# Topics in
# Physical Chemistry

vol 4

Edited by H. Baumgärtel, E. U. Franck, W. Grünbein
On behalf of Deutsche Bunsen-Gesellschaft für Physikalische Chemie

G. Brunner

# Gas Extraction

An Introduction to Fundamentals
of Supercritical Fluids and the
Application to Separation Processes

Springer-Verlag Berlin
Heidelberg GmbH

Author's address:
Prof. Dr.-Ing. Gerd Brunner
Technische Universität
Hamburg-Harburg
Arbeitsbereich Verfahrenstechnik II
Eißendorfer Straße 38
D-21073 Hamburg

Edited by:
Deutsche Bunsen-Gesellschaft
für Physikalische Chemie e.V.
General Secretary Dr. Heinz Behret
Carl-Bosch-Haus
Varrentrappstraße 40/42
D-60486 Frankfurt

Die Deutsche Bibliothek — CIP-Einheitsaufnahme
**Brunner, Gerd:**
Gas extraction : an introduction to fundamentals of
supercritical fluids and the application to separation processes
/ G. Brunner. [Ed. by: Deutsche Bunsen-Gesellschaft für
Physikalische Chemie e.V.]. — Darmstadt : Steinkopff ; New
York : Springer, 1994
   (Topics in physical chemistry ; Vol. 4)

   ISBN 978-3-662-07382-7      ISBN 978-3-662-07380-3 (eBook)
   DOI 10.1007/978-3-662-07380-3

NE: GT





Chemistry Editor: Dr. Maria Magdalene Nabbe — English Editor: James C. Willis
Production: Heinz J. Schäfer



Printed on acid-free paper

# Preface

Application of compressed gases as solvents has found widespread attention within the scientific community, and some processes have reached industrial, or near-industrial state.

This volume deals with the possibilities of supercritical gases as solvents for separation processes. It combines physico-chemical aspects with chemical engineering methods, and is intended to be used by students, practicing scientists, and engineers who want to make use of the techniques summarized as "Gas Extraction", or want to know what it is all about. Generalization goes as far as possible; on the other hand, examples are treated in detail. No comprehensive literature survey was intended.

Many persons have contributed to this book directly or indirectly. First, my teacher, Prof. Dr. S. Peter from the University of Erlangen-Nürnberg, who introduced me to the marvelous possibilities of supercritical gases. Then, my doctoral students whose efforts made this book possible. Their contributions are mentioned in the text. During the preparation of the manuscript useful discussions and remarks, which are acknowledged, have been contributed by Ralf Dohrn, Andreas Birtigh, Manuel Budich, Ariane Firus, Monika Johannsen, Jens-Torge Meyer, Klaus Nowak, Volker Riha, Carla Saure, and Jöran Stoldt.

Initiation and continuing support by the editors of the series "Topics in Physical Chemistry", Prof. E. U. Franck, Prof. W. Grünbein, and Prof. Baumgärtel are gratefully acknowledged. Dr. M. M. Nabbe and Mr. J. C. Willis helped to complete the manuscript and arranged its publication.

*Hamburg, February 1994*  	*G. Brunner*

# Contents

# 1 Introduction

This book is intended to be an introduction for people interested in the application of gaseous solvents, better known as supercritical fluids. It is not intended as a comprehensive treatment and review of the literature. In this work, I have tried to convey my experiences and my view of the matter.

The title of this book is "Gas Extraction"; therefore, the work is limited to gaseous solvents and the physical separation processes. There are interesting and important phenomena in the near-critical liquid region, which are not excluded, but also not especially stressed within this book. Chemical reactions in connection with solvent properties of supercritical gases are an important field of research and application, but are beyond the scope of this book.

This work deals with the possibilities of supercritical gases as solvents for technical processes. It combines physico-chemical aspects with chemical engineering methods in order to provide methods for the design of technical processes. It was the intention to generalize as far as possible. Illustrative examples will be given, including many from my work. Experimental techniques are discussed and data for some systems are presented, so that the information can be used for judging the applicability of gas extraction to special separation problems. The literature cited at the end of each chapter is by no means representative of the vast literature which is now available in this field.

Understanding and modeling a process and designing equipment for a process requires understanding and solution of a system of equations comprising:

1) theorems of conservation: conservation of energy, conservation of mass, and conservation of impulse;

2) information on the physico-chemical limits of changes, which can be achieved in a process: phase equilibrium and chemical equilibrium,

3) information on the timescale of changes in a process: mass transfer, heat transfer, and reaction rate of chemical reactions.

An important part of the tasks connected with gas extraction is concerned with component and mixture properties. Without meaningful data on real mixtures, the set of modeling and design equations can only be solved for simple cases, which may render the solution irrelevant. Therefore, the determination of component and mixture properties will also be treated.

Phase equilibrium is the most important basis for understanding the phenomena concerned with gas extraction. Importance is derived i.a. from the fact that phase equilibrium is independent on the equipment and the mode in which a process is carried out. An overview on phenomenological behavior and on thermodynamic calculation methods is given.

While equilibrium properties depend on conditions of state only, the terms and the solution of the rate equations depend on the mode of operation of the technical process and on the technical equipment for carrying out the process. Information in this field is still scarce.

In general, the design problem of a gas extraction process cannot be solved without some input from economical and other non-scientific data. These problems will not be covered in this book.

Separation processes are discussed in three chapters. First, extraction from solid is treated, which usually is carried out as a batch process, since solids are difficult to handle continuously in pressurized vessels, and as a one stage process, since separation factors are high. Fluid mixtures often have separation factors which make necessary the application of multistage contacting, which is carried out most effective in a countercurrent mode. Countercurrent multistage contacting is therefore treated in the second chapter on separation processes. If separation factors are approaching 1, many theoretical stages are necessary for separating the components. Chromatography is a magnificent tool for separating similar compounds. Chromatography with supercritical gases as solvents and mobile phases are presently applied in analytical separations. But the advantageous possibilities of gaseous solvents should also be applied in chromatographic processes on a process scale. Therefore, the third chapter on separation processes deals with chromatography and the problems connected with scale up.

# 2 Properties of Supercritical and Near-Critical Gases and of Mixtures with Sub- and Supercritical Components

Within this chapter properties of compounds in the near critical and supercritical state, as well as mixtures of subcritical compounds in combination with supercritical components are treated. An overview on the dependence of properties on conditions of state, and some methods and correlations for calculating properties are presented. Among the properties treated are: Density ($PVT$ behavior), viscosity, diffusion coefficient, thermal conductivity, dielectric constant, surface tension and Joule-Thompson coefficient.

## 2.1 What is a Supercritical Gas?

A pure component is considered, within the scope of this book, to be in a super-critical state if its temperature and its pressure are higher than the critical values (Fig. 2.1). Critical data for a number of components are listed in Table 2.1. At critical conditions for pressure and temperature, as indicated in Fig. 2.1 by the broken and hatched lines, there is no sudden change of component properties. The variation of properties with conditions of state is monotonous, when crossing the broken line in Fig. 2.1, with the exception of the critical point itself. Yet the magnitude of the variation can be tremendous, thereby causing different effects on solutes and reactants within neighboring conditions of state. Similar effects to that of the supercritical state can in some cases be achieved for $P > P_c$ and $T < T_c$, at near critical temperatures in the liquid state of a substance.

A mixture of components is considered to be supercritical with respect to pressure, temperature or concentration, if conditions of state of a mixture (pressure, temperature, composition) are beyond the critical point of a certain mixture. A binary mixture at constant temperature is supercritical for all pressures higher than the critical pressure of the binary mixture, as indicated in Fig. 2.2, part a, by the hatched area. The critical pressure is always the upper limit of the two-phase area at constant temperature for a binary system and, in addition, at constant composition in the case of a multicomponent system (see also Chapter 3). A gaseous mixture of the binary system, schematically represented in Fig. 2.2.a, as indicated by point A, is in the one-phase region, but not supercritical. There is no analogy with temperature, since in a binary system (and also in multicomponent systems), at constant pressure, the critical temperature does not limit the two-phase region, with respect to temperature. On the contrary, the two-phase area may well extend beyond the critical temperature, as

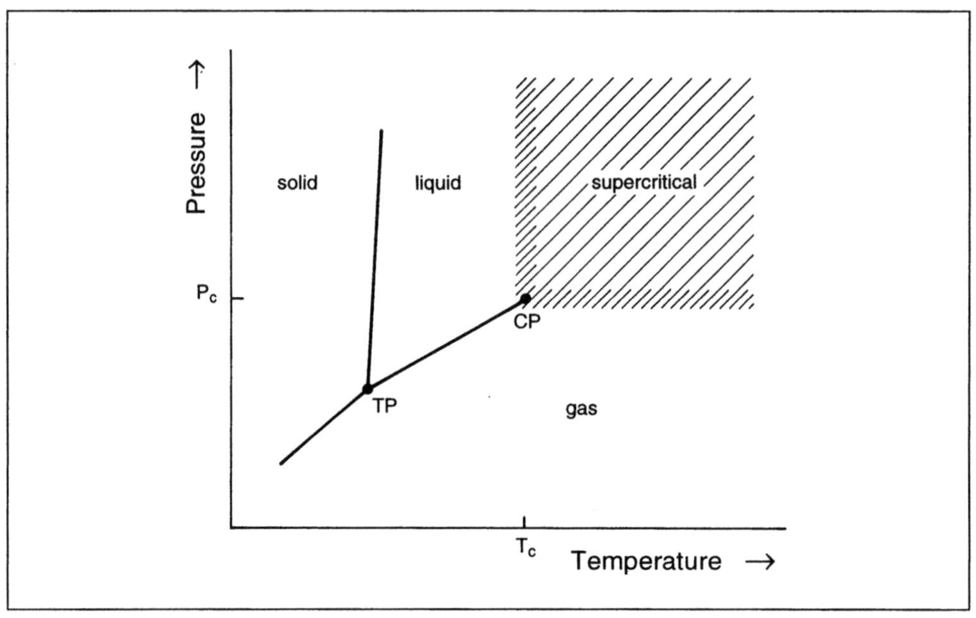



**Fig. 2.1.** Definition of supercritical state for a pure component. CP critical point, TP triple point, $T_c$ critical temperature, $P_c$ critical pressure.

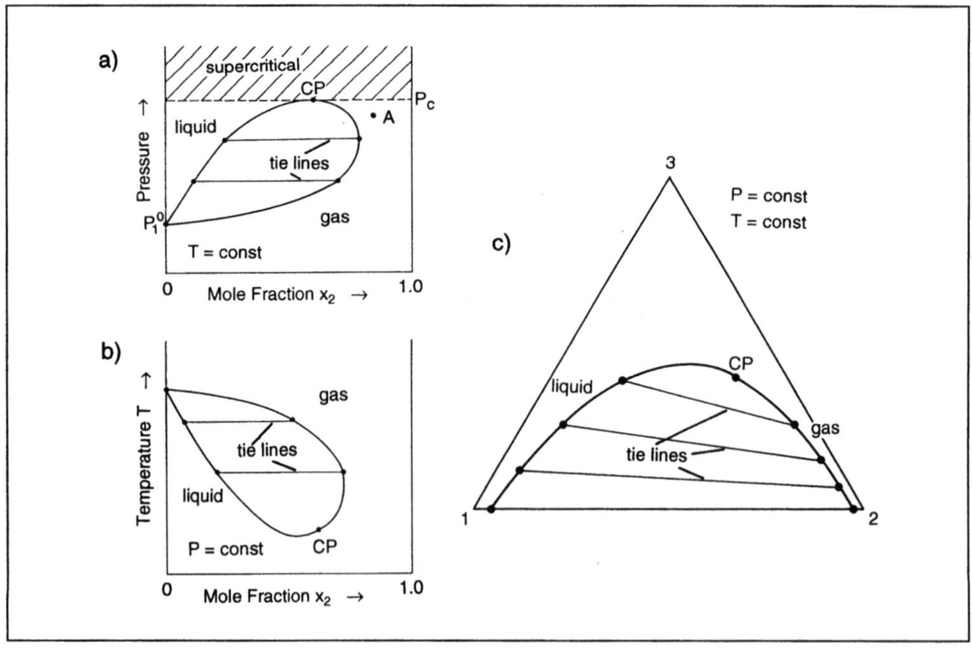



**Fig. 2.2.** Critical points in binary and ternary mixtures. $P_1^0$ vapor pressure of component 1.

is illustrated in Fig. 2.2, part b. In a ternary system at constant temperature and pressure, as illustrated in Fig. 2.2, part c, the critical point in general is not at a maximum value of the two-phase region with respect to one of the concentrations of the three components. The critical point may well be on one side of the two-phase area.

In conclusion, for mixtures it is more useful to use the notion "one-phase region" or "two-phase region" instead of "supercritical".

**Table 2.1.** Critical data for pure components [54]

| | $T_c$ [K] | $P_c$ [MPa] | | $T_c$ [K] | $P_c$ [MPa] |
|---|---|---|---|---|---|
| Helium | 5.19 | 0.23 | Chlorodifluoro-methane ($CHClF_2$) | 369.3 | 4.97 |
| Hydrogen | 33.0 | 1.29 | Propane | 369.8 | 4.25 |
| Neon | 44.4 | 2.76 | Hydrogen sulfide | 373.3 | 8.94 |
| Nitrogen | 126.2 | 3.39 | Ethyl fluoride | 375.31 | 5.02 |
| Carbon monoxide | 132.9 | 3.50 | Radon | 377.65 | 6.28 |
| Argon | 150.75 | 4.87 | Carbonyl sulfide (COS) | 378.8 | 6.35 |
| Oxygen | 154.6 | 5.04 | Dichlorodifluoro-methane ($CF_2Cl_2$) | 385.0 | 4.14 |
| Nitric oxide (NO) | 180.15 | 6.48 | Perfluorobutane | 386.35 | 2.32 |
| Methane | 190.4 | 4.60 | Propadiene ($C_3H_4$) | 393.15 | 5.47 |
| Krypton | 209.45 | 5.50 | Cyclopropane | 397.85 | 5.49 |
| Carbon tetrafluoride | 227.6 | 3.74 | Dimethyl ether | 400.0 | 5.24 |
| Silicon tetrafluoride | 259.1 | 3.72 | Ammonia | 405.55 | 11.35 |
| Silane ($SiH_4$) | 269.69 | 4.84 | Isobutane | 408.2 | 3.65 |
| Ethylene | 282.4 | 5.04 | Methyl chloride | 416.25 | 6.70 |
| Xenon | 289.7 | 5.84 | Chlorine | 416.9 | 7.98 |
| Hexafluoroethane | 293.0 | 3.06 | Hydrogen jodide | 424.0 | 8.31 |
| Trifluoromethane($CHF_3$) | 299.3 | 4.86 | n-Butane | 425.2 | 3.80 |
| Chlorotrifluoro-methane ($CClF_3$) | 301.95 | 3.87 | Methyl amine | 430.0 | 7.43 |
| 1,1Difluoroethene | 302.85 | 4.46 | Sulfur dioxide | 430.8 | 7.88 |
| Carbon dioxide | 304.15 | 7.38 | Diethylether | 466.7 | 3.64 |
| Ethane | 305.4 | 4.88 | n-Pentane | 469.7 | 3.37 |
| Chlorotrifluorosilane ($SiClF_3$) | 307.65 | 3.47 | Diethyl amine | 496.5 | 3.71 |
| Acetylene | 308.3 | 6.14 | n-Hexane | 507.5 | 3.01 |
| Nitrous oxide ($N_2O$) | 309.65 | 7.24 | Acetone | 508.1 | 4.70 |
| Monofluoromethane | 315.0 | 5.60 | Isopropanol | 508.3 | 4.76 |
| Sulfurhexafluoride ($SF_6$) | 318.75 | 3.76 | Methanol | 512.6 | 8.09 |
| Hydrogen chloride | 324.7 | 8.31 | Ethanol | 513.9 | 6.14 |
| Trifluorobromo-methane ($CBrF_3$) | 340.2 | 3.97 | Ethyl acetate | 523.25 | 3.83 |
| 1,1,1 Trifluoro-ethane ($CH_3$-$CF_3$) | 346.25 | 3.76 | n-Heptane | 540.3 | 2.74 |
| Chloropentafluoro-ethane ($C_2ClF_5$) | 353.15 | 3.23 | Acetonitrile | 545.5 | 4.83 |
| Hydrogen bromide | 363.15 | 8.55 | Cyclohexane | 553.5 | 4.07 |
| Propylene | 364.95 | 4.60 | Benzene | 562.2 | 4.89 |
| | | | Toluene | 591.8 | 4.10 |
| | | | Water | 647.3 | 22.12 |

# 2.2 Density (PVT Behavior)

## 2.2.1 Pure Components

(Mixtures are treated in Chapter 3 in connection with calculation of phase equilibria)

The interdependence of volume, temperature, and pressure are of utmost importance for gas extraction, since properties of supercritical compounds vary strongly with conditions of state and these variations are the basis for many applications.

A generalized *PVT* diagram (Fig. 2.3) shows for different temperatures the dependence of volume on pressure. At subcritical temperatures ($T < T_c$, isotherm a), the volume of a gas decreases rapidly with increasing pressure, till the phase boundary line is reached ($a_2$). Crossing this line by decreasing the total volume causes the formation of liquid drops of specific volume $a_3$. As long as the total volume of the two-

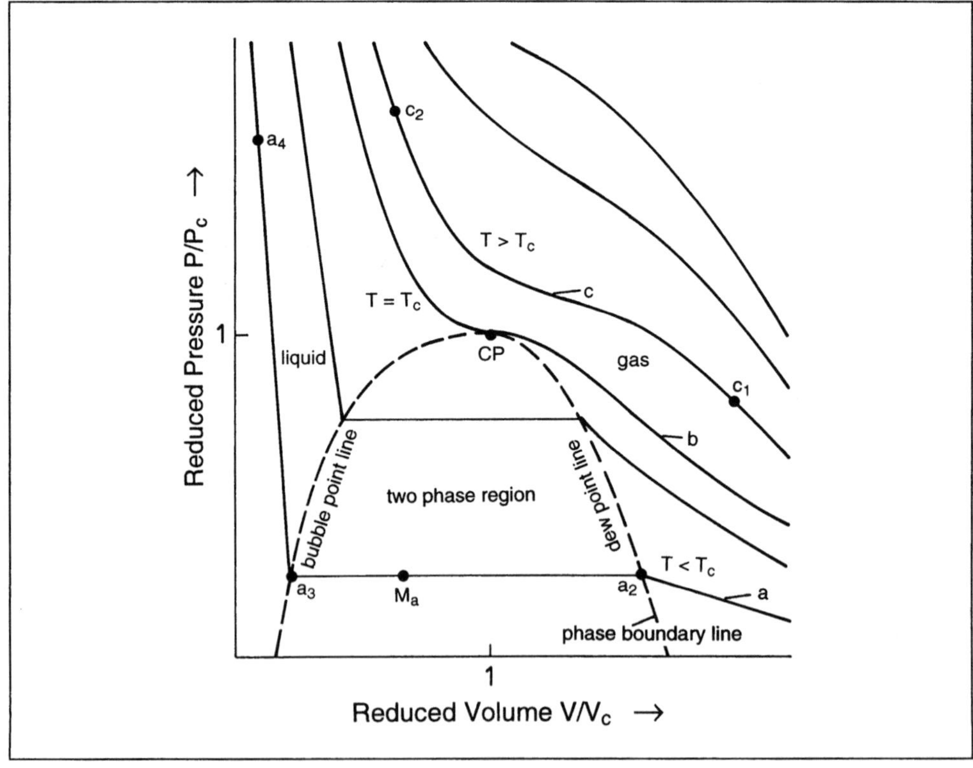


**Fig. 2.3.** *PVT* behavior of a fluid (e.g., carbon dioxide), as it may be used as supercritical solvent in gas extraction.

6

phase mixture lies in the two phase-region, a gas of volume $a_2$ and a liquid of volume $a_3$ coexist at constant pressure. Points within the two-phase region represent mixtures of gas and liquid, coexisting at constant pressure and temperature. The relative amounts of liquid and gas can be determined by the lever rule. At $a_3$ all the gas is condensed to liquid $a_3$. A further reduction of total volume results in a rapid increase in pressure $(a_3 \rightarrow a_4)$, since compressibility of a liquid is very low compared to that of a gas.

At temperatures higher than critical $(T > T_c$, isotherm c), a gas $(c_1)$ can be compressed to a liquid-like volume $(c_2)$ without visible change of phases. The critical isotherm $(T = T_c$, isotherm b), has a horizontal tangent at the critical point. At the critical point, the liquid and the gas branch of the phase boundary line meet and phases become identical. Isotherms at $T > T_c$, but near $T_c$ are flat in the vicinity of the critical point. Compressibility in this region is high. Small changes in pressure or temperature cause large variations of (specific) volume or density.

The simplest relation between pressure $P$, molar volume $V$, and temperature $T$ is that for the ideal gas:

$$P\,V \;=\; R\,T.$$

2.1

As is well known, it is applicable only at low pressures and high temperatures. It may seem not to be of much value for a high pressure application like gas extraction. But as a limiting relation for quickly estimating variations it is extremely useful.

More appropriate for quantitative representation of the $PVT$ behavior of supercritical fluids are equations of state (EOS), which take into consideration the properties of individual molecules. The corresponding-states principle and Van der Waals-type equations of state are appropriate for most substances and their mixtures applied in gas extraction, since these substances mostly are unpolar or moderately polar. The more strongly polar and associating compounds are involved, the more other relations must be used for calculating $PVT$ behavior. Typically, equations of state of the van der Waals-type represent gas-phase data well, but deviations from liquid phase densities may be substantial. There are EOS available, e.g., the Bender EOS, which represent $PVT$ data better than van der Waals-type EOS. But those EOS contain more parameters and are difficult to use with mixtures. Therefore, their application is limited to special cases, e.g., for calculations of the gas cycle of gas extraction (see also Chapter 5). Excellent reviews on EOS have been presented by Sandler [56] , and by Dohrn [16].


## 2.2.2 Principle of Corresponding States

At the critical point, coexisting phases, e.g., the gaseous and the liquid phase of a pure component become identical. The more remote conditions of state are from the critical point, the more different are properties of gas and liquid. The state of any substance can be defined with reference to the critical point. It is assumed (based on experience) that at same conditions with respect to the critical point, different sub-

7

stances behave equal. This assumption is called the "principle of corresponding states." *PVT* behavior can then be described by a generalized function. For representing this function, conditions of state are related to critical properties and the so-called reduced properties, $T_R$, $P_R$, $V_R$ are obtained:

$$T_R = \frac{T}{T_c}; \qquad P_R = \frac{P}{P_c}; \qquad V_R = \frac{V}{V_c}. \qquad\qquad 2.2$$

Index $c$ refers to critical properties.

Using reduced properties, a generalized EOS can be formulated:

$$P_R = F_1\,(T_R,\, V_R) \text{ or:} \qquad\qquad 2.3$$

$$Z = \frac{PV}{RT}; \quad Z = F_2\,(T_R,\, P_R), \qquad\qquad 2.4$$

where $F_1$ and $F_2$ are empirical functions, valid for any substance. Two substances are in corresponding state if reduced variables are equal. In this simple form, the corresponding principle can only be used for calculating *PVT* behavior of simple molecules with sufficient accuracy. All the influences of interaction, size, steric configuration and others, are assumed to be taken into account by the critical values of a substance. Several authors have calculated the functions $F_1$ and $F_2$ from measured *PVT* data of various substances. In Fig. 2.4 the result of Nelson and Obert [44] is presented. It is a two-parameter EOS with $T_c$ and $P_c$ as the parameters. From this diagram, *PVT* behavior of substances used in gas extraction can be calculated. Errors are reported to be about 5 %, but become larger in the critical region. The diagram is not to be applied to strongly polar substances and the light gases helium, hydrogen, and neon.

Compressibility, $Z = (PV)/(RT)$, in the supercritical region, decreases with increasing reduced pressure $P_r = P/P_c$ up to about a reduced temperature of $T_r = 2.2$, runs through a minimum, and then increases with pressure. Above a reduced temperature of $T_r = 2.2$, compressibility only increases with pressure. In the critical region, up to a reduced temperature of $T_r = 1.30$, compressibility rapidly decreases with pressure. In this range, density of a supercritical compound is strongly dependent on pressure and temperature, and solvent power, determined by density, can be varied with moderate changes in conditions of state.

Introduction of a third parameter to the corresponding states principle improves accuracy of the resulting EOS. Most widely accepted is the "acentric factor $\omega$", introduced by Pitzer and Curl [50]. The acentric factor considers deviation of a molecule from spherical shape. It is calculated from the vapor pressure $P^0$ of a substance at $0.7\,T_R$:

$$\omega = -\log(P_r)_{T_r = 0.7} - 1.0. \qquad\qquad 2.5$$

8

Then, the compressibility factor $Z$ can be calculated as follows:

$$Z = Z^{(0)} (T_R, P_R) + \omega \cdot Z^{(1)} (T_R, P_R) \qquad\qquad 2.6$$

The first term in Eq. 2.6 represents the behavior of spherical molecules, while the second term represents deviations by non-spherical molecules. In Table 2.2 acentric factors for some substances are tabulated. Functions $Z^{(0)}$ and $Z^{(1)}$ of the three parameter corresponding states principle can be calculated from a generalized equation of state. The most accurate values for $Z^{(0)}$ and $Z^{(1)}$ are based on the Lee-Kesler EOS and are tabulated in [54].

More widely used are analytical equations of state, because their parameters are related to the critical point of a substance, or can be individually fitted to the *PVT* behavior.

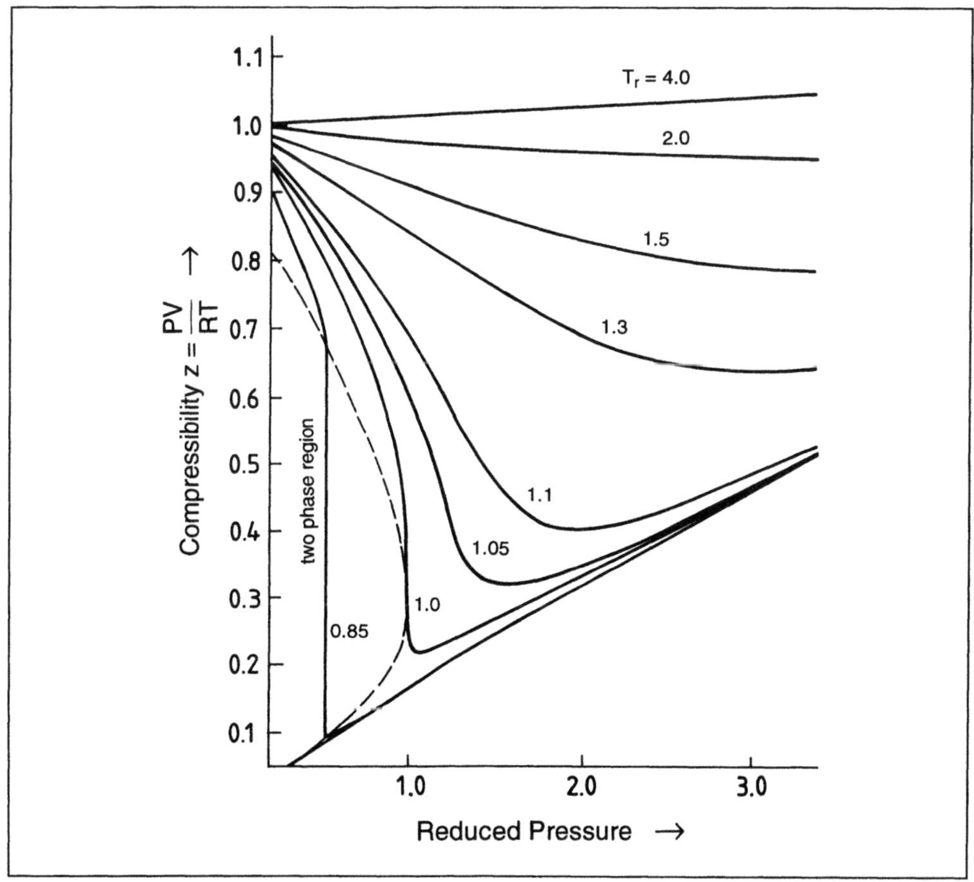


**Fig. 2.4.** *PVT* behavior of simple fluids ($\omega = 0$) according to a two parameter principle of corresponding states (after Nelson and Obert [44]).

**Table 2.2.** Acentric factors [54].

| Component | $\omega$ | Component | $\omega$ |
|---|---|---|---|
| Argon | 0.001 | Nitromethane | 0.310 |
| Chlorine | 0.090 | Methane | 0.022 |
| Sulfur hexafluoride | 0.286 | Methanol | 0.556 |
| Hydrogen chloride | 0.133 | Methyl amine | 0.292 |
| Hydrogen | −0.216 | Acetylene | 0.190 |
| Water | 0.344 | Vinyl chloride | 0.122 |
| Hydrogen sulfide | 0.097 | Acetonitrile | 0.327 |
| Ammonia | 0.250 | Ethylene | 0.089 |
| Krypton | 0.005 | Acetaldehyde | 0.303 |
| Nitric oxide (NO) | 0.588 | Ethane | 0.099 |
| Nitrogen dioxide ($NO_2$) | 0.834 | Ethanol | 0.644 |
| Nitrogen | 0.039 | Ethyl amine | 0.289 |
| Nitrous oxide ($N_2O$) | 0.165 | Propadiene | 0.313 |
| Neon | −0.029 | Propylene | 0.144 |
| Oxygen | 0.025 | Acetone | 0.304 |
| Sulfur dioxide | 0.256 | Propane | 0.153 |
| Xenon | 0.008 | Isopropyl alcohol | 0.665 |
| Carbon monoxide | 0.066 | n-Butane | 0.199 |
| Carbonyl sulfide | 0.105 | n-Butanol | 0.593 |
| Carbon dioxide | 0.239 | 2-Butanol | 0.577 |
| Dichloromethane | 0.199 | Benzene | 0.212 |

## 2.2.3 Analytical Equations of State

An analytical equation of state is an algebraic relationship between pressure, temperature, and molar volume. Equations of state should meet some conditions: The criterium of thermodynamic stability must be fulfilled at the critical point:

$$(\mathrm{d}P/\mathrm{d}V)_{T_c} = 0 \quad \text{and} \quad (\mathrm{d}^2 P/\mathrm{d}V^2)_{T_c} = 0. \tag{2.7}$$

Properties of a fluid should be represented by the equation of state from the liquid to the ideal gas state. At very low pressure the equation of state should approach the ideal gas law.

In 1873, van der Waals [64] published his famous equation of state (Eq. 2.8), based on the assumption of spherical molecules with attractive and repulsive forces depending on the relative distance of the molecules.

$$P = \frac{RT}{V-b} - \frac{a}{V^2}, \tag{2.8}$$

with:

$$V_c = 3b; \qquad P_c = \frac{a}{27\,b^2}; \qquad T_c = \frac{8\,a}{27\,Rb}. \tag{2.9}$$

10

The equation consists of the repulsion term with parameter $b$, the co-volume as parameter for representing repulsive forces, and the attraction term with parameter $a$ representing attractive forces. Both parameters are related to the critical point, as stated in Eq. 2.9, and are therefore dependent on the accuracy of these values.

It is useful for understanding EOS and thermodynamic properties, to visualize molecular interactions between two molecules as an interaction force depending on distance between the two molecules. This distance depends on pressure and temperature, since density of a substance is directly related to the average distance between the molecules. The interaction force can be expressed as a potential function. For an ideal gas, consisting of point-like molecules with no attractive or repulsive forces, there are no forces for distances $r>0$. For $r=0$, the repulsive force is infinite. The corresponding potential function is shown in Fig. 2.5 a. If spherical and hard molecules are assumed, exhibiting still no interaction force between different molecules, a hard sphere potential, Fig. 2.5 b applies to this model. If interaction forces are introduced, assumptions on their dependence on distance must be made. For example, an attraction varying inversely with the square of the distance ($\sim 1/r^2$) and a steep increase of repulsive forces when $r$ is in the range of the diameter of the molecules can be assumed. Best known of such a potential is the Lennard-Jones potential, Fig. 2.5 c. For molecules with dipole moments, the orientation of the electric charges influences

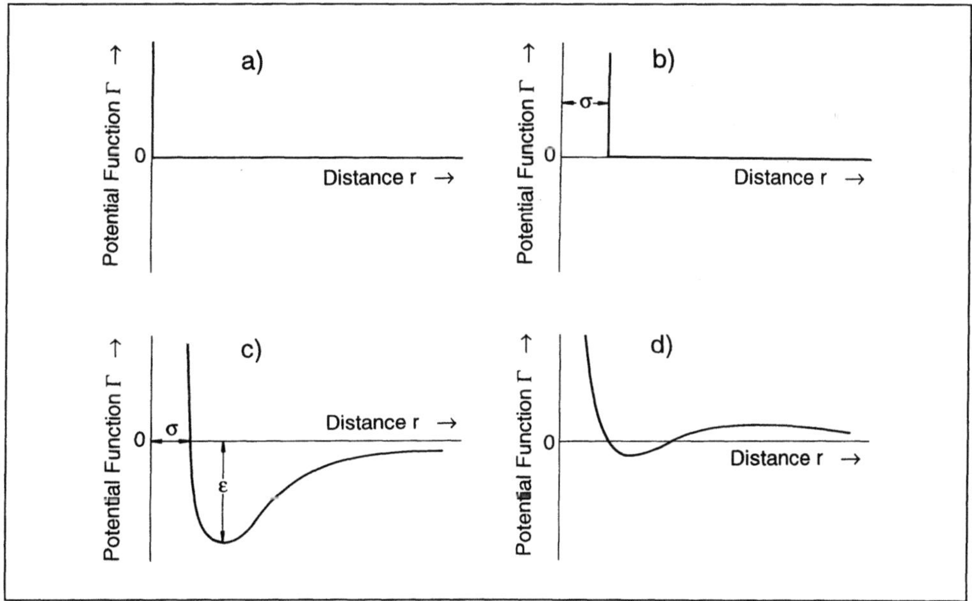


**Fig. 2.5.** Potential functions for different models of interaction forces of two molecules (σ = hard sphere diameter or collision diameter, ε = depth of energy well or minimum of potential energy). a) Point-like molecules, no interaction forces; b) Hard-sphere molecules, no interaction forces; c) Spherical molecules, repulsive and attractive forces; d) Dipoles lined up in repulsive position.

the potential function, e.g., causing a repulsive force if lined up in repulsive position, Fig. 2.5 d.

Still, calculation methods for thermodynamic properties from potential functions are not in a state whereby solute molecules for gas extraction can be treated. But a qualitative picture can be derived from potential functions.

The van der Waals-equation of state can qualitatively account for most phenomena in $PVT$ behavior and phase equilibria. Yet accuracy is not sufficient for practical purposes for most substances. Therefore various modifications have been proposed, but it was not until 1949 that Redlich and Kwong [53] brought forward a really good one, which is still in use today, with some modifications.

The Redlich-Kwong-EOS is presented in Eq. 2.10. The repulsive term remained unmodified, but the attractive term was modified by an explicit universal temperature dependence. Again, parameters $a$ and $b$ are related to the critical point, as written in Eq. 2.11.

$$P = \frac{RT}{V - b} - \frac{a}{T^{1/2}V(V+b)}; \qquad\qquad\qquad 2.10$$

$$a = \frac{0.4278\,R^2 \cdot T_2^{2,5}}{P_c}; \qquad b = 0.0867\,\frac{R \cdot T_c}{P_c}. \qquad\qquad 2.11$$

An excellent representation of $PVT$ data for the gaseous phase can be obtained with this EOS, even at high pressures.

The Redlich-Kwong EOS contains two parameters, $a$ and $b$, which are related to the critical point. Critical points are only known with limited accuracy or not determinable at all. Therefore, parameters $a$ and $b$ may be considered as adjustable parameters and fitted to $PVT$ data.

Parameters $a$ and $b$ can be generalized by introducing two coefficients, $\Omega_a$ and $\Omega_b$ (Eq. 2.12), which for the original Redlich-Kwong EOS have the values $\Omega_a = 0.4278$ and $\Omega_b = 0.0867$, but can be fitted to $PVT$ data or related to the acentric factor $\omega$.

$$a = \frac{\Omega_a R^2 T_c^{2.5}}{P_c}; \qquad b = \frac{\Omega_b R T_c}{P_c}. \qquad\qquad 2.12$$

The main disadvantage of the Redlich-Kwong EOS is the fixed temperature dependence of the attraction term. It is probable that different substances with different attractive forces, like, e.g., van der Waals forces of nonpolar molecules as compared to electrostatic forces of polar molecules, exhibit a different temperature dependence of their attractive forces, since these forces depend differently on distance between the molecules. Various modifications, taking into consideration individual temperature dependence for a substance, have been proposed. The most widely accepted stems from Soave [60] and is presented in Eq. 2.13 to Eq. 2.17.

12

$$P = \frac{RT}{V-b} - \frac{a(T)}{V(V+b)}, \text{ with} \qquad\qquad 2.13$$

$$b = 0.08664 \frac{RT_c}{P_c}, \qquad\qquad 2.14$$

where $a(T)$ depends on temperature:

$$a(T) = a'(T_c)\,\alpha(T) \qquad\qquad 2.15$$

$$a'(T_c) = 0.42748 \frac{R^2 T_c^2}{P_c} \qquad\qquad 2.16$$

$$\alpha(T) = [1 + (0.480 + 1.574\,\omega - 0.176\,\omega^2)(1 - (T/T_c)^{0.5})]^2. \qquad\qquad 2.17$$

Note that in this modification component parameters are also linked with the critical point, yet information on vapor pressure $(P^0)$ introduces some information on the interaction forces at conditions far from the critical point, according to the definition of the acentric factor $\omega$, at about 1/10 of the critical pressure.

Liquid densities as required in vapor-liquid equilibrium calculations are the weak point of these equations of state. Therefore, many modifications have been proposed. The Peng-Robinson equation of state [46], Eq. 2.18, is the most successful alternative, although problems with liquid density are not solved by it.

$$P = \frac{RT}{V-b} - \frac{a(T)}{V(V+b)+b(V-b)}. \qquad\qquad 2.18$$

$$b = 0.07780 \frac{RT_c}{P_c}; \qquad\qquad 2.19$$

$$a(T) = a(T_c)\,\alpha(T_R, \omega); \qquad\qquad 2.20$$

$$a(T_c) = 0.45724 \frac{R^2 T_c^2}{P_c}; \qquad\qquad 2.21$$

$$\alpha = [1 + \beta(1 - T_R^{0.5})]^2, \text{ with} \qquad\qquad 2.22$$

$$\beta = 0.37464 + 1.54226\,\omega - 0.26992\,\omega^2. \qquad\qquad 2.23$$

The Peng-Robinson EOS has the same characteristics with respect to parameter determination as the Soave modification of the Redlich-Kwong EOS: Parameters

**Table 2.3.** *PVT* data of carbon dioxide as calculated with different EOS, compared to IUPAC-data.

| T[K] | P [bar] | IG | PR | IUPAC | HPW(c) | HPW(d) | RKS |
|---|---|---|---|---|---|---|---|
| 313.15 | 80. | 135.225 | 284.444 | 279.334 | 278.802 | 598.004 | 281.291 |
|  |  | −.51590 | .01829 | .00000 | −.00191 | 1.14082 | .00700 |
| 313.15 | 100. | 169.032 | 564.624 | 629.815 | 538.620 | 699.651 | 542.471 |
|  |  | −.73162 | −.10351 | .00000 | −.14480 | .11088 | −.13868 |
| 313.15 | 200. | 338.064 | 829.980 | 840.371 | 755.540 | 861.392 | 756.850 |
|  |  | −.59772 | −.01236 | .00000 | −.10094 | .02501 | −.09939 |
| 313.15 | 300. | 507.096 | 928.271 | 910.335 | 840.522 | 938.962 | 841.454 |
|  |  | −.44296 | .01970 | .00000 | −.07669 | .03145 | −.07567 |
| 313.15 | 400. | 676.127 | 992.303 | 956.404 | 896.734 | 992.417 | 897.480 |
|  |  | −.29305 | .03754 | .00000 | −.06239 | .03765 | −.06161 |
| 313.15 | 80. | 127.107 | 196.448 | 191.759 | 194.136 | 199.427 | 195.438 |
|  |  | −.33715 | .02445 | .00000 | .01240 | .03999 | .01918 |
| 333.15 | 100. | 158.884 | 292.567 | 290.192 | 296.168 | 333.576 | 300.626 |
|  |  | −.45249 | .00818 | .00000 | .02059 | .14950 | .03595 |
| 333.15 | 200. | 317.769 | 694.380 | 724.366 | 658.841 | 754.770 | 663.288 |
|  |  | −.56131 | −.04140 | .00000 | −.09046 | .04197 | −.08432 |
| 333.15 | 300. | 476.653 | 831.370 | 830.215 | 770.833 | 861.230 | 773.890 |
|  |  | −.42587 | .00139 | .00000 | −.07153 | .03736 | −.06784 |
| 333.15 | 400. | 635.537 | 913.117 | 890.659 | 839.315 | 928.080 | 841.735 |
|  |  | −.28644 | .02521 | .00000 | −.05765 | .04201 | −.05493 |
| 353.15 | 80. | 119.909 | 163.632 | 160.529 | 162.870 | 165.244 | 163.906 |
|  |  | −.25304 | .01933 | .00000 | .01458 | .02938 | .02104 |
| 353.15 | 100. | 149.886 | 224.963 | 221.771 | 226.405 | 234.302 | 228.812 |
|  |  | −.32414 | .01439 | .00000 | .02089 | .05650 | .03175 |
| 353.15 | 200. | 299.772 | 562.945 | 594.892 | 560.416 | 637.917 | 567.430 |
|  |  | −.49609 | −.05370 | .00000 | −.05795 | .07233 | −.04616 |
| 353.15 | 300. | 449.659 | 734.407 | 745.064 | 700.960 | 780.978 | 705.988 |
|  |  | −.39721 | −.01549 | .00000 | −.06033 | .04694 | −.0535 |
| 353.15 | 400. | 599.545 | 833.739 | 823.601 | 782.222 | 862.949 | 786.222 |
|  |  | −.27204 | .01231 | .0000 | −.05024 | .04778 | −.04538 |
| 373.15 | 80. | 113.482 | 143.596 | 141.475 | 143.612 | 145.083 | 144.482 |
|  |  | −.19787 | 0.1499 | .00000 | .01510 | .02550 | .02125 |
| 373.15 | 100. | 141.853 | 191.107 | 188.800 | 192.591 | 196.438 | 194.336 |
|  |  | −.24866 | .01222 | .00000 | .02008 | .04045 | .02932 |
| 373.15 | 200. | 283.705 | 460.444 | 431.294 | 472.638 | 524.767 | 480.371 |
|  |  | −.41054 | −.04332 | .00000 | −.01798 | .09033 | −.00192 |
| 373.15 | 300. | 425.558 | 644.633 | 662.276 | 633.581 | 701.081 | 640.110 |
|  |  | −.35743 | −.02664 | .00000 | −.04333 | .05859 | −.03347 |
| 373.15 | 400. | 567.411 | 757.567 | 757.036 | 726.646 | 798.384 | 731.994 |
|  |  | −.25048 | .00070 | .00000 | −.04014 | .05462 | −.03308 |

Density in kg/m$^3$

IG:      ideal gas
PR:     Peng-Robinson EOS, using critical data
IUPAC:  IUPAC data, taken as reference
HPW(c):  Hederer-Peter-Wenzel EOS [26], using critical data
HPW(d):  Hederer-Peter-Wenzel EOS [26], parameter fitted to *PVT* data and vapor pressure
RKS:    Soave-modification of Redlich-Kwong EOS

14

are linked to the critical point. Some information on interacting forces and shape of the molecules is introduced by the acentric factor.

In Table 2.3 densities of carbon dioxide are tabulated as calculated with various EOS. Deviations from IUPAC-values are similar for the van der Waals type EOS (IUPAC: International Union for Pure and Applied Chemistry).

For some pure components, e.g., refrigerants, carbon dioxide and water, properties are relatively well known. Simple equations of state are not able to represent these properties very accurately over a large range of pressure and temperature. The deviations tend to be recognizable in the liquid state and around the critical point. In order to better represent the data, multi-parameter equations of state have been developed which are valid for one or some components. For phase equilibrium calculations needed in gas extraction, these equations of state are not very useful, since parameter determination for low volatile components is tedious or even impossible, because there are not enough data available to determine a great number of parameters. If the information on relations between components and the various parameters is generalized, then the method is similar to the theorem of corresponding states with all the limitations of a generalized method, but lacking simplicity and physical meaning of the equations of state as discussed above.

For *PVT* calculations of various states of a supercritical component, like carbon dioxide, as in a gas cycle of a separation process unit, the use of a multi-parameter equation of state may be useful. Therefore, the Bender EOS is mentioned in this context [2], Eq. 2.24.

$$P = T\varrho \left[ R + B\varrho + C\varrho^2 + D\varrho^3 + E\varrho^4 + F\varrho^5 + (G + H\varrho^2)\, \varrho^2 \exp\left(- a_{20}\varrho^2\right)\right], \qquad 2.24$$

with:

$$B = a_1 - a_2 / T - a_3 / T^2 - a_4 / T^3 - a_5 / T^4$$
$$C = a_6 + a_7 / T + a_8 / T^2$$
$$D = a_9 + a_{10} / T$$
$$E = a_{11} + a_{12} / T$$
$$F = a_{13} / T$$
$$G = a_{14} / T^3 + a_{15} / T^4 + a_{16} / T^5$$

$$H = a_{17} / T^3 + a_{18} / T^4 + a_{19} / T^5 \qquad\qquad 2.25$$

This equation has been generalized [51] in order to make it applicable for components, for which sufficient experimental data are not available to determine the coefficients in the equation. Yet with generalization, an increased deviation from experimental data must be taken into account, as mentioned before. Equations of this type are not to be used outside the range of experimental data with which the coefficients have been determined.

In Fig. 2.6, density of carbon dioxide is plotted, as calculated with the Bender EOS, using the coefficients listed in Table 2.4.

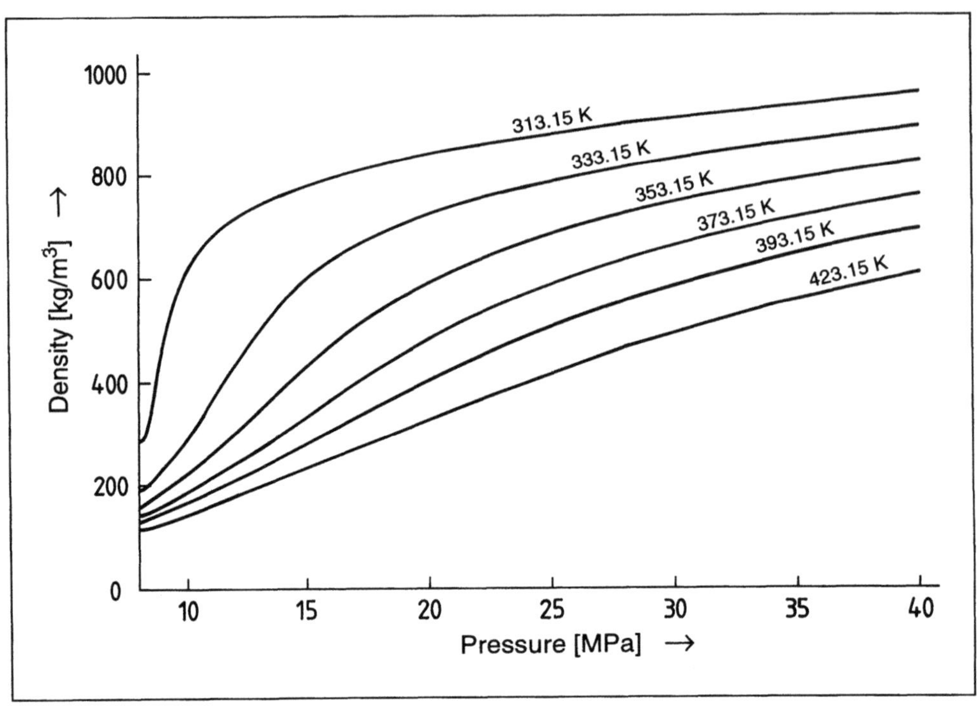

**Fig. 2.6.** Density of $CO_2$ calculated with the Bender-EOS.

**Table 2.4.** Coefficients of the Bender EOS for $CO_2$.

| | | | |
|---|---|---|---|
| $R =$ | $0.188918$ | $a_{10} =$ | $-0.39436077 \times 10^3$ |
| $a_1 =$ | $0.22488558$ | $a_{11} =$ | $0.12115286$ |
| $a_2 =$ | $0.13717965 \times 10^3$ | $a_{12} =$ | $0.10783386 \times 10^3$ |
| $a_3 =$ | $0.14430214 \times 10^5$ | $a_{13} =$ | $0.43962336 \times 10^2$ |
| $a_4 =$ | $0.29630491 \times 10^7$ | $a_{14} =$ | $-0.36505545 \times 10^8$ |
| $a_5 =$ | $0.20606039 \times 10^9$ | $a_{15} =$ | $0.19490511 \times 10^{11}$ |
| $a_6 =$ | $0.45554393 \times 10^{-1}$ | $a_{16} =$ | $-0.29186718 \times 10^{13}$ |
| $a_7 =$ | $0.77042840 \times 10^2$ | $a_{17} =$ | $0.24358627 \times 10^8$ |
| $a_8 =$ | $0.40602371 \times 10^5$ | $a_{18} =$ | $-0.37546530 \times 10^{11}$ |
| $a_9 =$ | $0.40029509$ | $a_{19} =$ | $0.11898141 \times 10^{14}$ |
| | | $a_{20} =$ | $0.50000000 \times 10^1$ |

Temperature $T$ in K, density $\varrho$ in g/cm$^3$. The resulting pressure $P$ is in MPa.

In gas extraction and related model systems, mixtures of components with very different size and properties are employed, e.g., mixtures of carbon dioxide and squalane, or ethylene and naphthalene. For the supercritical components, carbon dioxide and ethylene, the above discussed modifications of the Redlich-Kwong EOS, the Peng-Robinson EOS and parameter determination for these EOS works well, since vapor pressure and critical points are well known. Even for determination of the coefficients of the Bender EOS there is enough experimental information. But for the low volatile components it may not be possible to find enough experimental data, especially an experimentally determined critical point, since the substance may decompose before the critical point is approached. Even for substances where critical conditions can be determined, accuracy of the critical values deteriorates rapidly with increasing molecular weight of the compound. There are calculation methods available in the literature [54] for calculating critical constants from group contributions, i.e., from the chemical constitution of the component. While these calculation methods are very helpful in many cases, their value may be doubted if the resulting critical constants are used for the determination of parameters in an equation of state, which is meant to represent $PVT$ data accurately. Furthermore, there is no need to restrict oneself to such a method. The method of parameter determination can be easily modified, using information, which is more readily available, like $PVT$ data itself, vapor pressure, liquid densities or heat of vaporization. The parameters of an equation of state are then determined by fitting to known data. By choosing these data according to the intended calculations, the accuracy of the equation of state for representing data in the specified region can be improved.

An especially attractive possiblility for chemical engineering calculations is to determine parameters from heat of vaporization. Enthalpy calculations will then be more reliable than with parameters determined from $PVT$ data only.

This concept does not appear very attractive from a theoretical point of view and has therefore not gained much attention in the literature, but it is very effective and probably widely used in everyday practice. For phase equilibrium calculations, this concept was used by Hederer, Peter and Wenzel [26], Eq. 2.26. They modified the Redlich-Kwong EOS similar to the modification of Soave, but did not specify an analytical relationship between the temperature dependent attraction term and critical properties. Furthermore, they relied on the concept of determining all parameters by fitting to component properties.

$$P = \frac{RT}{V - b} - \frac{a\,T^{\alpha}}{V\,(V + b)}. \qquad\qquad 2.26$$

Parameters $a$, $b$, and $\alpha$ may be fitted to $PVT$ data and caloric data. A list of parameters of selected compounds is compiled in Table 2.5. Parameters can be correlated with simple component properties, like molar volume, vapor pressure or boiling temperature. Parameter correlations are obtained which may be used for components for which only few data are known, as is discussed in Section 2.2.4. In Chapter 3 the application of equations of state for calculation of phase equilibria is discussed.

Another well known and widely used EOS is the Peng-Robinson EOS [46], as presented in Eq. 2.27. Parameters $a_c$, $x_0$, and $b$, Eq. 2.28, also can be determined from component properties. Table 2.6 lists parameters of selected compounds. Parameter correlations have also been developed for this EOS [14, 15] and are presented below.

$$P = \frac{RT}{V - b} - \frac{a^+}{V(V + b) + b(V - b)}.$$

2.27

$$a^+ = a_c \left(1 + x_0 \left(1 - 2.4243 \frac{RTb}{a_c}\right)\right)^2.$$

2.28

**Table 2.5.** Parameter of modified Redlich-Kwong EOS for some substances, fitted to $PVT$ data [4, 27]

| Component | Parameter | | |
|---|---|---|---|
| | $a$ | $b$ | $\alpha$ |
| Methane | 13.02 | 0.02956 | $-0.31802$ |
| Ethane | 53.162 | 0.043506 | $-0.37913$ |
| Ethylene | 45.769 | 0.038821 | $-0.39684$ |
| Ethin | 88.05 | 0.03323 | $-0.51587$ |
| Carbon monoxide | 10.565 | 0.027424 | $-0.38829$ |
| Carbon dioxide | 89.651 | 0.027299 | $-0.55689$ |
| Nitrous oxide | 74.70 | 0.02798 | $-0.52381$ |
| Nitrogen | 6.7349 | 0.026526 | $-0.31575$ |
| Hydrogen | 0.30 | 0.01834 | $-0.01450$ |
| Propane | 121.90 | 0.05973 | $-0.42139$ |
| Monochlorotrifluoromethane | 86.67 | 0.05101 | $-0.43274$ |
| Monobromotrifluoromethane | 146.89 | 0.05829 | $-0.47849$ |
| Formic acid | 108.90 | 0.03301 | $-0.36374$ |
| Methanol | 412.20 | 0.03418 | $-0.61364$ |
| Ethanol | 2 152.00 | 0.05051 | $-0.82621$ |
| Acetone | 506.60 | 0.06130 | $-0.57049$ |
| 1-Propanol | 4 829.70 | 0.06561 | $-0.90775$ |
| Ethyl acetate | 1 092.00 | 0.08210 | $-0.62923$ |
| 2-Chlorphenol | 1 206.87 | 0.09163 | $-0.56989$ |
| Benzene | 531.30 | 0.07532 | $-0.52954$ |
| Vanillin | 72 293.98 | 0.13689 | $-1.10383$ |
| Caffeine | 56 804.50 | 0.14504 | $-1.13496$ |
| Naphthalene | 2 204.95 | 0.11593 | $-0.60953$ |
| Nicotene | 3 037.83 | 0.14786 | $-0.61205$ |
| Thymol | 25 224.82 | 0.14598 | $-0.96153$ |
| Decane | 5 880.10 | 0.17509 | $-0.74096$ |
| Phenanthrene | 5 821.50 | 0.15399 | $-0.66316$ |
| Palmitic acid | 4 908 225.50 | 0.28991 | $-1.59817$ |
| n-Hexadecane | 41 261.30 | 0.27492 | $-0.92199$ |
| 1-Hexadecanol | 112 689.00 | 0.28006 | $-1.05548$ |
| Stearic acid | 1 285 954.60 | 0.29264 | $-1.40091$ |
| Stearic acid methylester | 37 001.30 | 0.33418 | $-0.80218$ |
| Stearic acid monoglyceride | 46 385.30 | 0.34933 | $-0.82853$ |
| Cholesterol | 60 604.90 | 0.34877 | $-0.85881$ |
| Stearic acid triglyceride | 9 315 349.30 | 1.00950 | $-1.41920$ |
| Water | 114.60 | 0.01607 | $-0.47909$ |
| Ammonia | 42.04 | 0.01819 | $-0.40610$ |

**Table 2.6.** Parameters of the Peng-Robinson EOS for some selected substances, fitted to *PVT* data [14, 15]

| Component | Parameter | | |
|---|---|---|---|
| | $a_c$ | $b$ | $x_0$ |
| Methane | 2.49 | 0.02677 | 0.39157 |
| Ethane | 5.347 | 0.03245 | 0.56826 |
| Ethylene | 4.041 | 0.02549 | 0.6639 |
| Carbon dioxide | 3.355 | 0.02142 | 0.7603 |
| Nitrogen | 1.48 | 0.02408 | 0.43438 |
| Hydrogen | 0.27 | 0.01652 | 0.02560 |
| Monochlorotrifluoromethane | 7.45 | 0.05048 | 0.66943 |
| Nitrous oxide | 4.18 | 0.02766 | 0.62176 |
| Propane | 10.157 | 0.05667 | 0.48179 |
| n-Hexadecane | 97.631 | 0.26887 | 0.86464 |
| Formic acid | 10.282 | 0.03265 | 0.45394 |
| 2-chlorphenol | 29.816 | 0.09040 | 0.00000 |
| Triethyleneglycol | 44.658 | 0.12502 | 1.40940 |
| Salicylic acid | 32.352 | 0.08878 | 1.09248 |
| Vanillin | 49.636 | 0.13434 | 1.17754 |
| Caffeine | 41.597 | 0.14197 | 1.15279 |
| Tetralin | 42.039 | 0.12230 | 0.65972 |
| Nicotine | 53.771 | 0.14567 | 0.64571 |
| Thymol | 48.385 | 0.14328 | 1.00647 |
| Phenanthrene | 65.829 | 0.15179 | 0.71442 |
| Palmitic acid | 113.396 | 0.28412 | 1.71684 |
| 1-Hexadecanol | 104.734 | 0.27383 | 0.99562 |
| Stearic acid | 114.144 | 0.28552 | 1.29308 |
| Stearic acid methylester | 155.290 | 0.32803 | 0.76517 |
| Stearic acid monoglyceride | 160.106 | 0.34449 | 0.94692 |
| Cholesterol | 139.293 | 0.33605 | 0.77271 |
| Squalane | 215.557 | 0.49648 | 1.08108 |
| Stearic acid triglyceride | 501.540 | 0.99240 | 1.54588 |

With equations of state of the van der Waals-type, containing three parameters, it is in general possible to calculate *PVT* data for a gaseous phase with sufficient accuracy. In the liquid phase, the volume may not be calculated as accurately as needed. This has to be determined by the user for the individual case. If strongly polar and associating substances are involved, it is advisable to consult the literature or an expert. The application of EOS to calculation of phase equilibria, related to gas extraction, is covered in Chapter 3.

## 2.2.4 Correlations for Pure-Component Parameters

Usually, pure-component parameters of equations of state are calculated from the critical temperature, the critical pressure, and the acentric factor. Many substances which are of interest for supercritical fluid extraction processes decompose, when being heated, before the critical temperature is reached. For these substances, various estimation methods have been proposed. For a review see Reid et al. [54]. As mentioned above, calculation of pure-component parameters from critical values may not be adequate for the purposes of gas extraction calculations for several reasons:

- The critical point is unknown, since the compound decomposes;
- the critical point is not accurately known, because of experimental difficulties;
- the component actually is a mixture of compounds (impurities);
- several components are lumped together as a pseudo-component in order to calculate the phase envelope;
- the identity and chemical structure of the component or pseudo-component is unknown.

This leads to two conclusions: 1) Pure component parameters should be determined from other than critical properties, e.g., $PVT$ data, vapor pressure data, liquid molar volume and heat of vaporization; 2) Pure component parameters should be correlated with a physical property which is easy to measure. Brunner and Hederer developed correlations for pure component parameters of the Hederer-Peter-Wenzel modification of the Redlich-Kwong EOS. The correlations are based only on the molar volume at 20 °C, or at some other suitable temperature [4, 5]. The correlations are presented in Table 2.7.

This approach was further applied to the Peng-Robinson EOS by Dohrn and Brunner [14]. These correlations are presented in Table 2.8.

**Table 2.7.** Parameter correlations for the HPW-modification of the Redlich-Kwong EOS, after Hederer [27].

| Correlation | Correlation coefficient | Number of substances |
|---|---|---|
| $b = 0.991032\, v_{L20} - 0.01367$ | 0.99933 | 377 |
| $a = 2719624.208\, b^{3.264381736}$ | 0.92300 | 377 |
| $\alpha = -0.268042479 \ln b - 1.240725375$ | – | 377 |

$b$ in l/mole
$v_{L20}$ in l/mole at 293.15 K and 0.1013 MPa,
$a$ in (l$^2$ bar)/(mole$^2$).

**Table 2.8.** Parameter correlations for the Peng-Robinson EOS [14].

| Correlation | Average absolute deviation | Number of substances |
|---|---|---|
| $a_c = 459.905\, v_{L20} - 23.4643$ | 17.1 % | 531 |
| $b = 0.96371\, v_{L20} - 0.01199$ | 3.6 % | 531 |
| $x_0 = 1.19650\, v_{L20}^{0.25191}$ | 19.3 % | 531 |

$v_{L20}$ in l/mole

With these correlations, it is possible to calculate parameters for low volatile components and for impure components from their liquid density and their molecular weight, properties which can be determined in the laboratory by standard methods.

Accuracy of the correlations for representing $PVT$ data can be enhanced, if additional information is included in calculating the parameters. Using the normal boiling point $T_b$, Dohrn and Brunner [15] calculated another set of empirical correlations for the parameters of the Peng-Robinson EOS and the acentric factor. Representation of data can be improved, but on the other hand, an additional component property is needed ($T_b$ or another point on the vapor-pressure curve), rendering sometimes application to pseudo-components difficult, which was the main intention for developing parameter correlations.

Accuracy of $PVT$ data, calculated by the EOS using parameters from a correlation is lower than with individually fitted parameters. Especially parameters for low molecular weight compounds, obtained from the correlations, should not be used for calculation of $PVT$ behavior or phase equilibrium. If available, the fitted parameters from the tables, or personally obtained fitted parameters should be used. Parameters from correlations are only helpful if no other data are available. On the other hand, for complex mixtures of compounds they are in general the only way to make use of equations of state.

## 2.3 Thermodynamic Properties

Enthalpy, internal energy, entropy and others are important thermodynamic properties, which can be related to operating variables of processing equipment, e.g., enthalpy to a temperature increase in a heat exchanger or increasing pressure in an autoclave. These properties can be derived from $PVT$ behavior, as presented in textbooks on thermodynamics [52] and listed in Table 2.9.

Variations of these properties can be calculated in dependence of pressure, temperature and composition of a mixture using equations of state. Accuracy of the results depends on the accuracy of the experimental data on which the calculation is based, and on the accuracy of the equation of state, representing the $PVT$ behavior. In most cases, volume is subject to the greatest error in $PVT$ measurements. Pressure and temperature can be determined by carefully carried out measurements to an accuracy of 0.01 %. This corresponds to 0.1 bar or 0.01 MPa at a total pressure of 100 MPa (1000 bar), or 0.05 K at a temperature of 500 K (227 °C). Volumetric data at elevated pressures are seldom better known than with an error of 0.05 %, in most cases only 0.1 to 0.5 %, corresponding at 100 cm³/mole (nitrogen at 20 MPa and 320 K) to 0.1 to 0.5 cm³/mole. Values for thermodynamic state functions are derived from $PVT$ data by differentiation. In general, each differentiation reduces accuracy of the result by one order of magnitude. Then the resulting error is in the range of 1 % after the first differentiation and in the range of 10 % after the second differentiation. Therefore, thermodynamic state functions are only reluctantly calculated from $PVT$ data, since

**Table 2.9.** Derivation of thermodynamic properties from $PVT$ data [52].

$$U = \int_0^P [V - T \left( \frac{\partial V}{\partial T} \right)_{P,n}] \, dP - PV + \sum_i (n_i \, h_i^\circ)$$

$$H = \int_0^P [V - T \left( \frac{\partial V}{\partial T} \right)_{P,n}] \, dP + \sum_i (n_i \, h_i^\circ)$$

$$S = \int_0^P \left[ \frac{nR}{P} - \left( \frac{\partial V}{\partial T} \right)_{P,n} \right] dP - R \sum_i [n_i \ln (y_i P)] + \sum_i (n_i s_i^\circ)$$

$$A = \int_0^P \left[ V - \frac{nRT}{P} \right] dP + RT \sum_i [(n_i \ln (y_i P)] - PV + \sum_i [n_i (h_i^\circ - Ts_i^\circ)]$$

$$G = \int_0^P \left[ V - \frac{nRT}{P} \right] dP + RT \sum_i [(n_i \ln (y_i P)] + \sum_i [n_i (h_i^\circ - Ts_i^\circ)]$$

$$\mu_i = \int_0^P \left[ \bar{V}_i - \frac{RT}{P} \right] dP + RT \ln (y_i P) + h_i^\circ - Ts_i^\circ$$

$\bar{V}_i = (\delta V / \delta n_i)_{T,P,n_j}$, the partial molar volume of $i$;
$y_i$ = mole fraction of $i$;
$n$ = total number of moles;
$n_i$ = number of moles $i$;
$\mu_i^\circ = h_i^\circ - Ts_i^\circ$;
$s_i^\circ$ = molar entropy of pure $i$ as ideal gas at $T$ and $P = 0.1$ MPa;
$h_i^\circ$ = molar enthalpy of pure $i$ as ideal gas at $T$.

the underlying experimental data are available with sufficient accuracy only for a few substances.

We now consider for some thermodynamic properties dependency on pressure. In Fig. 2.7 entropy for supercritical water is shown in dependence of pressure. Increasing pressure leads to a higher degree of order in the molecular arrangement. Therefore entropy decreases with increasing pressure.

Enthalpy of a gas also decreases with pressure at constant temperature as shown in Fig. 2.8 for carbon dioxide. Enthalpy of the liquid phase at constant temperature remains constant for moderate pressures. Only at high pressures in the range of more than 100 MPa does influence of pressure on thermodynamic properties of most liquids become substantial.

Interdependence of thermodynamic properties is often represented in a $T,S$-diagram, as presented for carbon dioxide in Fig. 2.9. Thermodynamic work needed for a reversible cyclic process can be determined by the area in a $T,S$-diagram included by

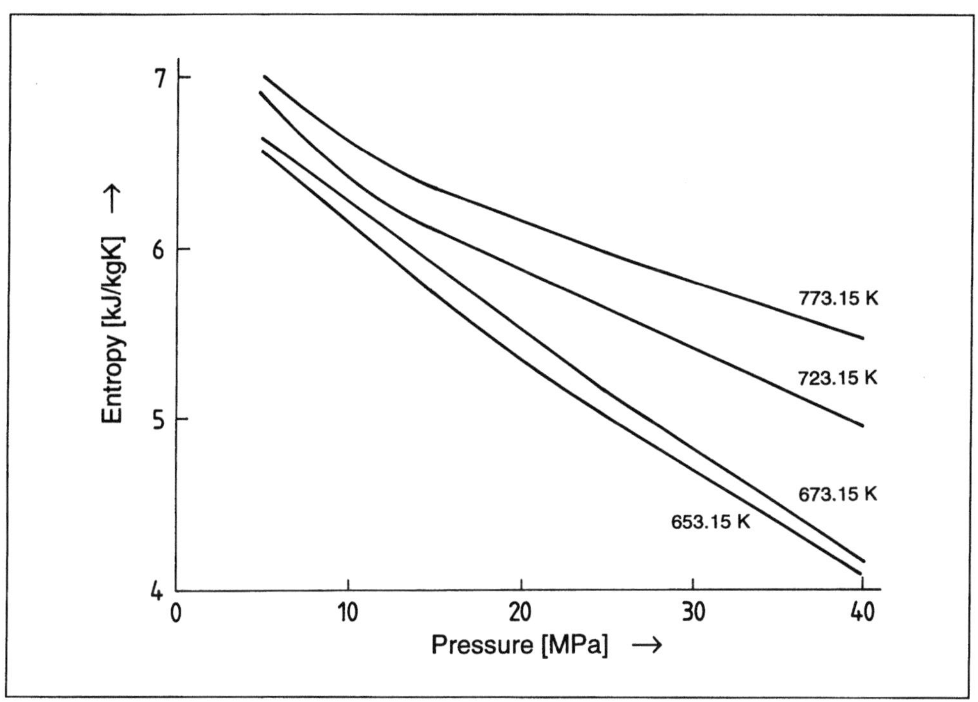



**Fig. 2.7.** Entropy dependence of gaseous water on pressure.

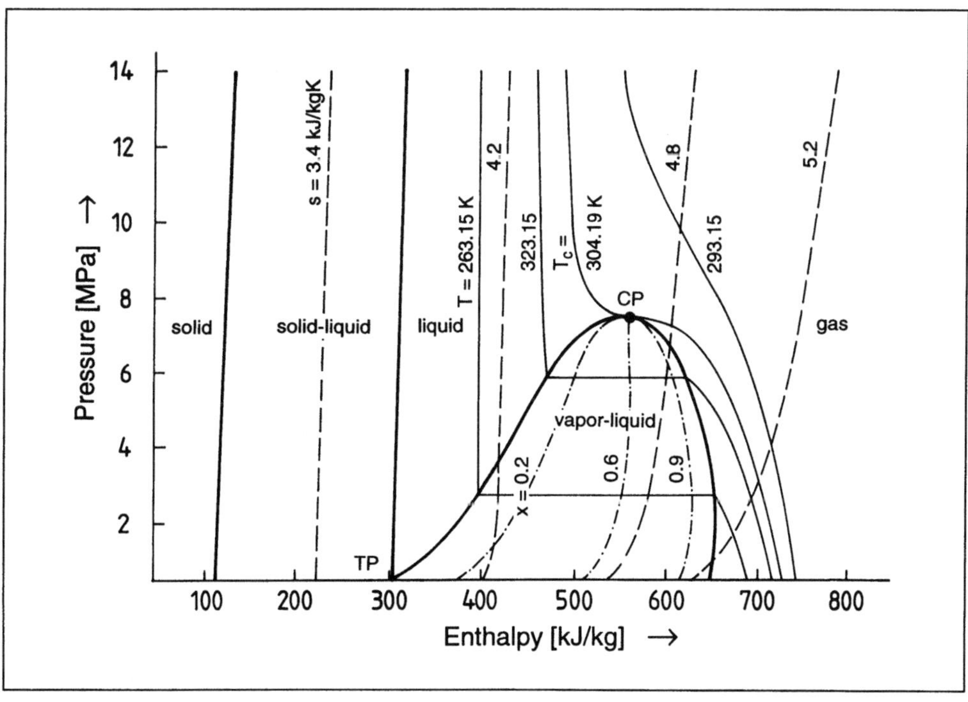



**Fig. 2.8.** Enthalpy dependence on pressure for carbon dioxide.

the process cycle. Such diagrams are similar for most substances. For substances, which may be used as solvents in gas extraction, diagrams or tables of thermodynamic properties can be found in data collections.

Heat capacity $c_p$ of gaseous nitrogen is shown in Fig. 2.10 [47]. Starting from low pressures, heat capacity $c_p$ increases with increasing pressure and passes at lower temperatures through a maximum in the range of $20-30$ MPa. At higher pressures, variation of $c_p$ with pressure is low. For carbon dioxide, $c_p$ in the region up to 22 MPa is plotted in Fig. 2.11.

In gases, heat capacity $c_p$ decreases with temperature and increases with pressure. Since supercritical solvents are applied while undergoing major changes of pressure and temperature, behavior of $c_p$ is better related to density, Fig. 2.12. On the other hand, since $P$ and $T$ are adjustable parameters in a process, dependence on $P$ and $T$ must be familiar.

An overview on the behavior of $c_p$ in the critical region is shown in Fig. 2.13 for water. Values for $c_p$ are reduced with the heat capacity of the liquid under boiling con-

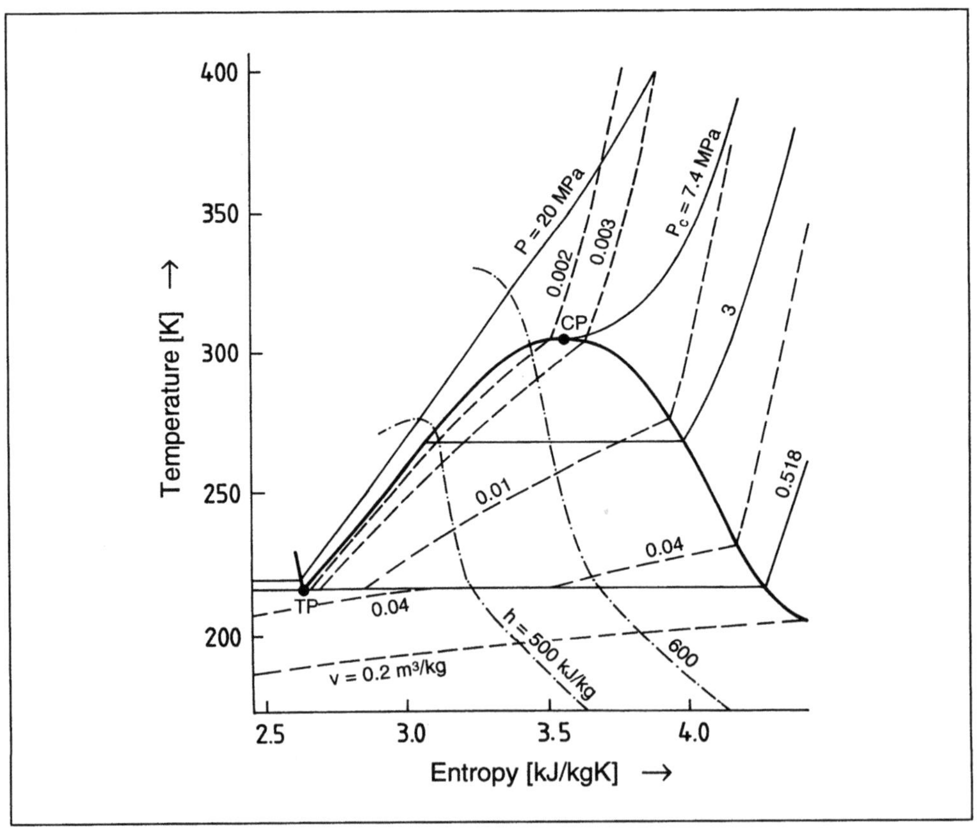


Fig. 2.9. $T,S$ diagram for carbon dioxide.

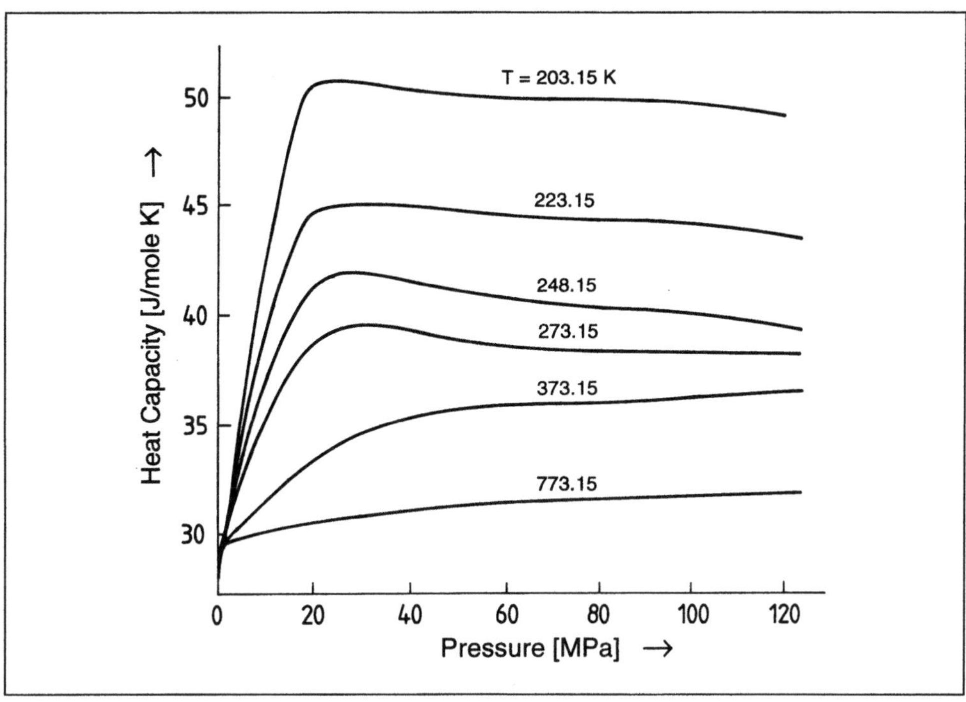

**Fig. 2.10.** Heat capacity $c_p$ of gaseous nitrogen.

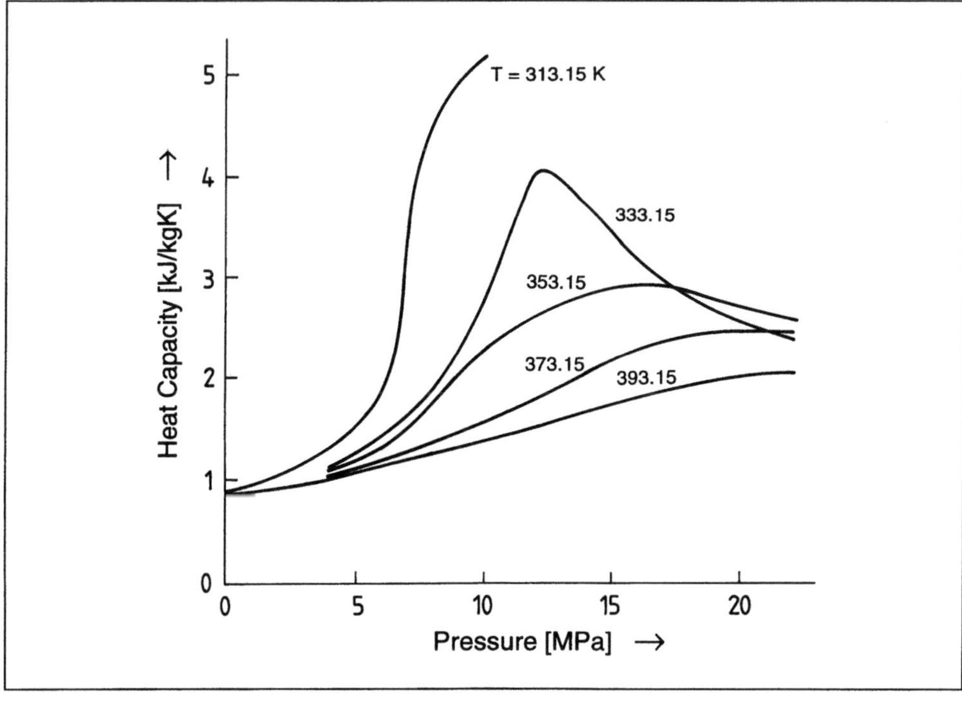

**Fig. 2.11.** Heat capacity $c_p$ of gaseous carbon dioxide.

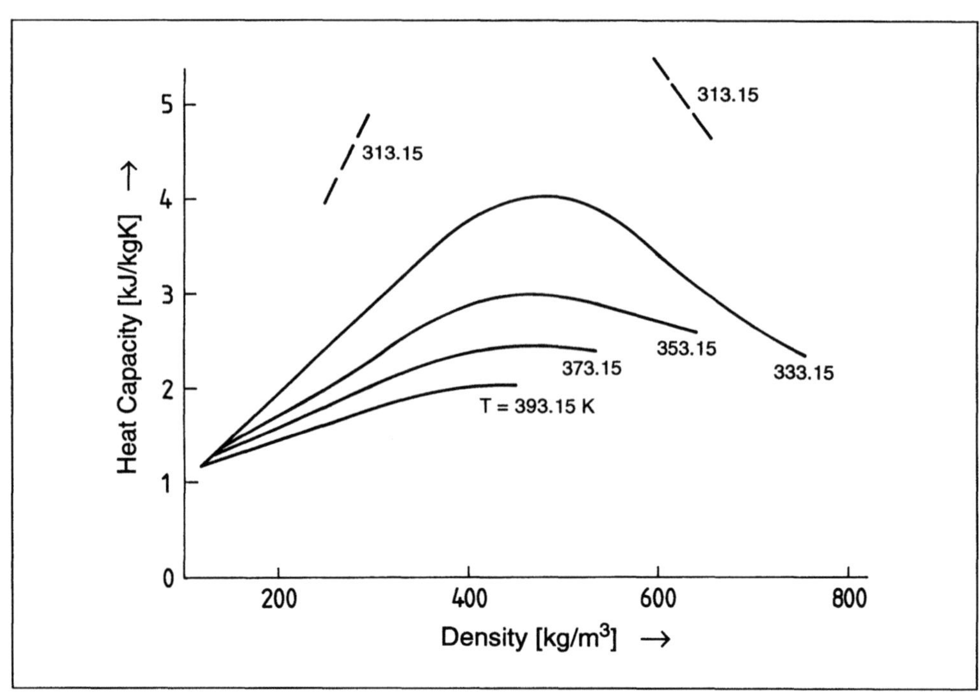

**Fig. 2.12.** Heat capacity $c_p$ of $CO_2$ as a function of density.

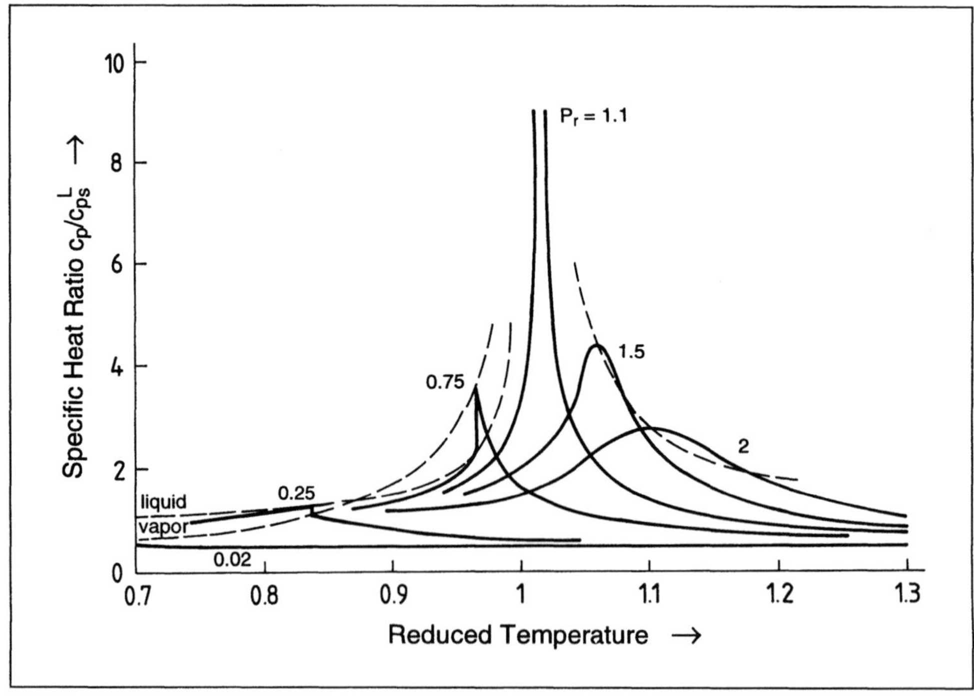

**Fig. 2.13.** Specific heat of water at constant pressure in the critical region. See text.

26

ditions at 1 bar for the liquid state, and for the gaseous state are reduced with the heat capacity of the ideal gas. Heat capacity of the liquid increases at constant pressure near boiling temperature. The increase is the more pronounced, the nearer the boiling temperature is to the critical temperature. Heat capacity of the gas exhibits the greatest deviations from ideal gas near the critical point. But these deviations decrease rapidly with increasing temperature. For pressures lower than critical, the ideal gas heat capacity is approached with increasing temperature. Between heat capacities of saturated gas and boiling liquid a gap occurs, which widens when the critical pressure is approached. At pressures higher than critical $(P_r > 1)$, $c_p$ runs through a maximum, which becomes flatter and is shifted to higher temperatures with increasing pressures. At the critical point heat capacity is infinite, $c_p \rightarrow \infty$.

The reader interested in thermodynamic and transport properties around the critical point, yet within the area of technical application, is referred to the work of Sengers [58] and Levelt-Sengers [40].

**The Joule-Thomson coefficient.** In high pressure processes like gas extraction, pressure changes are part of the process. From experience it is known (since Joule and Thomson carried out their experiments in the 19th century) that temperature of most gases decreases, when pressure is lowered in a flowing gas by throttling. This effect is called the Joule-Thomson effect and is easily experienced while experimenting with carbon dioxide. The throttling valve from which a sample is taken from a vessel, pressurized with $CO_2$, cools down to subzero temperature. Ice forms on the valve from the condensating humidity of the surrounding air. Solutes can solidify and block downstream lines. Therefore a throttling valve must be heated.

The Joule-Thomson coefficient $\eta_{JT}$ is defined as the variation of temperature with pressure at constant enthalpy:

$$\eta_{JT} = \left( \frac{\partial T}{\partial P} \right)_H \qquad 2.29$$

It can be derived from volumetric properties:

$$\eta_{JT} = \frac{T(\partial V/\partial T)_P - V}{c_p} \qquad 2.30$$

The Joule-Thomson coefficient of an ideal gas is zero. In real gases, the Joule-Thomson coefficient is different from zero and depends on pressure and temperature. For $\eta_{JT} > 0$, temperature decreases, and for $\eta_{JT} < 0$, temperature increases during an expansion. The temperature at which the sign of the Joule-Thomson coefficient changes, is the inversion- or Boyle temperature, which itself depends on pressure. The inversion temperature of most gases is above ambient temperature; for hydrogen the inversion temperature is about $-$ 80 °C. Values for the Joule-Thomson coefficient are listed in Table 2.10 for $CO_2$.

**Table 2.10.** Joule-Thomson coefficient $\eta_{JT}$ for $CO_2$ [47].

| | $\eta_{JT} = (\partial T/\partial P)_H$ in [K/(0.1 MPa)] | | | | | | |
|---|---|---|---|---|---|---|---|
| Tempera-ture [K] | Pressure [MPa] | | | | | | |
| | 0.1013 | 2.025 | 6.08 | 7.4 | 10.1 | 14.2 | 20.3 |
| 223.15 | 2.3814 | $-$ 0.0138 | $-$ 0.0148 | $-$ 0.0163 | $-$ 0.0158 | $-$ 0.0181 | $-$ 0.0245 |
| 273.15 | 1.2731 | 1.3837 | 0.0365 | 0.0306 | 0.0212 | 0.0113 | 0.0044 |
| 323.15 | 0.8833 | 0.8833 | 0.8685 | 0.8117 | 0.5497 | 0.1697 | 0.0918 |
| 373.15 | 0.6405 | 0.6292 | 0.6000 | 0.5843 | 0.5334 | 0.4264 | 0.2522 |
| 423.15 | 0.4826 | 0.4634 | 0.4372 | 0.4323 | 0.4101 | 0.3711 | 0.2872 |
| 473.15 | 0.3721 | 0.3528 | 0.3356 | 0.3282 | 0.3109 | 0.2852 | 0.2423 |

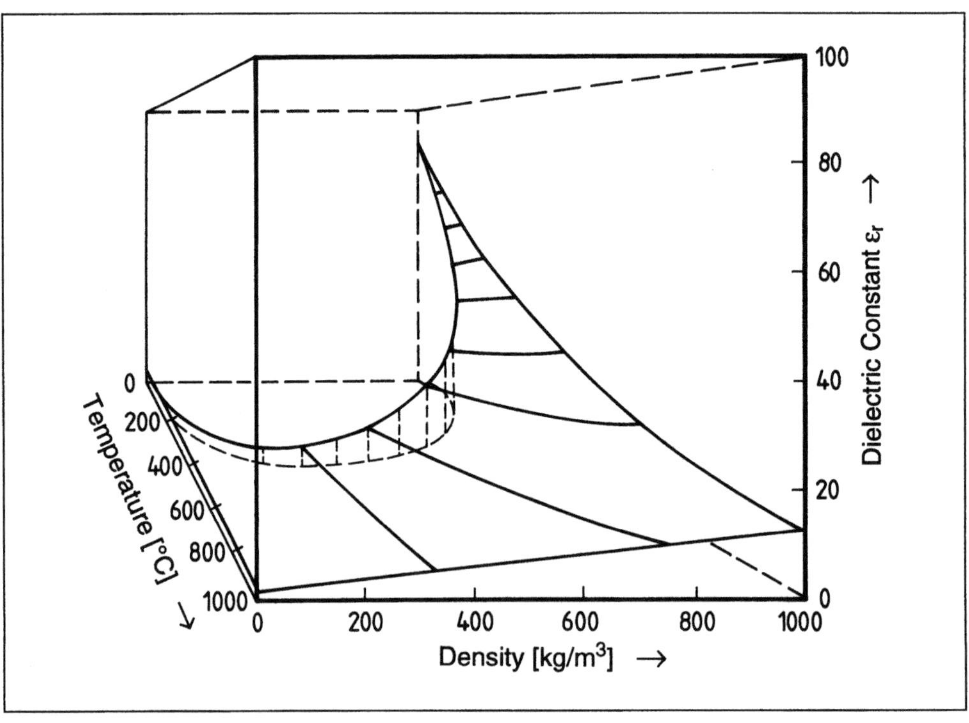


**Fig. 2.14.** Dielectric constant of water [11].

28

**Ionic and associating substances.** Ionic and associating substances change their properties when passing the critical temperature. Water changes its structure above the critical point, but remains a polar fluid, as can be seen from the dielectric constant $\varepsilon$, Fig. 2.14. At supercritical conditions $\varepsilon$ is still that of a strongly polar substance. Dissociation of water also remains at supercritical conditions, as illustrated in Fig. 2.15.

The addition of a polar substance like methanol to carbon dioxide should enhance polarity of the mixture, thus enabling the solvation of polar molecules. Roskar et al. [55] have determined the dielectric constant of carbon dioxide with methanol, as presented in Fig. 2.16. A substantial enhancement of $\varepsilon$ occurs only at concentrations for methanol higher than 20 mol-%. Since entrainer effects are observed at much lower concentrations, other effects dominate. Similar results were observed by Deul and Franck for benzene-water [11], Fig. 2.17. A substantial concentration of water has to be added to benzene, in order to enhance the value of the dielectric constant of the unpolar molecule. But the dielectric constant for water decreases rapidly, when benzene is added.

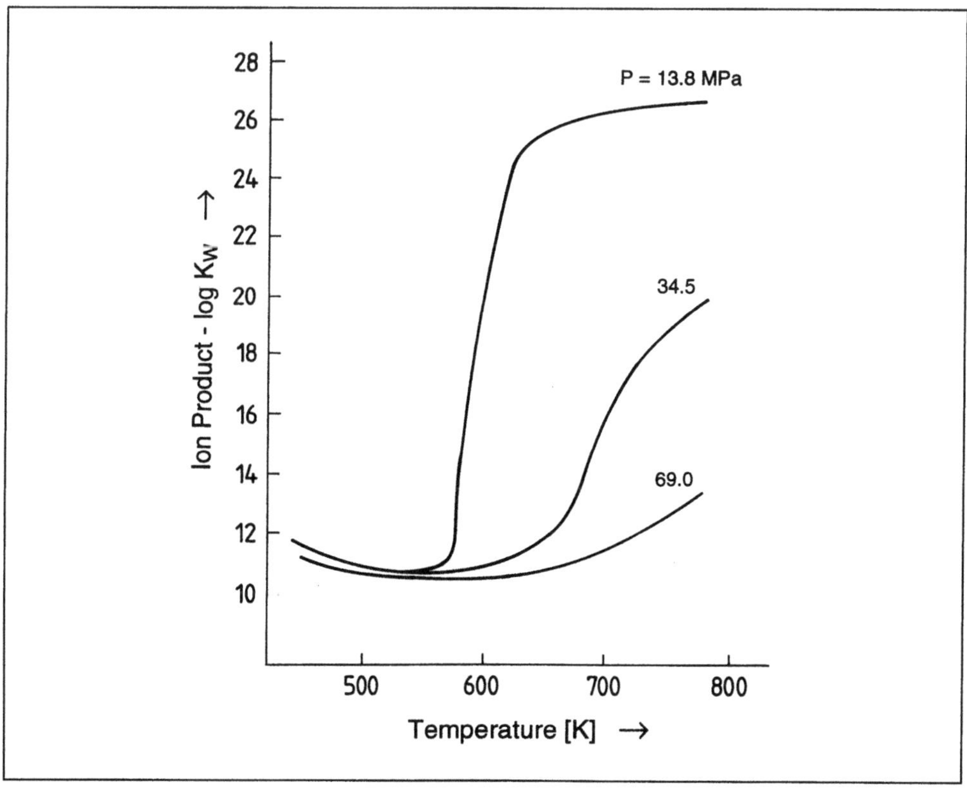


**Fig. 2.15.** Ion product of water.

29

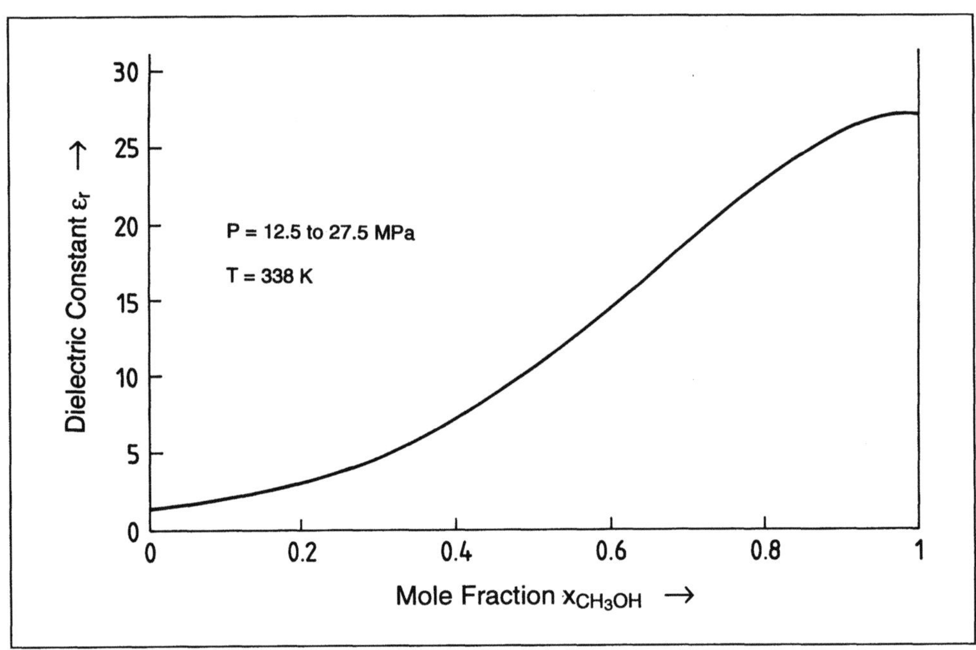

**Fig. 2.16.** Dielectric constant for $CO_2$ – methanol mixtures [55].

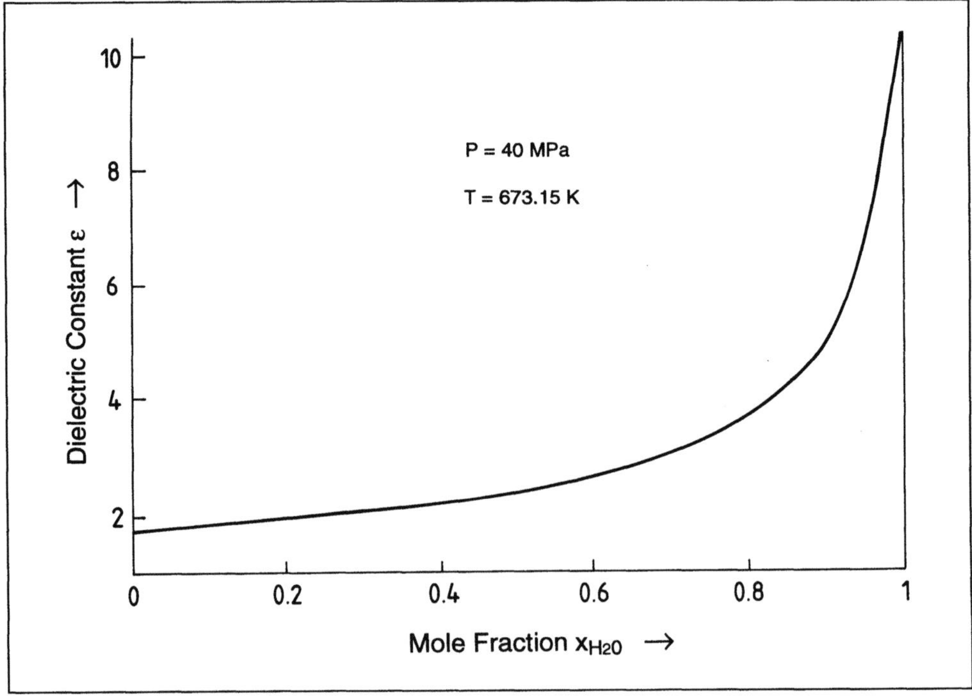

**Fig. 2.17.** Dielectric constant for benzene – water [11], mole fraction.

30

# 2.4 Transport Properties

In gas extraction, as in other technical processes, fluids and solids are transported, their temperature is changed by heat transfer and composition of mixtures is altered by mass transport via phase boundaries. Bulk transport of matter is determined by viscosity, transport of thermal energy by thermal conductivity and molecular transport is characterized by the diffusion coefficient. Enhanced pressure, as necessary for gas extraction processes, strongly influences these transport properties. Contrary to equilibrium properties, the influence of pressure on nonequilibrium properties like viscosity, thermal conductivity and diffusion coefficient cannot be calculated with thermodynamic relations.

## 2.4.1 Viscosity

Viscosity is defined by a fluid, flowing laminar or free of turbulences, with a velocity gradient vertical to the direction of flow. This gradient in velocity induces a shear stress. By relating shear stress to an area and dividing it by the velocity gradient, the viscosity of the fluid is obtained, as given by

$$\eta = \frac{\tau}{d\omega/dy}, \qquad\qquad 2.31$$

with:
$\tau$ = shear stress per unit of area
$\omega$ = flow velocity
$y$ = coordinate vertical to direction of flow

Fluids are discerned by their behavior of viscosity in dependence on velocity gradient and shear stress. Viscosity of so-called Newtonian fluids is independent on shear stress and velocity gradient and depends only on conditions of state *(P, V, T)*. Most pure liquids, simple mixtures and gases are Newtonian fluids. Non-Newtonian fluids, on the other hand, are characterized by a viscosity depending either on shear stress or on shear velocity.

### 2.4.1.1 Influence of Pressure on the Viscosity of Gases

Influence of pressure on the viscosity of gases is strong only in certain ranges of temperature and pressure. In general, influence of pressure is low at high reduced temperatures and at low reduced pressures. As shown in Fig. 2.18, influence of pressure is low for temperatures well above the critical temperature [33]. For higher pressures than those shown in Fig. 2.18, viscosity increases more strongly with pressure.

Influence of pressure can be derived from a generalized presentation of viscosity, as shown in Fig. 2.19 [63]. Viscosity at supercritical conditions decreases at constant pressure with temperature to a minimum, and then increases with temperature. The viscosity minimum is shifted to higher temperatures with increasing pressure. At temperatures below the minimum, the supercritical fluid behaves, with respect to viscosity, like a liquid: Viscosity decreases with temperature. At temperatures above the minimum, the supercritical fluid behaves, with respect to viscosity, like a gas: Viscosity increases with temperature. In the region of pressure and temperature, which is mostly exploited for gas extraction purposes, viscosity of the supercritical fluid decreases with temperature.

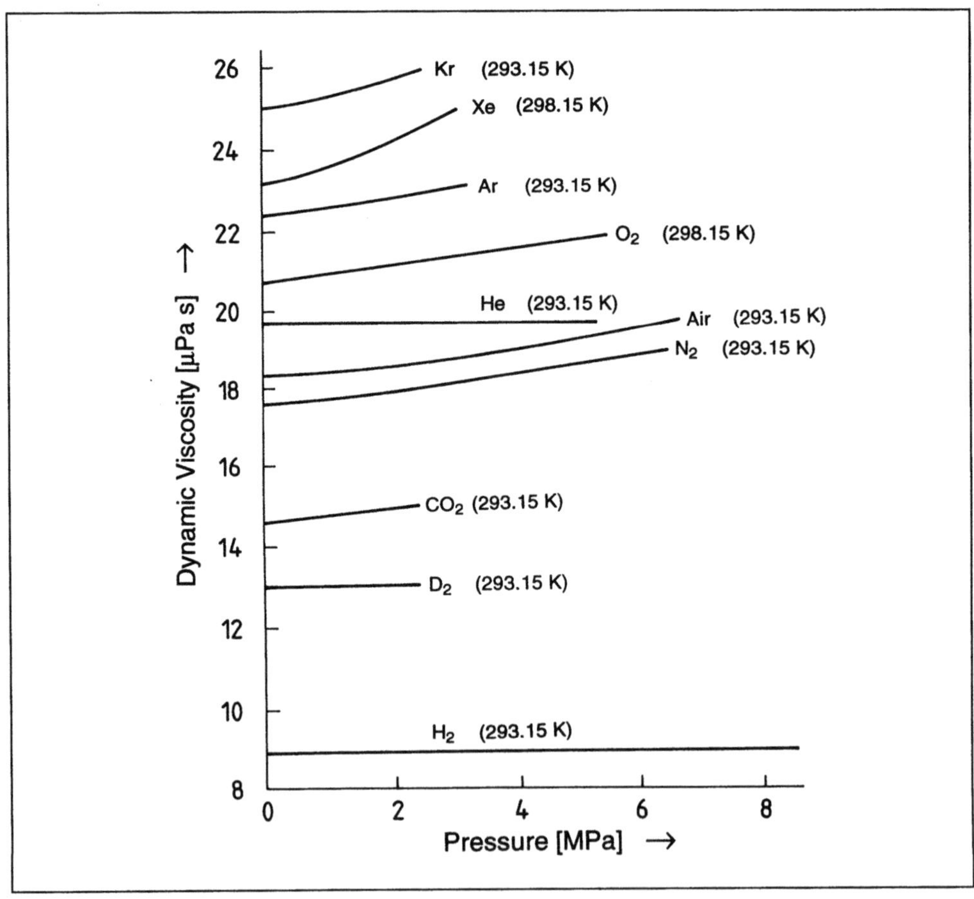


**Fig. 2.18.** Viscosity of gases at low reduced pressures and high reduced temperatures (after Kestin and Leidenfrost [33]).

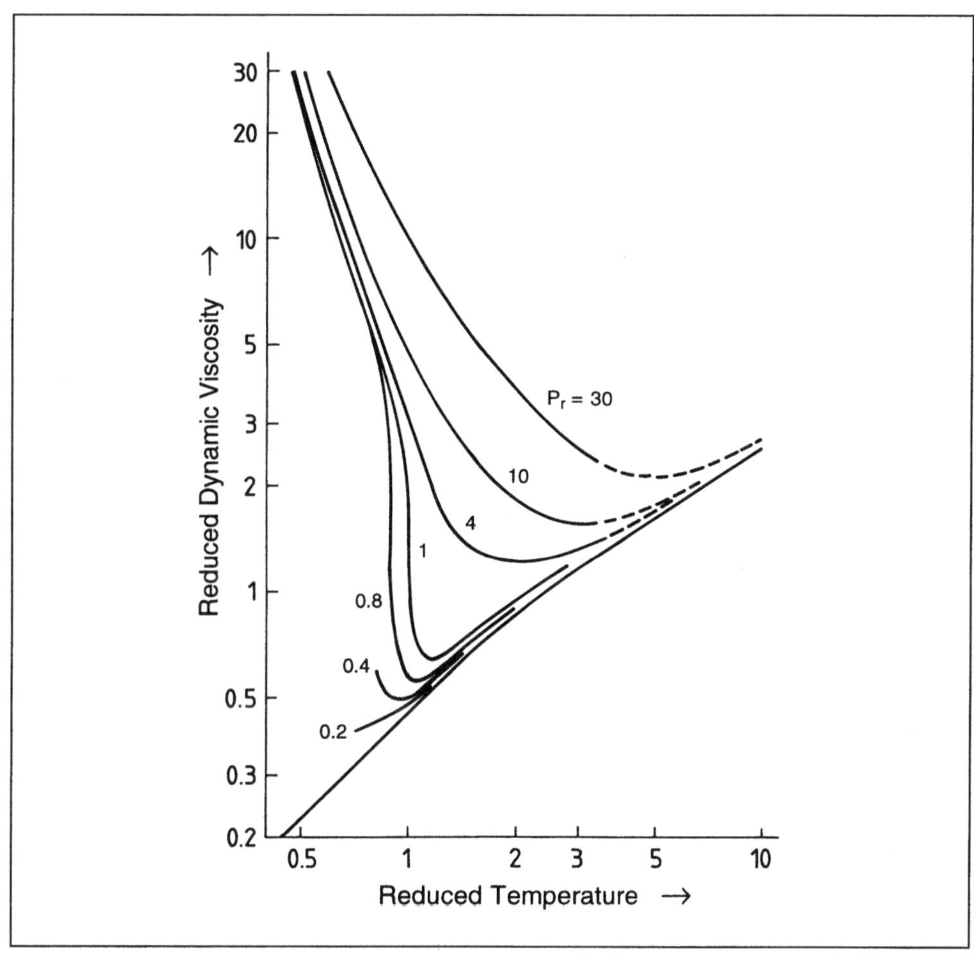


**Fig. 2.19.** Generalized viscosity behavior (after Uyehara and Watson [63]).

According to Childs and Hanley [6], a criterium, presented in Eq. 2.32, can be formulated for dense gas regions, where the influence of pressure on viscosity must be taken into account. Viscosity for gases at pressures lower than that given by Eq. 2.32 can be determined by normal correlations without considering pressure. Influence of pressure is less than 1 % on the value of viscosity. Otherwise, a correction for pressure is considered necessary. The correction is applied as a departure function, relating viscosity at high pressure $(\eta)$ to viscosity at low pressure $(\eta^o)$ at the same temperature. Data for $\eta^o$ can be found from experimental data compilations, by correlations for calculation of compound properties [54] published in textbooks, or by the correlations presented below.

$$P/P_c > 0.183 \, (T/T_c) - 1. \tag{2.32}$$

### 2.4.1.2 Nonpolar Gases

For nonpolar gases, Jossi et al. [31] have published the following correlation:

$$[(\eta - \eta^o)\,\xi + 1]^{1/4}$$
$$= 1.0230 + 0.23364\,\varrho_r + 0.58533\,\varrho_r^2 - 0.40758\,\varrho_r^3 + 0.093324\,\varrho_r^4, \qquad 2.33$$

with
$\eta$ = viscosity of the dense gas, $\mu$P;
$\eta^o$ = viscosity of the gas at low pressure, $\mu$P;
$\varrho_r$ = reduced density, $\varrho/\varrho_c = v_c/v$;
and for $\xi$:

$$\xi = \frac{T_c^{1/6}}{M^{1/2} \cdot P_c^{2/3}}, \qquad 2.34$$

with $T_c$ in K, and $P_c$ in atm and $M$ for molecular mass. The range of application is:
$0.1 \le \varrho_r < 3$.

### 2.4.1.3 Polar Gases

For polar gases, Stiel and Thodos [61] published the following set of equations for
correlating viscosity at high pressures from viscosity at low pressures:

$$(\eta - \eta^o)\,\xi = 1.656\,\varrho_r^{1.111} \qquad \text{for } \varrho_r \le 0.1 \qquad 2.35$$

$$(\eta - \eta^o)\,\xi = 0.0607\,(9.045\,\varrho_r + 0.63)^{1.739} \qquad \text{for } 0.1 \le \varrho_r \le 0.9 \qquad 2.36$$

$$\log\{4 - \log[(\eta - \eta^o)\xi]\} = 0.6439 - 0.1005\,\varrho_r - \triangle \quad \text{for } 0.9 \le \varrho_r < 2.6 \qquad 2.37$$

with

$$\triangle = \begin{cases} 0 & \text{for } 0.9 \le \varrho_r < 2.2 \qquad 2.38 \\ 4.75 \cdot 10^{-4}\,(\varrho_r^3 - 10.65)^2 & \text{for } 2.2 \le \varrho_r < 2.6 \qquad 2.39 \end{cases}$$

$$(\eta - \eta^o)\,\xi = 90.0 \qquad \text{for } \varrho_r = 2.8 \qquad 2.40$$

$$(\eta - \eta^o)\,\xi = 250 \qquad \text{for } \varrho_r = 3.0 \qquad 2.41$$

### 2.4.1.4 Mixtures of Gases

For mixtures of dense gases a departure function similar to that for pure gases can be applied. Such a function was presented by Dean and Stiel [9]:

$$(\eta_m - \eta_m{}^o)\,\xi_m = 1.08\,[\exp(1.439\varrho_{rm}) - \exp(-\,1.111\varrho_{rm}^{1.858})] \qquad\qquad 2.42$$

with
$\eta_m$ = viscosity of the mixture at high pressure, $\mu$P
$\eta^o{}_m$ = viscosity of the mixture at low pressure, $\mu$P
$\varrho_{rm}$ = reduced density of the mixture, $\varrho_m/\varrho_{cm}$
$\varrho_{cm}$ = pseudo critical density of the mixture, mol/cm$^3$

$$\varrho_{cm} = \frac{P_{cm}}{z_{cm}\,RT_{cm}} \qquad\qquad 2.43$$

$$T_{cm} = \sum y_i T_{ci} \quad z_{cm} = \sum y_i z_{ci}; \quad v_{cm} = \sum y_i v_{ci}; \qquad\qquad 2.44$$

$\xi_m$ as defined in Eq. 2.34.

This correlation should only be applied to nonpolar gases or mixtures of non-polar gases. In this case, calculated values compare well with experimental values. Errors for low molecular weight gases are said to be below 10 %. For mixtures of polar gases no adequate correlation could be found so far.

### 2.4.1.5 Liquids

Viscosities of liquids and gases are quite different. Viscosities of liquids are far higher than those of gases, but decrease rapidly with increasing temperature. Viscosity of gases under ambient conditions is about $\eta \approx 10^{-5}$ Ns/m$^2$. Dynamic viscosities of liquids cover a broad range. The viscosity of liquid n-heptane at 20 °C is: $\eta = 4 \cdot 10^{-4}$ Ns/m$^2$, for n-hexadecane $35 \cdot 10^{-4}$ Ns/m$^2$, for water $10 \cdot 10^{-4}$ Ns/m$^2$, for methylene chloride $1{,}8 \cdot 10^{-4}$ Ns/cm$^2$ (all at 20 °C).

Viscosity of gases at low pressure is due to transfer of impulse energy by individual impacts between molecules, moving at random between layers of different velocity in the flowing medium. In liquids, an impulse transfer by a similar mechanism is also effective, but is superimposed by the intermolecular forces of the densely packed liquid molecules. Density of liquids is high enough, that the average distance of two molecules is in the range of the force fields of the molecules. Based on molecular properties so far no correlation could be developed for calculation of the viscosity of liquids. Therefore, experimental data are the only data base.

At low temperatures, below normal boiling point, viscosity is nearly independent on pressure for moderate values of pressure. Up to 4 MPa no influence of pressure can be determined, but at high pressures, viscosity covers a broader range of values than other component properties. Experimental results show that the decrease in viscosity with temperature is enhanced with increasing pressure, an unusual behavior compared to equilibrium properties. In Fig. 2.20 viscosity behavior for benzene is shown for example [45].

An expressive model for the pressure dependence of the viscosity of liquids is given by Bridgman. He observed that the influence of pressure is greatest on complex molecules. Therefore, he suggested that the increase in viscosity with increasing pressure is due to the interhooking of molecules of complex structure, which thus are hindered in moving relative to one another. According to this model, an increase in temperature causes an increase of internal energy and of average distance of molecules, thereby reducing interhooking of the molecules and thus decreasing viscosity. So far,

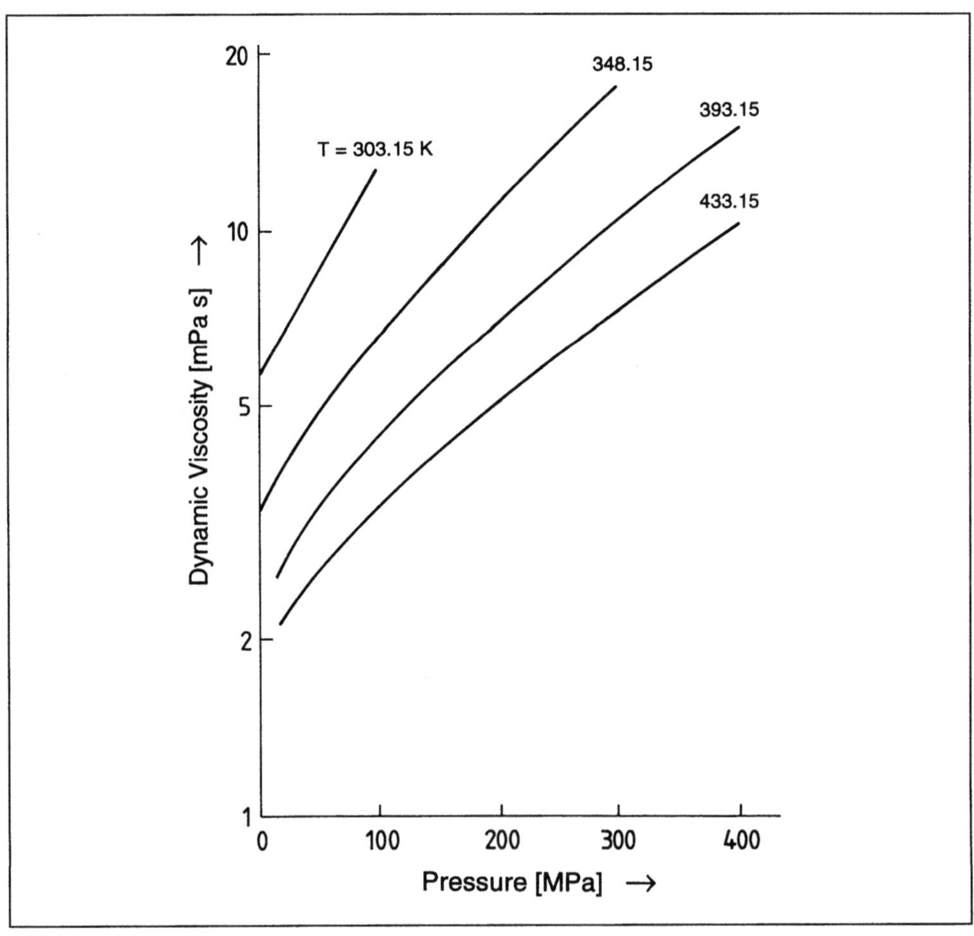


**Fig. 2.20.** Dependence of viscosity of liquid benzene on pressure (after Parkhurst and Jonas [45]).

36

this model has not been transferred into a correlation for viscosity of liquids at high pressures.

For calculating pressure dependence of viscosity of liquids, a satisfactory correlation is not available. Since in gas extraction, liquids in general are saturated with the supercritical solvent, the pressure effect on pure liquids is not very important for this application.

### 2.4.1.6 Viscosity of Systems with Supercritical Fluids

Viscosity is an important parameter for operating gas extraction process equipment, since it influences pressure drop, mass transfer and capacity. It is well known that dissolved supercritical carbon dioxide reduces viscosity of a substance in the liquid phase, since Kuss and Golly published their results [37]. A few experimental investigations have been carried out since then [12, 29, 32, 36, 38, 48]. Components involved are restricted to mineral oil components, fatty acids, fatty acid esters, glycerides and similar components.

Viscosity in the saturated liquid phase decreases with increasing pressure, which corresponds to increasing amount of dissolved supercritical fluid. Figure 2.21 pre-

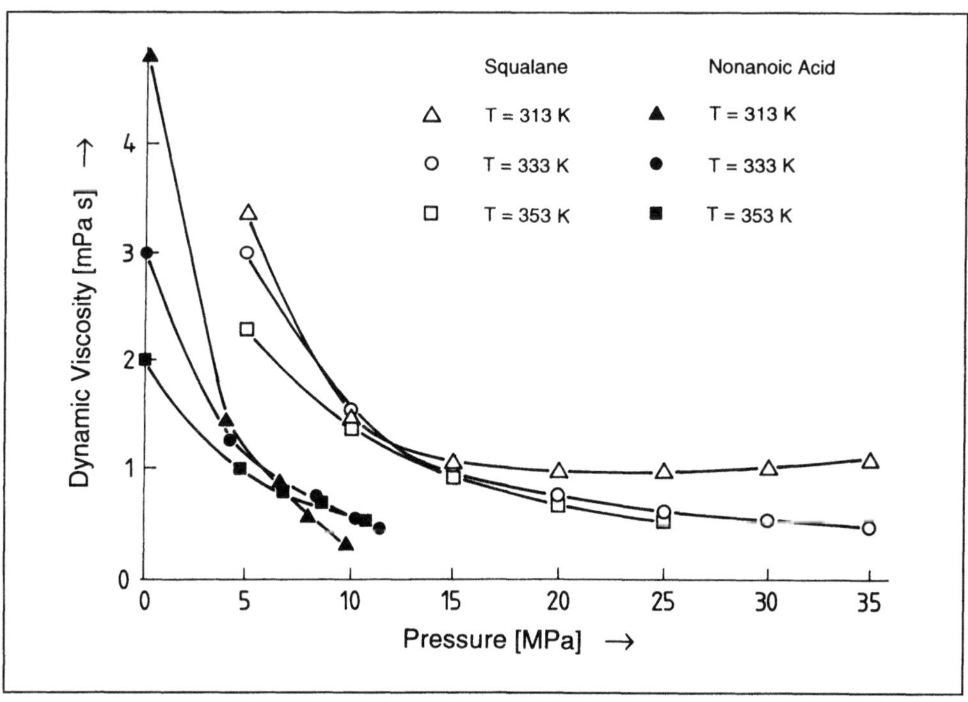


**Fig. 2.21.** Viscosity of liquid phases saturated with carbon dioxide. Squalane – $CO_2$ after Dilchert [12], nonanoic acid – $CO_2$ after Jakob [29], cited from [21].

sents data from Jakob [29] on the dynamic viscosity of the saturated liquid phase of pelargonic acid ($CH_3(CH_2)_7COOH$) – carbon dioxide and data from Dilchert [12] on the system squalane – $CO_2$.

Jakob further determined viscosities in systems carbon dioxide – oleic acid ($CH_3(CH_2)_7CH = CH(CH_2)_7COOH$), ethane – nonanoic acid, ethane – oleic acid, kerosene – $CO_2$ and kerosene – ethane. Peter et al. [48] published data on mixtures of oleic and stearic acid with $CO_2$ and ethane, and soybean oil with $CO_2$ and propane. Kashulines et al. [32] investigated liquid phase viscosities of oleic acid, linoleic acid, milk fat, methyl oleate and methyl linoleate with $CO_2$. The following tendencies are common to all data:

– Viscosity of the liquid phase, saturated with the supercritical compound, decreases rapidly with pressure;
– liquid phase viscosity decreases with temperature;

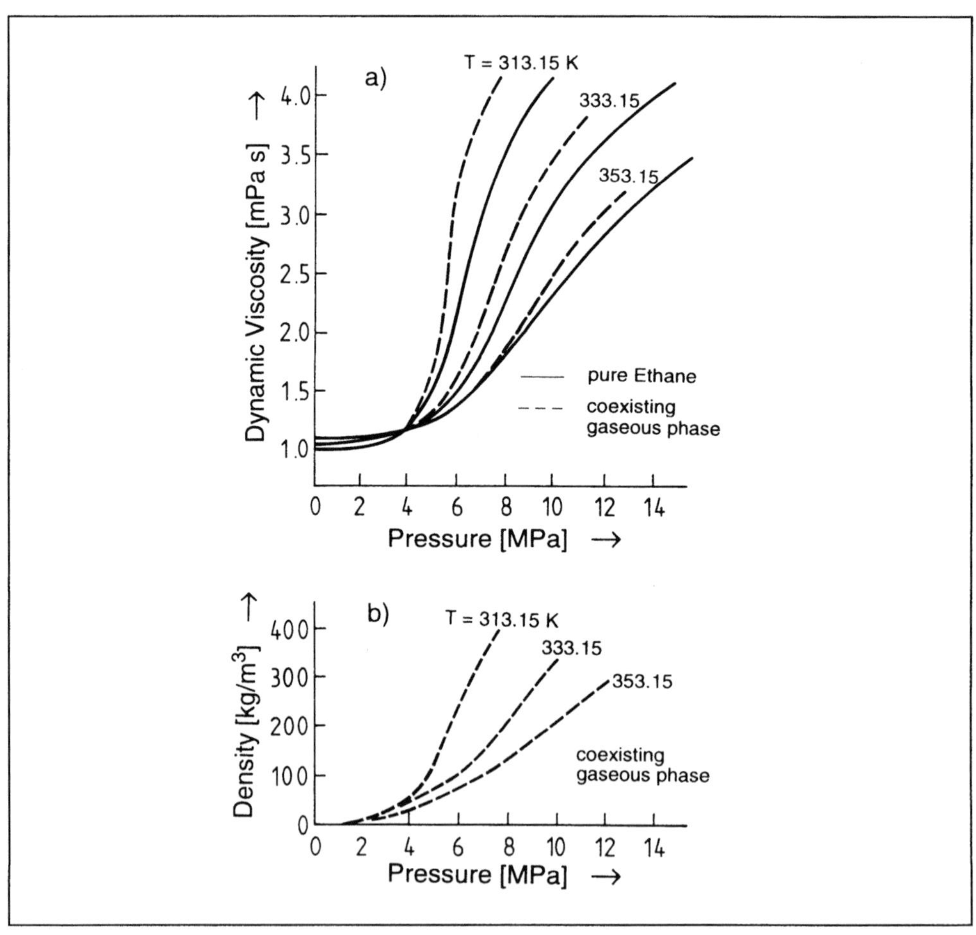


**Fig. 2.22.** Gas phase viscosity (a), and density (b) of nonanoic acid – ethane (after Jakob [29]).

– decrease of liquid phase viscosity is higher at lower temperatures.

The pure components are Newtonian fluids. The gaseous phases also behave like Newtonian fluids. The same behavior was found for the liquid phase in systems oleic acid – $CO_2$, oleic acid – ethane and linoleic acid/oleic acid – ethane for shear velocities up to $1000\,s^{-1}$ [29]. For systems nonanoic acid – ethane, nonanoic acid – carbon dioxide and kerosene – $CO_2$, viscosity increases with shear velocity, which is characteristic for shear-thickening or dilatancy.

Viscosity of the gaseous phase depends on the viscosity of the supercritical solvent and the concentration of the low volatile components. Since this concentration is mostly low in gas extraction, viscosity of the gaseous phase is approximately that of the supercritical compound. But with increasing concentration, gas phase viscosity also deviates from viscosity of the pure supercritical compound. In Fig. 2.22 data on gas phase viscosity of pelargonic acid – ethane are plotted from the work of Jakob

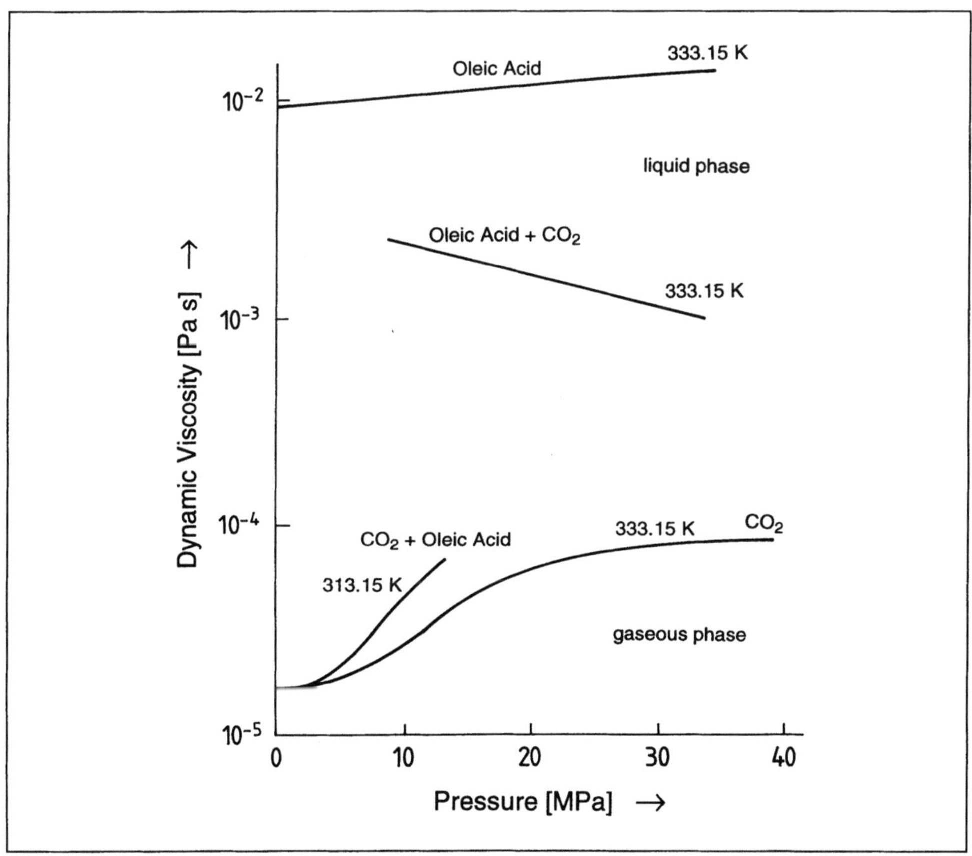


**Fig. 2.23.** Dynamic viscosity of saturated phases for the system oleic acid – $CO_2$. Data from Kashulines et al. [32] and Jakob [29].

[29]. Viscosity of the saturated gas increases with increasing pressure, corresponding to increasing amount of low volatile compound, dissolved in the gas.

Viscosities of both phases are compared to pure component viscosities in Fig. 2.23 for the system oleic acid – $CO_2$. Viscosity of the liquid phase decreases with pressure, while viscosity of the gaseous phase increases with pressure.

As the supercritical compound dissolves in the liquid phase, properties of the liquid are changed. Decrease in density might be an explanation for decreasing viscosity. But experimental evidence shows that the main influence is due to the number of molecules dissolved, since density, as shown in Fig. 2.24 for the system nonanoic acid – $CO_2$, may even increase. If viscosity of both phases is plotted against mole fraction, a relatively simple picture is obtained. Viscosity of the liquid decreases with con-

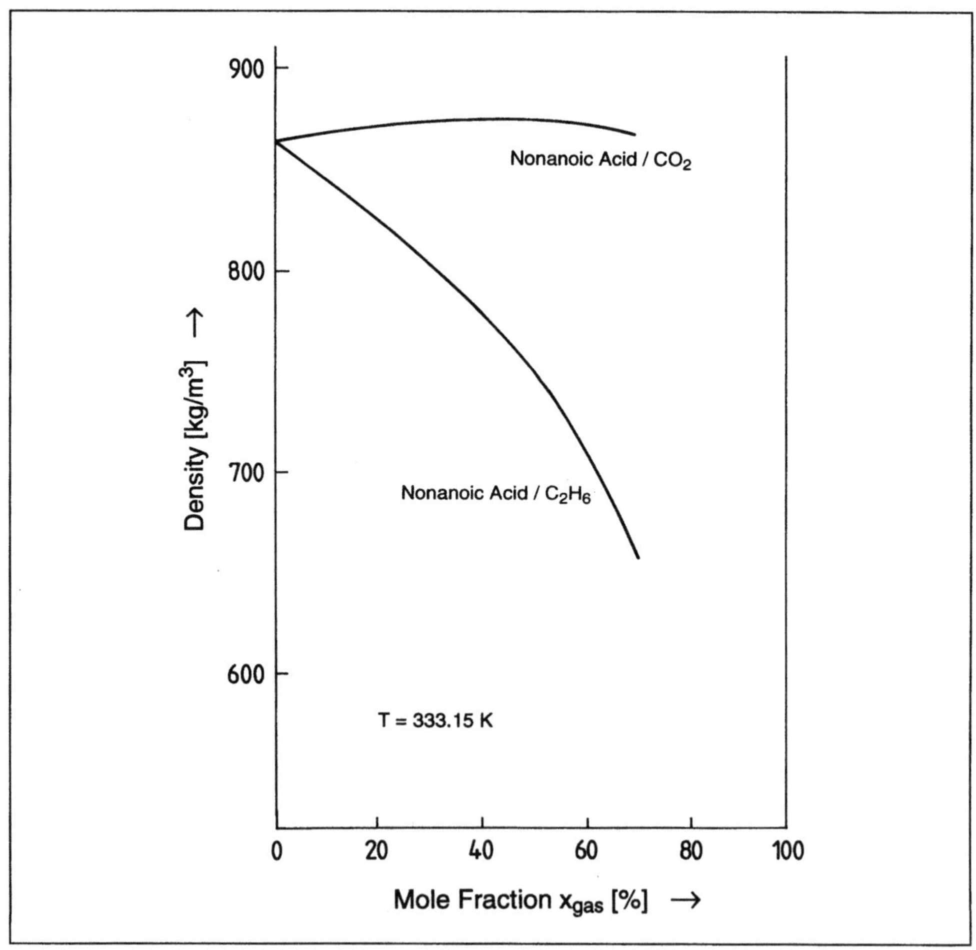


**Fig. 2.24.** Density of coexisting phases of nonanoic acid – $CO_2$ and nonanoic acid – ethane as function of composition. Data from Jakob [29].

40

centration of the supercritical compound, from the viscosity of the pure liquid, to the viscosity of the pure supercritical compound (Fig. 2.25).

For calculation of viscosities of saturated phases, the Grunberg-equation [22, 23] may be employed, which is for a binary mixture:

$$\eta_m = \eta_1^{x_1}\, \eta_2^{x_2}\, \exp\left(G_{12} x_1 x_2\right)$$

with:

$\eta_m$     = viscosity of the mixture,
$\eta_1$, $\eta_2$ = viscosity of pure component 1 and 2 at system temperature,
$x_1$, $x_2$ = concentration of components 1 and 2 (mole or mass fraction),
$G_{12}$    = interaction coefficient.

Equation 2.45 can be used for correlating viscosity data, using $G_{12}$ as fitting parameter.

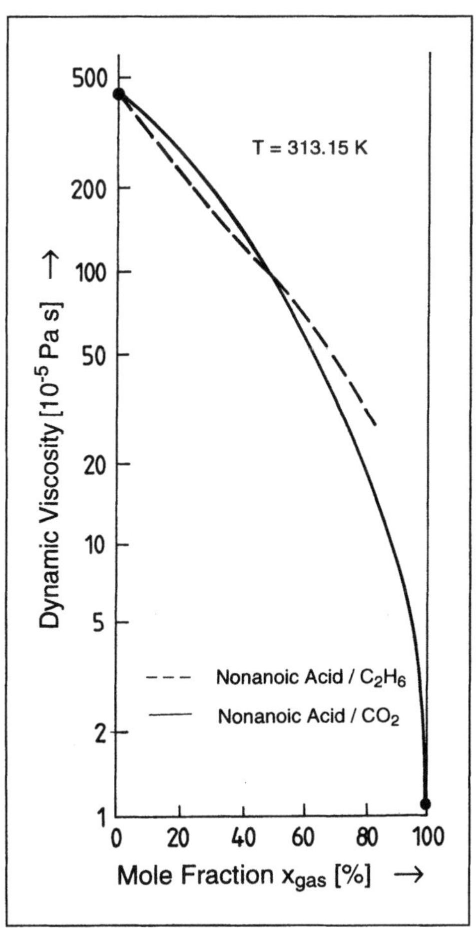


Fig. 2.25. Dynamic viscosity of coexisting phases in dependence of composition at constant shear stress ($D_w = 1000\ \mathrm{s}^{-1}$). Data from Jakob [29].

## 2.4.2 Thermal Conductivity

### 2.4.2.1 Gases

Thermal conductivity of gases is connected with the molecular transport of energy and is therefore also connected with viscosity and heat capacity. Under ambient conditions, thermal conductivity of gases is in the range of 0.01 to 0.025 W/(mK), with exceptions for helium and hydrogen. The thermal conductivity $\lambda$ of helium is: $\lambda = 0.150\,W/(mK)$, and of hydrogen: $\lambda = 0.181\,W/(mK)$, both at 0.1 MPa and 25 °C.

With increasing pressure, thermal conductivity of gases increases at constant temperature. The influence of pressure is low up to moderate pressures. An overview is presented in Fig. 2.26, which has been calculated by Schäfer and Thodos [57] for biatomic gaseous molecules.

Three regions can be determined for the influence of pressure: At very low pressures ($P < 1$ mbar) the mean free distance between molecules is of the same order as, or greater than the available space. Influence of pressure is strong. At pressures below 0.1 mbar, thermal conductivity $\lambda$ is approximately proportional to pressure. In the low pressure range (1 mbar $< P <$ 10 bar), thermal conductivity increases with

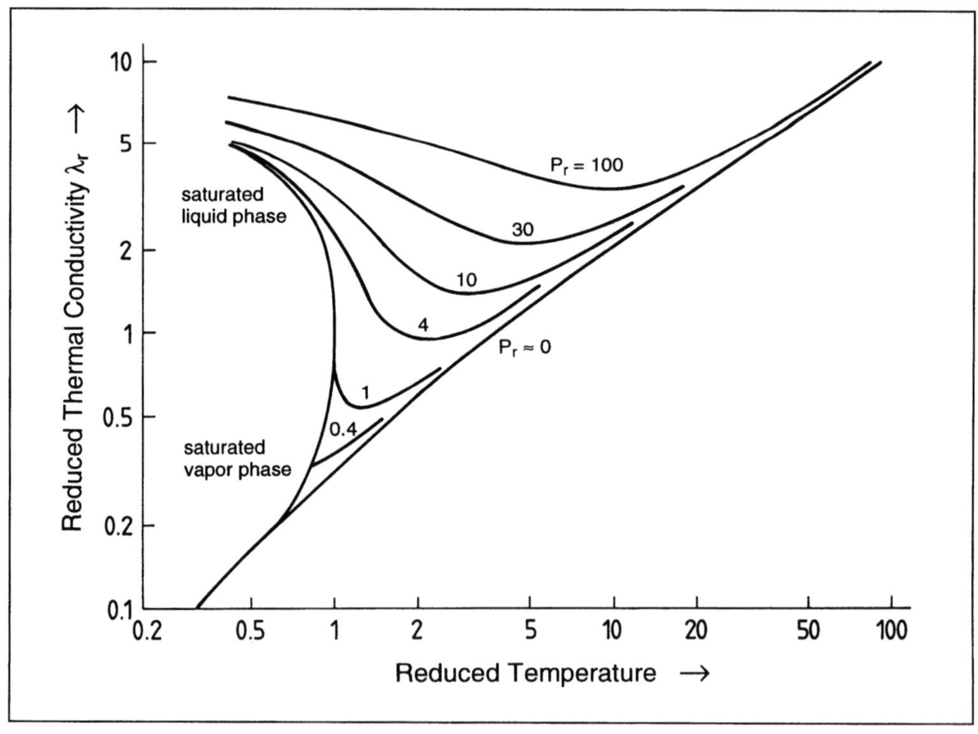


**Fig. 2.26.** Reduced thermal conductivity for bi-atomic gases (after Schäfer and Thodos [57]).

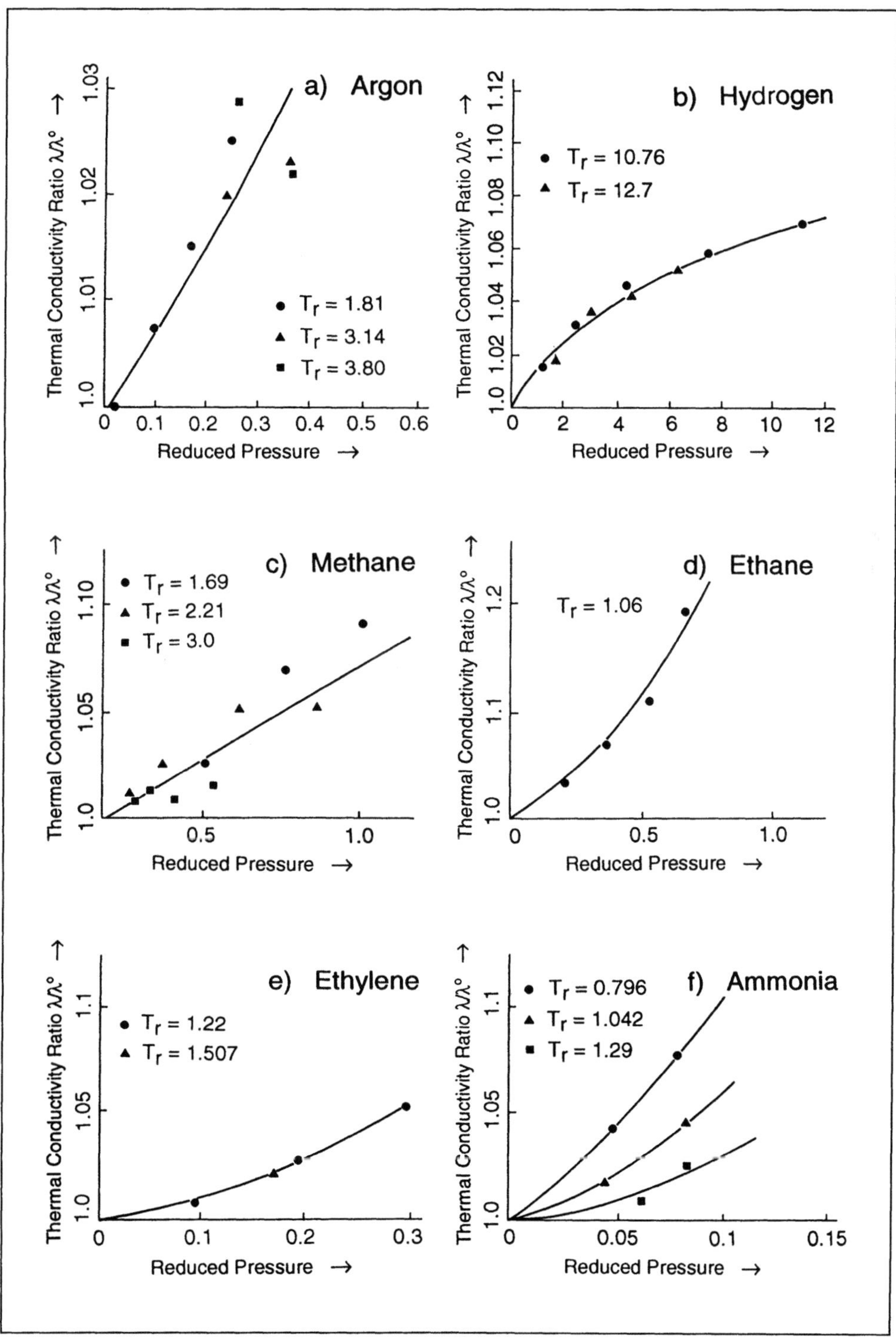

**Fig. 2.27.** Influence of pressure on thermal conductivity of gases at low pressure. Data from Keyes [34, 35].

pressure. The increase is about 1 % with 0.1 MPa increase in pressure, but is often neglected. Within this pressure range, correlations from the literature for thermal conductivity can be applied. In Fig. 2.27 pressure dependence of heat conductivity is presented for this region of pressure. For high pressures ($P > 10$ bar), it has been tried to correlate thermal conductivity according to the principle of corresponding states. With an estimated average error of 20 %, the generalized chart for pressure dependence of thermal conductivity from Lenoir, Junk and Comings [39] may be used, which is shown in Fig. 2.28.

Beyond normal variation with pressure and temperature values for thermal conductivity show great deviations in the region of the critical point. In Fig. 2.29 thermal conductivity for carbon dioxide is presented for illustration [25]. The great deviations are tentatively explained by transitions in the molecular structure or by convective transport of molecule clusters.

In general, at supercritical conditions, thermal conductivity decreases with increasing temperature, runs through a minimum and then increases with temperature (Fig. 2.26). The minimum is shifted with increasing pressure to higher temperatures. At constant temperature, thermal conductivity increases with pressure.

For nonpolar gases, dependence of thermal conductivity on pressure can be correlated by a departure function, determined by Stiel and Thodos [61].

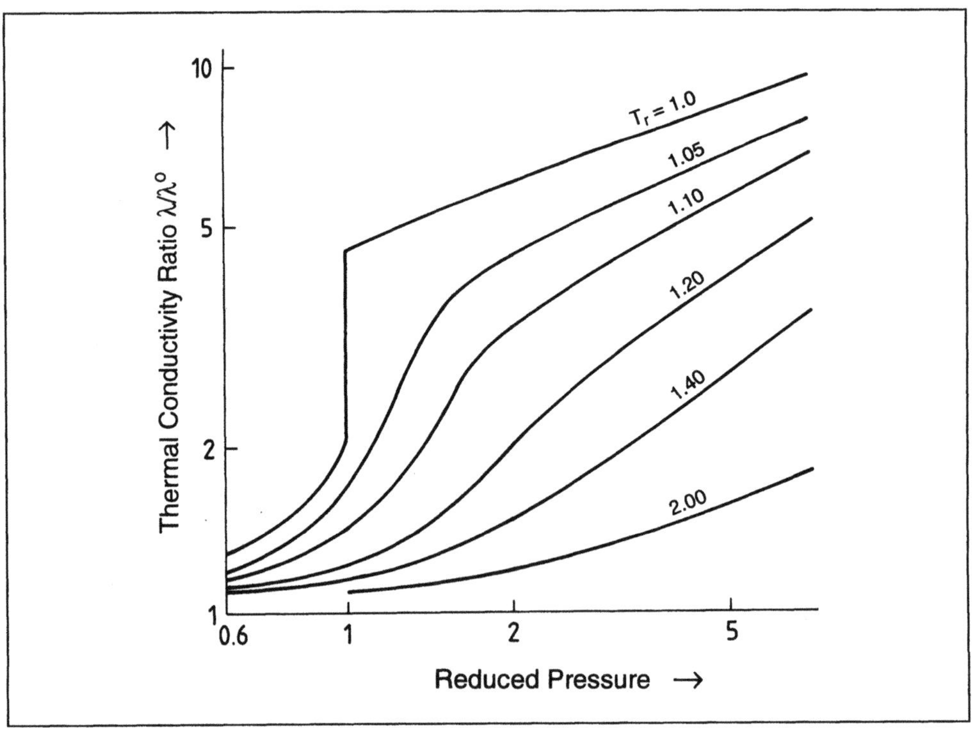



**Fig. 2.28.** Effect of pressure on thermal conductivity (after Lenoir, Junk and Comings [39]).

44

$$\lambda - \lambda^o = f(\varrho)$$

$$(\lambda - \lambda^o)\, \Gamma \cdot Z_c^5 = 14.0 \cdot 10^{-8}\, (e^{0.535\varrho r} - 1), \text{ for } \varrho_r < 0.5 \qquad\qquad 2.46$$

$$(\lambda - \lambda^o)\, \Gamma \cdot Z_c^5 = 13.1 \cdot 10^{-8}\, (e^{-0.67\varrho r} - 1.069), \text{ for } 0.5 < \varrho_r < 2.0 \qquad\qquad 2.47$$

$$(\lambda - \lambda^o)\, \Gamma \cdot Z_c^5 = 2.976 \cdot 10^{-8}\, (e^{1.155\varrho r} + 2.016), \text{ for } 2.0 < \varrho_r < 2.8 \qquad\qquad 2.48$$

with

$$\Gamma = \frac{T_c^{1/6}\, M^{1/2}}{P_c^{2/3}}, \qquad\qquad 2.49$$

$Z_c$ = compressibility factor at the critical point.

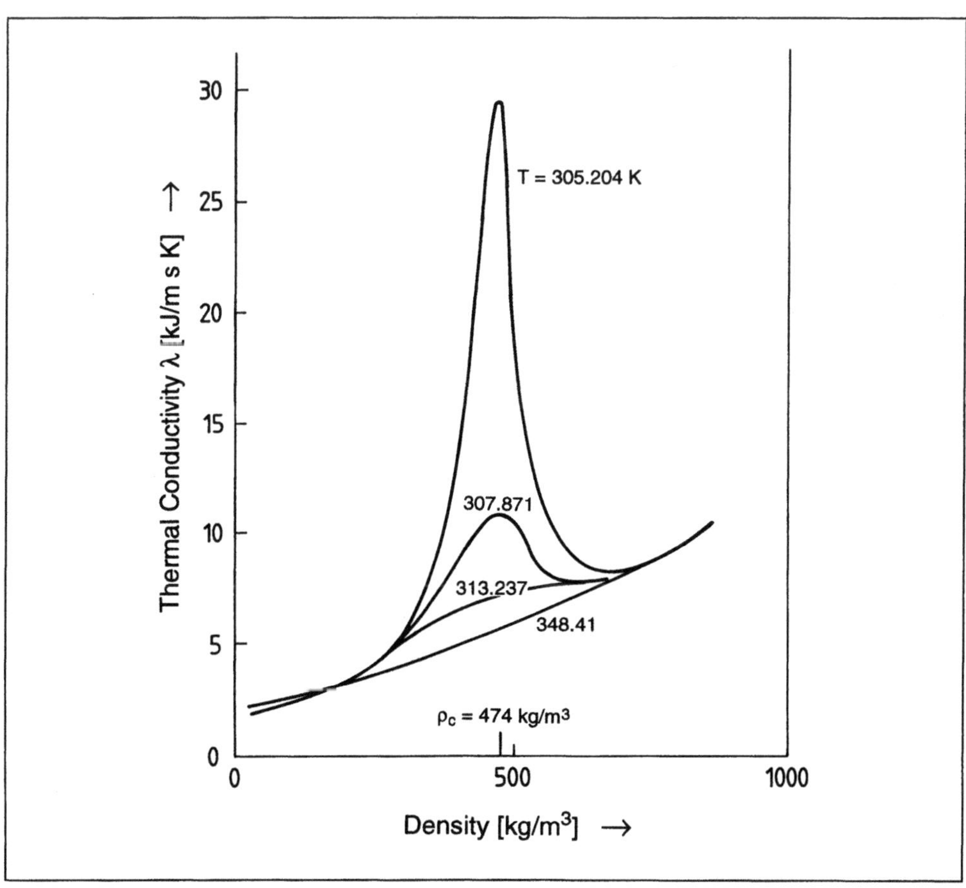


**Fig. 2.29.** Thermal conductivity of $CO_2$ in the vicinity of the critical point (after Guildner [25]).

The correlations can be used for mixtures, but not for polar gases, and the gases helium and hydrogen. Accuracy cannot be stated, even for unpolar gases. Errors may be in the range of 10 to 20 % [54].


### 2.4.2.2 Liquids

Thermal conductivity of most organic liquids is in the range of 0.1 to 0.2 W/(mK) at temperatures below boiling temperature. Water, ammonia, and strongly polar molecules have a higher thermal conductivity by a factor of 2 to 3. Highly viscous liquids often also have a relatively high thermal conductivity. Liquid metals and some organic silica compounds have very high thermal conductivities.

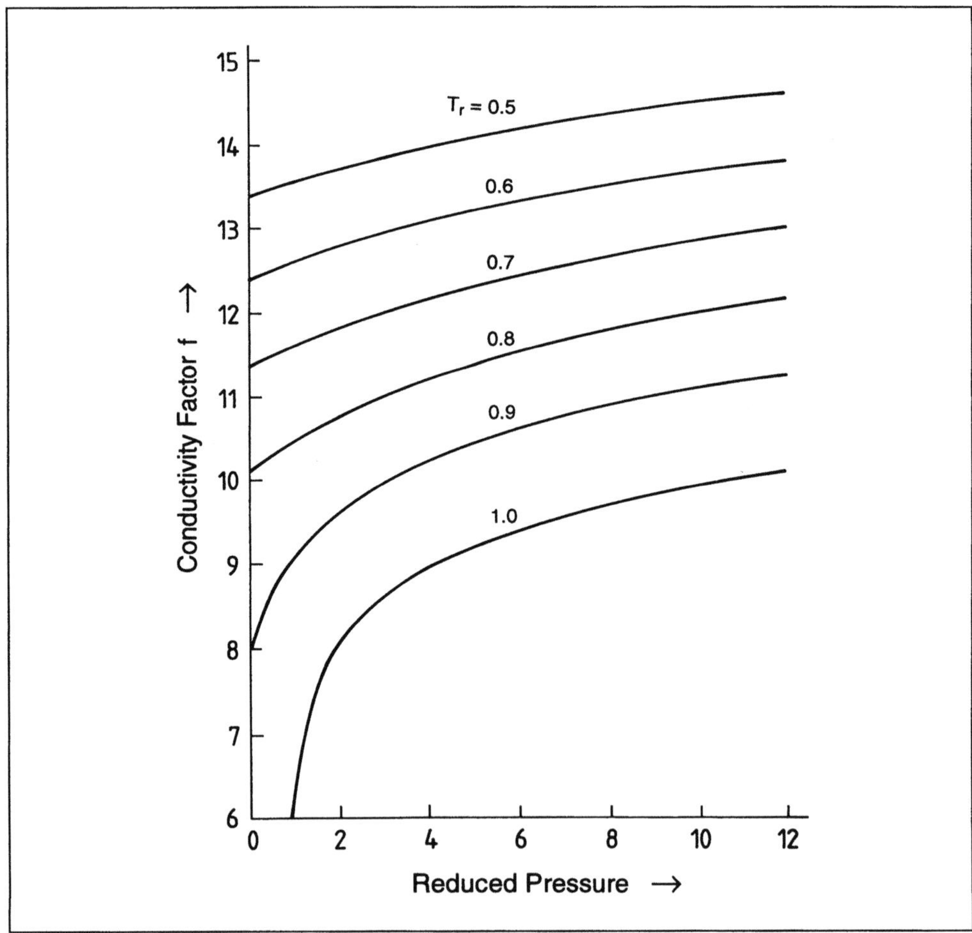


**Fig. 2.30.** Influence of pressure on thermal conductivity of liquids near the critical point (modified after Reid, Prausnitz and Poling [54]).

Thermal conductivity of liquids is not strongly influenced by pressure. In general, thermal conductivity of liquids increases up to 1000 MPa by a factor of 2 to 3. A compilation of data has been published by Jamieson et al. [30].

At moderate pressures, up to about 4 MPa, the influence of pressure on heat conductivity of liquids can be neglected. An exception is the critical region, where liquids behave more like dense gases. The influence of pressure can be considered by referring to values at low pressure:

$$\frac{\lambda_2}{\lambda_1} = \frac{f_1}{f_2}, \qquad\qquad 2.50$$

with $\lambda_1$ and $\lambda_2$ the thermal conductivities at temperature $T$ and pressures $P_1$ and $P_2$. $f_1$ and $f_2$ depend on reduced pressure and temperature, as shown in Fig. 2.30. For polar and nonpolar liquids deviations from experimental values are reported to be in the range of 2 to 4 % [54].

## 2.4.3 Diffusion Coefficient

Diffusion is transport of matter without convection or mechanically induced mixing. If we restrict for our purposes diffusion as being induced only by concentration differences, then the diffusion coefficient is the proportionality factor between the diffusion potential (concentration difference) and diffusive flow (mass flow). While viscosity and thermal conductivity of gases and liquids differ by one to two orders of magnitude, diffusion coefficients of gases and liquids differ by four orders of magnitude:

$$\frac{\lambda_L}{\lambda_G} = 10 - 100; \quad \frac{\eta_L}{\eta_G} = 10 - 100; \quad \frac{D_L}{D_G} = 10^{-4}. \qquad 2.51$$

### 2.4.3.1 Gases

Up to moderate pressures, the diffusion coefficient of gases is reverse proportional to pressure or density. At higher pressures mostly self diffusion coefficents have been determined by experiment. They have been correlated by Dawson, Khoury and Kobayashi [8]:

$$\frac{D \cdot \varrho}{(D\varrho)^o} = 1 + 0.053432 \, \varrho_r - 0.030182 \, \varrho_r^2 - 0.029725 \, \varrho_r^3 \qquad 2.52$$

$$0.8 < T_r < 1.9; \qquad 0.3 < P_r < 7.4, \qquad\qquad 2.53$$

with:

$D$     = self diffusion coefficient at $T$ and $\varrho$;
$\varrho$     = density;
$(D\varrho)^o$ = $D\varrho$ at $T$, but low pressure;
$\varrho_r$     = reduced density.

At higher densities, $D$ decreases markedly. In this range, the presented correlation is only a rough approximation.

### 2.4.3.2 Diffusion in Dense Fluids

**Gaseous Phase**

Diffusion in dense fluids can be connected with viscosity by the Stokes-Einstein-equation [24]. According to this equation, the self-diffusion coefficient for molecules with diameter $\sigma$ in a medium of viscosity $\eta$ is obtained from:

$$D = (kT)/(C\,\pi\,\sigma\,\eta), \qquad\qquad 2.54$$

where $k$ is the Boltzmann factor and $C$ a coefficient with a value between 3 and 6. In Fig. 2.31 experimental self diffusion coefficients for tetramethylsilane and values for the viscosity are plotted [45]. In this case the Stokes-Einstein relation is fulfilled. The widely used Wilke-Chang equation seems not to be adequate for gas extraction systems. Feist and Schneider [17] therefore proposed an exponential relation between the binary diffusion coefficient and the viscosity of the supercritical component, Eq. 2.55.

$$D_{12} \sim \eta^q, \text{ with } q \approx 0.66. \qquad\qquad 2.55$$

Diffusion coefficients for systems containing supercritical gases have been experimentally determined. Some are listed in Table 2.11. A review is given by Liong et al. [41]. Some of the experimental results for binary diffusion coefficients are shown in Fig. 2.32.

Binary diffusion coefficients in mixtures of a supercritical gas and a low volatile component are of the order of $10^{-8}$ m$^2$/s. They are about one order of magnitude higher than that for liquids and about two orders of magnitude lower than that for gases. Components of similar structure and molecular weight also have similar diffusion coefficients (see C16- and C22-fatty acid ethylester). Binary diffusion coefficients decrease with density (at constant temperature) and with molecular weight.

Diffusion coefficients may be influenced by an entrainer, or a third component. Smith et al. [59] found a decreasing diffusion coefficient in the system $CO_2$ – benzoic acid, if methanol was added as entrainer, while the diffusion coefficient of phenanthrene was not affected by methanol. Different interactions account for this effect.

**Table 2.11.** Binary diffusion coefficients in selected systems [21].

| Low volatile component | Supercritical solvent | Reference |
| --- | --- | --- |
| Benzoic acid | $CO_2$/methanol | Smith et al. [59] |
| Benzene | $CO_2$ | Funazukuri et al. [18] |
| | | Feist and Schneider [17] |
| Caffeine | $CO_2$ | Feist and Schneider [17] |
| Fatty acid ethyl ester | $CO_2$ | Liong et al. [41] |
| Fatty acid methyl ester | $CO_2$ | Liong et al. [41] |
| | | Funazukuri et al. [18, 19] |
| Ketones | $CO_2$ | Dahmen et al. [7] |
| Squalane | $CO_2$ | Dahmen et al. [7] |
| $\alpha$-Tocopherol | $CO_2$ | Funazukuri et al. [18] |
| | | Zehnder [66] |

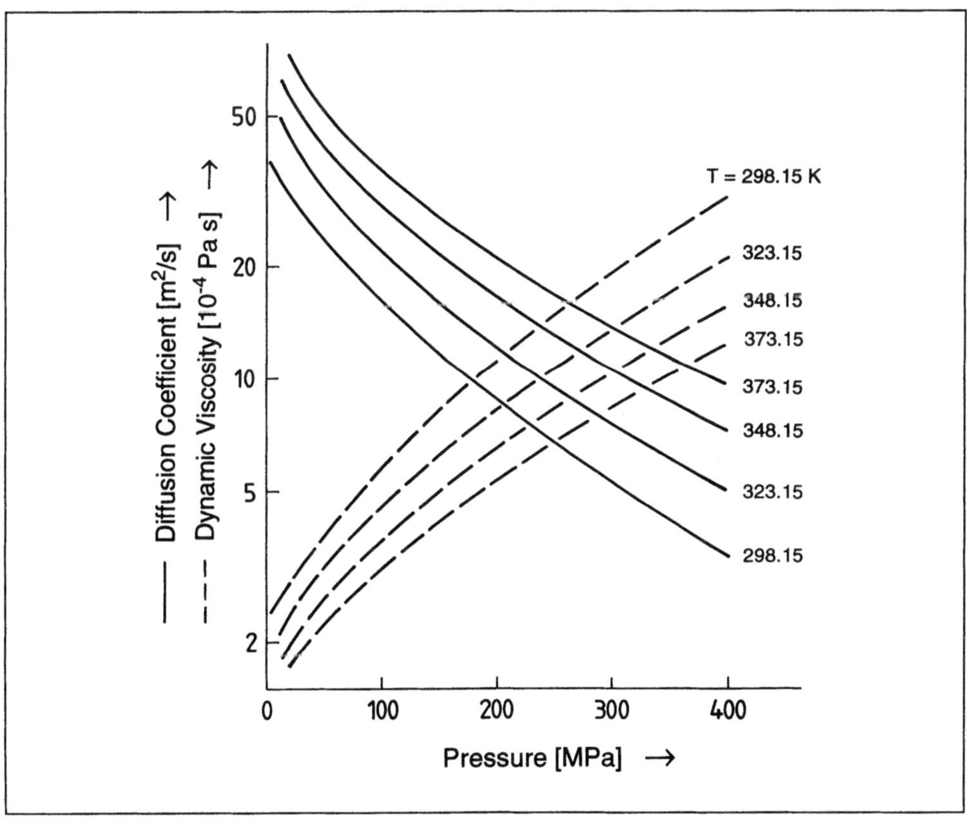


**Fig. 2.31.** Diffusion coefficient (——) and viscosity (-----) for tetramethyl silane (after Parkhurst and Jonas [45]).

For calculation of binary diffusion coefficients in systems, typical for gas extraction, Funazukuri et al. [20] proposed:

$$Sc/Sc^* \approx 1 + 2.45\,(M_s/M)^{-0.089}\,F_v^{1.12}, \qquad\qquad 2.56$$

with

| | | |
|---|---|---|
| $Sc$ | $=$ | $\eta_2/(\varrho_2 D_{12})$, Schmidt-number at system conditions; |
| $M_s$ | $=$ | molecular weight of solvent; |
| $M$ | $=$ | molecular weight of solute; |
| $F_v$ | $=$ | $x/(x-1)^2$; |
| $x$ | $=$ | $V_2/[1.384(\tilde{V}_2)_o]$; |
| $(\tilde{V}_2)_o$ | $=$ | $N\sigma_2^3/\sqrt{2}$, m$^3$/mol; |
| $N$ | $=$ | Avogadro-number; |
| $\sigma_2$ | $=$ | hard-sphere diameter of solvent, m; |

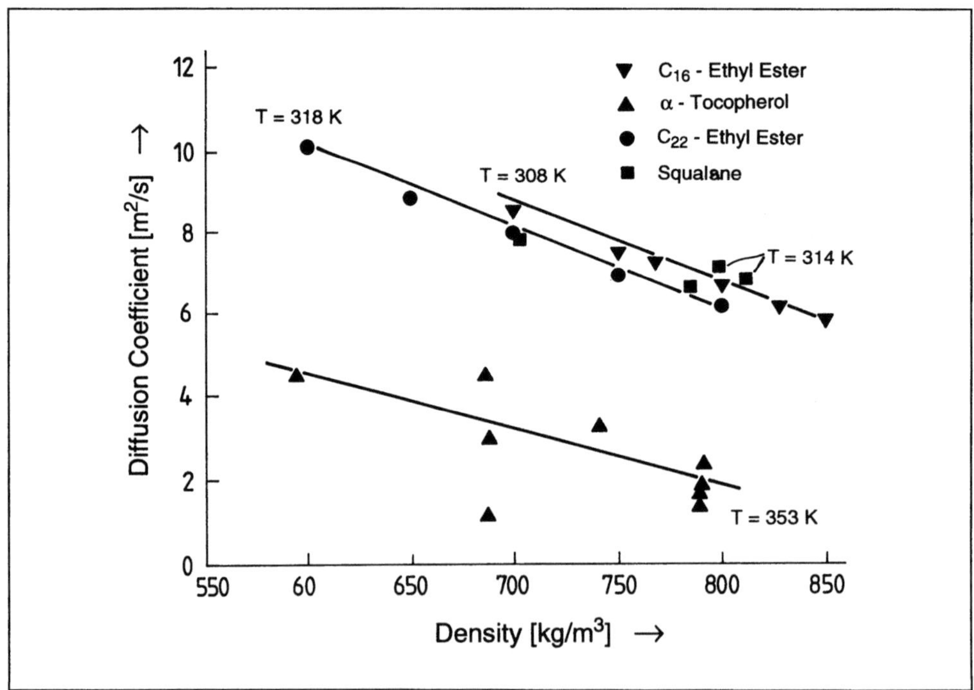


**Fig. 2.32.** Binary diffusion coefficients for fatty acid ethyl esters, squalane and $\alpha$-tocopherol in supercritical $CO_2$. Data from Dahmen et al. [7] (squalane), Liong et al. [41] (fatty acid ester) and Zehnder [66] ($\alpha$-tocopherol). Adapted from Gottschau [21].

50

$V_2$ = molar volume of the supercritical component, $m^3/mol$;

$Sc^*$ = $\eta_2^*/(\varrho_2^* D_{12}^*)$, Schmidt-number at system temperature and ambient pressure.

$D_{12}^*$ can be determined with Fuller's method [54]. The equation has been tested by the authors for 12 low volatile components with $CO_2$ up to pressures of 25 MPa and compared to the available data from literature. General representation of the data is sufficient, though in regions with scattering data deviations of up to 20 % may be expected.

## Liquid Phase

Recently, Lockemann and Schlünder [42] published some data on diffusion coefficients of carbon dioxide in the liquid phase. Their data are presented in Table 2.12.

**Table 2.12.** Diffusion coefficients $D$ of carbon dioxide in oleic acid, myristic acid methylester and palmitic acid methylester. $D$ in $10^{-9}\,m^2/s$. $D^\infty$ – diffusion coefficient at infinite dilution, $D^*$ – diffusion coefficient at equilibrium concentration of the liquid phase. Data from [42]

| | Oleic acid: $D^\infty/D^*$ | | |
|---|---|---|---|
| $P$[MPa] | $T$:313.15 [K] | 323.15 [K] | 333.15 [K] |
| 7.5 | 0.9/1.6 | | |
| 10 | 0.9/1.5 | 1.1/2.0 | 1.3/2.5 |
| 12.5 | 0.9/1.6 | 1.1/2.1 | |
| 15 | 0.9/1.6 | | |

| | Myristic acid methylester: $D^\infty/D^*$ | | |
|---|---|---|---|
| $P$[MPa] | $T$:313.15 [K] | 323.15 [K] | 333.15 [K] |
| 7 | 0.8/1.6 | 1.5/2.2 | |
| 8 | 0.8/1.5 | 1.5/2.2 | 2.0/3.2 |
| 9 | 0.8/1.6 | | |

| | Palmitic acid methylester: $D^\infty/D^*$ | | |
|---|---|---|---|
| $P$[MPa] | $T$:313.15 [K] | 323.15 [K] | 333.15 [K] |
| 7 | 0.8/2.1 | 0.93/2.7 | |
| 8 | 0.8/2.1 | 0.93/2.5 | 2.0/4.8 |
| 9 | 0.8/1.8 | 0.93/2.4 | |
| 10 | | 0.93/2.0 | |
| 11 | | 0.93/1.9 | |
| 12 | | 0.93/1.95 | |
| 13 | | 0.93/2.0 | |

# 2.5 Surface Tension

At the boundary between two phases properties change within probably several molecular diameters from that of one phase to those of the other phase. The boundary layer between a liquid and a gaseous phase may then be considered as a separate third phase with intermediate properties. Molecules in the boundary layer experience strong attraction on the liquid side, but weak attraction on the gaseous side. Thus, a tension tends to contract the surface layer to the smallest area compatible with constraints on the system. Stability of phase boundaries is dependent on surface tension. With the variation of surface tension, a continuous liquid film may disintegrate to droplets.

Surface tension $\sigma$ is defined by the work $W_\sigma$ needed for enlarging surface $A$:

$$dW_\sigma = \sigma dA \qquad\qquad 2.57$$

Surface tension of liquids is in the range of $20-40$ dyn/cm, for water $\sigma = 72.8$ dyn/cm at 20 °C. Values for the surface tension of pure liquids can be calculated by an empirical equation, proposed by Macleod [43]:

$$\sigma^{1/4} = [P] \, (\varrho_L - \varrho_V) \qquad\qquad 2.58$$

with

$\varrho_L$ = density of liquid phase;
$\sigma_v$ = density of gaseous phase;
$[P]$ = coefficient, called parachor.

For the calculation of the parachor structural contributions have been calculated and are given in [54]. Using this method, densities are in gmole per cubic centimeter and surface tension in dynes/cm. Equation 2.58 can also be used for correlating data. Then $[P]$ may be considered as adjustable parameter.

For mixtures, Eq. 2.58 may be applied in the form:

$$\sigma_m^{1/4} = \sum_{i=1}^{n} \left( [P_i] \, (\varrho_{Lm} x_i - \varrho_{Vm} y_i) \right) \qquad\qquad 2.59$$

with

$\sigma_m$ = surface tension of mixture;
$[P_i]$ = parachor of component $i$;

$x_i, y_i$ = mole fractions of component $i$ in liquid and gas;
$\varrho_{Lm}$ = density of liquid mixture;
$\varrho_{Vm}$ = density of gaseous mixture.

Experimental data relevant to gas extraction are known for gases with water [65], hydrocarbons, and fatty acids. Surface tension for methane – nonane is shown in Fig. 2.33 and for $CO_2$ – squalane in Fig. 2.34.

Surface tension of the liquid decreases with increasing pressure, corresponding to increasing dissolution of the supercritical component in the liquid. Figure 2.35 shows this effect for different gases. Surface tension of the gas increases with pressure, corresponding to the increasing concentration of the less volatile component in the gas. Surface tension approaches zero at the critical point, where density difference between the phases vanishes, as can be seen from Fig. 2.36.

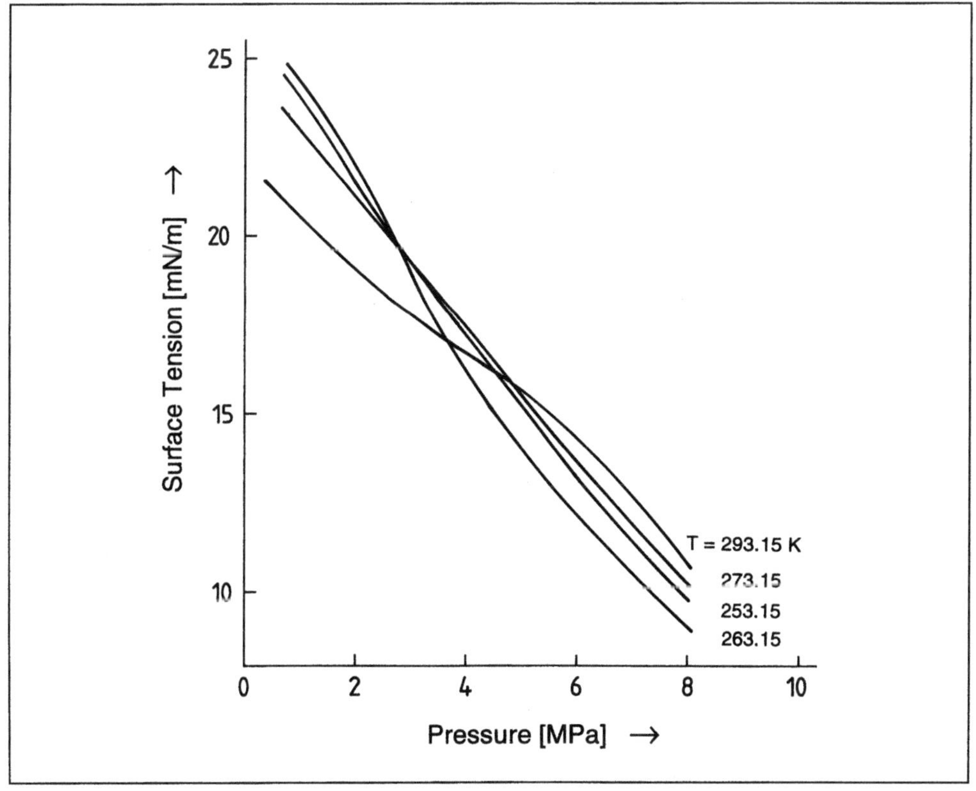


**Fig. 2.33.** Surface tension for the system methane – nonane after Deam and Mattox [10].

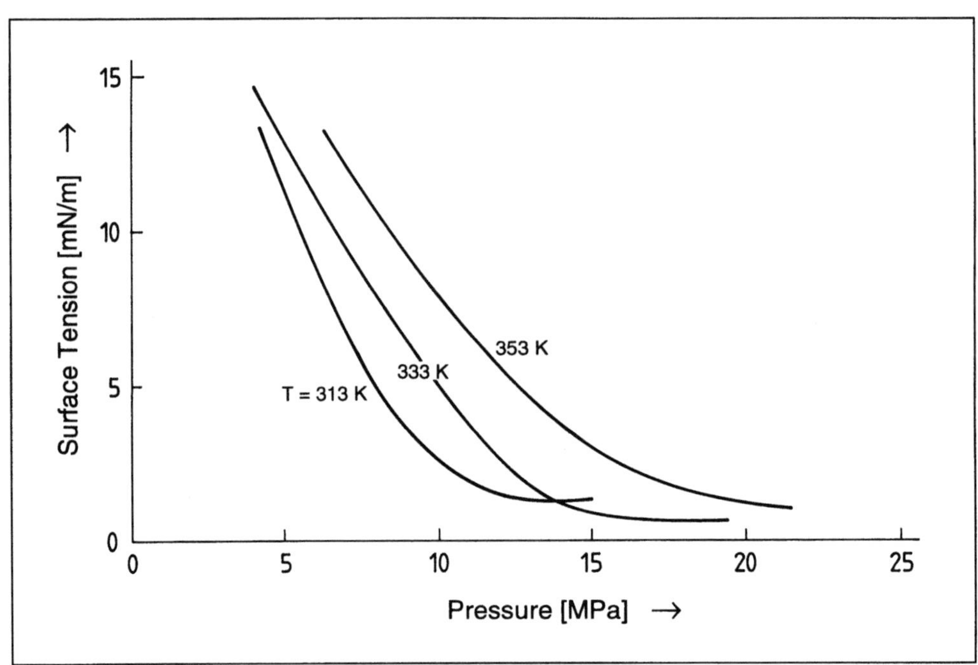

**Fig. 2.34.** Surface tension for the system squalane – $CO_2$ (after Hiller [28]).

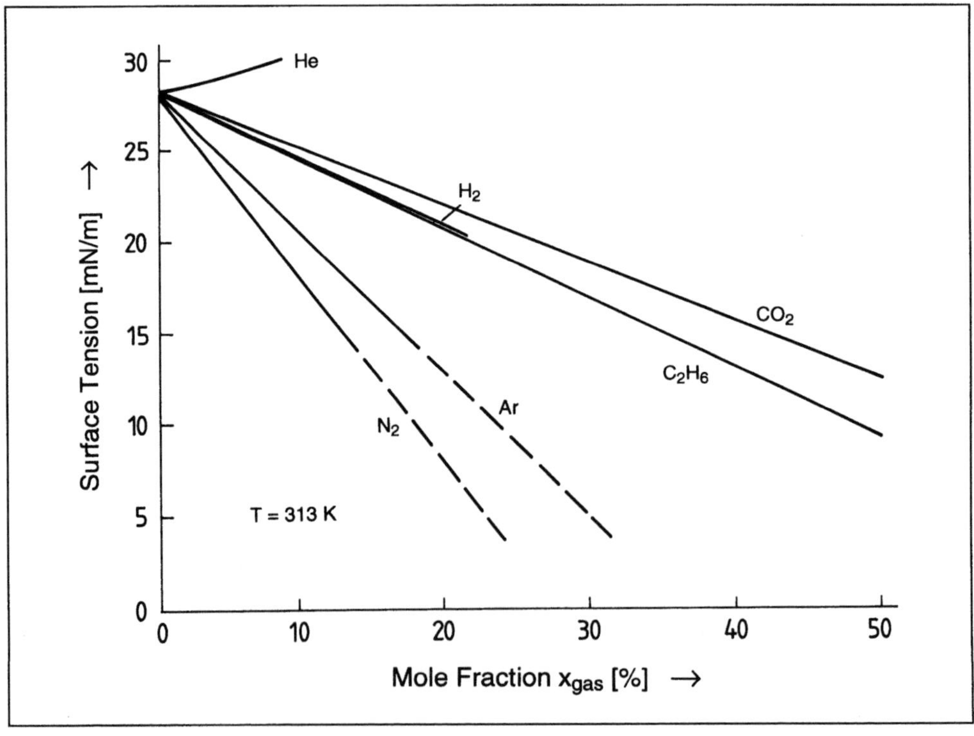

**Fig. 2.35.** Surface tension as a function of solvent concentration in the liquid phase (adapted from Peter [49]).

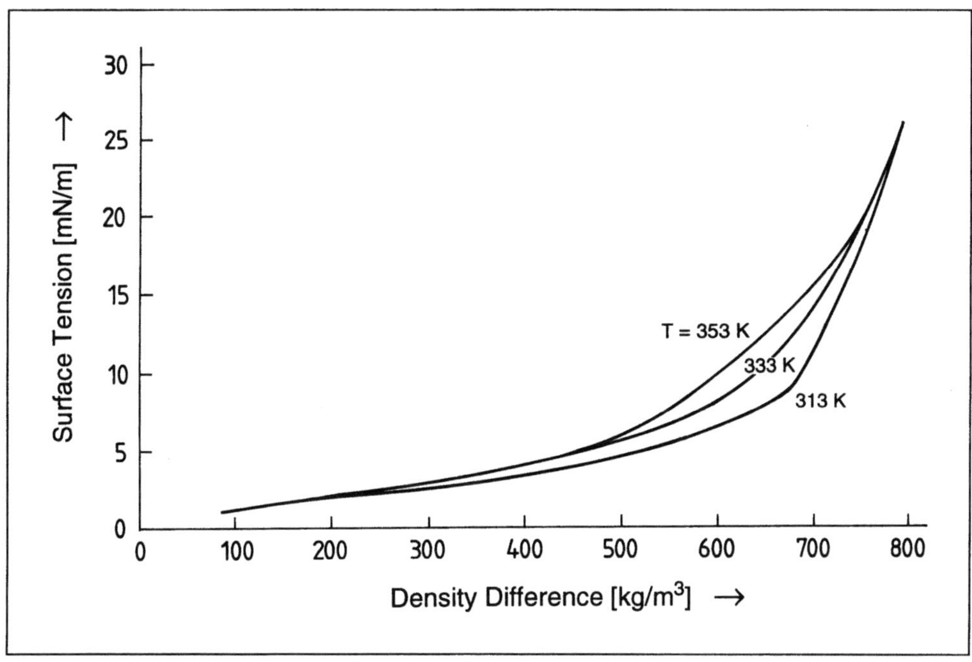


**Fig. 2.36.** Surface tension for the system squalane – $CO_2$ in dependence on the density difference between coexisting phases (adapted from Hiller [28]).

## References


1. Balenovic Z, Meyers MN, Giddings JC (1970) Binary diffusion in dense Gases to 1360 atm by chromatographic peak-broadening method. J Chem Phys 52: 915 – 922
2. Bender E (1971) Die Berechnung von Phasengleichgewichten mit der thermischen Zustandsgleichung dargestellt an den reinen Fluiden Argon, Stickstoff, Sauerstoff und an ihren Gemischen. Habilitationsschrift, Ruhr-Universität Bochum
3. Bender E (1975) Cryogenics 15: 667
4. Brunner G (1978) Phasengleichgewicht in Anwesenheit komprimierter Gase und ihre Bedeutung bei der Trennung schwerflüchtiger Stoffe. Habilitationsschrift, Universität Erlangen-Nürnberg
5. Brunner G, Hederer H (1979) The calculation of vapour-liquid phase equilibria of low volatile substances in systems with compressed gases. In: Timmerhaus KD, Barber MS (eds) Proc 6th AIRAPT Conf, Boulder 1977, vol 1. Plenum Press, New York, pp 527 – 534
6. Childs GE, Hanley HJM (1968) Cryogenics 8: 94
7. Dahmen N, Kordikowski A, Schneider GM (1990) Determination of binary diffusion coefficients of organic compounds in supercritical carbon dioxide by supercritical fluid chromatography. J Chromatography 505: 169 – 178
8. Dawson RF, Khoury F, Kobayashi R (1970) Self-diffusion measurements in methane by pulsed nuclear magnetic resonance. AIChE J 16: 725 – 729
9. Dean DE, Stiel LI (1965) The viscosity of nonpolar gas mixtures at moderate and high pressures. AIChE J 11: 526 – 532
10. Deam JR, Mattox RN (1970) Interfacial tension in hydrocarbon systems. J Chem Eng Data 15: 216 – 222

11. Deul R, Franck EU (1991) The static dielectric constant of the water-benzene mixture system to 400 °C and 2800 bar. Ber Bunsenges Phys Chem 95: 847 – 853

12. Dilchert G (1979) Der Einfluß gelöster Gase auf das Viskositäts-Druckverhalten flüssiger Kohlenwasserstoffe im gesättigten und untersättigten Lösungsbereich. Dissertation, Universität Hannover

13. Dohrn R, Brunner G (1988) Programmsystem zur Berechnung von Hochdruckphasengleichgewichten – Anwendung auf Stoffe mit unbekannten kritischen Daten. Chem Techn 41: 65 – 68

14. Dohrn R, Brunner G (1988) Empirische Korrelationen zur Bestimmung der Reinstoffparameter der Peng-Robinson-Zustandsgleichung und ihre Anwendung zur Phasengleichgewichtsberechnung. Chem Ing Tech 60: 1059 – 1061

15. Dohrn R, Brunner G (1991) Correlations for pure-component parameters of the Peng-Robinson equation of state. In: McHugh MA (ed) Proc 2nd International Symposium on Supercritical Fluids, Boston, p 471

16. Dohrn R (1994) Berechnung von Phasengleichgewichten mit Hilfe von Zustandsgleichungen. Habilitationsschrift, Technische Universität Hamburg-Harburg

17. Feist R, Schneider GM (1982) Determination of binary diffusion coefficients of benzene, phenol, naphthalene and caffeine in supercritical $CO_2$ between 308 and 333 K in the pressure range of 80 to 160 bar with supercritical fluid chromatography (SFC). Sep Sci Technol 17: 261 – 270

18. Funazukuri T, Hachisu S, Wakao N (1989) Measurements of diffusion coefficients of C18 unsaturated fatty acid methyl esters, naphthalene and benzene in supercritical carbon dioxide by a tracer response technique. Anal Chem 61: 118 – 122

19. Funazukuri T, Hachisu S, Wakao N (1991) Measurements of binary diffusion coefficients of C16 – C24 unsaturated fatty acid methyl esters in supercritical carbon dioxide. Ind Eng Chem Res 30: 1323 – 1329

20. Funazukuri T, Ishiwata Y, Wakao N (1992) Predictive correlation for binary diffusion coefficients in dense carbon dioxide. AIChE J 38: 1761 – 1768

21. Gottschau T (1994) Zur Anreicherung von Tocopherolen aus Dämpferdestillat. Dissertation, Technische Universität Hamburg-Harburg

22. Grunberg L (1976) An equation for predicting the viscosity of liquid mixtures, NL L Report No. 626, National Engineering Laboratory, East Kilbride, Glasgow, Scotland

23. Grunberg L, Nissan AH (1949) Mixture law for viscosity. Nature 164: 799

24. Gubbins KE (1973) in: Singer K, (ed) Statistical Mechanics, vol 1. The Chemical Society, London

25. Guildner LA (1958) Proc Natl Acad Sci 44: 1149

26. Hederer H, Peter S, Wenzel H (1976) Calculation of thermodynamic properties from a modified Redlich-Kwong equation of state. Chem Eng J 11: 183 – 190

27. Hederer H (1981) Die Berechnung von Phasengleichgewichten, insbesondere bei hohen Drücken, mit einer modifizierten Zustandsgleichung nach Redlich und Kwong. Dissertation, Universität Erlangen-Nürnberg

28. Hiller H (1990) Bestimmung der Grenzflächenspannung in Systemen aus schwerflüchtigen und überkritischen Komponenten unter hohen Drücken. Dissertation, Universität Erlangen-Nürnberg

29. Jakob H (1988) Das Fließverhalten koexistierender Gas- und Flüssigphasen bei hohen Drücken. Dissertation, Universität Erlangen-Nürnberg

30. Jamieson DT, Irving JB, Tudhope JS (1975) Liquid thermal conductivity: A data survey to 1973. H M Stationery Office, Edinburgh

31. Jossi JA, Stiel LI, Thodos G (1962) The viscosity of pure substances in the dense gaseous and liquid phases. AIChE J 8: 59 – 63

32. Kashulines P, Rizvi SSH, Harriott P, Zollweg JA (1991) Viscosities of fatty acids and methylated fatty acids saturated with supercritical carbon dioxide. JAOCS 68: 912 – 921

33. Kestin J, Leidenfrost W (1959) The effect of pressure on the viscosity of $N_2 - CO_2$ mixtures. Physica 25: 525 – 536

34. Keyes FG (1954) Thermal conductivity of gases. Trans ASME 76: 809 – 816

35. Keyes FG (1955) Thermal conductivity of gases. Trans ASME 77: 1395 – 1396

36. Killesreiter H (1984) Continuous viscosity measurements in order to observe the solubility and miscibility of carbon dioxide on crude oil. Ber Bunsenges Phys Chem 88: 838 – 841

37. Kuss E, Golly H (1972) Das Viskositäts-Druckverhalten von Gas-Flüssigkeitslösungen. Ber Bunsenges Phys Chem 76: 131 – 138

38. Kuss E (1982) PVT-Daten und Druckviskosität von ausgesuchten Mineralölen. DGMK-Bericht Projekt 198, Deutsche Gesellschaft für Mineralölwissenschaft und Kohlechemie e.V., Hamburg

39. Lenoir, Junk, Comings (1953) Chem Eng Progr 49: 539

40. Levelt-Sengers JMH (1994) Critical behavior of fluids: Concepts and applications. In: Kiran E, Levelt-Sengers JMH (eds) Supercritical Fluids – Fundamentals for Application. NATO ASI, Kluwer

41. Liong KK, Wells PA, Foster NR (1991) Diffusion in supercritical fluids. Journal of Supercritical Fluids 4: 91 – 108

42. Lockemann CA, Schlünder EU (1993) An experimental method to determine the diffusion coefficient of carbon dioxide in liquids at high pressures. J Energy Heat and Mass Transfer 15: 125 – 133

43. Macleod DB (1923) Trans Faraday Soc 19: 38

44. Nelson LC, Obert EF (1954) Generalized PVT properties of gases. Trans ASME 76: 1057 – 1066

45. Parkhurst HJ, Jonas J (1975) Dense liquids. II. The effect of density and temperature on viscosity of tetramethylsilane and benzene. J Chem Phys 63: 2705 – 2709

46. Peng DY, Robinson DB (1976) A new two-constant equation of state. Ind Eng Chem Fundam 15: 59 – 64

47. Perry RH, Chilton CH (1984) Perry's chemical engineers' handbook, 6th edn. McGraw-Hill, New York

48. Peter S, Weidner E, Jakob H (1987) Die Viskosität koexistierender Phasen bei der überkritischen Fluidextraktion. Chem Ing Tech 59: 59 – 62

49. Peter S (1994) Interfacial tension in binary systems containing a dense gas. In: Kiran E, Levelt-Sengers JMH (eds) Supercritical Fluids – Fundamentals for Application. NATO ASI, Kluwer

50. Pitzer KS, Curl RF (1955) The volumetric and thermodynamic properties of fluids. I. Theoretical basis and virial coefficients. J Am Chem Soc 77: 3427 – 3433

51. Platzer B (1990) Eine Generalisierung der Zustandsgleichung von Bender zur Berechnung von Stoffeigenschaften unpolarer und polarer Fluide und deren Gemische. Dissertation, Universität Kaiserslautern

52. Prausnitz JM, Lichtenthaler RN, Azevedo EG de (1986) Molecular thermodynamics of fluid-phase equilibria, 2nd edn. Prentice-Hall, Englewood Cliffs, NJ

53. Redlich O, Kwong JNS (1949) On thermodynamics of solutions V: an equation of state. Fugacities of gaseous solutions. Chem Rev 44: 233 – 244

54. Reid RC, Prausnitz JM, Poling BE (1989) The properties of gases and liquids, 4th edn. McGraw-Hill, New York

55. Roskar V, Dombro RA, Prentice GA, Westgate CR, McHugh MA (1992) Comparison of the dielectric behavior of mixtures of methanol with carbon dioxide and ethane in the mixture-critical and liquid regions. Fluid Phase Equilibria 77: 241 – 259

56. Sandler SI (1994) Equations of state for phase equilibrium computations. In: Kiran E, Levelt-Sengers JMH (eds) Supercritical Fluids – Fundamentals for Application. NATO ASI, Kluwer

57. Schäfer CA, Thodos G (1959) Thermal conductivity of diatomic gases: Liquid and gaseous states. AIChE J 5: 367 – 372

58. Sengers JV (1994) Effects of critical fluctuations on the thermodynamic and transport properties of supercritical fluids. In: Kiran E, Levelt-Sengers JMH (eds) Supercritical Fluids – Fundamentals for Application. NATO ASI, Kluwer

59. Smith SA, Shenai V, Metthews MA (1990) Diffusion in supercritical mixtures: $CO_2$ + Cosolvent + Solute. Journal of Supercritical Fluids 3: 175 – 179

60. Soave GS (1972) Equilibrium constants from a modified Redlich-Kwong equation of state. Chem Eng Sci 33: 225 – 229

61. Stiel LI, Thodos G (1964) The thermal conductivity of nonpolar substances in the dense gaseous and liquid regions. AIChE J 10: 26 – 30

62. Stiel LI, Thodos G (1964) The viscosity of polar substances in the dense gaseous and liquid regions. AIChE J 10: 275 – 277

63. Uyehara OA, Watson KM (1944) A universal viscosity correlation. Natl Pet News 36: R-714

64. Waals JD van der (1873) Over de continuiteit van den gas- en vloestoftoestand. Dissertation, University of Leiden

65. Wiegand G, Franck EU (1994) Interfacial tension between water and non-polar fluids up to 473 K and 2800 bar. Ber Bunsenges Phys Chem 98: 809 – 817

66. Zehnder BH (1992) NIR-spektroskopische Bestimmung von Stoffaustauschkoeffizienten und Gleichgewichtslöslichkeiten in Fluid-Fluid-Systemen unter überkritischen Bedingungen. Dissertation ETH Nr. 9657, ETH Zürich

# 3 Supercritical Gases as Solvents: Phase Equilibria

## 3.1 The Need for Phase Equilibria

Gas extraction and related processes are characterized by using a dense gas as solvent. In processes with solvents, the driving potential for mass and heat transfer is determined by the difference from the equilibrium state, in most cases the difference to heterogeneous thermodynamic equilibrium, at given conditions of state. Equilibrium determined processes, like gas extraction, are analyzed with respect to the equilibrium state.

The equilibrium state provides information about:
- the capacity of a supercritical (gaseous) solvent, which is the amount of a substance dissolved by the gaseous solvent at thermodynamic equilibrium,
- the amount of solvent, which dissolves in the liquid process streams, and the equilibrium composition of the liquid phase;
- the selectivity of a solvent, which is the ability of a solvent to selectively dissolve one or more compounds, expressed by the separation factor $\alpha$ as defined by:

$$\alpha = \frac{y_i / y_j}{x_i / x_j}, \qquad\qquad 3.1$$

with $\alpha$ – separation factor, $x_i$, $x_j$ – equilibrium concentrations of component $i$ and $j$ in the liquid phase, in mole or mass fractions, $y_i$, $y_j$ – equilibrium concentrations of component $i$ and $j$ in the gaseous phase, in mole or mass fractions,
- the dependence of these solvent properties on conditions of state $(P, T)$,
- the extent of the two-phase area, as limiting condition for a two-phase process like gas extraction.

If capacity and selectivity are known, a good guess can be made about whether a separation problem can be solved with the separation process of gas extraction and on its mode of operation. To get more information on these parameters in general, it is necessary to know phase equilibrium behavior of typical compounds and their mixtures encountered in gas extraction.

What types of heterogeneous systems are of importance in gas extraction?

The systems applied for gas extraction are characterized by a great difference in volatility, vapor pressure, and critical temperature between solvent and dissolved compounds. The main component of the solvent is a fluid at supercritical conditions of state. Its properties can be intentionally modified by additional compounds. The solvent is discussed in more detail in Section 3.5. The dissolved components are pure components or mixtures of substances with a low volatility at the critical temperature of the main component of the solvent. Examples of systems typical in gas extraction are: ethylene – naphthalene, or carbon dioxide – ethanol – triglycerides.

How many components must be considered?

Solubility of a substance in a pure supercritical fluid can be extracted from information on a binary system. Separation of two components by means of a supercritical fluid occurs in a ternary system. If the supercritical solvent consists of two components, in order to modify its solvent properties, a system of four components (quaternary system) is the basis for the process and has to be investigated. Often the feed mixture consists of more than two compounds, so multicomponent, even complex mixtures, have to be taken into account. Equilibrium phases may occur in the gaseous, liquid or solid states. Regions of immiscibility may occur in the liquid state, and the solid state may interfere with gas-liquid equilibrium. In this sequence phase equilibria relevant to gas extraction will be discussed below.


## 3.2 Solubility in Supercritical Solvents

Compounds of low volatility dissolve in compressed supercritical gases to such an extent that the concentration in the gaseous phase is enhanced by several orders of magnitude, as compared to the concentration that would occur if only their vapor pressures would determine the concentration in the gaseous phase by using Raoult's law. This enhanced volatility is the starting point for deliberately exploiting the solvent properties of supercritical fluids for separation processes and other processes.

Historically the process of investigating supercritical fluids began in the 19th century, when Cagniard de la Tour first observed a critical point [14], Andrews carried out basic measurements [1], and Hannay and Hogarth [27] and Villard [64], among others, observed that the solvent power of a liquid solvent is not totally lost, when crossing the critical point of the solvent. Up to then it had been assumed that a dissolving component lowers the vapor pressure of the solvent. This is true for dissolving components with a very low volatility. If there is some volatility of the dissolving component, two effects determine the equilibrium concentration: On one hand, the dissolving component lowers the partial pressure of the solvent, on the other hand the solvent enhances the volatility of the dissolving component. In open vessels and at ambient pressure only the first effect can be observed. But at higher pressures, around 10 to 20 MPa, the enhanced volatility emerges.

In this context the enhancement of vapor pressure of a condensed, usually solid phase, by hydrostatic pressure is of interest. This effect is called the Poynting-effect. The Poynting-effect is far smaller than the above discussed enhancement of the volatility by the solvent effects of a supercritical gas. Poynting calculated the effect of hydrostatic pressure $P$ on vapor pressure $P^0$ [46]:

$$\left( \frac{dP^0}{dP} \right)_T = \frac{V_L}{V_G},$$

3.2

Under the assumption that $V_L$ is incompressible ($V_L \neq f(P)$) and $V_G$ is the volume of an ideal gas, the integration can be carried out and Eq. 3.3 is obtained.

$$\ln \frac{P^{01}}{P^{02}} = \frac{V_L}{RT} (P - P^{01}),$$
3.3

with:
$P^{01}$ = vapor pressure of the condensed substance at ambient pressure;
$P^{02}$ = vapor pressure of the condensed substance at system pressure;
$P$  = system pressure (total pressure);
$V_L$ = molar volume of condensed phase;
$V_G$ = molar volume of gaseous phase.

Since molar volumes of condensed phases are small, only very high pressures enhance volatility (vapor pressure) substantially. For 100 MPa, the enhancement is still less than by a factor of 2, at 25 °C, as illustrated in Fig. 3.1.

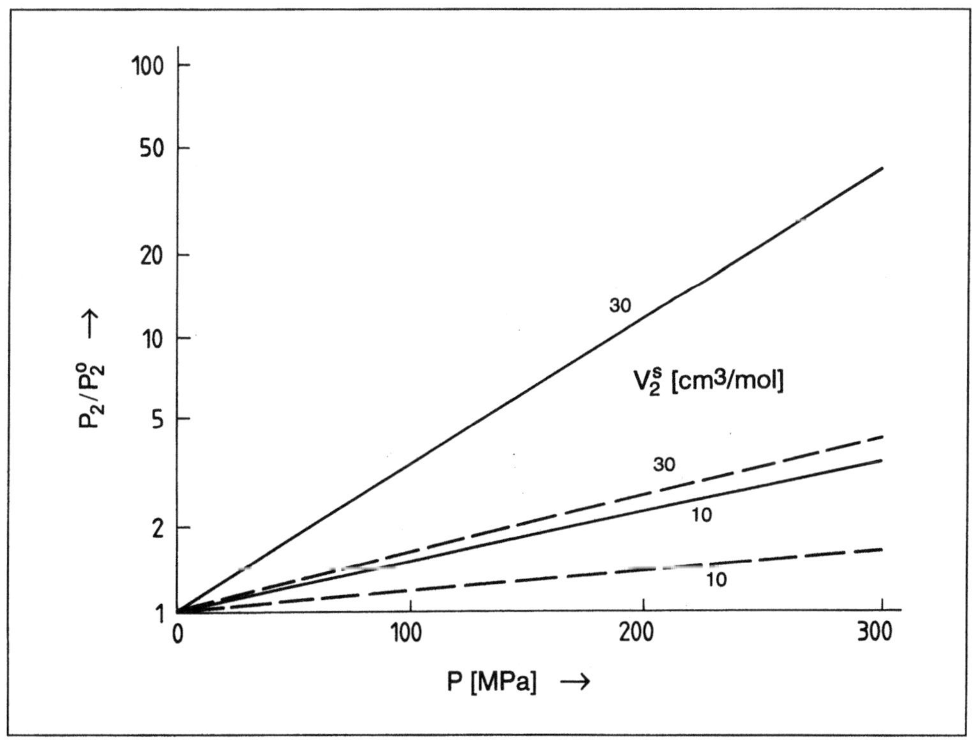


**Fig. 3.1.** Vapor pressure enhancement due to the Poynting-effect (after Franck [22]).
—— at 25 °C, ------ at 500 °C

In practical processes the aim is a high concentration of the dissolved component in the supercritical fluid. For separation processes the concentration is chosen as high as possible with regard to the separation to be achieved.

Solubility of a low volatile component in a supercritical solvent exhibits a characteristic scheme: At high pressures solubility increases with temperature and at lower presssures solubility decreases with temperature. This is illustrated in Fig. 3.2 on the example of the solubility of $SiO_2$ in water, after Kennedy [34].

In Fig. 3.2 the solubility of quartz ($SiO_2$) in water is shown in dependence of pressure, temperature, and constant specific volume. In the lower left corner of the diagram, three phases coexist: solid quartz, a liquid solution of quartz in water, and a gaseous phase. Outside that area, below the critical temperature of water, solid quartz coexists with one liquid phase. Above the critical temperature of water, solid quartz coexists with a gaseous phase. The critical temperature (broken line in Fig. 3.2) represents no sudden change in properties. Properties of phases in the regions where quartz coexists with a liquid and regions where quartz coexists with a gaseous phase merge gradually.

The area around the critical point of a solvent (or a mixture) may be called "near-critical" or "critical region," comprising sub- and supercritical conditions of state for

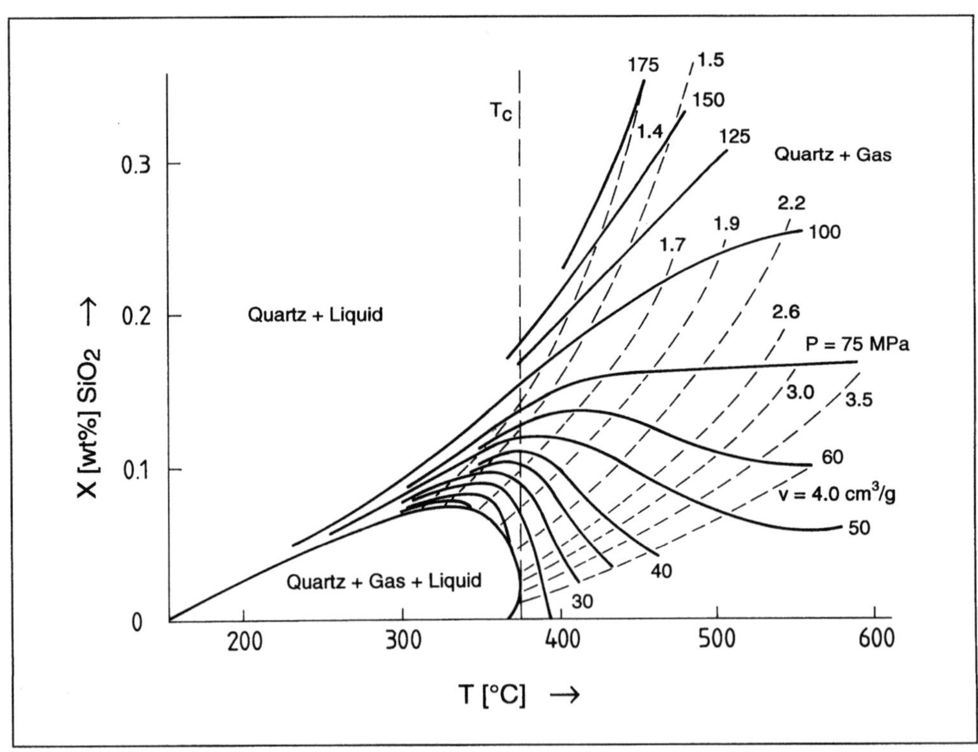


Fig. 3.2. Solubility of quartz in water. Data from Kennedy [34].

the solvent. Properties of a solvent in this region of temperature and pressure are different from those of a solvent in the liquid or gaseous state far from the critical temperature. In the critical region, properties vary drastically within narrow ranges of pressure and temperature, thus producing quite different effects at neighboring conditions of state.

Figure 3.2 shows that solubility of $SiO_2$ in liquid water increases at constant pressure up to temperatures about 30 K below the critical temperature of water. A further increase in temperature leads at low pressures to a decrease of the dissolved amount of quartz in the subcritical liquid water and at high pressures still to an increase. "High" and "low" pressures refer to the "medium" pressure level applied in this system of about 70 MPa. The same dependence of solubility of quartz in water remains at temperatures higher than the critical temperature. At low pressures, solubility of quartz in supercritical water decreases with temperature. Only at much higher temperatures, e.g., 200 to 300 K above the critical temperature, does solubility increase again at moderate pressures. This can be explained by two effects:
– solvent power of a solvent increases with increasing density;
– vapor pressure of the solute increases exponentially with temperature.

In the region beyond the critical temperature, at about 375 °C to 450 °C, density of water decreases rapidly with increasing temperature, if pressures are low. At high pressures, density changes with temperature are far more moderate, so that the increase of vapor pressure is the dominating factor, while at low pressures loss in

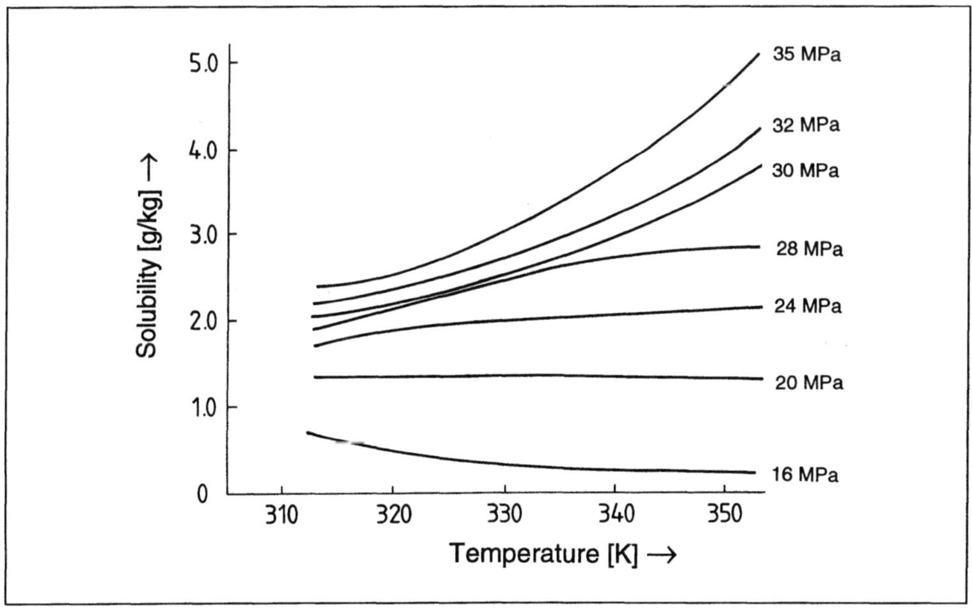



**Fig. 3.3.** Solubility of caffeine in carbon dioxide. Data from various authors: Gährs [24], Ebeling and Franck [18], Johannsen and Brunner [30].

solvent power induced by lower density prevails. At constant density, represented in Fig. 3.2 by the isochores $v$ = const., solubility of quartz in supercritical water always increases with increasing temperature. It is remarkable that the solvent power of supercritical water for silicium dioxide can be higher than that of subcritical liquid water.

Analogous solubility behavior can be found in other systems of a supercritical fluid with a solid substance or a low volatile liquid. The following figures present solubilities of caffeine in $CO_2$ (Fig. 3.3), naphthalene in ethylene (Fig. 3.4), and oleic acid in ethylene (Fig. 3.5).

If the condensed phase is liquid, the supercritical solvent dissolves in the liquid. For such cases, solubility is discussed using the system oleic acid – ethylene as an example. The solubility of oleic acid in supercritical ethylene exhibits a similar

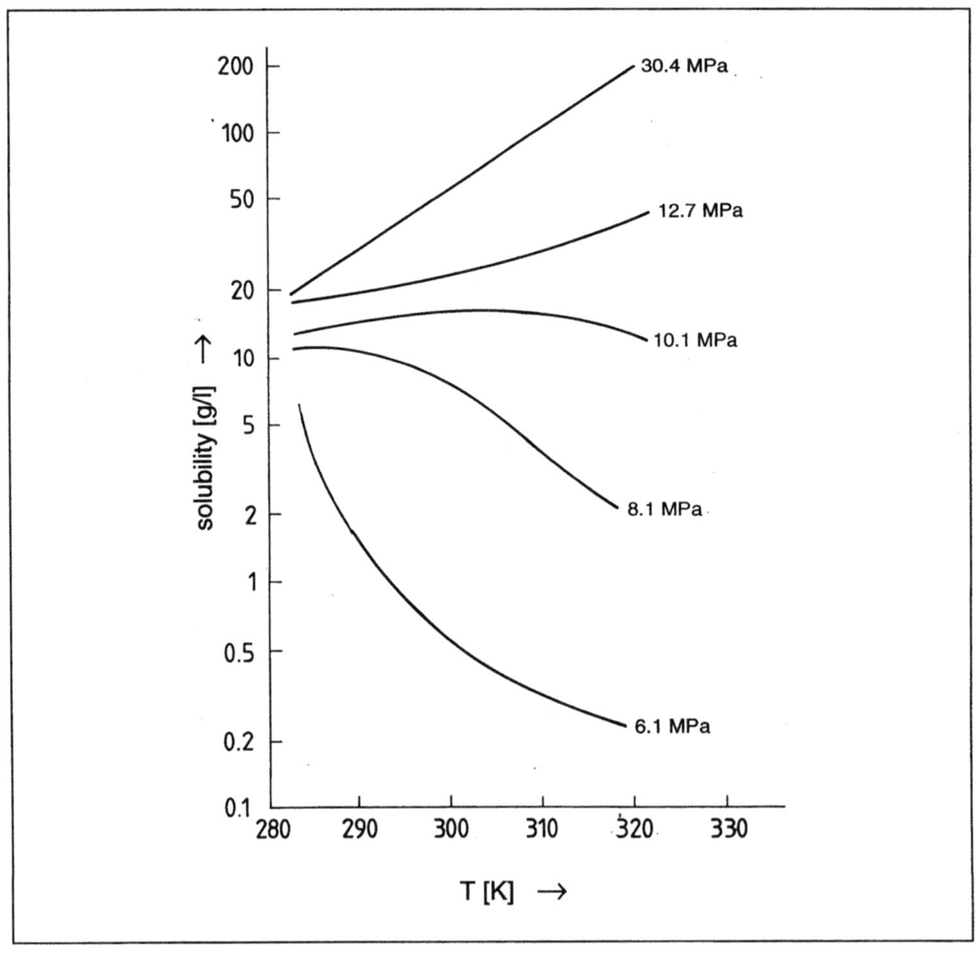


**Fig. 3.4.** Solubility of naphthalene in ethylene. Data from Tsekhanskaja [63].

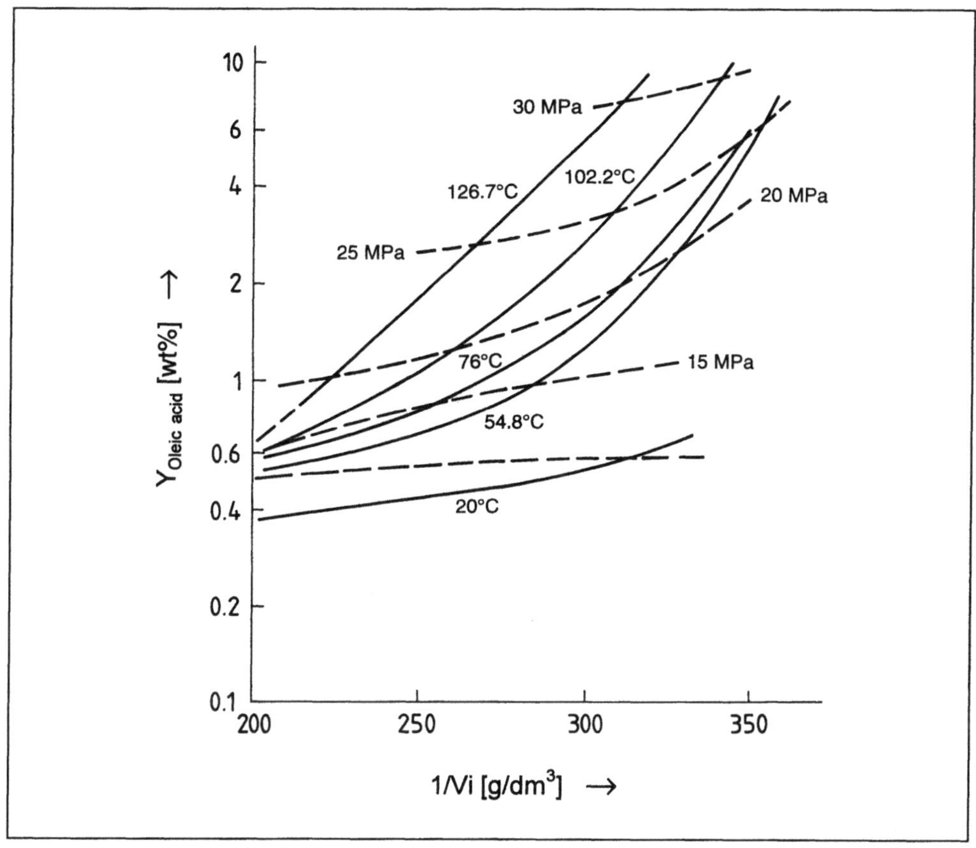



**Fig. 3.5.** Solubility of oleic acid in ethylene [4].

behavior as binary systems of a solid with a supercritical compound. With increasing density, the amount of oleic acid, dissolved in supercritical ethylene increases (see Fig. 3.5). On the other hand, a substantial amount of ethylene dissolves in oleic acid. Figure 3.6 presents several isotherms in a $P,x$-diagram. The lefthand side of the diagram represents the liquid phase, i.e. the solubility of ethylene in oleic acid at the temperatures specified. The righthand side of the diagram represents the gaseous phase, i.e., the solubility of oleic acid in ethylene at the specified temperatures. The two-phase region extends to far higher pressures than the critical pressure of ethylene (5.12 MPa).

The phase diagram, plotted in mole fractions, shows the great number of solvent molecules, dissolving in the liquid. Relatively few solute molecules dissolve in the gaseous phase, therefore concentrations on a molar basis are small. For mass fractions, the picture is somewhat different (Fig. 3.7), since molar masses of the two components differ by an order of magnitude ($M_w = 28.05$ for ethylene, $M_w = 282.45$ for oleic acid). Solubility values for ethylene in the liquid phase are much lower, solubil-

65

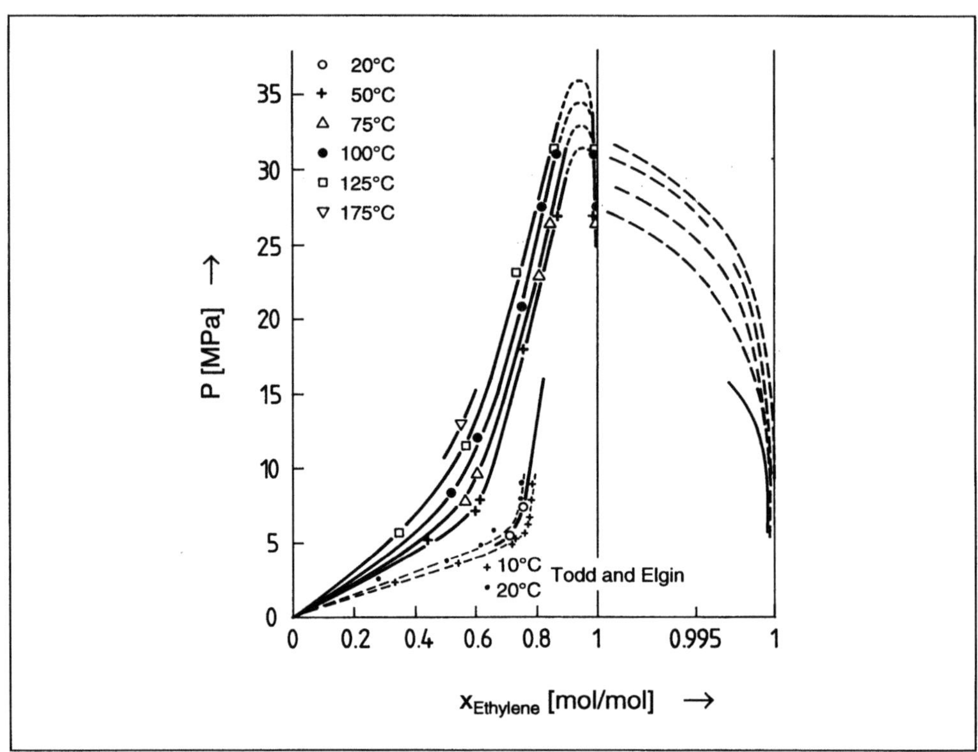


**Fig. 3.6.** Phase equilibria of oleic acid – ethylene. Mole fractions Data interpolated [4].

values for oleic acid in the gaseous phase are much higher, if presented as mass fractions, according to the relation between mole fraction $x_i$ and mass fraction $x'_i$:

$$x'_i = \frac{x_i M_{wi}}{\sum\limits_i (M_{wi} x_i)} \qquad\qquad 3.4$$

The dissolution of the supercritical solvent in the liquid phase changes properties of the liquid phase and influences the amount of substance dissolved in the supercritical solvent. This influence is connected to the density of the liquid phase in the system oleic acid – ethylene, which is determined by the dissolved amount of supercritical solvent. Great variations in density of the liquid phase result in great variations of the dissolved amount in the gaseous phase and vice versa. This effect is superimposed by the variation of vapor pressure of the solute with temperature, and the variation of density of the supercritical solvent with pressure and temperature. Furthermore, the effect of solvation varies with different compounds, so that the resulting effect of dissolution of the supercritical compound in the liquid phase on concentration of the low volatile compound in the gaseous phase may be more complicated in other systems.

66

Different supercritical gases behave differently against a certain compound with respect to mutual solubility. This is again discussed according to the example oleic acid. In Fig. 3.8, equilibrium solubilities in the liquid and the gaseous phase (Fig. 3.9) are plotted for oleic acid with different supercritical solvents at a temperature of 100 °C. The amount of gas dissolved at constant pressure in the liquid phase increases from nitrogen, monochlorotrifluoromethane ($CF_3Cl$), carbon dioxide, ethylene, monobromotrifluoromethane ($CF_3Br$) up to propane. The solubility of oleic acid in the gaseous phase, at constant conditions, increases in the same sequence of gases. Triglycerides (palm oil, a mixture of triglycerides of mainly palmitic, stearic and oleic acid, containing about 5 wt.-% of the free fatty acids) behave similar, as shown in Fig. 3.10 and 3.11.

At given conditions, the solubility of a low volatile substance in a supercritical gas, to a first approximation, increases with increasing critical temperature of the gas, or with decreasing temperature difference between critical temperature of the supercrit-

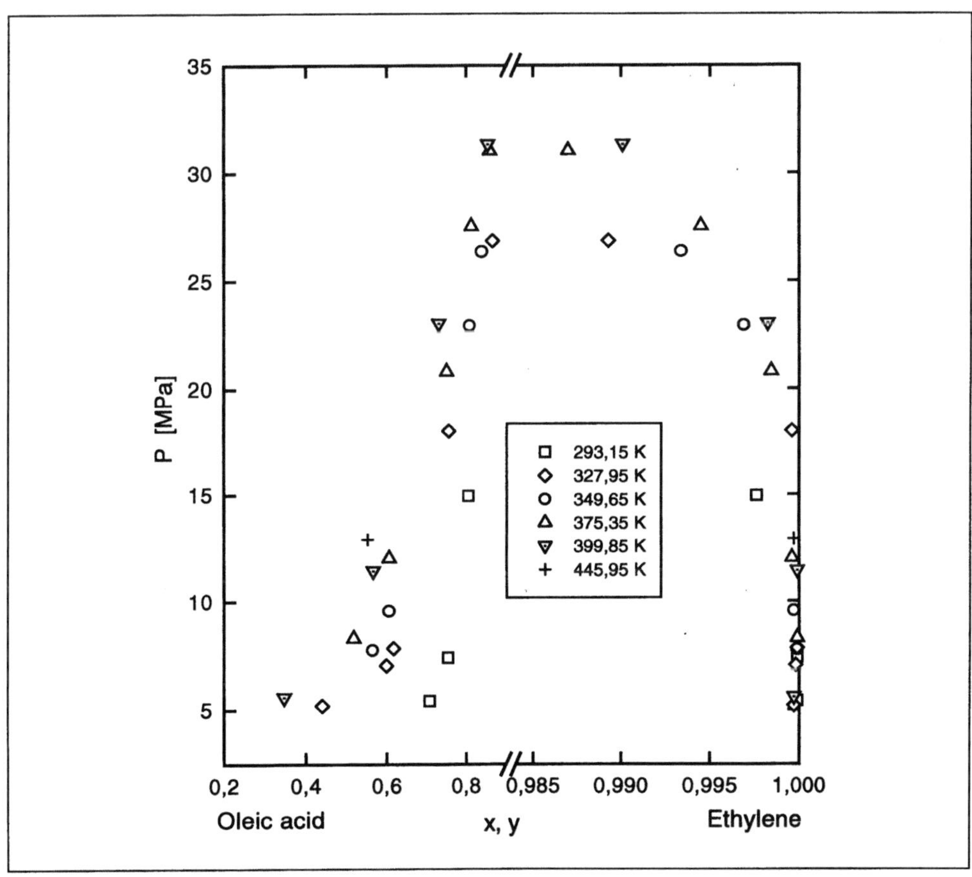


Fig. 3.7. Phase equilibria of oleic acid – ethylene. Mass fractions. Data from Brunner [4].

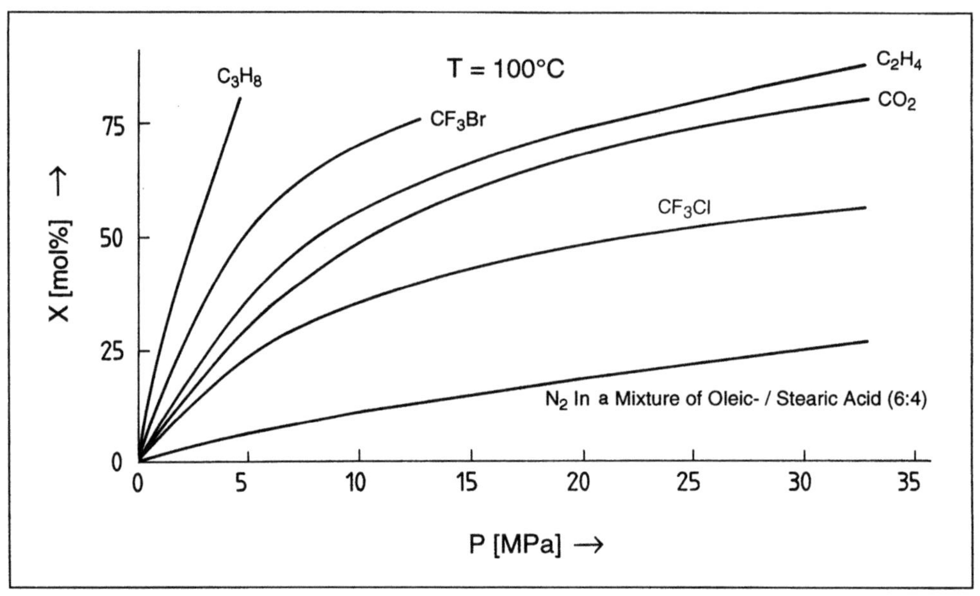



**Fig. 3.8.** Solubility of various gases in oleic acid. $T = 100\,°C$ [4].

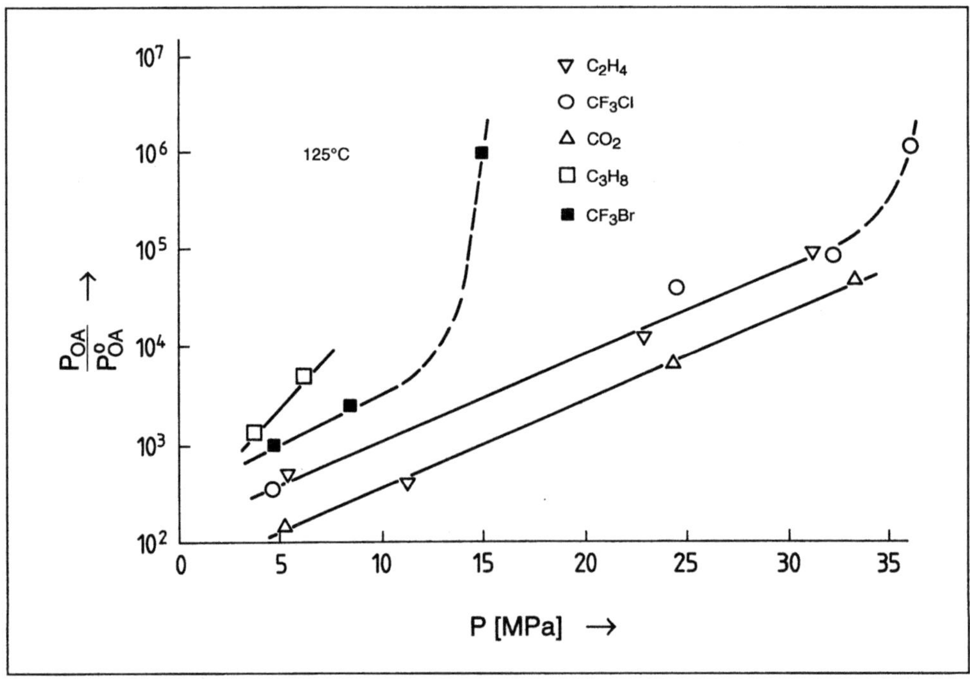



**Fig. 3.9.** Solubility of oleic acid (OA) in various gases. $T = 125\,°C$ [4].

ical solvent and system temperature. But there are also interaction forces of importance. The solubility of oleic acid (Fig. 3.9) increases at constant pressure with
decreasing difference between the critical temperature of the gas and the system temperature of 125 °C, with the exception of ethylene. Palm oil dissolves in ethane,
nitrous oxide, carbon dioxide, monochlorotrifluoromethane, at same conditions, to
quite different amounts, though the critical temperatures of these gases are nearly
equal (Fig. 3.12), see also Table 2.1.

Solubility of low volatile substances in unpolar supercritical gases decreases with
increasing molecular mass and with increasing polarity and number of polar functional groups. Fig. 3.13 shows the dependency of solubility on molecular weight. At
higher molecular weight the influence of the unpolar part of the molecules, which is
proportional to molecular weight, dominates solubility, while at lower molecular
weight, the influence of the functional groups dominates solubility. The concentration of unpolar hydrocarbons in the gaseous phase increases with decreasing molecular weight until complete miscibility is reached. The solubility of low molecular
weight carboxylic acids in supercritical $CO_2$ is dominated by the carboxylic group and
is far lower than for higher molecular weight acids, where the unpolar chain dominates. Stahl and Glatz [59] reported that with three and more polar functional
groups, like -OH and -COOH, the solubility of tetracyclic steroid compounds in

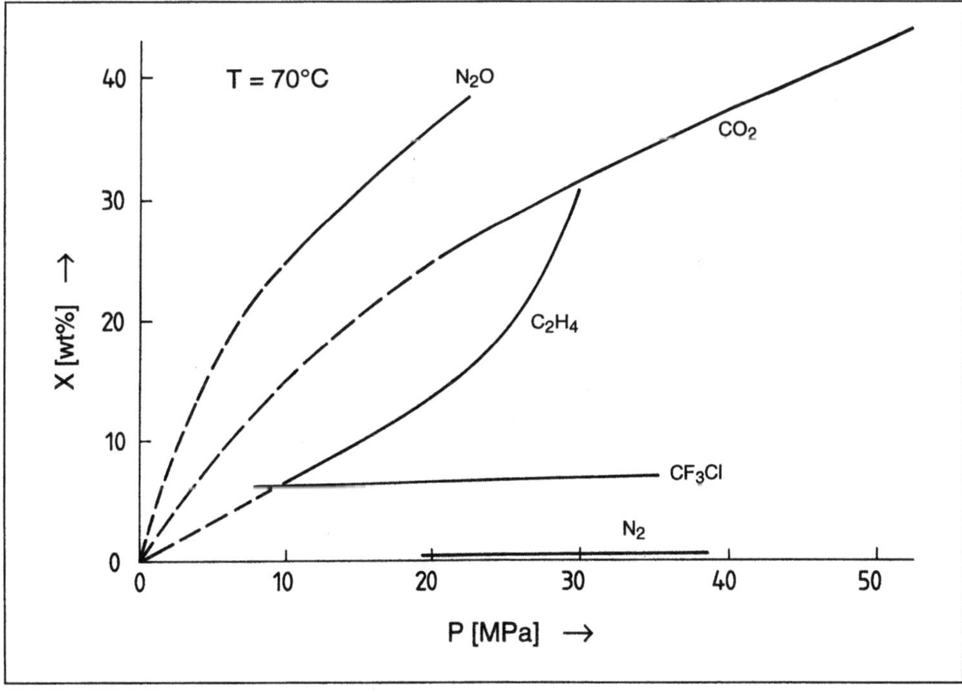


Fig. 3.10. Solubility of various gases in triglycerides of palm oil [4].

supercritical carbon dioxide is very low. Carbohydrates and glycosides are nearly insoluble in pure supercritical carbon dioxide.

The influence of chemical structure is clearly shown by the solubility of different xanthines in supercritical $CO_2$ (Fig. 3.14). Though the xanthines differ only by one methyl group and the position of the methyl groups, solubilities in supercritical $CO_2$ are different by orders of magnitude.

The calculation of solubility is part of the discussion of calculating phase equilibria in Section 3.8, yet it may be useful to have a simple relation at hand for interpolating solubility data. Such a relation is discussed in the following part.

Solubility depends on the variation of partial molar volumes with pressure. According to Eq. 3.5, which can be derived from equilibrium conditions, the vari-

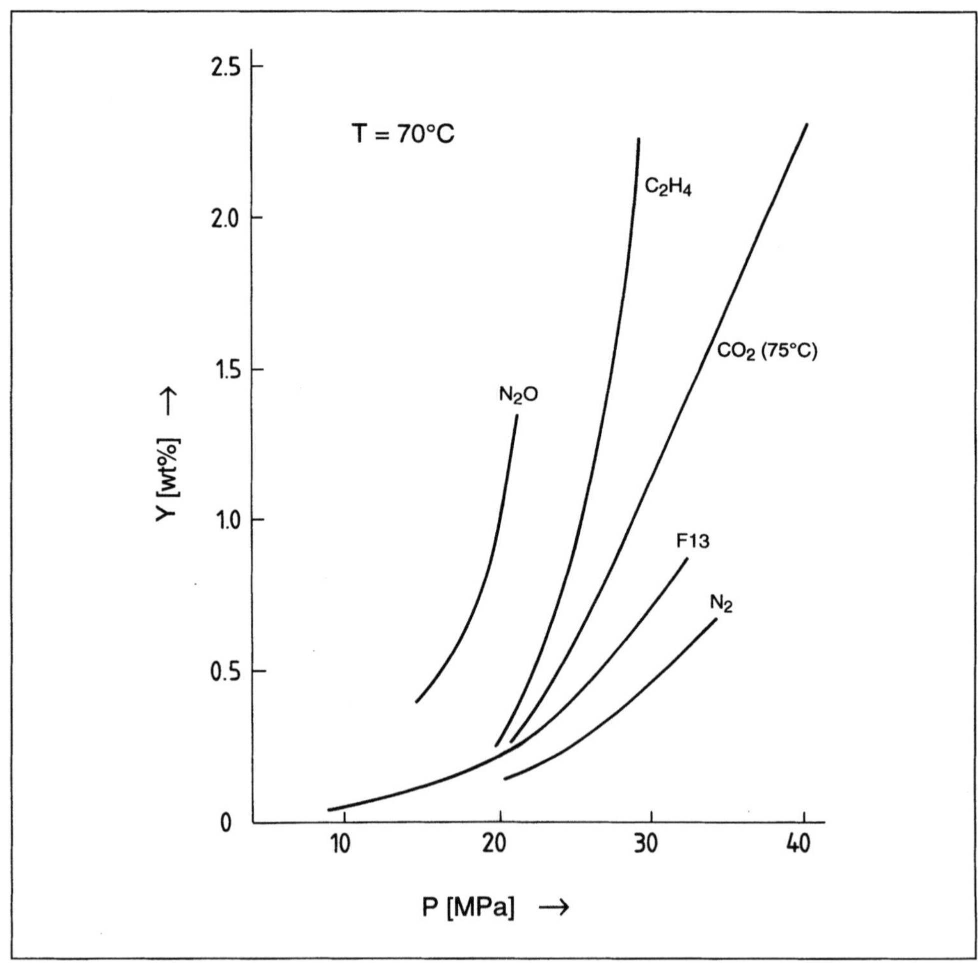



Fig. 3.11. Solubility of triglycerides of palm oil in various gases [4].

70

ation of concentration $y_2$ of the low volatile compound in the gas depends on the difference of the partial molar volumes of the dissolved compound in the gaseous phase $(\bar{V}_2^G)$ and in the condensed phase $(\bar{V}_2^L)$.

$$\left(\frac{\partial \ln y_2}{\partial P}\right)_T = \frac{\bar{V}_2^G - \bar{V}_2^L}{RT}.$$

$$3.5$$

Various semi-empirical equations have been derived for calculating solubilities in the gaseous phase, based on different expressions for the partial molar volumes and considering molecular interactions of different molecules, e.g. (Ewald [20,21] and Franck [22]). The resulting equations are useful, but are now replaced by methods employing equations of state for calculating equilibrium solubilities in the gaseous and condensed phase (see Section 3.8). The interested reader is therefore referred to the literature, which can be derived from the references. On the other hand, Mitra and Wilson [40] have reported a simple empirical equation (Eq. 3.6), relating solubility of some substances and density of the supercritical solvent. The coefficients for several binary systems, of which some are listed in Table 3.1, were calculated by non-linear regression from experimental data and should only be used in the ranges of pressure and temperature from which they are derived.

$$\ln y_2 = A\varrho_s + BT + C,$$

$$3.6$$

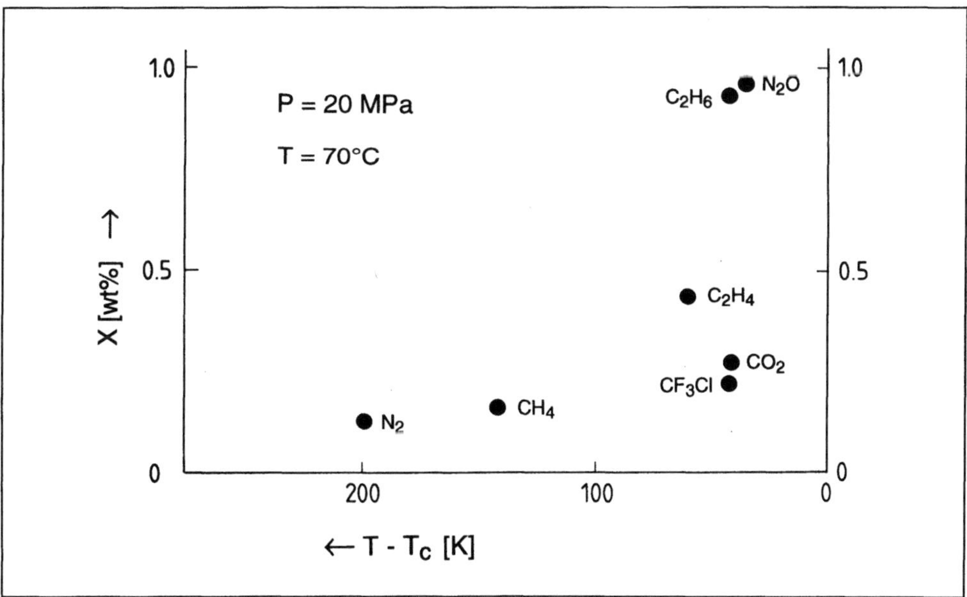


**Fig. 3.12.** Solubility of triglycerides of palm oil in various gases as a function of the difference between system temperature and the critical temperature of the supercritical solvent [4].

with

$y_2$ = concentration of dissolved component, mole-% (wt.-%, depending on experimental data used for correlating the coefficients);

$\varrho_s$ = density of the solvent, g/cm³;

$T$ = temperature, K;

$A, B, C$ = coefficients, obtained by fitting Eq. 3.6 to experimental data.

**Table 3.1.** Coefficients for solubility correlation. Data from Mitra and Wilson [40].

| Solute | $A \cdot 10^3$ | B | C | $r^2$ | Range of application max. P [MPa] | T [K] |
|---|---|---|---|---|---|---|
| Solvent: Carbon dioxide | | | | | | |
| Hexamethylbenzene | 43.11 | 5.995 | −24.607 | 0.93 | | |
| Triphenylmethane | 27.01 | 6.879 | −20.870 | 0.96 | | |
| Naphthalene | 46.24 | 4.684 | −19.55 | 0.88 | 8−30 | 300−330 |
| Anthracene | 65.35 | 6.676 | −35.875 | 0.96 | | |
| Phenanthrene | 32.52 | 7.138 | −22.647 | 0.90 | | |
| Fluorene | 47.03 | 6.551 | −26.398 | 0.94 | | |
| Pyrene | 45.98 | 7.958 | −29.534 | 0.91 | | |
| $\alpha$-Naphthol | 42.39 | 4.938 | −23.733 | 0.98 | | |
| $\beta$-Naphthol | 60.72 | 5.986 | −31.161 | 0.97 | | |
| Soybean Triglycerides | 61.22 | 16.409 | −34.393 | 0.99 | 13−62 | 300−380 |
| Solvent: Ethylene | | | | | | |
| Hexamethylbenzene | 48.87 | 15.589 | −26.406 | 0.97 | | |
| Naphthalene | 65.59 | 15.376 | −24.678 | 0.98 | | |
| Anthracene | 46.20 | 14.542 | −24.473 | 0.98 | | |
| Phenanthrene | 48.95 | 17.481 | −23.29 | 0.99 | | |
| Fluorene | 47.01 | 16.941 | −26.573 | 0.97 | | |
| Pyrene | 40.85 | 19.005 | −27.577 | 0.96 | | |
| Solvent: Ethane | | | | | | |
| Triphenylmethane | 58.61 | 18.187 | −31.214 | 0.97 | | |
| Naphthalene | 49.37 | 13.555 | −23.913 | 0.92 | | |
| Anthracene | 50.15 | 10.54 | −28.730 | 0.98 | | |
| Phenanthrene | 34.0 | 13.678 | −21.830 | 0.98 | | |
| Solvent: Water | | | | | | |
| Silicium dioxide | 9.09 | .0042 | −10.74 | 0.99 | 20−175 | 650−730 |

To achieve a high concentration of a low volatile substance in a supercritical gas, density and system temperature should be as high as possible. For practical purposes, these parameters are limited, i.a. by the stability of the compounds against elevated temperatures, investment costs for high pressure processing equipment and hydro-dynamics, if countercurrent contact of two phases under gravity is necessary.

Solvent power of a supercritical gas and solubility of a component in the supercritical gas are the most important aspects of a solvent/solute system to be used in gas extraction. But these aspects are only part of the picture. In order to select a solvent and appropriate conditions of state for a process, a more comprehensive view is necessary, comprising all existing phases at equilibrium and their behavior with respect to temperature and pressure. Therefore, in the next sections the phenomenological phase behavior of mixtures consisting of a supercritical compound and one or more substances of medium or low volatility will be discussed.

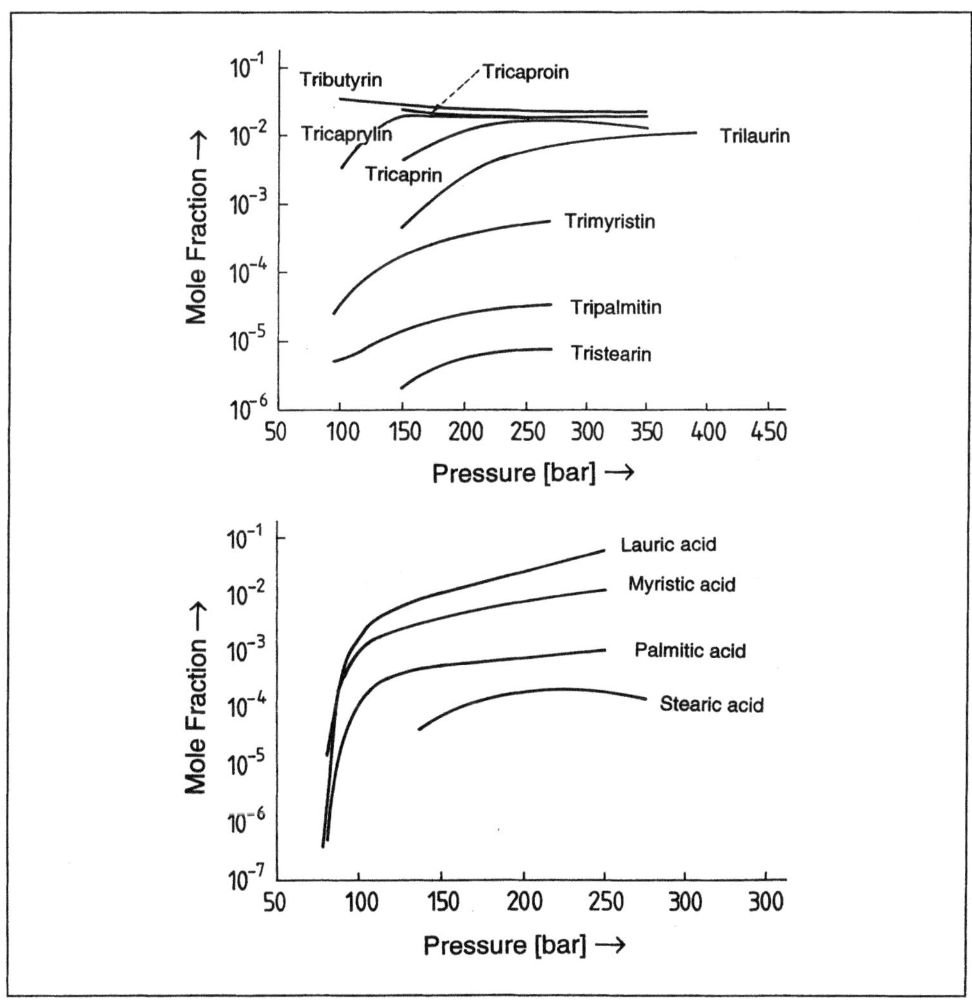


**Fig. 3.13.** Solubility of components of homologous series in supercritical carbon dioxide.

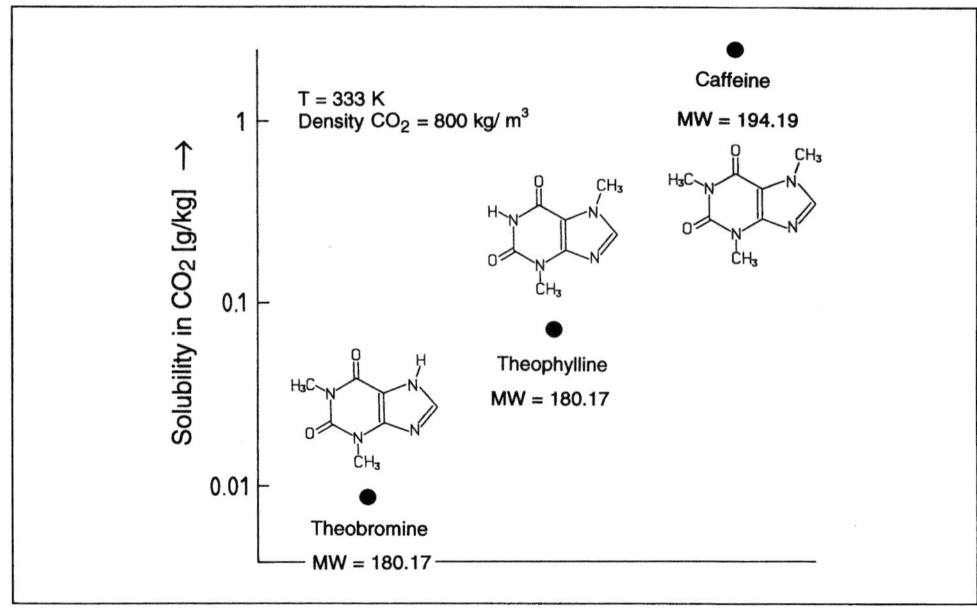


**Fig. 3.14.** Solubility of xanthines in supercritical carbon dioxide (dry $CO_2$) at constant density. Data from Johannsen and Brunner [30].

# 3.3 Phase Equilibrium in Binary Systems

Within this section an introduction to equilibrium phase behavior of binary systems consisting of a supercritical component and a subcritical component of much lower volatility is presented. In the next section (Section 3.4) phenomenological phase behavior of binary systems is treated more generally.

The phase rule for non-reacting systems (Eq. 3.7), determines the number of degrees of freedom $F$ for a mixture of $m$ components and $\pi$ coexisting phases:

$$F = m + 2 - \pi \tag{3.7}$$

According to Eq. 3.7, a binary system may exist in up to four phases at equilibrium. For an introduction, first gas-liquid phase equilibria and then gas-liquid-solid equilibria will be discussed.

## 3.3.1 Gas-liquid Equilibria of Systems, Completely Miscible in the Liquid Phase

Figure 3.15 schematically represents the behaviour of a binary system with respect to temperature, pressure, and composition in a three-dimensional plot in the region of complete miscibility in both phases. The phase equilibrium shown is characteristic for systems like nitrogen – n-heptane or carbon dioxide – ethanol, or in general, binary systems of a supercritical component and a compound of medium volatility. At the limiting values for the composition ($x = 0$ and $x = 1$), vapor-pressure curves of components $A$ and $B$ separate regions of a gas from that of a liquid. The vapor-pressure curves end at the critical point of each component ($K_A$, $K_B$). The critical curve, represented by the broken line, connects the critical points and limits the two-phase region, which extends between the vapour-pressure curves and below the critical line. The two-phase space is made visible by cuts at constant temperature (isotherms), one cut at constant pressure (isobar) and one cut at constant composition (isopleth). For quantitative representation of phase equilibria, these cuts are projected to the respective plane of two variables as coordinates.

Phase equilibrium data are often represented graphically. For binary systems, often a $P,x$-diagram is used, as schematically shown in Fig. 3.16. In such a diagram, composition of coexisting phases can be shown. According to the phase rule, Eq. 3.7, for two coexisting phases, a liquid and a gaseous one, pressure and temperature must be fixed at equilibrium. In a $P,x$-diagram, composition is plotted on the horizontal axis and pressure on the vertical axis. Equilibrium compositions of coexisting phases are plotted for a certain temperature. If pressure is varied at this temperature, compositions of the coexisting phases form a closed loop, as shown in Fig. 3.16. At a certain pressure $P_1$, a liquid phase $l_1$ and a gaseous phase $v_1$ coexist at thermodynamic equilibrium for a total composition $x_1$ of the binary system. The line $\overline{l_1v_1}$ is a tie line,

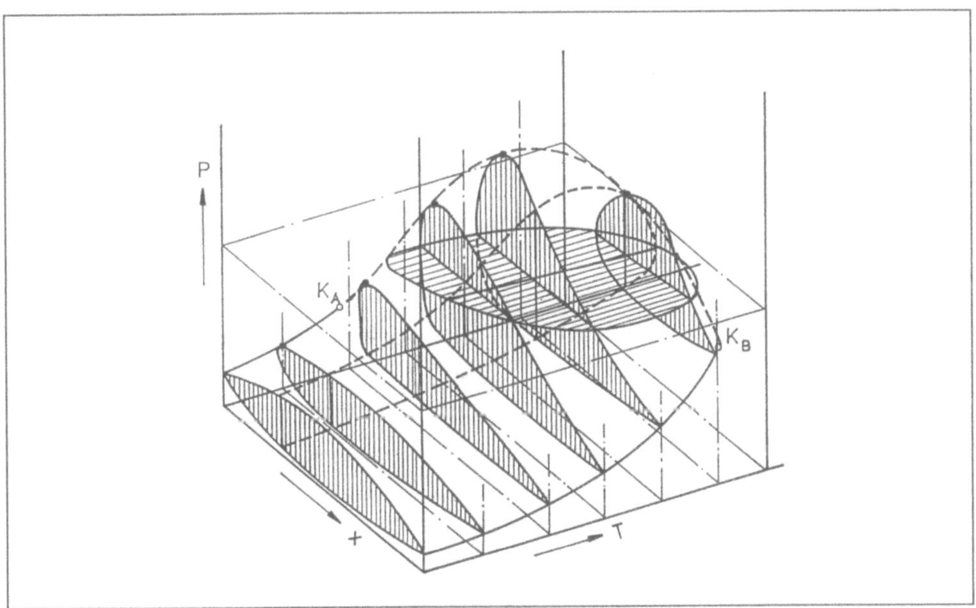


**Fig. 3.15.** $P, T, x$-diagram of a binary system with one gaseous and one liquid phase [4].

connecting coexisting equilibrium phases. The line has no further physical meaning. In a $P,x$-diagram, tie lines are horizontal lines. The amount of phases can be derived from the lever rule:

$$l_1 = \frac{\overline{x_1 v_1}}{\overline{l_1 v_1}} \quad \text{and} \quad v_1 = \frac{\overline{l_1 x_1}}{\overline{l_1 v_1}},$$

3.8

with

$l_1$ = number of moles in the liquid phase;
$v_1$ = number of moles in the gaseous phase;

$\overline{x_1 v_1}$, $\overline{l_1 x_1}$, $\overline{l_1 v_1}$ = distance between respective points in the $P,x$-diagram.

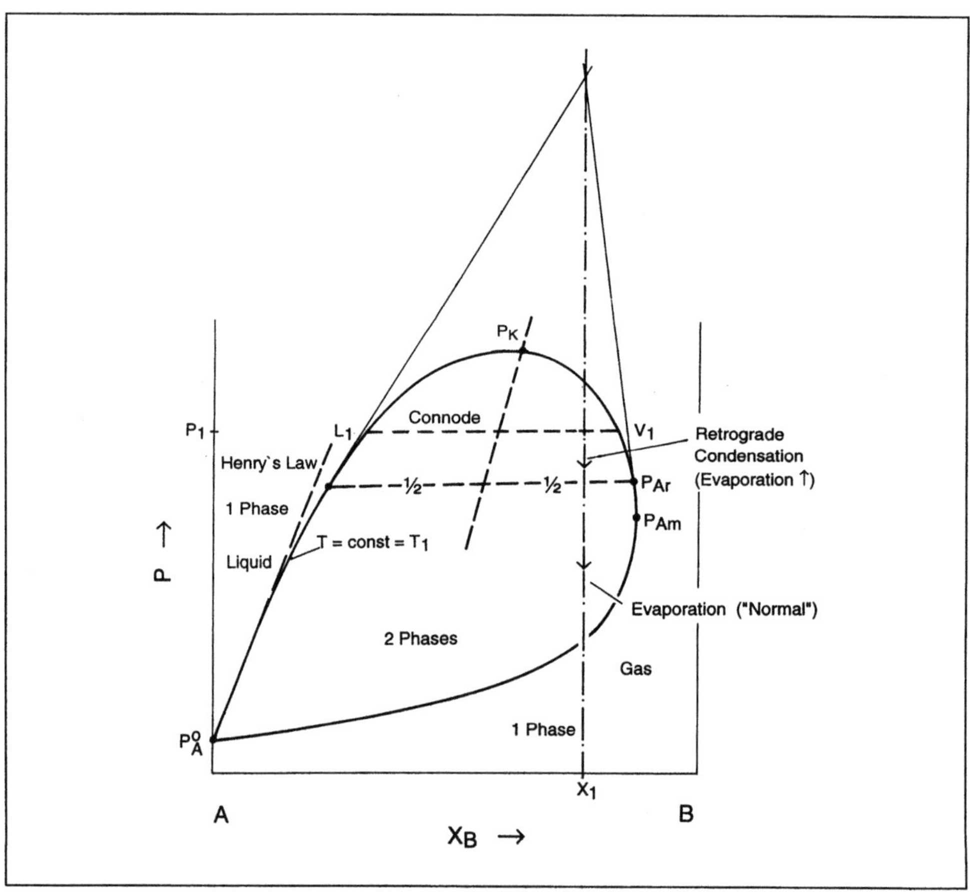


**Fig. 3.16.** Schematic representation of phase equilibrium of a binary system with one supercritical component in a $P,x$-diagram.

The closed loop, representing composition of coexisting phases at a constant temperature $T_1$, encloses the two-phase area. Each mixture with a total composition within this loop, at equilibrium splits into two coexisting phases, with concentrations lying on the enclosing phase boundary line. Compositions of the coexisting phases start from the vapor pressure of component $A$ ($P_A{}^O$), the subcritical component and meet again at the critical point, at which liquid and gaseous phase become identical. The critical point very closely lies on a straight line which divides the two phase area, see Fig. 3.16. Component $B$ is supercritical. Therefore, no vapor pressure of $B$ exists at $T_1$, and the part of the phase boundary line, representing the gaseous phase, does not reach the ordinate for pure component $B$. At pressures higher than the vapor pressure of $A$, and concentrations for the binary system left of the phase boundary line, a homogeneous liquid mixture of $A$ and $B$ exists. At pressures below the vapor pressure of $A$ and on the righthand side of the phase boundary line, homogeneous gaseous mixtures of $A$ and $B$ exist. At pressures higher than the critical pressure, components $A$ and $B$ are completely miscible. The character of the mixture gradually changes in this region from gaseous, at or nearby ordinate $B$, to liquid, at or nearby ordinate $A$. For total compositions of mixtures of $A$ and $B$, containing more $B$ than at the critical point, and above the pressure of minimum concentration of $A$ in the gaseous phase $(P_{Am})$, a pressure reduction induces condensation, called retrograde condensation. The point of inversion, $P_{Ar}$, from normal to retrograde condensation lies

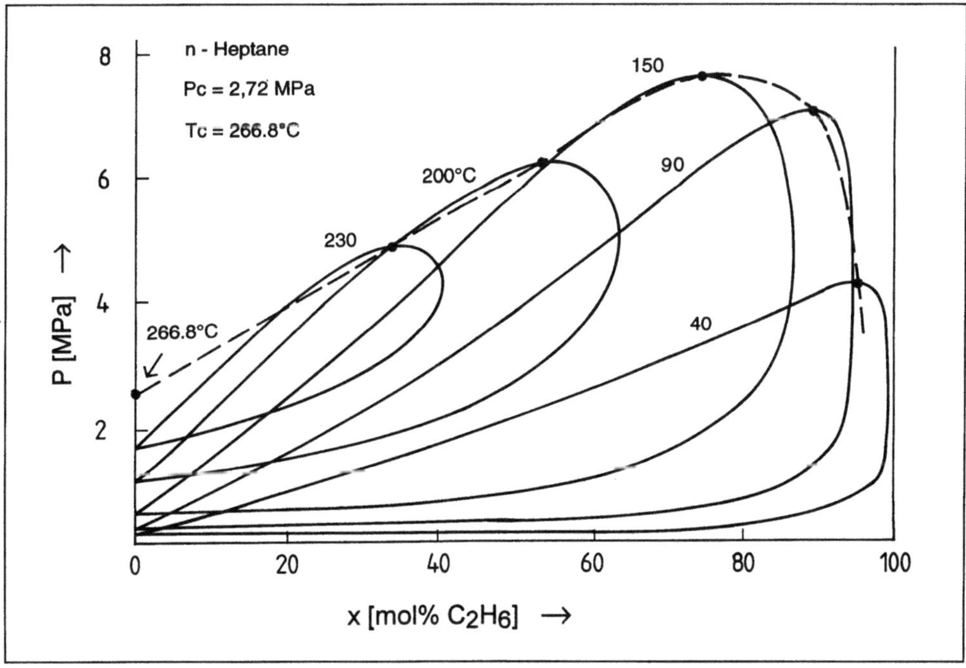



**Fig. 3.17.** $P,x$-diagram of ethane-n-heptane. After Kay [31].

somewhat above $P_{Am}$. The exact location of $P_{Ar}$ for a given total composition depends on the tangential point of the tangent from the line of total composition to the phase envelope, as indicated in Fig. 3.16. Normally, $P_{Ar}$ is very close to $P_{Am}$. Below $P_{Ar}$, reducing the pressure causes evaporation, as expected.

Projections of phase equilibrium data of binary systems are shown in the following figures. Isotherms, the $P,x$-projection of the binary system ethane – n-heptane are presented in Fig. 3.17. Full lines represent the boundary line of the two-phase region for a certain temperature. At the temperatures shown, ethane is supercritical. Therefore the two-phase region extends not to pure ethane, since this component cannot exist in the liquid state at supercritical temperatures. The maximum pressure (which

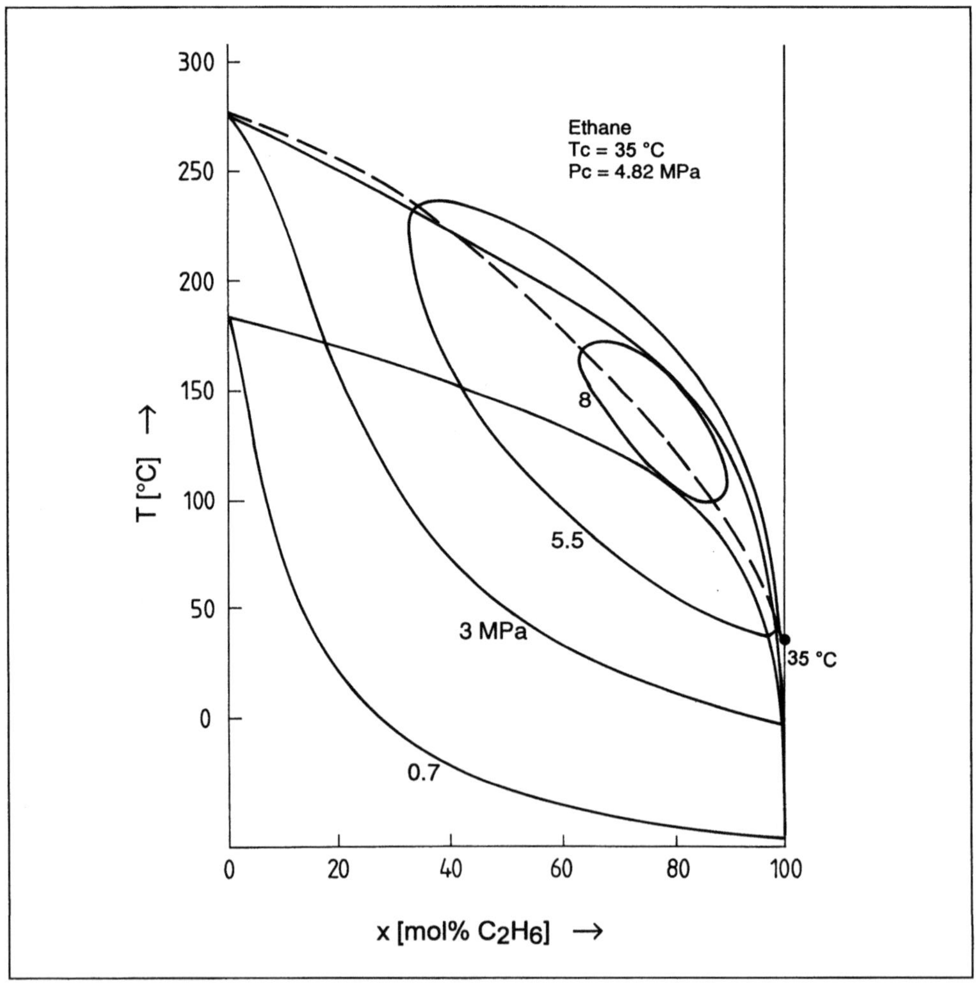


**Fig. 3.18.** $T,x$-diagram of ethane-n-heptane. After Kay [31].

78

is the critical pressure) of each isotherm increases if the temperature is raised from near critical temperatures of ethane (40 °C), reaches a maximum ($\approx 150$ °C) and then decreases until the critical pressure of n-heptane, the component of lower volatility is reached at its critical temperature.

Figure 3.18 presents the $T,x$-projection of the binary system ethane – n-heptane. Full lines are the boundary lines for the two-phase regions at constant pressure. At subcritical conditions for both components, the two-phase regions extend over the whole range of concentration. At higher pressures than the critical pressure of one component, the two-phase area detaches from the ordinate. If both components are supercritical with respect to pressure, an isolated two-phase region exists.

Projections of isopleths on the $P,T$-plane, shown in Fig. 3.19 for ethane – n-heptane, are similar to vapor-pressure curves. Each full line represents the envelope of the two-phase region for a binary mixture of certain total composition. The lines ending at points $K_A$ and $K_B$ are the corresponding vapor pressure curves for pure components $A$ (ethane) and $B$ (n-heptane). The broken line is the projection of the critical curve to the $P,T$-plane. This enveloping curve meets each isopleth at the critical point, where gaseous and liquid phases become identical.

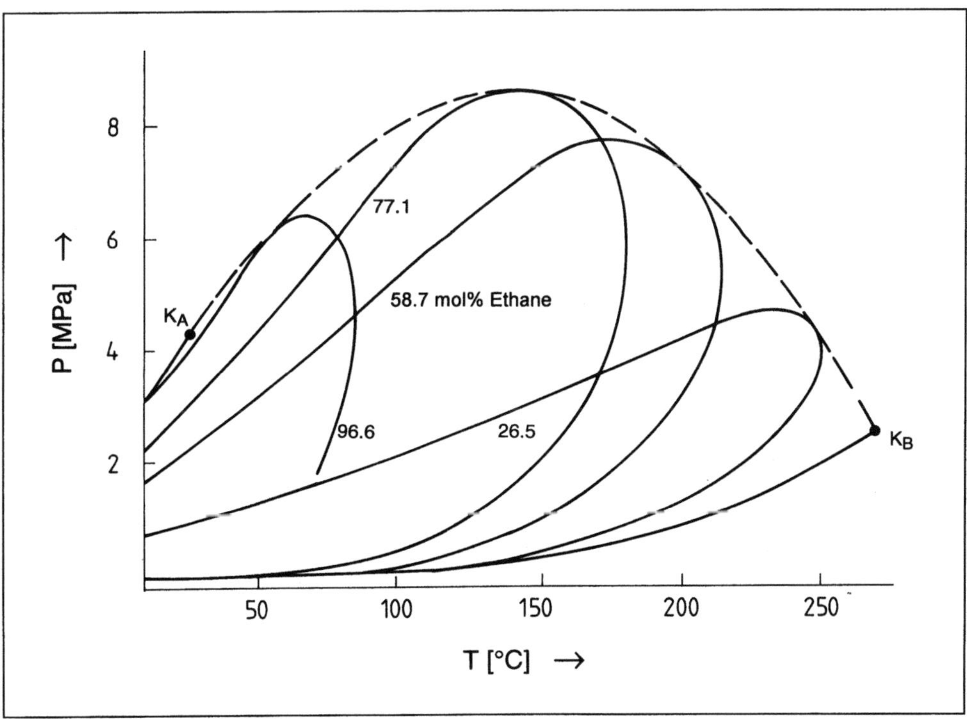


**Fig. 3.19.** $P,T$-diagram of ethane-n-heptane. After Kay [31].

## 3.3.2 The Critical Curve of Mixtures

The extent of the two-phase region depends on the size of the molecules, the molecular structure and the chemical properties of the components of the mixture. Conditions at which gaseous and liquid phases become identical characterize critical points. All the critical points line up to a critical curve, which is characteristic for a mixture. Critical curves have been plotted in Figs. 3.17 to 3.19 as broken lines. The critical curve of a binary mixture, with respect to composition, depends on the ratio of critical pressure to critical temperature of the pure components. Maximal critical pressures of mixtures increase with the ratio of critical temperatures of the compounds, as shown in Fig. 3.20. If components of a mixture are chemically similar, e.g.

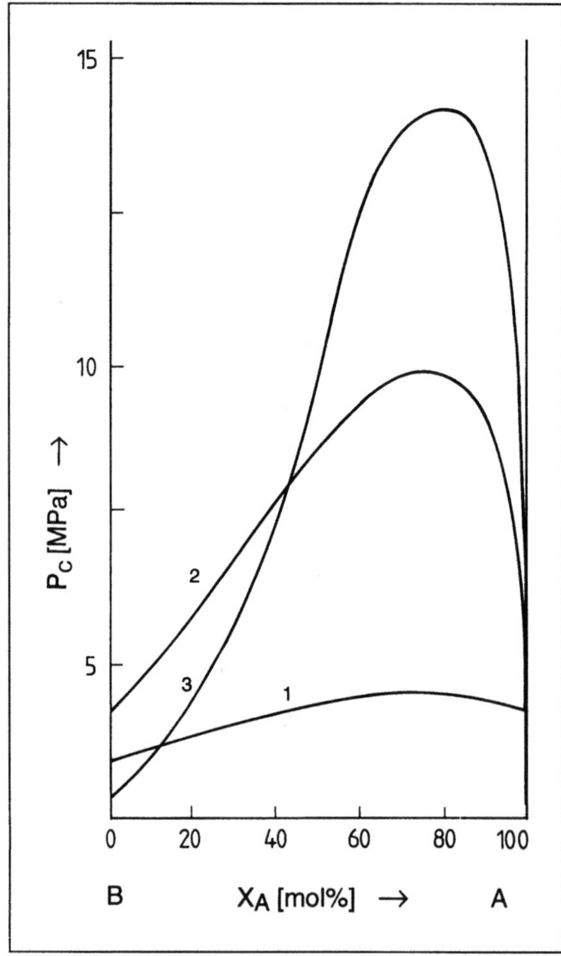


**Fig. 3.20.** Critical pressures of binary mixtures (after King [35]).
1) propane – n-heptane: $T_{cB}/T_{cA} = 1.46$;
2) methane – propane: $T_{cB}/T_{cA} = 1.94$;
3) methane – n-heptane: $T_{cB}/T_{cA} = 2.83$.

a series of homologues, and do not differ much in size, the critical curve is nearly a straight line. With increasing difference in molecular size, the maximum critical pressure increases and can reach much higher values than the critical pressures of both pure components. Lower members of a homologous series are of much more influence on the critical curve than higher members. This is due to the greater differences of critical temperatures between lower members of a homologous series. In Fig. 3.21 this is illustrated for mixtures of acetone with paraffines.

If pure-component critical temperatures differ appreciably between solvent and solute, as is the normal case in gas extraction, critical pressures increase rapidly with the amount of solute in the gaseous phase. Therefore, the concentration of the solute in the gaseous phase remains at relatively low values (compare the solubility of oleic acid in ethylene, Figs. 3.6, 3.7).

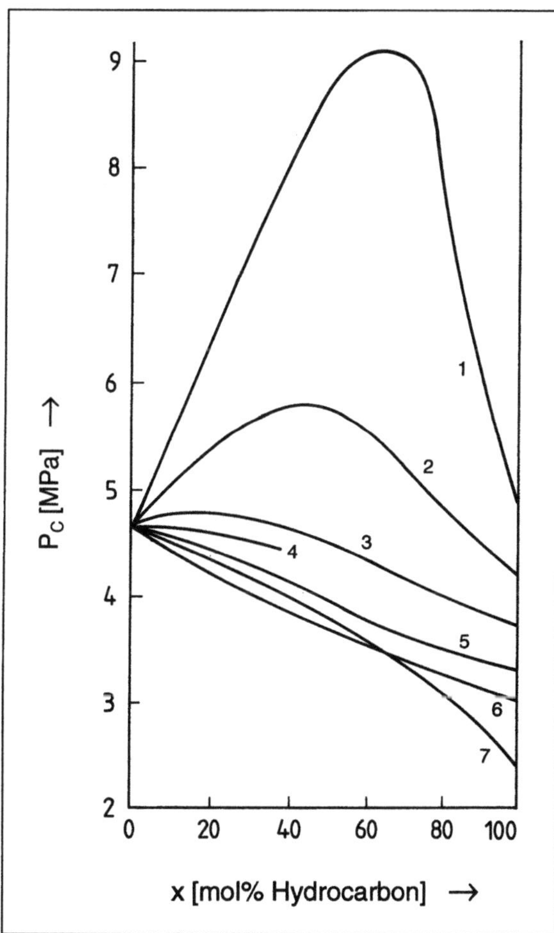


**Fig. 3.21.** Critical pressures of binary mixtures of acetone and alkanes (after Kay [32]).
1) ethane; 2) propane; 3) n-butane; 4) n-decane; 5) n-hexane; 6) n-heptane; 7) n-octane.

Critical temperatures of binary mixtures often are intermediate to critical temperatures of the pure components, as illustrated in Fig. 3.22. If critical temperatures of the two pure components are similar, there may be a maximum or minimum value for the critical temperature of the mixture, which corresponds to azeotropic behavior (see Fig. 3.23).

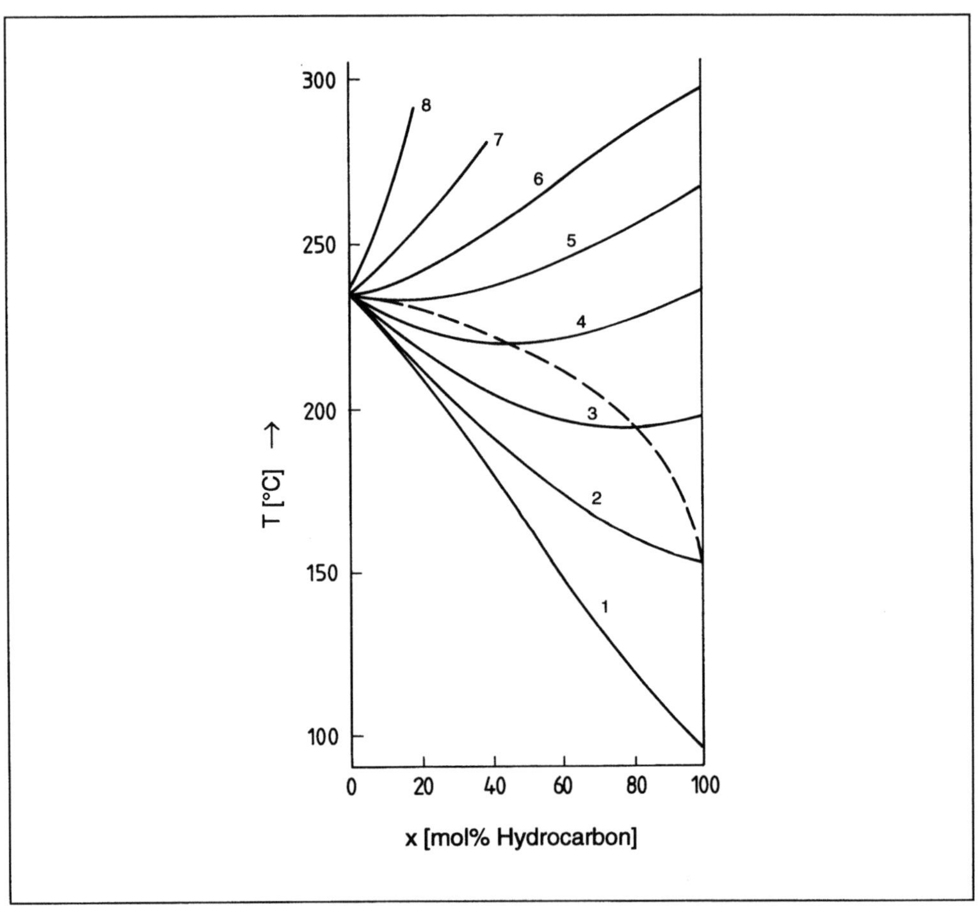



**Fig. 3.22.** Critical temperatures of binary mixtures of acetone and alkanes (after Kay [32]).
1) propane; 2) n-butane; 3) n-pentane; 4) n-hexane; 5) n-heptane; 6) n-octane; 7) n-decane; 8) n-tridecane.

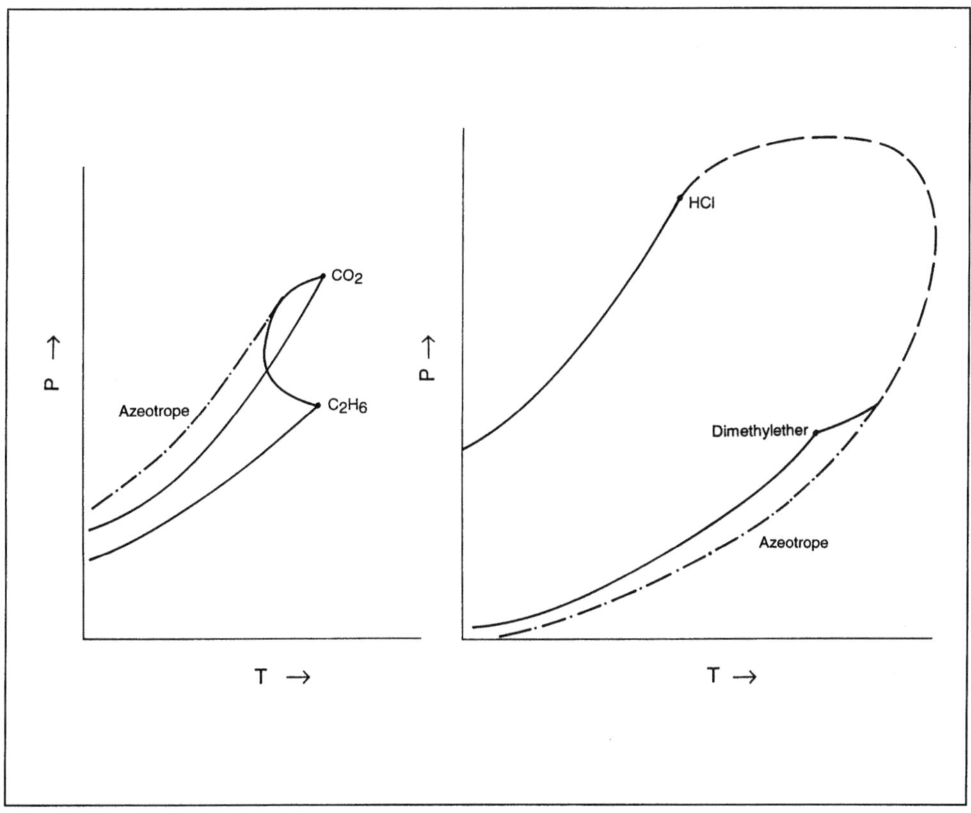


**Fig. 3.23.** Azeotropic behavior in mixtures of similar critical temperatures [48].


### 3.3.3 Systems with Miscibility Gaps in the Condensed Phase

As component properties become increasingly different, limited miscibility may occur in condensed phases. Liquid-liquid immiscibility often occurs in the immediate vicinity of the critical point of the solvent. This region of phase equilibrium will be discussed in Section 3.4. More important for gas extraction processes is the occurrence of a solid phase in binary mixtures. Such equilibria exist in systems where the melting point temperature of the low volatile component is higher than the critical temperature of the solvent. Phase behavior of such a binary mixture is presented in Fig. 3.24 for the case where the solid phase does not extend to the critical curve, and in Fig. 3.26 for the case where the solid phase extends beyond the critical curve and divides it into two sections. To simplify understanding, it is assumed that there is no immiscibility in the liquid phases. Furthermore, the discussion is restricted to near critical and higher temperatures for component $A$, the solvent, excluding solid phases of $A$.

Figure 3.24 presents the three-dimensional plot and the $P,T$-projection of the phase behavior of a binary system, consisting of a volatile component (component $A$, which at supercritical conditions is the solvent in gas extraction), and a nonvolatile component (component $B$). The melting point temperature of the nonvolatile component is higher than the critical temperature of the volatile component. Three cuts at constant temperature illustrate phase equilibrium in different regions. Curve $(V+L)_A$ is the vapor-pressure curve of component $A$ and ends at its critical point $K_A$ and at lower temperatures at the triple point $A_t$ of component $A$ (not shown). The three-phase line $V+L+B$, the vapor pressure curve of saturated solutions of $B$, runs to a maximum

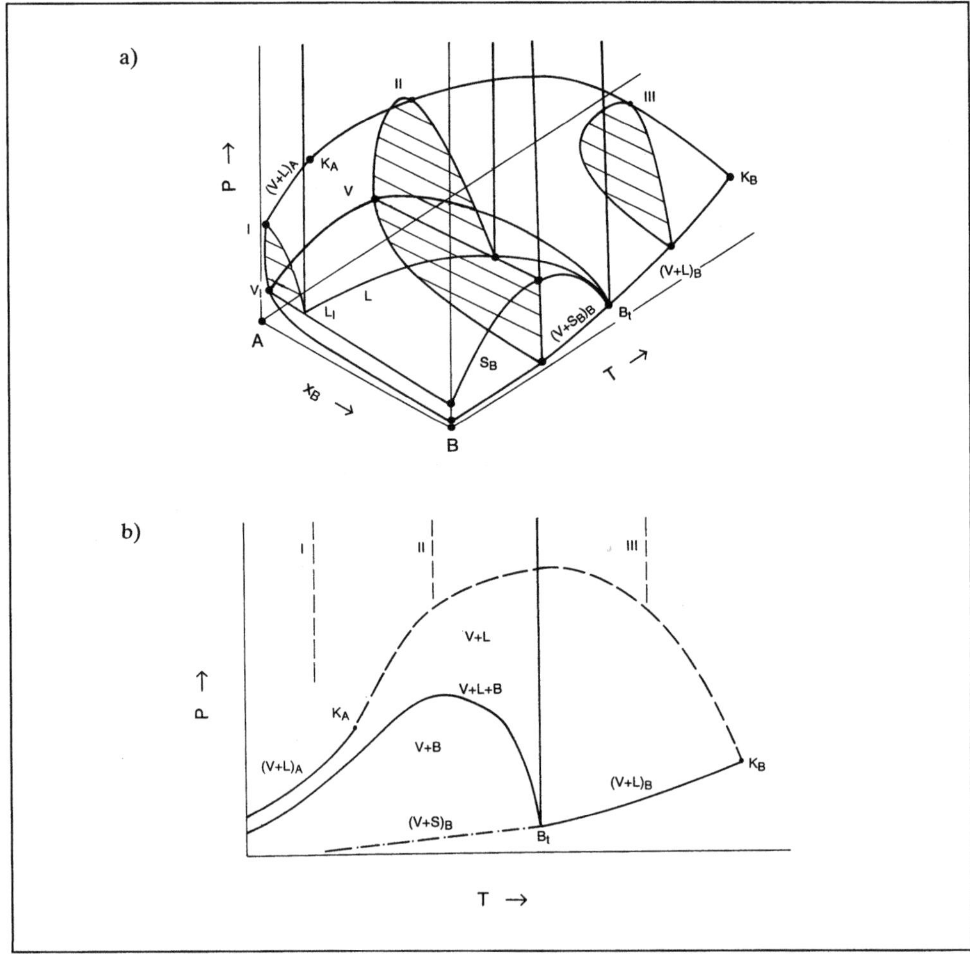



**Fig. 3.24.** Phase equilibrium in a binary system. Three dimensional plot (part a) and $P, T$-diagram (part b). The melting point temperature of the low volatile component is higher than the critical temperature of the solvent.

pressure and then declines to the vapor pressure of $B$ at its triple point $B_t$. From this point the vapor-pressure lines $(V+L)_B$ and $(V+S)_B$ proceed. The vapor-pressure curve of liquid component $B$ ends at the critical point $K_B$. The critical points of both pure components are connected by the critical curve $K_A K_B$, which limits the two-phase area for vapor-liquid equilibrium.

Lines $V_t B_t$ and $L_t B_t$ represent the gas-liquid equilibrium of the three-phase equilibrium $V+L+B$, which proceeds to the triple point of component $B$. In the region marked $V+B$, a gaseous phase coexists with solid $B$. If the pressure is raised at constant temperature in this region to values at which solid $B$ melts, the gas-liquid equilibrium region marked $V+L$ is reached.

An example of such a type of phase behavior is shown in Fig. 3.25 for the mixture carbon dioxide – methane. Similar behavior show mixtures of water and many inorganic salts.

Figure 3.26 presents the three-dimensional plot and the $P,T$-projection of the phase behavior of a binary system, in which the solid region extends beyond the critical curve. Four cuts at constant temperature illustrate phase equilibrium in different regions. Line $L_t P_{VL}$ represents the solubility of $B$ in liquid $A$, which decreases in the region of the critical temperature of $A$. Line $V_t P_{VL}$ represents the solubility of $B$ in gaseous $A$, which increases in the same region until both lines meet at the critical endpoint $P_{VL}$. In the $P,T$-projection the critical endpoint is represented by a single point, while in the three-dimensional plot it can be seen that there are two endpoints

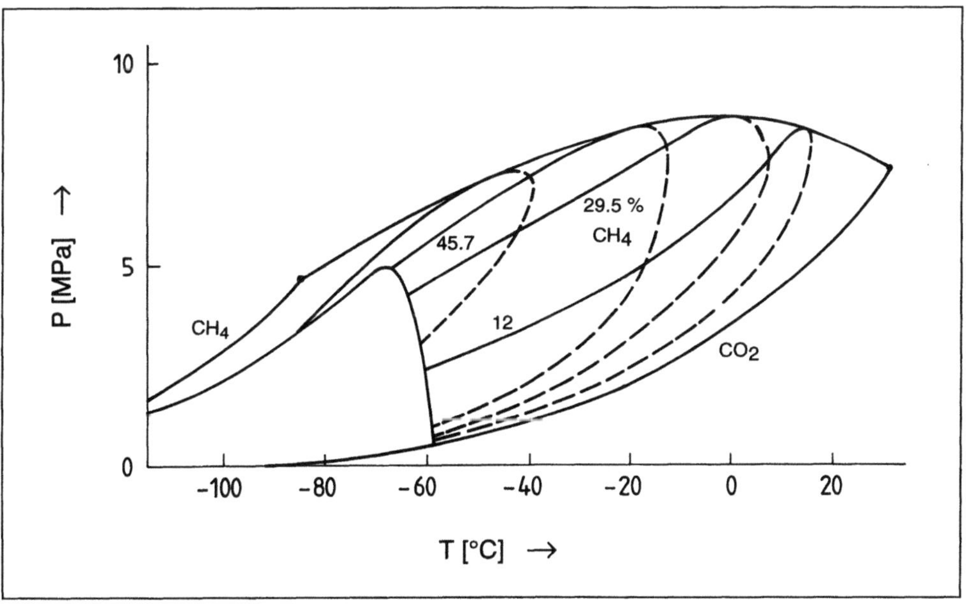


**Fig. 3.25.** Phase behavior of the binary system carbon dioxide – methane. $P,T$-diagram with isopleths (after Donnelly and Katz [17]).

of different composition, namely, $P_{VL}$, representing the endpoint of vapor-liquid equilibrium and $P_{SB}$, representing the endpoint for solid $B$. Lines $V_lP_{VL}$, $L_lP_{VL}$ and $S_lP_{SB}$ fall on a single line $(V+L+B)_l$ in the $P,T$-projection.

The melting point of component $B$ is lowered by component $A$, as line $(V+L+B)_2$ shows, starting from the triple point $B_t$ of pure $B$. The pressure necessary to keep the volatile component in solution rapidly increases, as can be seen from the righthand side of part $b$ of Fig. 3.26. The coexisting gaseous phase increases in density. More

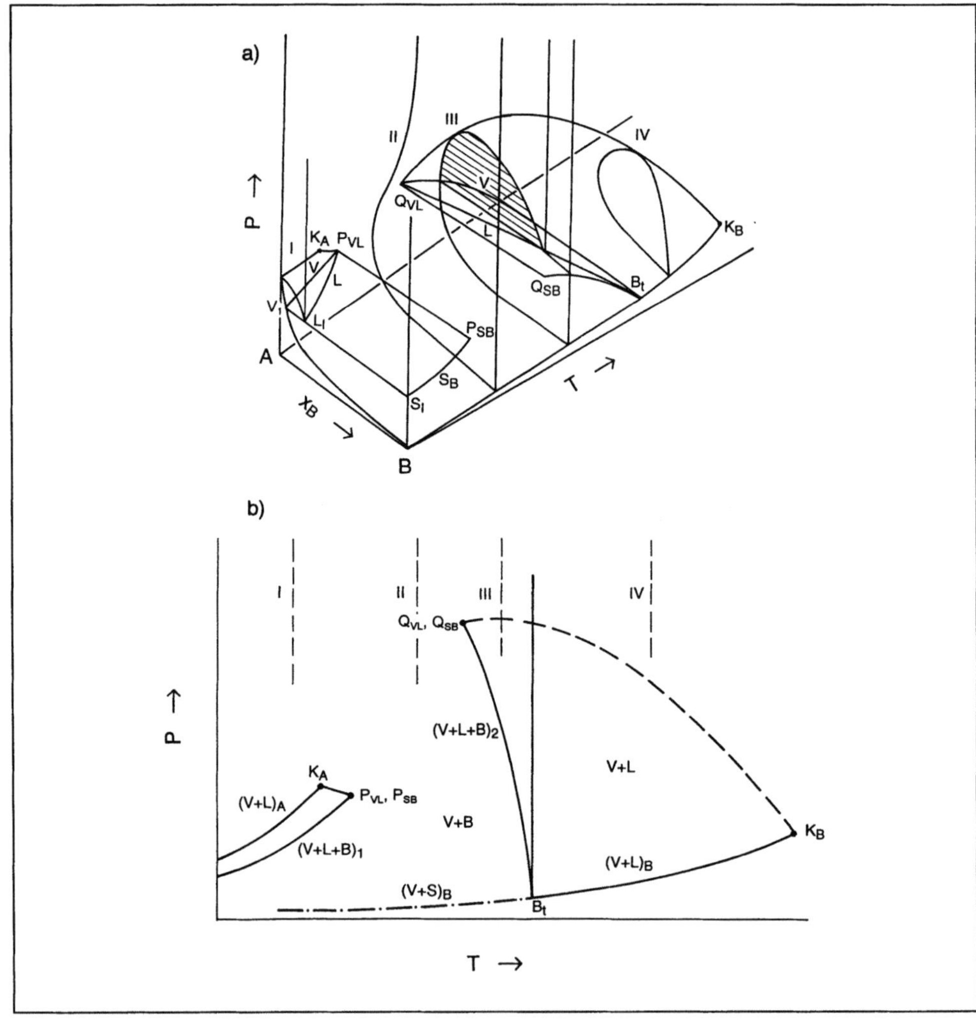


**Fig. 3.26.** Phase equilibrium in a binary system. Three dimensional plot (part a) and $P,T$-diagram (part b). The melting point temperature of the low volatile component is higher than the critical temperature of the solvent. The solubility curve divides the critical line into two parts.

86

and more of component $A$ dissolves in the coexisting liquid phase, until the compositions become equal at the critical endpoint $Q$. At the critical endpoint a single fluid phase $Q_{VL}$ is in equilibrium with solid $B$ at $Q_{SB}$. Lines of the coexisting phases fall on a single line $(V+L+B)_2$ in the $P,T$-projection, as was discussed above for the lower temperature part of the diagram.

For such a type of phase behavior, the binary system anthrachinone – diethylether, as presented in Fig. 3.27 is an example.

In this type of systems, equilibrium between a gaseous and a solid phase is not only found at pressures below the three-phase line $V+L+B$, but also at temperatures between $P$ and $Q$. In this region no liquid phase exists. Only a gaseous and a solid phase coexist at equilibrium, see Fig. 3.26, cut II. Pressure can be enhanced to very high values in order to achieve high solubilities of the low volatile component in the supercritical solvent. In the pressure region, to where an extension of the critical lines would lead, a relatively large increase in solubility of $B$ in $A$ with pressure can be achieved. This is due to the transformation of phase equilibrium corresponding to cut I and III to that of cut II. At temperatures higher than the melting point of $B$, a gaseous and a liquid phase coexist. For gas extraction the region between $P$ and $Q$ can be used for extraction of solid compounds from solid substrates, insoluble in the solvent, and the region at higher temperatures than the melting temperature, for gas-liquid separations.

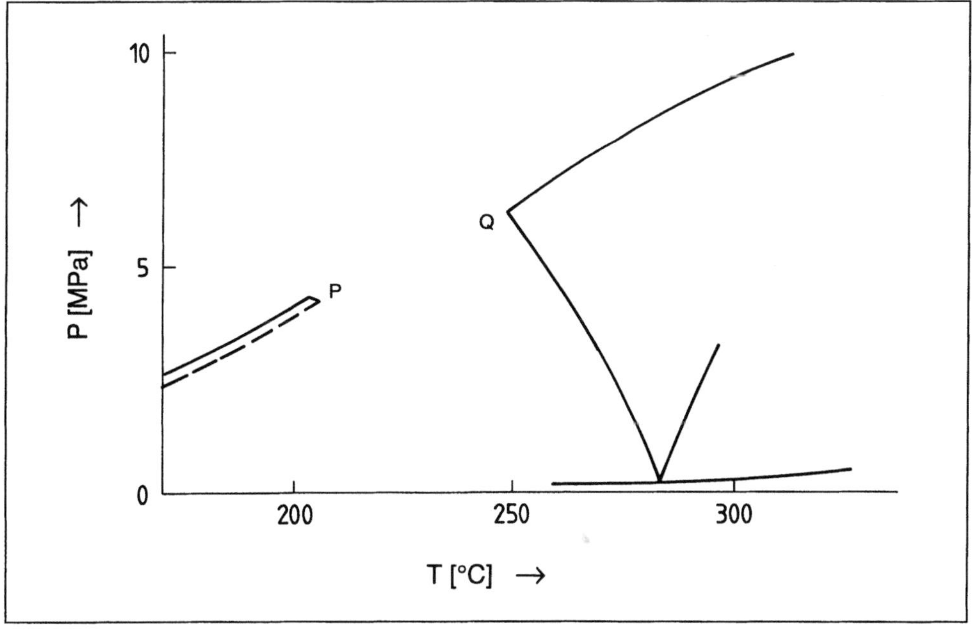

Fig. 3.27. Phase behavior of the system anthrachinone – diethylether. $P,T$-diagram (after Smits [57, 58]).

The formation of solid phases including component $A$, the solvent, has been omitted in this discussion, since this region is of minor importance for gas extraction processes. The interested reader is referred to review articles, e.g., the one by Koningsveld and Diepen [36].

In order to know solubility for the normal operation of a gas extraction separation process, from the phase behaviour, shown in the $P,T$-diagrams, only cuts at constant temperature or pressure at near critical conditions of the solvent are needed. On the other hand, understanding of phenomenological phase behavior is fundamental for selecting operating conditions for new solvents, solutes and operational modes. Phase behavior of multicomponent systems can often be explained by the binary systems which are the basis for multicomponent systems. It is these reasons, beside the desire to understand in general, which leads to the next chapter, wherein the phenomenology of binary systems is treated.

# 3.4 Phenomenology of Binary Gas-Liquid Systems

As differences in the physical and chemical properties of components become greater, immiscibility in the liquid phase may occur, and phase behavior exhibits some new features. The effect of increasing differences in the properties of compounds can be demonstrated for a series of homologous compounds. For paraffins with ethane as the more volatile compound in each binary, liquid immiscibility comes up with components of more than 19 carbon atoms. In many cases miscibility gaps extend to the critical region. Schneider systematically investigated and reviewed this phase behavior and the following discussion is partly based on his work.

There have been successful attempts to calculate the different types of phase behavior. Van Konynenburg and Scott [37] suggested five types which can be defined by characteristic values formulated with the parameters $a$ and $b$ of the van der Waals equation of state. This simple equation is able to represent the various types of phase behavior, although coincidence between experimental and calculated results is not good. Rowlinson and Swinton [48] reviewed this work and added a 6th class, which can only be predicted by angle-dependent potential functions. The six types of phase behavior of binary systems are shown in Fig. 3.28 and are shortly discussed below. Only $P,T$ projections are presented, since with the aid of the previous figures, the other projections can be derived.

If the components of a binary system are not too different in molecular weight and chemical structure, they are miscible in the liquid phase, and the vapor-liquid coexistence region is limited by a continuous critical curve, leading from the critical point of one component to the critical point of the other component. Such a system is classified as type I. Its $P,T$ diagram is shown in Fig. 3.28.a, and has been discussed above in Section 3.3.1.

Type I: Continuous, uninterrupted critical curve gas-liquid. Azeotropic behavior possible.

Examples:
- no maximum or minimum in the critical curve:    $CO_2 - C_3H_8$;
- maximum with respect to pressure:    $C_2H_6 - n\text{-}C_7H_{16}$;
- minimum with respect to pressure:    $SO_2 - CH_3Cl$;
- maximum with respect to temperature:    $HCl - (CH_3)_2O$;
- minimum with respect to temperature:    $CO_2 - C_2H_6$.

In systems with an uninterrupted gas-liquid critical curve, at lower temperatures immiscibility in the liquid phase can emerge. This phase behavior is classified as type II. The upper critical solution temperature *(UCST)* may increase with pressure or decrease with pressure. At temperatures below the upper critical solution temperature liquid-liquid equilibria exist, which extend to the-three phase line *VLL*. At pressures below the *VLL*-line, a gaseous and a liquid phase coexist at equilibrium. At temperatures higher than the *UCST* and pressures higher than the vapor-pressure of *A* or higher than pressures of the critical curve, there is complete miscibility between components *A* and *B*.

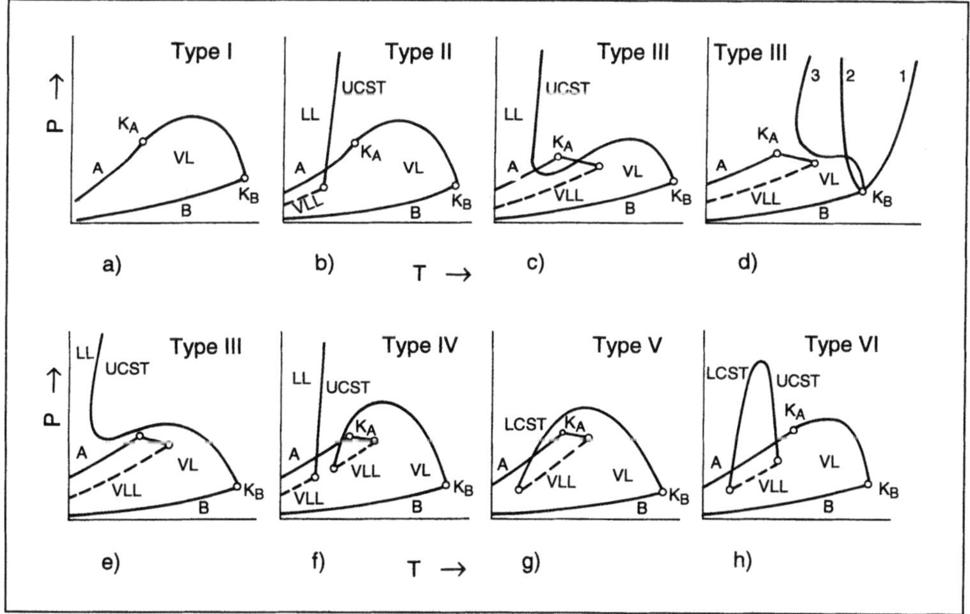


**Fig. 3.28.** *P, T*-projections of phase boundary lines (*P, T*-diagrams) for different classes of phase behavior of binary systems. The six types of binary phase behavior of gas-liquid-systems. See text for examples.

Type II: Continuous, uninterrupted critical curve gas-liquid. Azeotropic behavior possible. Three-phase line liquid-liquid-gas ends below critical temperature of the more volatile component. The critical curve liquid-liquid is nearly vertical.
Examples:
n-octane – carbon dioxide;
n-hexane – nitrobenzene.

With decreasing tendency of mutual miscibility, the region of immiscibility extends to higher temperatures, as does the critical curve (see Fig. 3.28, part $c$). The critical curve is divided into two branches. This phase behavior is classified as type III. The critical curve gas-liquid does not meet the gas-liquid critical point of the more volatile component, but merges with the former upper critical solution temperature. Example: n-dodecane – carbon dioxide.

The critical curve can tend to higher temperatures until in curve 1, Fig. 3.28, part $d$, the total critical curve extends to temperatures beyond the critical temperature of the less volatile compound. At pressures above this curve, and temperatures above this critical temperature, so-called gas-gas equilibria exist. Examples: $d1$: Helium – ammonia, helium – xenon, $d2$: ammonia – nitrogen, water – carbon dioxide, $d3$: n-hexadecane – carbon dioxide.

In some systems, there is a pressure minimum of the critical curve at relatively low temperatures (see Fig. 3.28, part $e$); example: methane – methylcyclohexane.

Type III: Systems with a divided gas-liquid critical curve. One leg starts at the critical point of the less volatile component. The other leg begins at the critical point of the more volatile component and ends on the three-phase line liquid-liquid-gas. Type III systems are subdivided according to Fig. 3.28, parts $c$, $d$ and $e$.

If the mutual solubility of both the components is slightly higher than for a type III system shown in Fig. 3.28.e, then the critical curve may extend to the liquid-liquid line and divide it into two parts, as shown in Fig. 3.28.f. This phase behavior is classified as type IV [48].

Type IV: Systems with three critical curves according to Fig. 3.28.f. The third critical curve begins at the critical point of the component with the higher critical temperature and ends at an critical endpoint of the three-phase line.
Example:
Methane – n-hexene.

If no liquid-liquid immiscibility at low temperatures occurs before crystallization, then phase behavior is as shown in Fig. 3.28.g, which is classified as type V. The lower part of the critical curve between component $B$ and the $VLL$-line has the character of a lower critical solution temperature $(LCST)$.

Note that the character of the critical curve changes gradually with increasing temperature to a *VL*-critical curve, which is also true for some of the critical curves of type III and the critical curve of type IV phase behavior.

Type V: Systems with two critical curves as type IV, but without liquid-liquid critical curve
Examples:
Ethane – ethanol, methane – n-hexane, methane – i-octane.

At low temperatures, liquid-liquid immiscibility is possible in a type-I system. Such a phase diagram is classified as type VI. Type-VI systems are only found in systems where one or both components exhibit self-association due to hydrogen bonding and in the mixture there are additional strong inter-component hydrogen bonding [48]. Such mixtures are not normally encountered in gas extraction, using nonpolar supercritical solvents, but may become of more importance with the application of supercritical water in gas extraction processes.

Type VI: Systems with liquid immiscibility over a certain range of temperature and critical curves like that shown in Fig. 3.28.h.
Examples:
$D_2O$ – 2-methylpyridine;
Water – n-butanol
Water – 2-butanol

Between all the types shown in Fig. 3.28, there are intermediate cases of phase behavior.

Critical curves of binary systems provide information on the extent of the two- or three-phase region and can be used to derive estimates of the phase behavior of multicomponent systems. Critical curves on binary systems are compilated i.a. in the following references: With carbon dioxide [54, 55], nitrogen [55], ammonia [39], ethane [54], water [3, 55].

# 3.5 Phase Equilibria of Ternary Systems

In the context of this book, processes are of interest where the solvent power of dense supercritical gases is exploited. For the dissolution and precipitation of one pure component by a supercritical gas, binary systems, as discussed above, provide sufficient information on equilibrium. Even a mixture of more than two components of similar volatility may be treated with sufficient accuracy as a pseudo-binary system. But in many cases, phase behavior of three different components must be considered, e.g., for the separation of mixtures into compounds, the modification of the properties of a solvent, or the common case, that a mixture of compounds is in contact with a supercritical solvent. Therefore, in the following phase behavior of three-component systems will be treated.

### 3.5.1 Phenomenology of Ternary Systems

In a three-component system, used in gas extraction, one component always is a supercritical gas. The other two components may be relatively similar in their properties, especially with respect to the solubility in the gas. Then phase behavior of such a system is very similar to that of a binary system and Section 3.4 can be used to explain or predict the system's behavior. On the other hand, solubility in the gas of the two subcritical compounds may be very much different, one component even may be practically insoluble in the gas. Then, phase behavior is determined only by the more volatile or soluble component. Equilibrium properties of the system are that of a binary system, consisting of the volatile component and the supercritical gas. Well known examples of application of such type of systems are the extraction of caffeine or nicotine from plant materials. In all the other cases, where properties of the subcritical components differ more or less, all possible types of phase behavior, as discussed above, may occur in each binary system, containing the supercritical compound. In addition, phase behavior of binary systems consisting of two subcritical compounds, as well as phenomena occurring in the three-component area, have to be considered. Discussions on the phase behavior of ternary systems, including the critical region of the system, have been worked out by Sadus [49] and Deiters [13]. They will not be reviewed here, because relevance to gas extraction processes is limited.

A three-component system for discussing phenomena, relevant to gas extraction, consists of the supercritical component, a low volatile component, and a component of intermediate volatility. Such a system provides a generalization for many types of mixtures found at various processes. This type of system also stands for the modification of a solvent with an entrainer or modifying agent, or for mixtures of gases used as solvent, one supercritical, the other subcritical, or for multicomponent systems where groups of compounds can be lumped to a volatile and a nonvolatile fraction.

Such a ternary system, consisting of the supercritical compound, a nonvolatile compound, and a compound of intermediate volatility is represented in a three-dimensional plot of triangular diagrams in Fig. 3.29 in dependence of pressure, and in Fig. 3.30 in dependence of temperature. First the pressure dependence is discussed.

Temperature for phase equilibria, shown in the diagram of Fig. 3.29, is constant and higher than the critical temperature of the gas (the supercritical solvent). The subcritical components are completely miscible. The two-phase areas of the binary systems can be seen as $P,x$ diagrams on the lateral faces of the triangular prism. Cuts through this prism at constant pressures show the phase behavior in the ternary region in the well-known form of the Gibbs' triangular diagrams. Some equilibrium tie lines are included for illustration.

92

The two-phase area in the ternary region shrinks with increasing pressure. At lower pressures ($P_1$, $P_2$) both binary systems of the subcritical components with the supercritical gas show a two phase region. A cut at constant pressure in this region reveals a phase equilibrium of ternary type II (ternary type II: two binary systems with limited miscibility; ternary type I: one binary system with limited miscibility). It is characteristic for such a system that the gradient of the equilibrium tie lines gradually changes from one side line of the triangle, to the other. There is no critical point. At higher pressures ($P_3$, $P_4$), the supercritical component and the component of medium volatility are completely miscible. At these pressures, the two-phase area, as revealed by a cut at constant pressure, is of ternary type I. Such an equilibrium is characterized by a critical point, where two phases become identical. In a ternary type I system, higher solubilities in the gaseous phase can be reached than in a ternary type-II system at comparable conditions. If the pressure is lowered, a ternary type-I system develops into a ternary type-II system.

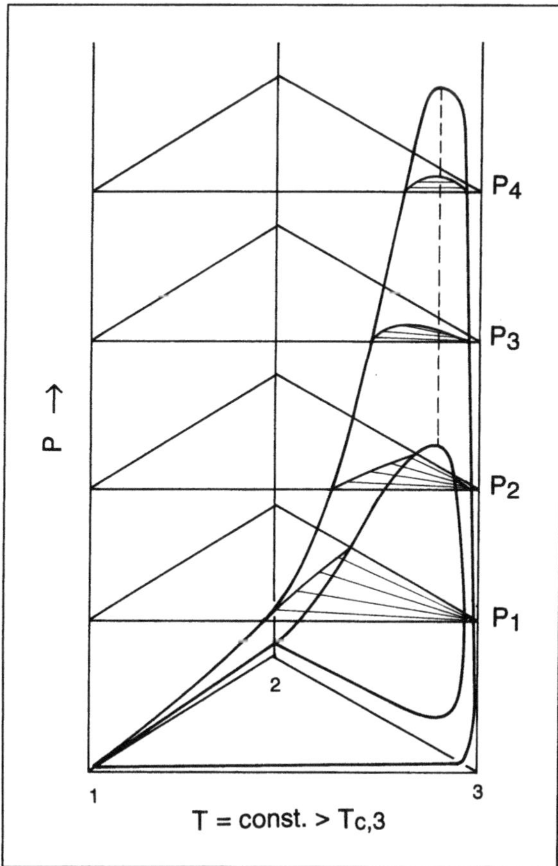



**Fig. 3.29.** Phase behavior of a ternary system in dependence of pressure. Component 1: low volatile substance; 2: substance with intermediate volatility (entrainer or modifier); 3: supercritical gas. System temperature higher than the critical temperature of the supercritical gas.

In Fig. 3.30, phase equilibria of a ternary system are shown in dependence of temperature. Pressure is constant and higher than the critical pressure of the supercritical component. On the lateral faces of the triangular diagram, the $T,x$-diagrams of the binary systems are visible. Cuts at constant temperature reveal the phase behavior in the ternary region. Equilibrium tie lines are added for illustration.

For this system, at temperatures in the range of about $T_{c3} < T < 1.2 * T_{c3}$, where $T_{c3}$ is the critical temperature of the supercritical solvent, ternary type-I phase equilibrium exists. Equilibrium tie lines have a positive gradient relative to the base line $(T_1)$. With increasing temperature, the two-phase area expands, and the slope of the tie lines changes, until it is negative with respect to the base line $(T_2, T_3)$. Ternary type-I phase equilibrium occurs up to the lower critical temperature (for the given pressure) of the binary system consisting of supercritical gas and medium volatile

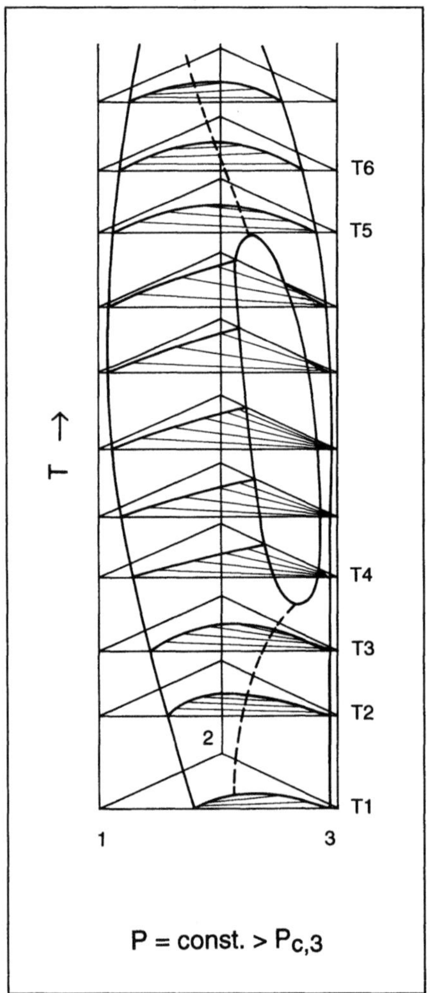


Fig. 3.30. Phase behavior of a ternary system in dependence of temperature. Component 1: low volatile substance; 2: substance with intermediate volatility (entrainer or modifier); 3: supercritical gas. System pressure higher than the critical pressure of the supercritical gas.

component. At still higher temperatures, a ternary type-II equilibrium develops ($T_4$). Immiscibility between supercritical gas and medium volatile component first increases with temperature and then decreases, as shown in the $T,x$-diagrams of Fig. 3.30. At the upper critical temperature, the binary system of supercritical gas and medium volatile component becomes totally miscible. At still higher temperatures, phase equilibrium is of ternary type I again ($T_5$, $T_6$).

Critical curves in a ternary system may also exhibit maximal or minimal values with respect to pressure and temperature [29].

Phase behavior according to ternary type I and type II can be achieved by temperature changes at constant pressure. Solubility of a component of low volatility in the gaseous phase may be essentially higher in a ternary type-I system, than that in a type II system. This can be used in both ways: enhance solubility or precipitate a component by decreasing the solubility with temperature changes.

Examples for the types of systems described in this section are presented in the following figures: $CO_2$ – benzene – oleic acid (Fig. 3.31), ethylene – dimethylformamide – C18-fatty acids (Fig. 3.32) and $CO_2$ – propane – palm oil (Fig. 3.33). A quasi ternary system for glycerides in equilibrium with a solvent mixture is shown in Fig. 3.34.

The effect of adding a modifying compound to the supercritical solvent is discussed below.

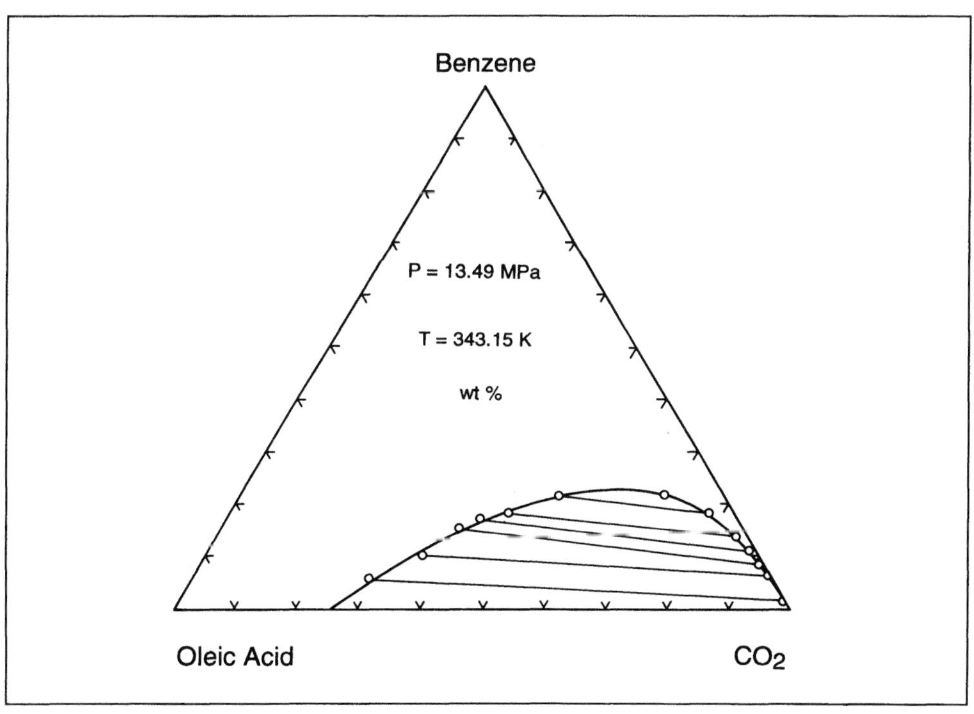


**Fig. 3.31.** Phase equilibrium in the ternary system $CO_2$ – benzene – oleic acid at $P = 13.49$ MPa and $T = 343.15$ K.

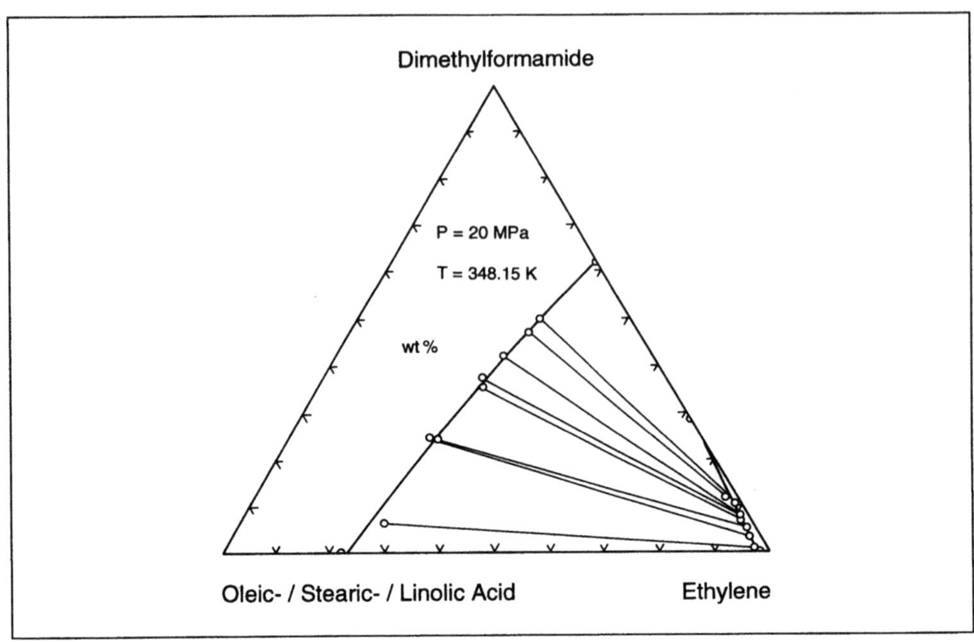



**Fig. 3.32.** Phase equilibrium in the ternary system ethylene – dimethylformamide – C18-fatty acids at $P = 20$ MPa and $T = 348.15$ K.

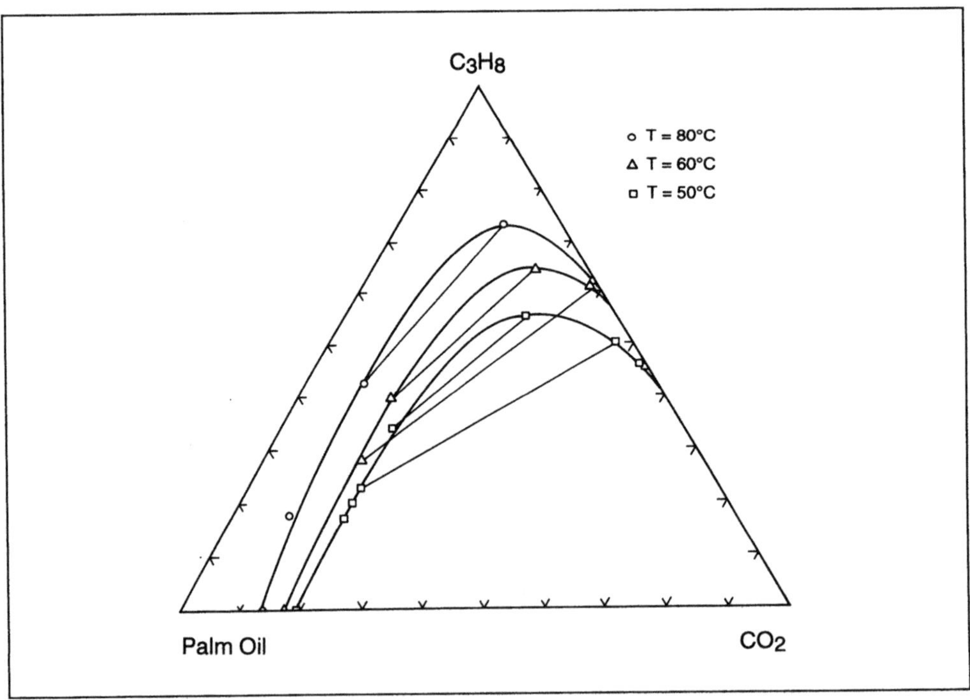



**Fig. 3.33.** Phase equilibrium in the ternary system $CO_2$ – propane – palm oil at $P = 9$ MPa and $T = 323.15$, 333.15 and 353.15 K. Data from Peter et al. [45].

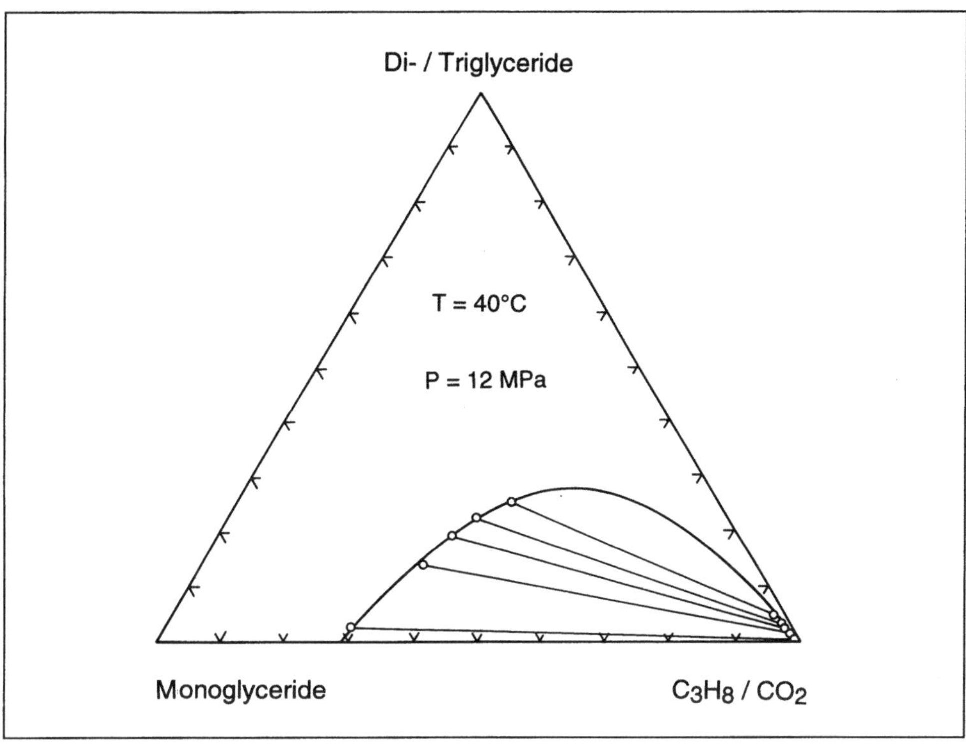


**Fig. 3.34.** Phase equilibrium in the ternary system $CO_2/C_3H_8$ – di-/triglycerides – monoglycerides at $P = 12$ MPa and $T = 313.15$ K (after Ender [19]).

### 3.5.2 The Effect of a Modifier or Entrainer on Solvent Power, Selectivity, and their Pressure and Temperature Dependence

If an additional component is added to a supercritical gas, it is with the intention to impose properties on the solvent that none of the pure component has alone. A modifying component or entrainer is added to a supercritical gas, in order to achieve the following solvent effects:
– enhance solvent power of the supercritical gas;
– enhance temperature and pressure dependence of solvent power;
– enhance the separation factor.

Because of the higher solvent power of a mixed gaseous solvent, a desired high loading of the gaseous phase can be achieved at appreciably lower pressures. The enhanced dependence of solubility on temperature and pressure may allow regeneration of the solvent through temperature changes and only minor pressure changes. An influence on the separation factor makes feasible separation processes, that would not be possible otherwise.

The idea of intentionally adding a further component to the supercritical gas in order to change solvent power and selectivity was applied by Peter, Brunner, and Riha [43] in 1973, in an investigation of the separation of C18-fatty acids. Other systematic investigations have been carried out by various authors, e.g., Brunner 1978 [4], 1982 [7], Brunner and Peter 1979 [5], 1982 [6], Schmitt and Reid 1985 [52], Sunol 1985 [60]. In all of these investigations, an enhanced solvent power of the gaseous solvent mixture was observed by an additional component with higher critical temperature than the supercritical gas. Figure 3.35 shows enhancement of the solubility of

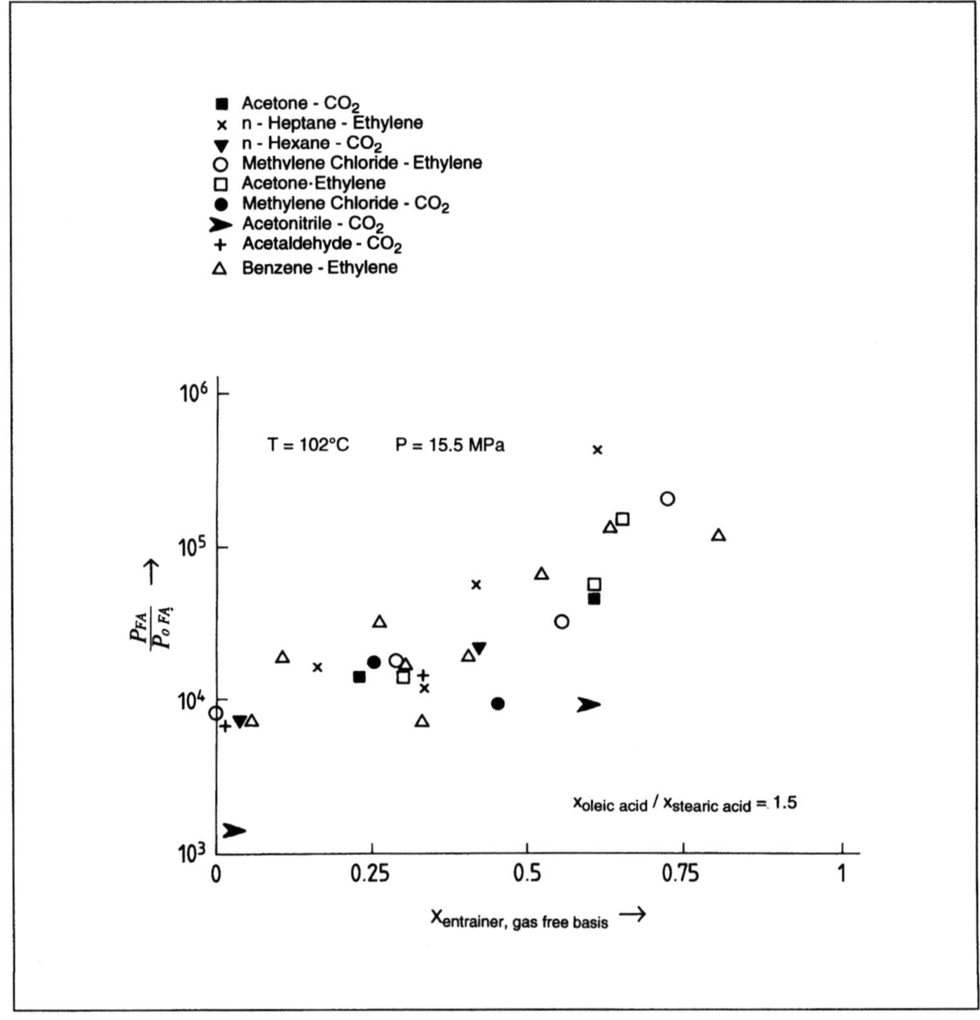


**Fig. 3.35.** Enhancement of concentration of oleic acid in the gas phase by the addition of components of medium volatility (entrainer, modifier). $x$ is the mole fraction in the liquid phase.

98

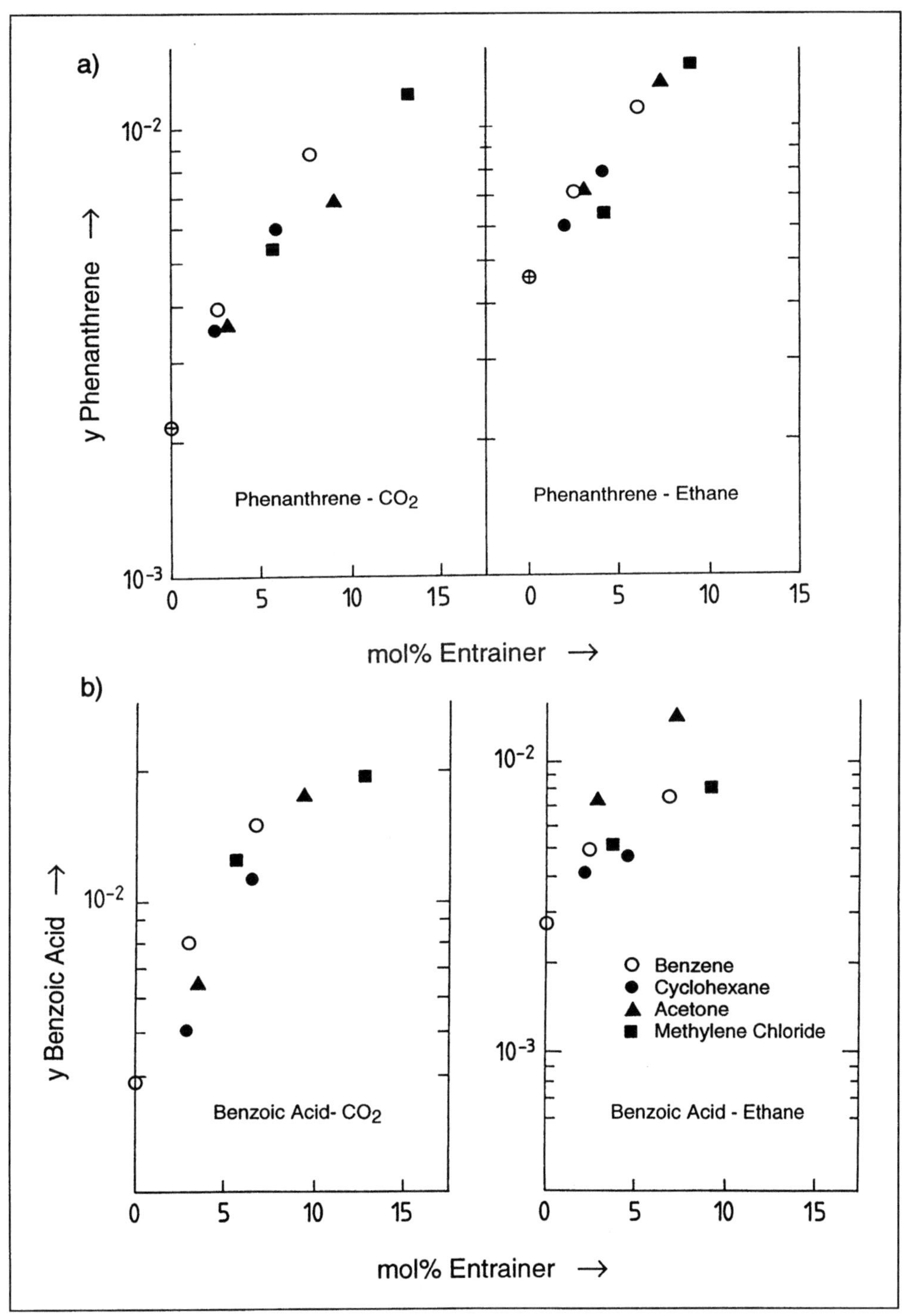


**Fig. 3.36.** Enhancement of phenanthrene- and benzoic acid-concentration in the gaseous phase by entrainers (after Schmitt and Reid [52]).

C18 fatty acids in carbon dioxide by one order of magnitude, caused by different entrainers. Despite the scatter of the experimental values, the trend is significant. Schmitt and Reid [52], who investigated the solubility of phenanthrene and benzoic acid in mixtures of $CO_2$ and ethylene, arrived at the same conclusion. Different quantities of benzene, cyclohexane, acetone, and methylene chloride were added. At 55 °C and 20 MPa, addition of 2.5 mol.-% benzene raised the concentration of phenanthrene in the gas by 60 %, whereas 6 mol-% benzene raised it by 160 % (Fig. 3.36). In the system of benzoic acid – ethane, phenanthrene — $CO_2$, and phenanthrene – ethane with four entrainers, a nearly linear enhancement of the concentration was found with increasing mole fraction of the entrainer in the gas.

Gährs [24] reported the change in the equilibrium concentration of caffeine in $CO_2$ when different amounts of $N_2$ were added (Fig. 3.37). A nitrogen content as low as 5 mol-% reduces the solubility by a factor of 2. Similar results were reported for nicotine. In a theoretical analysis, Prausnitz and Joshi [47] were able to show that the phenanthrene concentration in $CO_2$ or ethylene would be increased by two orders of magnitude, if propane were added (Fig. 3.38). The concentration enhancement by an entrainer was also shown for benzoic acid as solute and $SO_2$, $NH_3$, and dimethylether as entrainers.

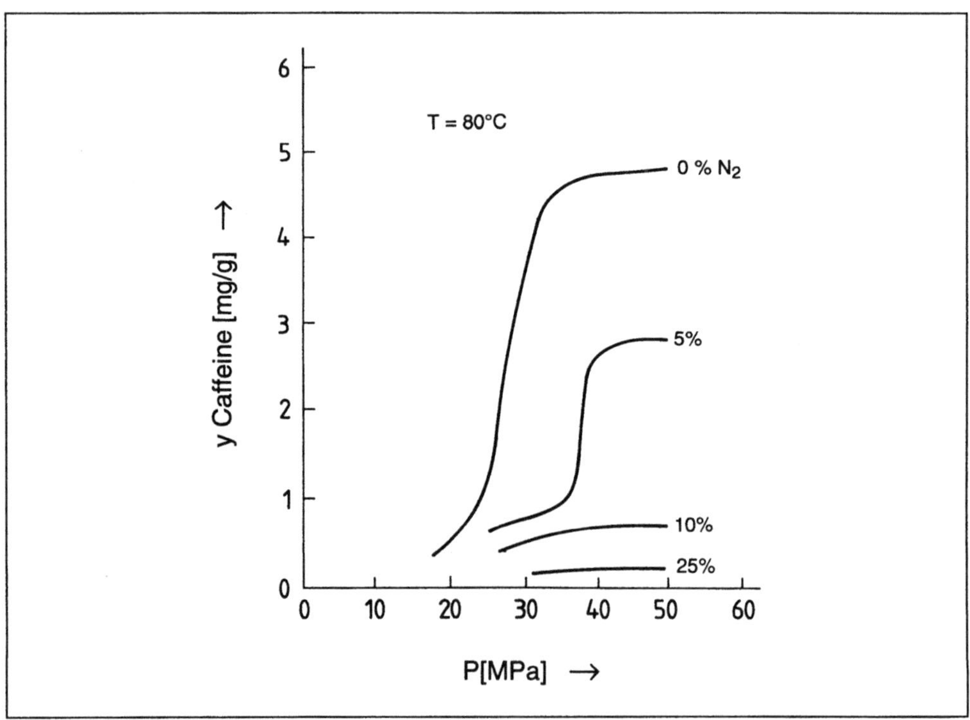



**Fig. 3.37.** Influence of nitrogen on the solubility of caffeine in carbon dioxide (after Gährs [24]).

100

In general, a modifier that has a lower critical temperature than the supercritical gas causes a decrease in the solubility of a low volatile component in the gas, whereas a modifier of higher critical temperature causes an increase in solubility. Yet concentration of the modifier (or the component of medium volatility) is significant, as is illustrated in Fig. 3.39. In this figure, the supercritical gas corner of a triangular diagram is shown, representing phase equilibria of a system consisting of supercritical gas, modifier, and solute. Corresponding diagrams show the concentration of the solute as a function of the concentration of the modifier. A monotonous rise of the solute concentration with the concentration of the modifier is possible only in the case of a ternary type-I equilibrium. Temperature and pressure must be chosen appropriately.

It is possible to use an additional component in the gaseous solvent, in order to achieve complete miscibility with a solute, and thereby effectively raise the loading of the solvent. This may be applied advantageously for the extraction of substances from solid substrates.

The effects of an entrainer are further discussed on model systems [7].

Low-volatile model components were hexadecanol $[CH_3(CH_2)_{15}OH]$, octadecane $[CH_3(CH_2)_{16}CH_3]$, and salicylic acid phenylester $[C_{13}H_{10}O_3]$, which have the same vapor pressure at 70 °C. Phase equilibrium for an equimolar mixture of hexadecanol

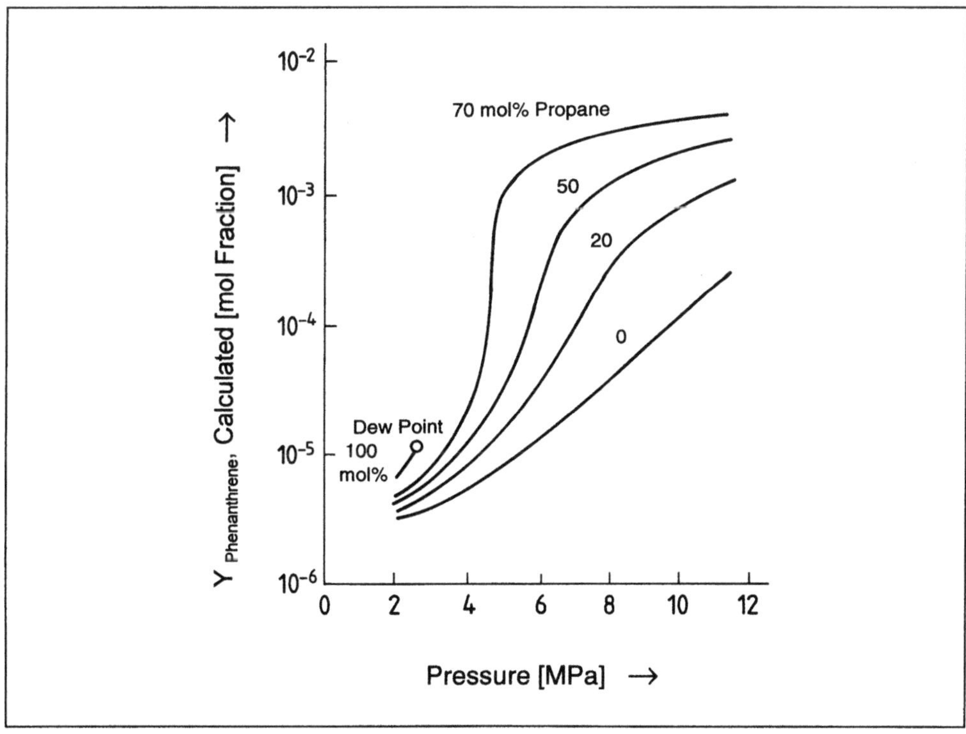


**Fig. 3.38.** Influence of propane on the solubility of phenanthrene in carbon dioxide (after Prausnitz and Joshi [47]).

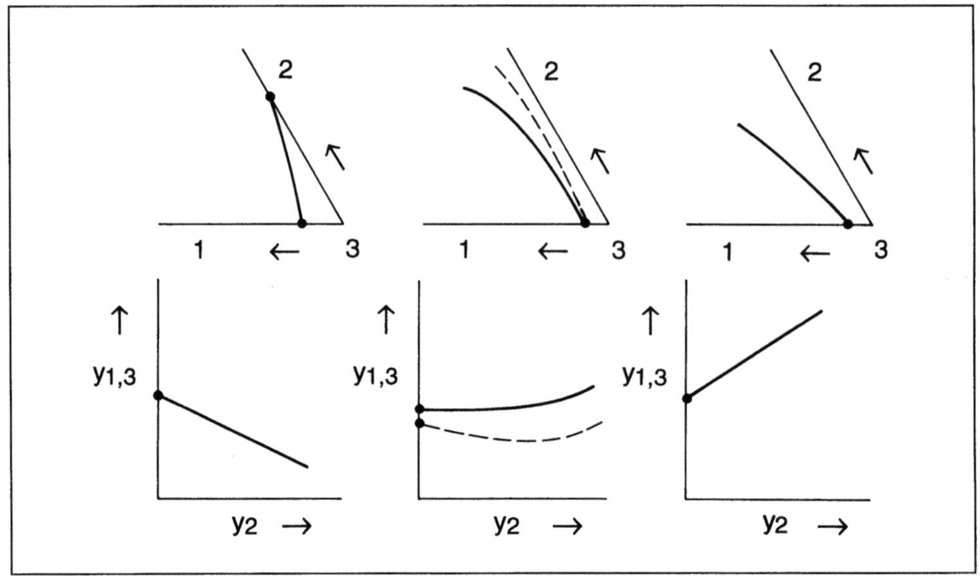


**Fig. 3.39.** Solubility of a low volatile compound (1) in a supercritical gas (3) in dependence of concentration of the entrainer (2).

and octadecane with carbon dioxide as supercritical solvent and hexane as entrainer are presented in Fig. 3.40. At higher pressures the mutual miscibility is enhanced and the two-phase area shrinks. The equilibrium concentration of the low volatile components in the gaseous solvent mixture ($CO_2$ – hexane) increases with pressure and concentration of the entrainer. Temperature dependence of the solubility in the gaseous phase is shown in Fig. 3.41. This dependence may even be strong enough for a regeneration of the solvent to be accomplished by a temperature change. The investigation on the model systems confirmed that some entrainers enhance the separation factor: For hexadecanol – octadecane in nitrous oxide as the supercritical solvent, these entrainers are acetone and methanol (Fig. 3.42), because of the specific interaction with the hydroxyl group of hexadecanol. The separation factor decreases with increasing amount of solute in the gas, from about $\alpha_{21} = 4$ (without entrainer $\alpha_{21} = 2$) to $\alpha_{21} = 1.4$ at an equilibrium concentration of 4 wt.-% in the gaseous phase. If hexadecanol is replaced by salicylic acid phenylester, separation factors are lower.

For a separation process, it must be considered that the separation factor varies also with concentration of the low volatile components. For the octadecane – hexadecanol – $CO_2$-system, the separation factor is high at low concentrations of octadecane and decreases substantially at higher concentrations, as shown in Fig. 3.43. On the other hand, for fatty acid methylesters and squalene – tocopherol mixtures with pure $CO_2$, a nearly constant separation factor was observed over a wide range of concentration of the low volatile substances (see also Chapter 8).

The entrainer dissolves to a high degree in the liquid phase. Since the density of the liquid phase is generally higher than that of the gas phase, stronger intermolecular

102

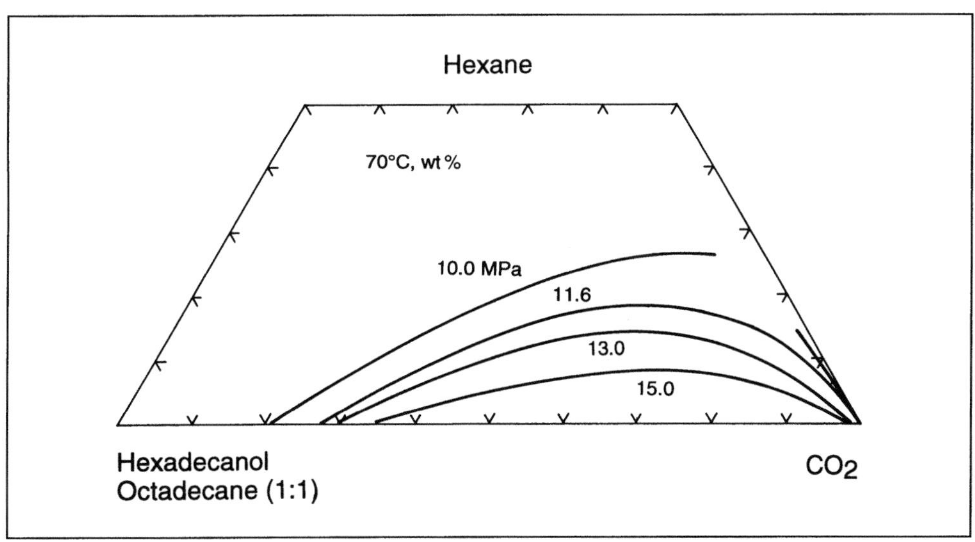



**Fig. 3.40.** Phase equilibrium of $CO_2$ – n-hexane – hexadecanol/octadecane (see text).

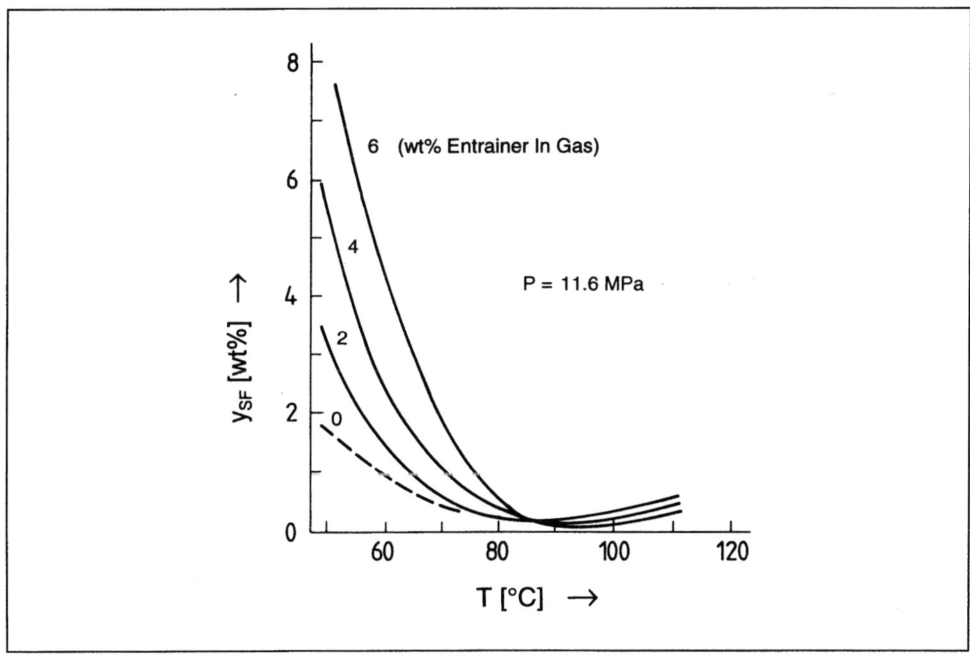



**Fig. 3.41.** Temperature dependence of the solubility of hexadecanol/octadecane in supercritical $CO_2$ with hexane as entrainer [7].

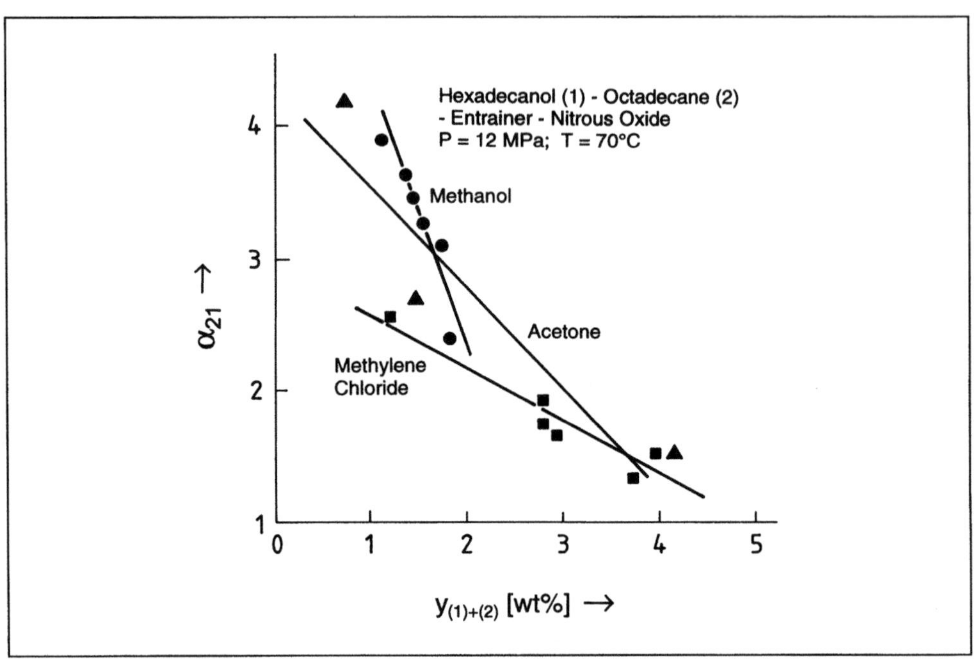



**Fig. 3.42.** Separation factor for hexadecanol and octadecane in supercritical $CO_2$ with different entrainers [7].

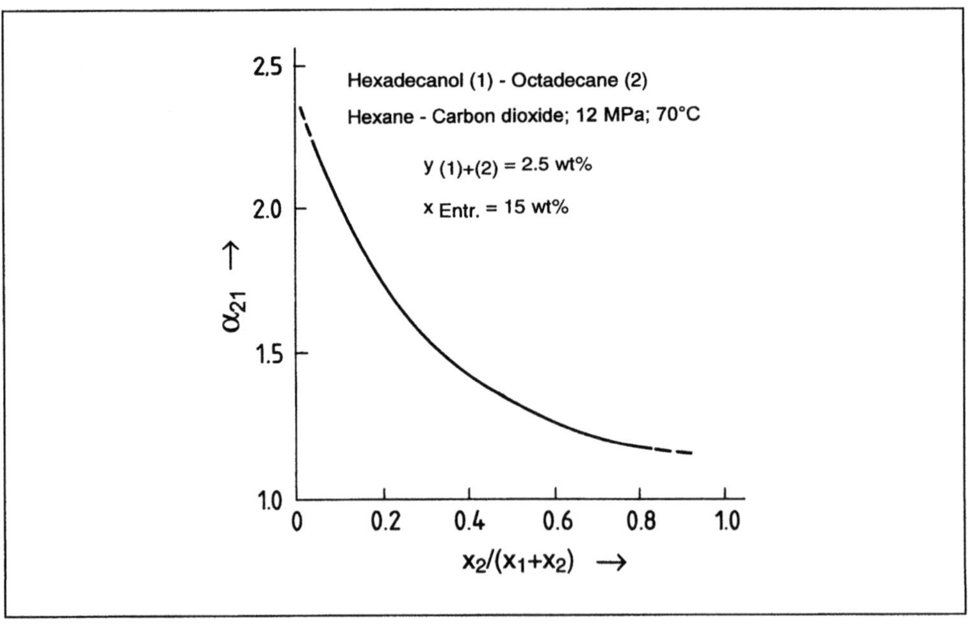



**Fig. 3.43.** Separation factor for hexadecanol and octadecane in supercritical $CO_2$ with hexane as entrainer for different concentrations of hexadecanol [7].

104

forces are present which exert a greater effect on the distribution of the components than they do in the gaseous phase. Therefore, it is not surprising that hexadecanol is preferentially held in the liquid phase, if methanol is applied as an entrainer.

The influence of additional components in the gaseous solvent can be summarized as follows:
- The critical temperature of the solvent and therefore the process temperature can be chosen so as to be optimally suited for the feed mixture.
- The solvent power is generally increased to an appreciable extent, but is diminished if the second component has a much lower critical temperature than the supercritical solvent.
- The selectivity can be influenced by properly chosen additional components.
- The temperature and pressure dependence of the solvent power are enhanced.

# 3.6 Phase Equilibria of Quaternary Systems

Four-component equilibria must be considered when the four components or pseudo-components have different properties, so that they cannot be further lumped together. Normally these equilibria are represented numerically. Since more information can be transmitted, equilibrium concentrations are often represented in diagrams as solubilities of certain compounds in the respective phases, as distribution coefficients, or as separation factors. It is useful to acquire an overview on phase equilibrium of a mixture when it has to be processed in gas extraction. Yet phase behavior of a four-component system in dependence of temperature and pressure is not easily represented graphically. Therefore, it is necessary to restrict diagrams to narrow ranges of temperature and pressure. Since gas extraction is carried out mostly at constant pressure and temperature, this is no serious restriction.

For representation of four-component equilibria, the equilateral pyramid is chosen. Phase equilibria of four components can thus be shown at constant temperature and pressure. Cuts at constant composition will help to understand phase equilibria and to demonstrate the effect of individual components. The following cases of four component equilibria of interest in gas extraction processes are discussed [5]:

I) Two components (or mixtures of components) of low volatility and similar solubility in the gaseous phase, a supercritical gas, and a component of medium volatility or another gas. Complete miscibility in the liquid phase.

II) Two components (or mixtures of components) of low volatility, but different solubility in the gaseous phase, a supercritical gas and a component of medium volatility or another gas. Complete miscibility in the liquid phase.

III) Two components (or mixtures of components) of low volatility. The solubility of the two components in the gaseous phase is extremely different. A supercritical gas and a component of medium volatility or another gas. Immiscibility in the liquid or solid phase.

IV) Four components (or mixtures of components) of different properties. Immiscibilities in the liquid phase occur.

## Case I

In Fig. 3.44, phase equilibrium of a four-component system according to case I is represented at constant temperature and pressure. It is an example of phase equilibrium conditions which may be the basis for a separation of two components of low volatility with a supercritical gas and an entrainer. System temperature $T$ is about $1.0$ $T_c < T < 1.2\ T_c$ and system pressure $P$ is about $2P_c < P < 4P_c$.

In part a of Fig. 3.44 equilibrium concentrations are plotted in an equilateral pyramid. Part b depicts two of the triangular diagrams (projected into one) which are the lateral faces of the pyramid. The triangular diagrams represent the two ternary systems with gas-liquid equilibrium. The full and the broken line, representing phase boundaries of the two-phase area for the two low volatile components, are nearly equal, since these two components behave similarly.

Parts c, d, and e represent cuts I, II, and III of the pyramid at constant concentrations of the entrainer, or in general, the component of intermediate volatility. The cuts are normalized to the size of a lateral face triangle of the pyramid. Cut I is entirely in the one-phase area. The components of the systems are completely miscible. Cuts II and III, at lower concentrations of the medium-volatile component, cross the two-phase region. The higher the concentration of the medium volatile component, the higher is also the concentration of the low-volatile component in the gaseous phase. A separation of the two low volatile compounds can be achieved if the gradient of the tie lines is different from the gradient of the broken lines, representing a constant ratio between the two low-volatile components. Molecular interaction forces between supercritical gas and intermediate volatile component are important for achieving a different gradient. The intermediate volatile component may be chosen so that these interacting forces are introduced. In that case the component may be called an *entrainer.*

An example for a separation in such a four-component system is the separation of mono- and diglycerides with carbon dioxide as supercritical gas and propane or acetone as entrainer [8, 44].

In Fig. 3.45, phase equilibrium of a four-component system according to case I is represented for a temperature at which there is limited misciblility between the component of medium volatility and the supercritical gas. At higher concentrations of the medium volatile component, the solubility of the low volatile components in the gaseous phase is low and approaches zero with increasing concentration of the medium volatile component. These conditions of temperature and pressure can be used for precipitating low volatile compounds which have been dissolved under different conditions and regenerate the gaseous solvent.

## Case II

In Fig. 3.46 phase equilibrium of a four-component system according to case II is plotted at constant temperature and pressure. It is also an example for phase equilib-

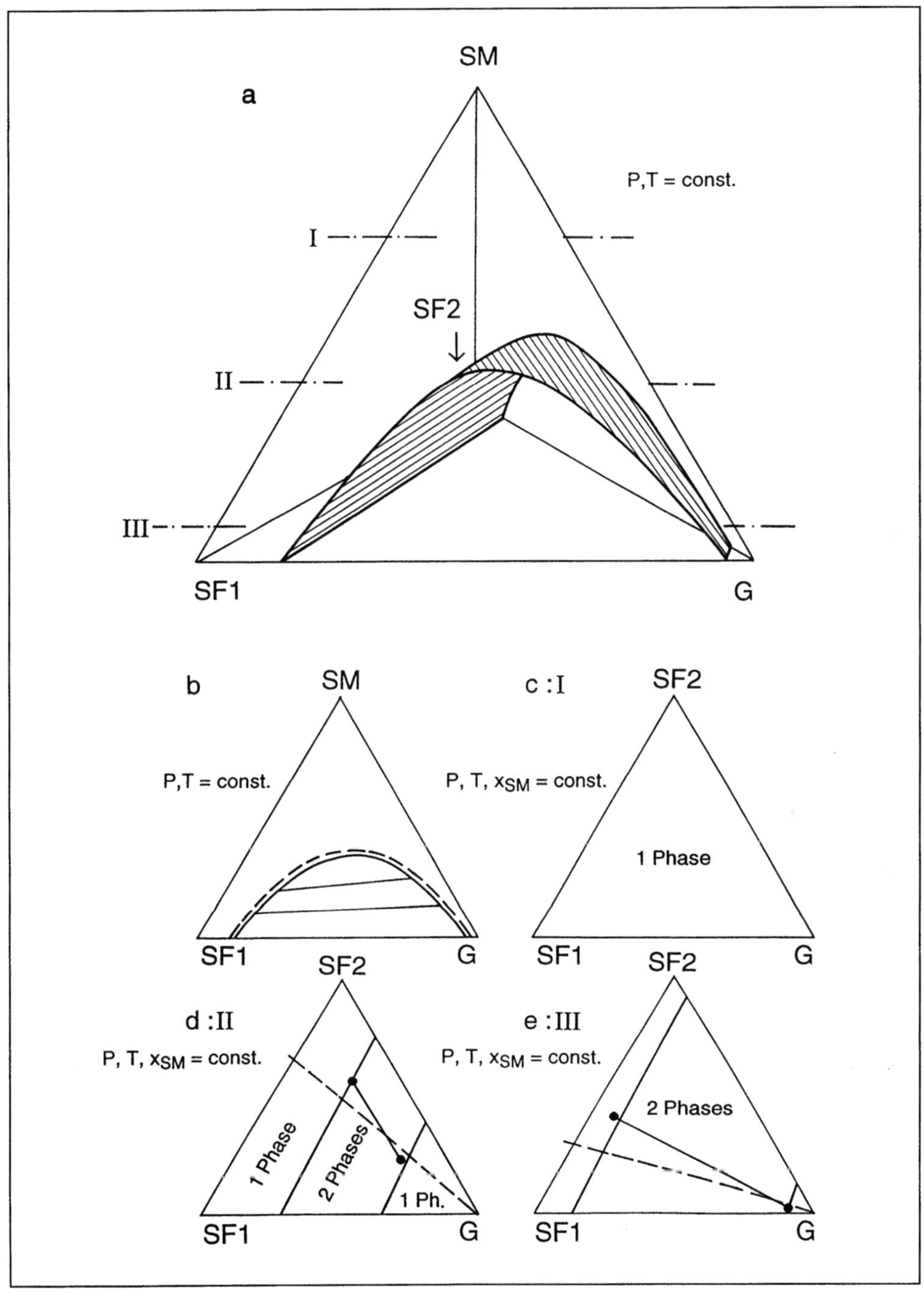


**Fig. 3.44.** Phase equilibrium in a quaternary system. System temperature and pressure higher than the critical temperature and pressure of the gas. Component 1 (SF1): low volatile substance; 2 (SF2): low volatile substance, similar solubility in the gas as component 1; 3 (SM): component of intermediate volatility (entrainer); 4 (G): supercritical gas.

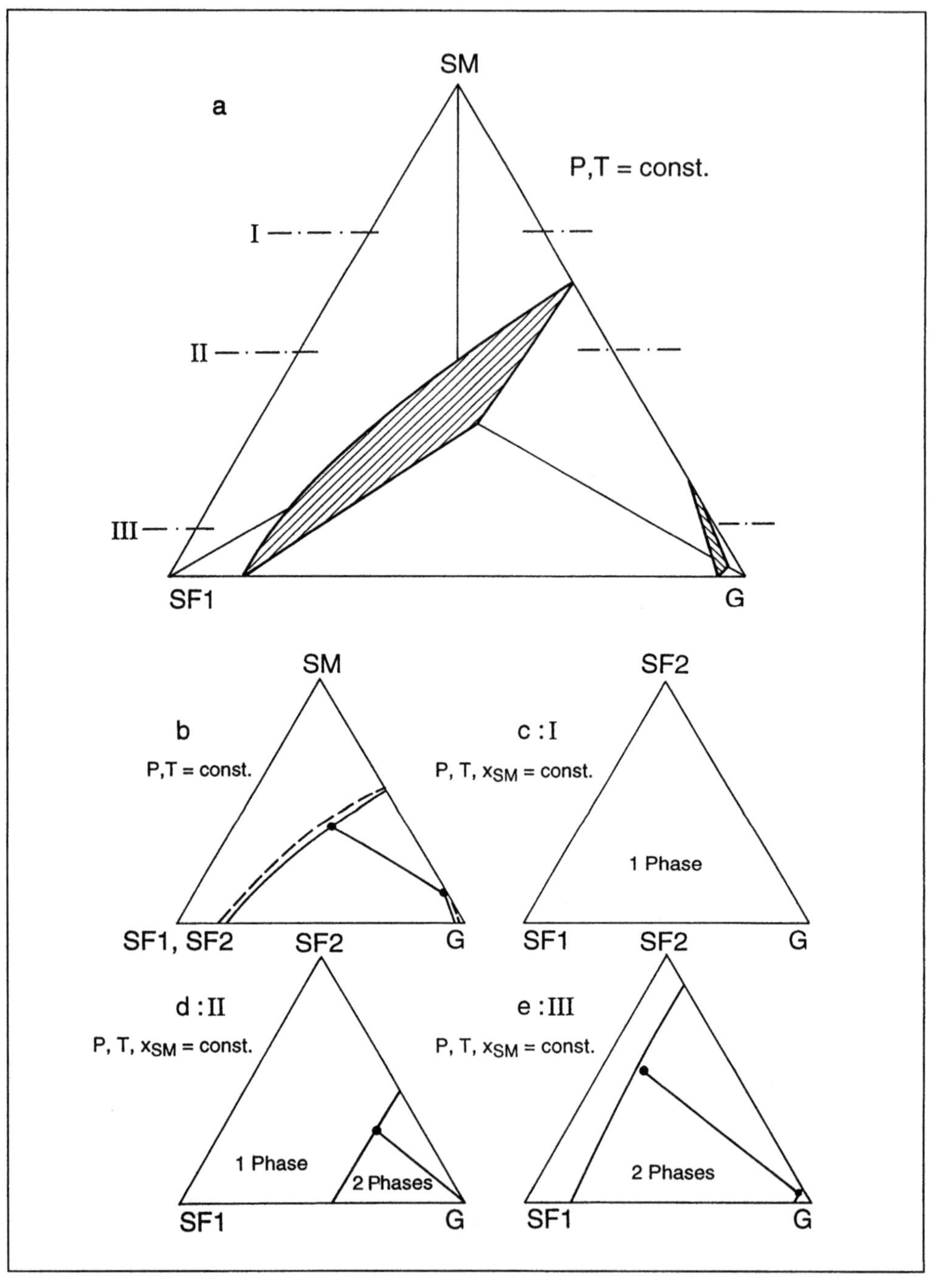


**Fig. 3.45.** Phase equilibrium in a quaternary system. System temperature and pressure higher than the critical temperature and pressure of the gas. Contrary to Fig. 3.44, *P* and *T* are chosen so, that immiscibility between entrainer and supercritical gas occurs. Component 1 (SF1): low volatile substance; 2 (SF2): low volatile substance, similar solubility in the gas as component 1; 3 (SM): component of intermediate volatility (entrainer); 4 (G): supercritical gas.

rium conditions which may be the basis for a separation of two components of low volatility with a supercritical gas and an entrainer. System temperature $T$ is about $1.0\,T_c < T < 1.2\,T_c$ and system pressure $P$ is about $2P_c < P < 4P_c$.

Part a of Fig. 3.46 illustrates equilibrium concentrations in an equilateral pyramid. Part b depicts two triangular diagrams, projected into one, from the lateral faces of the pyramid, where gas-liquid equilibrium occurs in the ternary systems. Since the two low-volatile components are dissolved in the gaseous phase to different quantities, the two lines, representing phase boundary lines, do not coincide, and the two-phase areas in the ternary systems are of different size.

Parts c, d and e of Fig. 3.46 represent cuts I, II, and III of the pyramid at constant concentrations of the component of intermediate volatility. The cuts are normalized to the size of a lateral face triangle. Cut I is entirely in the one-phase area. The components of the systems are completely miscible. Cuts II and III, at lower concentrations of the medium volatile component, cross the two-phase region. For cut II, the concentration of the medium volatile component is in such a range that supercritical gas and medium volatile components are completely miscible with one of the low-volatile components. In this case relatively high concentrations of the low-volatile component in the gaseous phase can be achieved. At lower concentrations of the medium-volatile component, as shown in cut III, again both components of low volatility are only partially miscible with the supercritical gas and the medium volatile component. In the two-phase area of the cuts, equilibrium tie lines have been plotted for example. They do not end on the phase boundary lines of the cuts because they are tilted against the plane of cut III. Liquid mixture $A$, chosen on the phase boundary line, is in equilibrium with gaseous mixture $B$. The end of the tie line may be inside or outside of the two-phase area on the plane of cut III, depending on the slope of the tie lines.

An equilibrium as shown in Fig. 3.46 part d, may be useful in separation processes, since a relatively high loading of the gaseous solvent at reasonable separation factors is possible. An example is the removal of the free fatty acids from edible oils [8] or the purification of lecithin, where the oil is removed from the lecithin [45].


## Case III

In this case the four-component system may be reduced to a three-component system. If one of the components is nonvolatile and insoluble in the supercritical gas, as is the case with many solid substances, it may be considered not to influence the equilibrium of the ternary system.

If the volatile component is distributed within the insoluble compound, as, for example, the triglycerides in oil seeds, or the caffeine in coffee beans, equilibrium solubility in the gaseous solvent may be influenced by the cellulosic structure which represents the nonvolatile component. If equilibrium solubility is known, in dependence of the many parameters, which determine the influence of the solid component, the system may be treated on the basis of four components. Since in many cases such equilibrium solubilities are unknown, equilibrium solubilities of the pure low volatile components in the gaseous solvent may be used for design purposes. Then the four component system is again reduced to a three-component system.

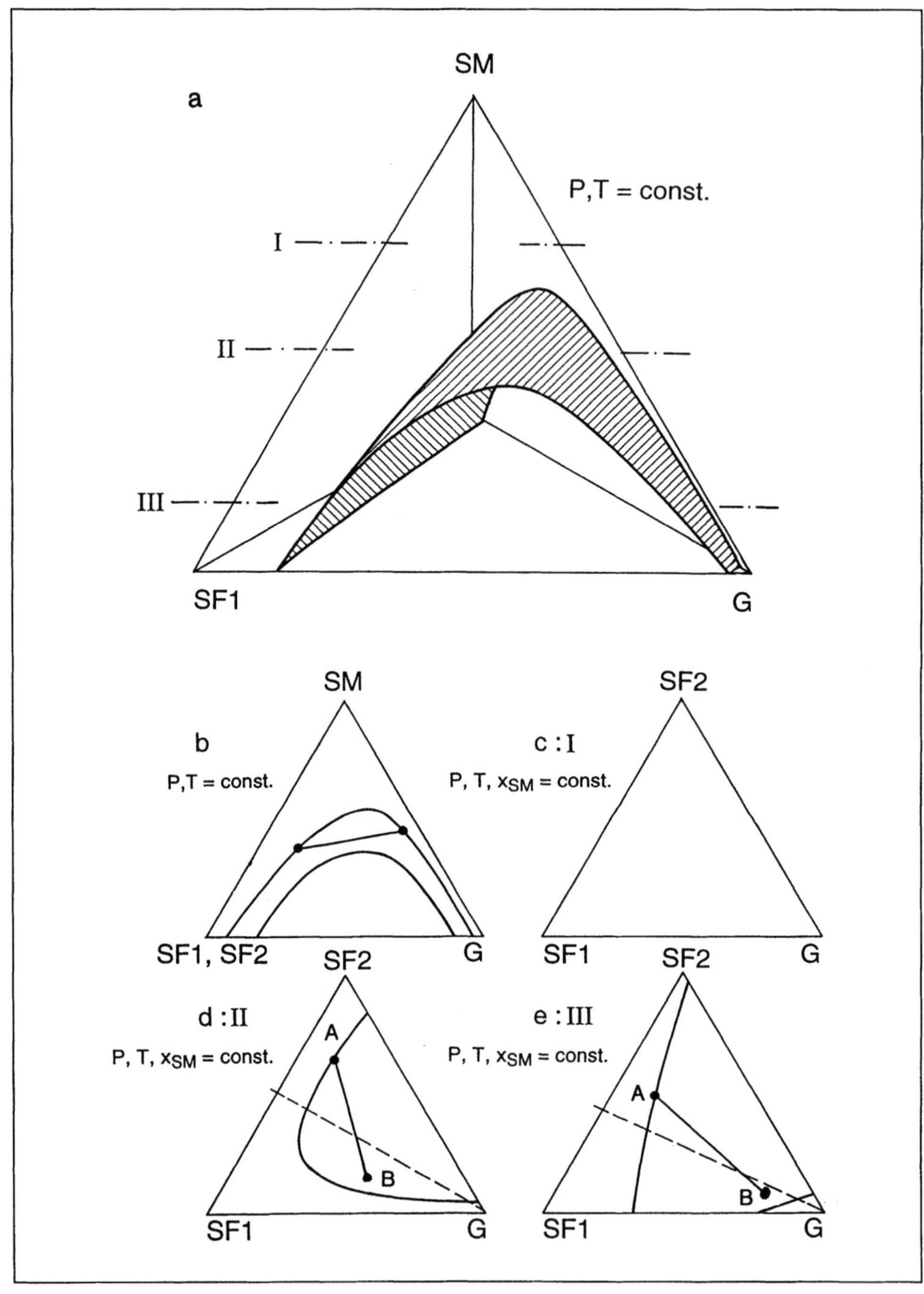


**Fig. 3.46.** Phase equilibrium in a quaternary system. System temperature and pressure higher than the critical temperature and pressure of the gas. Component 1 (SF1): low volatile substance; 2 (SF2): low volatile substance, different solubility in the gas as component 1; 3 (SM): component of intermediate volatility (entrainer); 4 (G): supercritical gas.

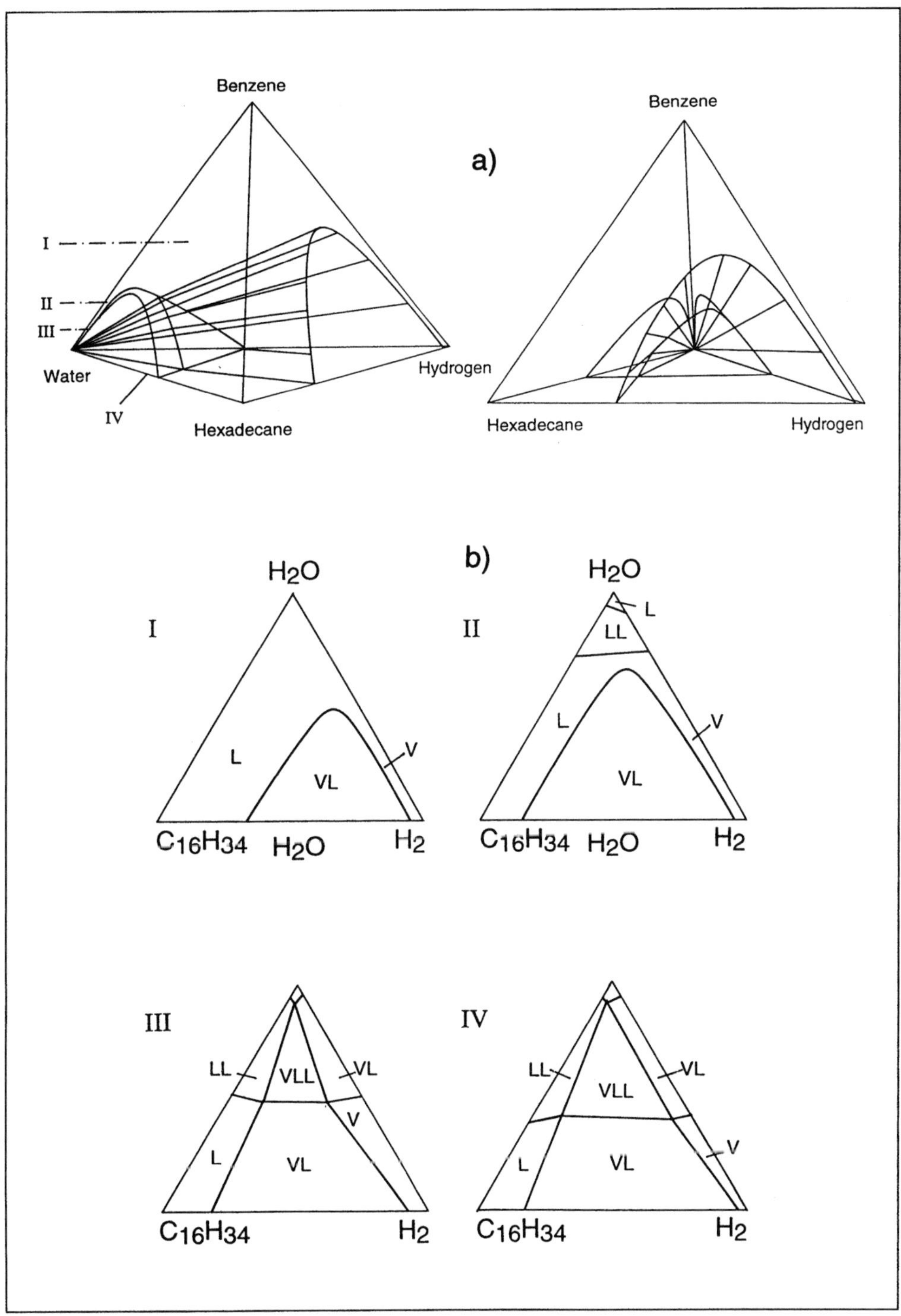



**Fig. 3.47.** Phase equilibrium in the quaternary system hydrogen – water – benzene – n-hexadecane at $P = 20$ MPa and $T = 573.15$ K. The water corner is enlarged [15].

111

**Case IV**

In the most general case, there may be immiscibility between all the four components of a quaternary mixture. In Fig. 3.47 a concentration pyramid is shown with limited miscibility in the liquid phase. Conditions of state and projections to the plane are similar as explained for case I: Two components of low volatility, a supercritical gas and an entrainer. System temperature $T$ is about $1.0\ T_c < T < 1.2\ T_c$ and system pressure $P$ is about $2P_c < P < 4P_c$. For clarity, a special system is selected [15]. Cuts at constant concentration of benzene are drawn, as in previous cases, to show the influence of this compound, which at higher concentrations acts as solubilizing compound (at 200 bar, 300 °C).

At a certain concentration range of the medium-volatile component, there is limited miscibility in the liquid phase and therefore a three-phase equilibrium region exists, Fig. 3.47 part III, IV. This effect may occur even with completely miscible low volatile components, as illustrated in Fig. 3.48. Starting in the concentration triangle, Fig. 3.48, from the binary side of the two low-volatile components, increasing amounts of the supercritical gas are added. With increasing concentration of the supercritical component, the liquid splits into two liquid phases in which all the gas is dissolved. At higher concentrations of the supercritical gas, the two liquid phases no

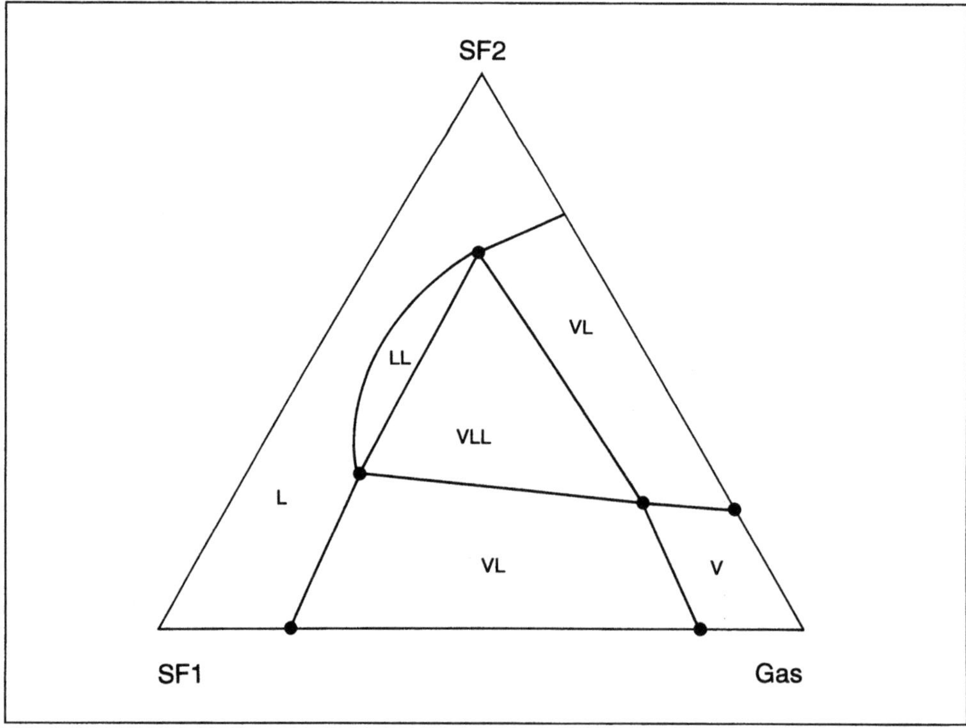


**Fig. 3.48.** Three-phase equilibrium in a ternary system (schematic drawing, see text).

112

longer are able to dissolve all the gas. A gaseous phase appears in which some amount of the low-volatile components is dissolved. Now there exist three phases in equilibrium, two liquid phases and one gaseous phase. The compositions of these three phases remain constant, until the total concentration of the supercritical gas reaches the point where one liquid phase disappears because of dissolution in the gas, and only one liquid phase remains in equilibrium with the gaseous phase.

Such a phase behavior is similar to the three-phase equilibrium liquid-liquid-gas of a binary system in the neighborhood of the critical temperature of the supercritical gas. Since the induced immiscibility is also a partial separation of the low volatile components, such conditions of three-phase equilibrium might be of interest for a separation process. A single equilibrium stage may well be used for partial separation.

Liquid-liquid immiscibility may be induced by a supercritical gas in a mixture of completely miscible compounds. This effect is similar to the salting-out effect experienced if salts are added to liquid solutions.

# 3.7 Multicomponent and Complex Systems

Mixtures containing many compounds are often encountered when real systems have to be dealt with. The extracts of plant materials, mineral oil fractions or liquids from pyrolysis of hydrocarbon or carbohydrate material are examples. In many cases, concentrations of most of the components may be low, compared to the main components of interest. In such cases, phase behavior may be sufficiently accurately represented by the main components only. During processing of multicomponent mixtures, compounds of low concentration may accumulate in certain fractions to such an extent that the minor components have to be considered for phase behavior. For example, edible oils from oil seeds contain tocopherols, carotinoids, and sterols at a concentration level of about 0.05 wt.-%, beside the by far dominating triglycerides. During the refining process a condensate fraction is removed, which may contain these compounds at a level of up to about 5 wt.-% and more. If treatment by gas extraction is considered, phase behavior of these side streams clearly is dependent on those compounds, while phase behavior of the initial raw oil is not. On the other hand, the minor components may be the components of interest. Then, distribution between equilibrium phases is important, since it determines the possibility of enhancing the concentration by extraction. The phase behavior with respect to the amount of liquid or gaseous phases, is determined only by the main components. Use of individual compounds for representing phase equilibrium depends on the purpose for which the data are needed.

The distinction between multicomponent and complex mixtures is somewhat arbitrary. A multicomponent mixture shall be a mixture of more than three components, wherein all the components are identified. A complex mixture shall be a mixture of far more than three components, where the total number of components is unknown and many of the compounds are not identified. Identification of compounds, and characterization of groups of compounds, depend on the problem to be solved.

### 3.7.1 Phase Behavior of Complex Mixtures

Phase behavior of complex mixtures is discussed in the following. For simplicity only mixtures completely miscible in the liquid phase are considered. The overall phase behavior of a complex mixture with respect to dew- and bubble-point curves and the amount of liquid or gas is best represented in a $P, T$-diagram, as shown in Fig. 3.49. Dew-point curve and bubble-point curve limit the two-phase gas-liquid region. At the critical point gaseous and liquid phases become identical. Pressure and temperature at the critical point are not necessarily extreme conditions for the two-phase region. The critical point is normally on the upper or lower phase boundary curve and is situated only by coincidence at an extreme value of temperature and pressure. Curves of constant liquid content are also shown in the diagrams.

If conditions of state of a complex mixture (in general, of any mixture of two or more compounds) are changed isobarically or isothermally, e.g., from $A$ to $B$ in Fig. 3.49.a, a pressure reduction causes in a certain region (from $d$ to $m$) condensation of liquid *(retrograde condensation)*. Similar retrograde condensation can occur if the temperature of a complex mixture is enhanced at constant pressure along the line $DE$ in Fig. 3.49.a. By reversing the change of conditions, retrograde evaporation can be obtained by increasing pressure or decreasing temperature.

The extent of the two-phase area in the $P, T$-plane depends on the volatility of the most and least volatile compounds. This can be easily explained on binary systems, as has been done by King [35]. The two-phase area of an equimolar mixture of carbon dioxide and nitrous oxide is very small, since volatility of these both components is very similar, as expressed by the ratio of the critical temperatures: $T_c(N_2O)/T_c(CO_2)$ = 1.02; $V_c(N_2O)/V_c(CO_2)$ = 1.04. The two-phase area of an equimolar mixture of hydrogen and n-hexane is much wider, since volatility of the two components is different, and the critical constants are also much different: $T_c(H_2)/T_c(\text{n-hexane})$ = 11.7; $V_c(H_2)/V_c(\text{n-hexane})$ = 7.4.

Multicomponent and complex mixtures show a phase behavior on the $P, T$-plane similar to binary systems. The extent of the two-phase area is mainly determined by the most and the least volatile compounds. In gas extraction, the most volatile compound is the supercritical solvent. The compounds of the mixture to be separated are much less volatile. Therefore, the two-phase area is wide on the $P, T$-plane. If a complex low volatile mixture is considered as a pseudocomponent, a pseudo-binary system is formed with the supercritical gas. In this binary system, all phenomena of binary phase equilibria may occur, e.g., liquid-liquid immiscibility and interrupted critical curves. The formation of additional phases results in different distributions of the compounds of the low volatile mixture between the phases and may offer possibilities for a separation.

Typical examples (beyond edible oils and mineral oil fractions) where complex mixtures are involved are the enhanced oil recovery by carbon dioxide flooding and conversion processes of carbonaceous material, like coal gasification, coal liquefaction, or oil shale retorting, if carried out under elevated pressures, especially under a hydrogen atmosphere.

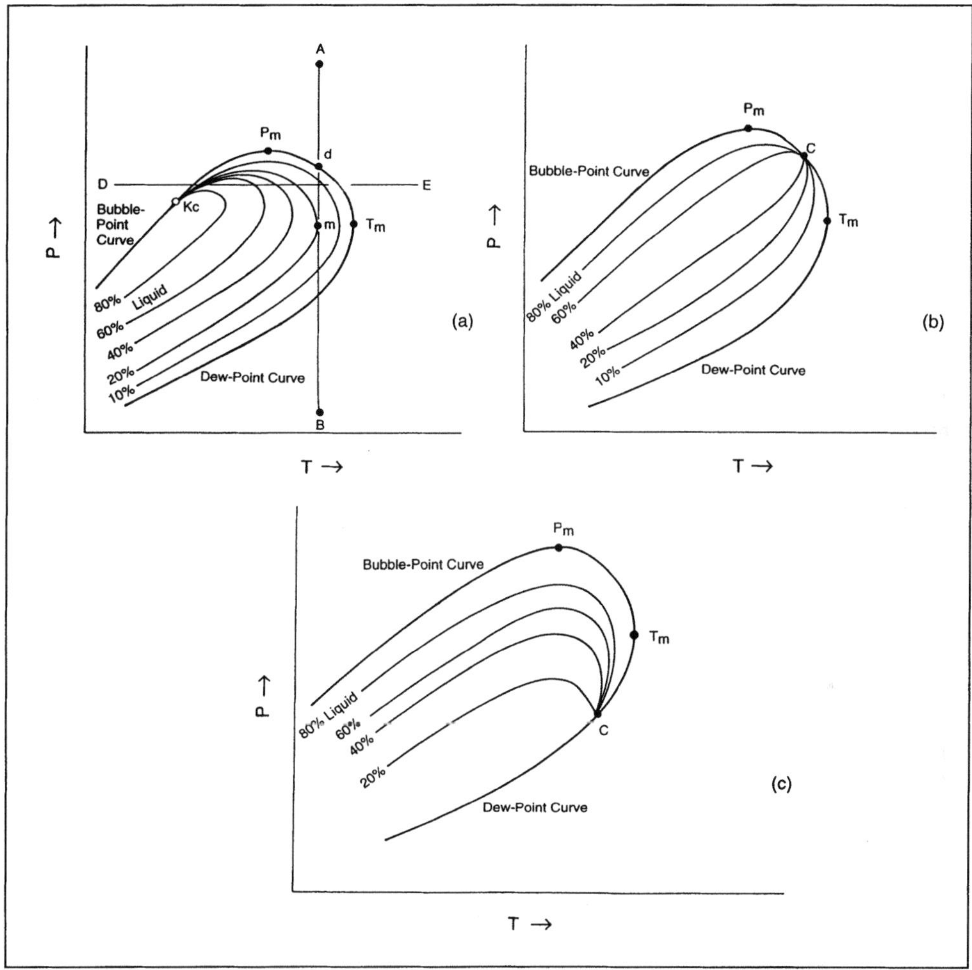


**Fig. 3.49.** Phase equilibrium in complex mixtures. *P,T*-diagrams with different position of the critical point.

## 3.7.2 Lumping and Pseudo-Components

Information on complex systems, which shall be treated by gas extraction, must be prepared so that the information needed can be provided at the minimum necessary effort. One way is to transform the complex system into a system with as few as possible components. For the purpose of gas extraction a complex system often can be handled as a pseudo-binary system consisting of the supercritical solvent and the low volatile mixture. The information which can be transferred is the total solubility in each of the phases and the extent of the two- or three-phase area, as part of the phase behavior of the pseudo-binary system. In addition, the solute in the gaseous phase and the remaining part of the low volatile mixture in the liquid phase, may be further characterized by well known and often standardized methods which characterize product qualities, like acid- and iodine-number in edible oils. Equilibrium distribution of identified compounds of interest may be determined by analyzing the concentration of these individual compounds in the different phases.

Often, a pseudobinary treatment of the complex system is not sufficient. An entrainer, a modifying compound, or compounds of medium volatility, which have a similar effect on phase behavior as an entrainer, should be treated as individual components.

The complex mixture of low volatility itself must be split into as many pseudocomponents, and into such groups of pseudocomponents, that the goal of characterizing phase equilibria, distribution of interesting compounds, and the effect of the supercritical solvent in the gas extraction process can be evaluated. This information may represent experimental data, e.g., equilibrium distribution coefficients, which are used for process design calculations without modification. If phase equilibrium calculation shall be carried out, pseudocomponents for the complex mixture must be chosen, so that parameters necessary for the calculation procedures can be obtained. Normally, splitting the complex mixture into three to five pseudocomponents will be sufficient for modeling phase behavior with respect to solubilities and distribution of groups of similar compounds. In addition, certain individual compounds may be considered. At least the compounds of the gaseous solvent (supercritical compound, entrainer, modifying agent) have to be treated individually. For most cases, a total number of 10 components or pseudocomponents will be more than sufficient. The addition of each pseudocomponent must be justified by the effect this pseudocomponent has to represent for modeling or characterizing the complex mixture. Furthermore, each component must be characterized by physical meaningful parameters, according to the applied method.

Complex mixtures of similar compounds may be treated as a mixture of an indefinite number of components, which may be characterized by continuous parameter functions [11, 12, 33]. These methods may be useful for special applications, as, for example, reservoir fluid calculations. In gas extraction it may be difficult to define parameter functions which make calculation of the distribution coefficients for certain compounds possible. A comparison [26] has shown that for less than 10 components, methods based on continuous thermodynamics have no advantage.

A difficulty with handling data on complex mixtures is that data depend on the individual treatment of the mixture, with respect to the concept of pseudocomponents. Pseudocomponents are mostly chosen after a chromatographic analysis of the complex mixture. Depending on separation efficiency of this analysis and on component identification, different pseudocomponents may be obtained. Therefore, the results of experiments on complex mixtures are somewhat difficult to compare. Conclusions on agreement or disagreement of results can only be drawn after extensive compari-

116

son of the original complex mixture, the analytical procedures, and the mixture of arbitrarily chosen pseudocomponents. The concept of pseudocomponents for complex mixtures is further treated in Chapter 8 in context with countercurrent separations.

# 3.8 Experimental Determination of Phase Equilibria for Gas Extraction Systems

## 3.8.1 General Remarks on Measuring Phase Equilibrium

The objective of measuring phase equilibria is to determine the values of the intensive variables *temperature, pressure,* and *equilibrium compositions* of all phases under the conditions of thermodynamic equilibrium.

In this section a general discussion of experimental methods for measuring phase equilibria is given. Due to the scope of this book, the discussion is limited to phase equilibria at elevated pressures and to mixtures containing at least one supercritical or near critical component. Detailed examples of experimental equipment are presented for each method. Applicability to multicomponent and real systems, as compared to model systems, is a condition. There are methods for binary systems, for volatile and easy to handle substances, which are quicker, more accurate and can be carried out with smaller equipment than that presented here, but which are not adequate for most systems of practical importance in gas extraction.

Determination of high-pressure phase equilibria is carried out by experiment. In most cases the experimental data are then correlated with the models and methods discussed in Section 3.9. Phase equilibrium calculations may be used to limit experiments to the minimum number necessary for determining dependency on temperature, pressure, and composition, or for determining the parameters in a calculation method. Interpolating calculations are quite safe with the models discussed in Section 3.9. Careful extrapolations, especially at constant temperature, can be used for quantitative purposes, but mostly extrapolations yield only qualitative results. Development of predictive calculations will make such extrapolations more realistic in the near future. But a design procedure for process equipment will always be based on experimental data of the real mixtures.

Multicomponent and complex mixtures of low volatile substances combined with supercritical compounds impose special problems on phase-equilibrium experiments by the great number of components, the low vapor-pressures, and the enormous differences in compound properties.

At thermodynamic equilibrium, temperature, pressure, and concentrations in the coexisting phases remain constant (see Section 3.9). In general, in an equilibrium experiment a constant pressure over an extended period of time, or more generally, the invariance of conditions of state, is used as an indication for established equilibrium in high-pressure phase equilibrium experiments. According to the phase rule (Eq. 3.7), in a binary sytem at constant temperature and constant pressure the compositions of the phases are fixed. Therefore, from measurements of pressure and temperature the equilibrium concentrations of the phases can be determined if total composition is known. For mixtures of three and more components, in addition to temperature and pressure, the equilibrium concentrations of the coexisting phases must be measured.

Experimental methods for measuring phase equilibria at elevated pressure are static and dynamic methods. Static methods comprise all methods where the components are placed within a closed volume and equilibrium is established by waiting until equilibrium conditions have been reached. The approach of equilibrium can be hastened by enhancing mass transfer with appropriate methods, like stirring or circulating the phases with pumps. Dynamic methods comprise all methods by which one phase is passed once through the other phase, thereby approaching equilibrium under preset conditions. At the outlet, equilibrium is assumed to be established.

All the mentioned methods have advantages and disadvantages. There is no single method and no single set-up for determination of phase equilibrium over the whole range of pressure, temperature, or number, type and concentration of components. In the following, three methods and examples of equipment are presented and discussed: The static-analytical method, the (static)-synthetic method, and the dynamic method.


## 3.8.2 The Static Analytical Method

The arrangement of an experimental set-up for measuring phase equilibria of multicomponent systems at elevated pressures according to the static-analytical method is presented in Fig. 3.50. The apparatus consists essentially of a pressure resistant equilibrium cell, the installations for establishing, controlling and monitoring pressure and temperature, and for removing and analyzing samples of the phases.

According to the measuring procedure, the components are brought into the equilibrium cell and are intensively mixed at constant temperature. Normally pressure is adjusted by pumping in the most volatile (in our case supercritical) compound. For binary systems it is not necessary to establish a certain pressure exactly, since there is only one combination of equilibrium concentrations of the phases for each pressure value. In ternary systems it is more advisable to measure all different concentrations at constant values of temperature and pressure, since data are represented mostly in triangular diagrams and equilibrium conditions for any binodal curve require pressure and temperature to be constant. It is possible to correct data by relating them to common values of temperature and pressure. But for this procedure the dependency on pressure and temperature must be known from experiments or an appropriate model, e.g., an equilibrium calculation by an equation of state must be used (see Section 3.9).

After equilibrium is established, the stirrer is shut off and the phases separate according to their density. When this process is considered to be completed, samples

118

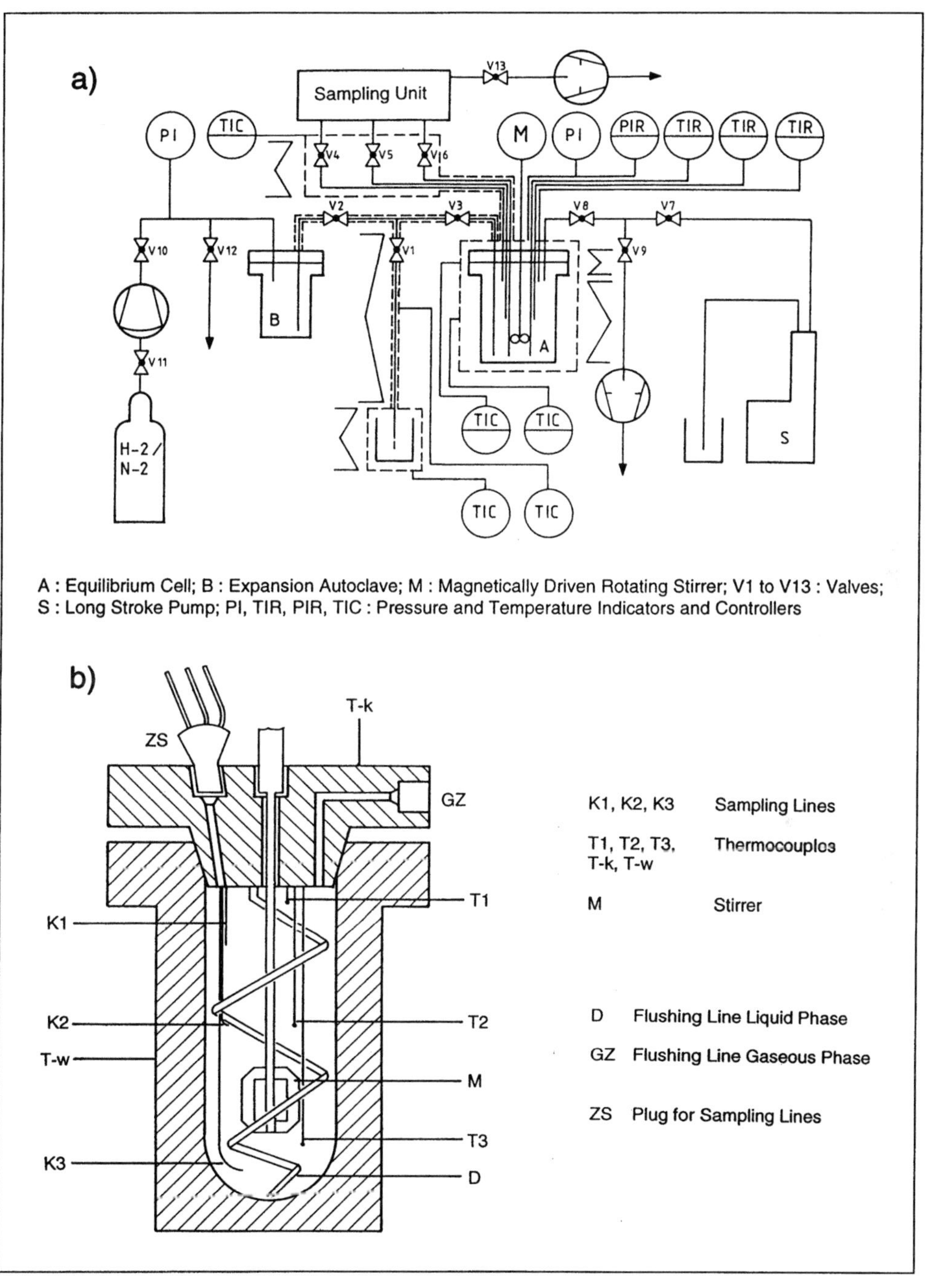


A : Equilibrium Cell; B : Expansion Autoclave; M : Magnetically Driven Rotating Stirrer; V1 to V13 : Valves; S : Long Stroke Pump; PI, TIR, PIR, TIC : Pressure and Temperature Indicators and Controllers



**Fig. 3.50.** Experimental equipment for determining phase equilibrium according to the static-analytical method. Part a) A: equilibrium cell; B: expansion autoclave; M: magnetically driven rotationg stirrer; V1 –V13: valves; S: long-stroke pump; PI, TIR, PIR, TIC: pressure and temperature indicators and controllers. Part b) K1, K2, K3: sampling lines; T1, T2, T3, T-k, T-w: thermocouples; M: stirrer; D: flushing line for liquid phase; GZ: flushing line for gaseous phase; ZS: plug for sampling lines.

are drawn from the different phases by removing a small quantity of the phase through a thin capillary (0.1 to 0.5 mm inner diameter) and expanding it through a valve into sample flasks. The sample separates into a condensed and a gaseous phase. The quantity of the non-condensible part is determined by its gas volume, either by measuring the pressure in a known expansion volume or by quantitatively pumping the gas into gas burettes. The quantity of the condensed part of the sample is determined by weighing the sample flasks. If one or both parts of the sample consist of more than one compound, then further analysis is carried out, e.g., by gas chromatography. If the condensible part of the sample comprises substances of very different volatility, these substances can be quantitatively separated in a sequence of sample flasks held at different temperatures. For a mixture of triglycerides and entrainers like ethanol, the first sample flask was maintained at a temperature around $0°$ to $20\,°C$ and the second flask at $T = -78\,°C$ by a dry ice/acetone mixture. After removal of the gas in the flasks to a pressure of about 1 mbar, a complete separation between glycerides and ethanol was established. Yet this part of the sampling procedure must be adjusted to the specific mixtures used.

It proved useful to extend the sampling capillary into the sampling flask, so that it is only interrupted along a small distance in the valve. Due to expansion, temperature during sampling in most cases will drop (dramatically for carbon dioxide). Precipitation of solid particles in the sampling line must be avoided. Otherwise, the sampling line may be blocked and if the valve is still held open, the increasing pressure in the sample line will press out the plug and the following rapid expansion will blow the sampling flask to pieces. Therefore, sampling valves and lines should be carefully heated. Only with hydrogen or helium will the temperature increase during expansion, since the Boyle temperature of these substances is below room temperature (see Section 2.3).

The sampling procedure is the critical operation in a static-analytical method. Ideally, a small volume of an equilibrium phase is removed from the cell, so as not to disturb equilibrium, and is brought quantitatively into the sample flask. Since this is not possible in practice, the sampling procedure must make sure that the ideal process is approached as nearly as possible. The following rules apply:

1) Remove only small quantities. Open valves slowly and carefully. For samples from the liquid phase, the critical quantity is the amount of gas. 10 to 20 $cm^3$ gas at ambient pressure may be sufficient, depending on the accuracy of the burettes. For samples from the gaseous phase, the condensible part is more critical, the lower the concentration of low volatile compounds in the gaseous phase. If there is only one condensible compound, the quantity may be as low as some mg (the total weight of sample flask and handling equipment weighs about 100 to 200 g!), but if a gas chromatographic analysis is necessary, the quantity of the sample should not be smaller than 20 mg. Handling of the sample flasks for weighing must be carried out especially carefully. Pressure and temperature of the flask must be adjusted to ambient values. The flasks must be carefully and reproducibly cleaned from sealing fat and stored in an exsiccator. Even better are flasks with seals from polytetrafluoroethene. Make sure that the weighing equipment is in best working order.

2) Sample lines, valve, and sample flask must be absolutely clean and gas tight. Pressure increase in this part after evacuation to about 0.01 mbar should be not more

than 0.1 mbar in 10 min! It makes no sense to flush the sample lines outside the equilibrium cell, since substances of low volatility may precipitate and the cleansing problem is the same again.

3) Furthermore, there must be taken several samples, usually three, in order to ensure representative average values. Then there is the problem with the content of the sampling capillary which lies within the equilibrium cell and may not be at equilibrium conditions, since mass transfer in the capillary is slow. Flushing of this part of the sampling line would be appropriate, but in most cases is avoided, because it would complicate the apparatus too much. The content of the sampling line in the equilibrium cell is supposed to be at least in approximate equilibrium with its surrounding phase.

4) Pressure drop during sampling cannot be avoided totally. Yet experience has shown that a reasonable pressure drop of around 2 % of the total pressure for the whole sampling procedure of taking six samples does not change the composition values beyond the error of analysis. The composition values are attributed to the ini-

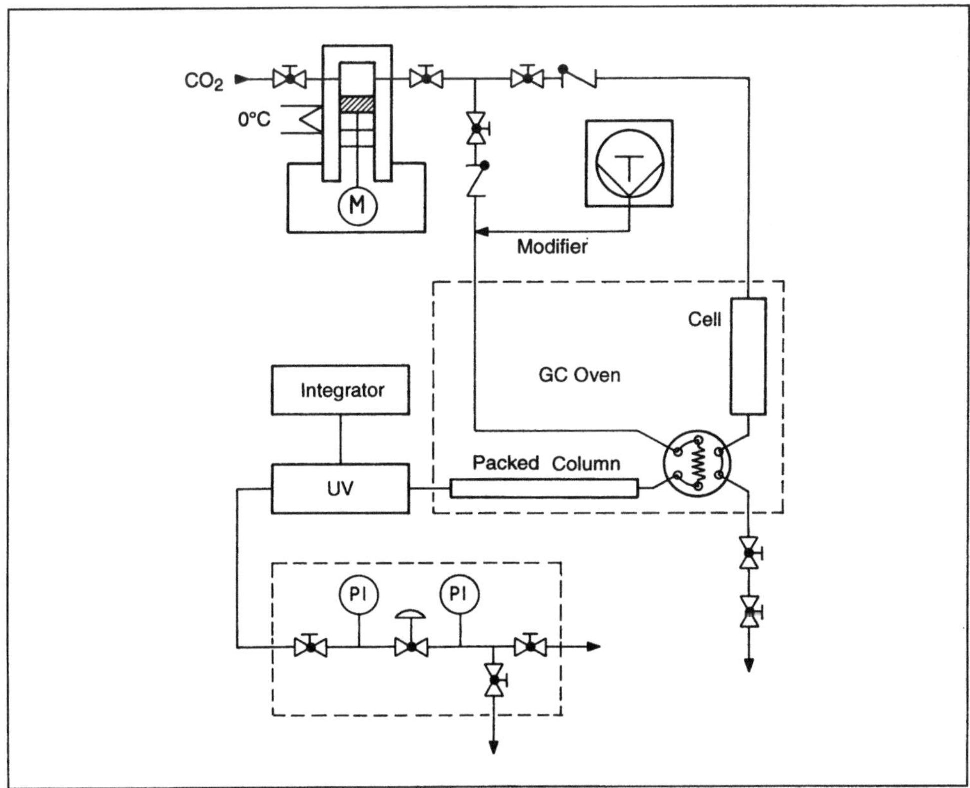


**Fig. 3.51.** Apparatus, consisting of standard chromatographic equipment, for measuring solubilities of solids and low volatile liquids in supercritical gases [30].

tial value of pressure. This reasonable pressure drop can be maintained in systems with non-volatile and supercritical compounds if the equilibrium cell has a volume of at least 500 cm$^3$ and small samples are taken as discussed above. Furthermore, a certain sequence is recommended: First take samples from the phase causing the least pressure drop during sampling, usually the liquid phase. Then take samples from the other phase, in most cases the gaseous phase.

The static-analytical method is simple in principle and its results are relatively reliable. Quality of the results depends strongly on the skill, the expertise, and the conscientiousness of the operator. For determination of solubilities of low-volatile components in supercritical gases, a much simpler installation can be used, which is derived from chromatographic equipment [30]. The equilibrium cell is a common HPLC-column in a GC-oven. See Fig. 3.51 for a flow diagram of the apparatus. The

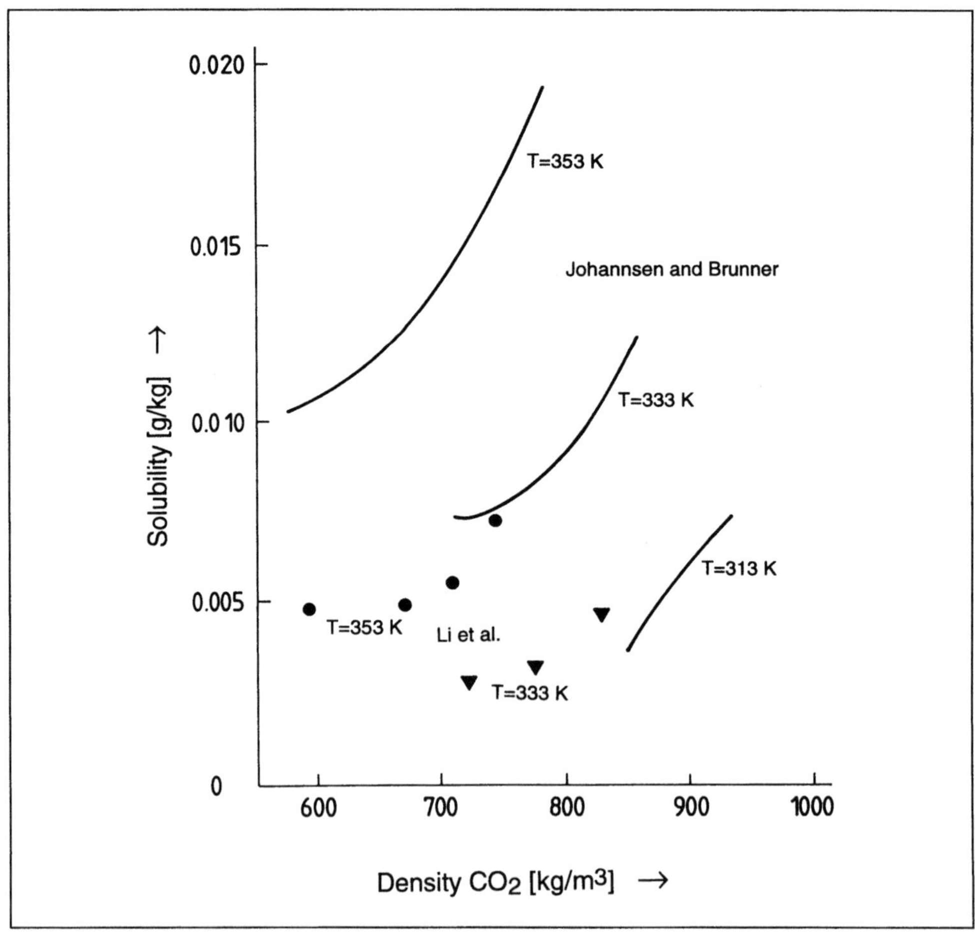


**Fig. 3.52.** Solubility of theobromine in supercritical $CO_2$. Measured with the apparatus of Fig. 3.51 [30].

122

substances of which solubility is to be measured are inserted into the column together with some glass beads. Then the column is fixed in the oven and connected to the gas supply line. The supercritical gas is pumped into the column by an HPLC-syringe pump. Equilibrium is established by waiting for about 30 min and shaking the column from the outside of the oven in order to fasten establishing of equilibrium. Then a sample is withdrawn by means of a multiport-valve and injected into a chromatographic column, which is in the same GC-oven as the equilibrium cell. The eluted peaks are detected by common gas chromatographic detectors. Peaks are calibrated by an external standard.

Results for solubilities in a supercritical solvent are excellent, if solubility is low. The time needed for one data point is about 1/10th of the time necessary for the static-analytical method discussed above. An example of experimental data determined by this method are the solubilities of theobromine in carbon dioxide, presented in Fig. 3.52.

### 3.8.3 The Synthetic Method

The synthetic method is very convenient for determining binary phase equilibria and phase envelopes in multicomponent systems. A schematic drawing of an experimental setup for the synthetic method is shown in Fig. 3.53. The apparatus consists of

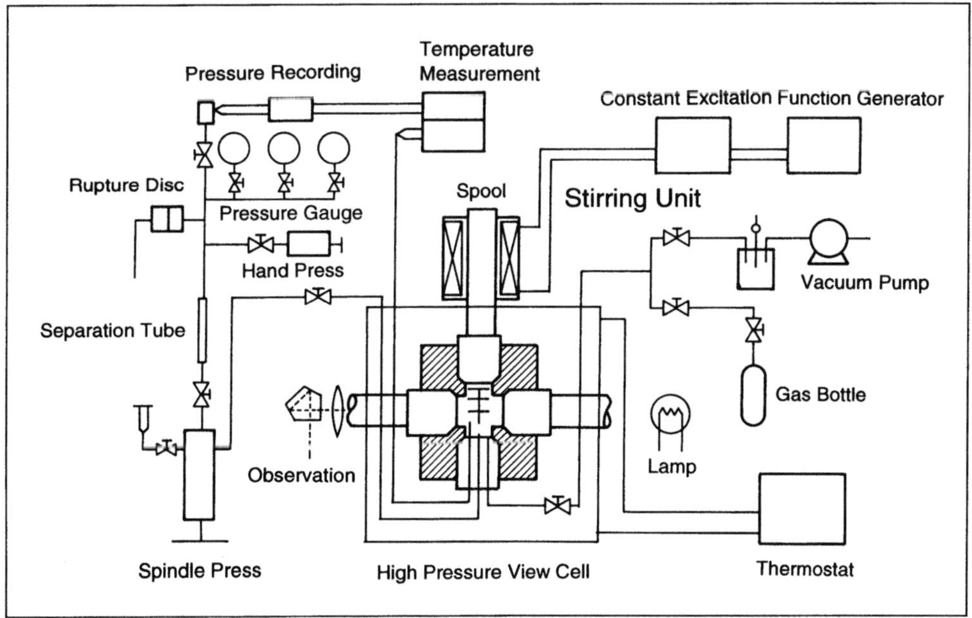


**Fig. 3.53.** Flow scheme of an apparatus for measuring phase equilibrium according to the synthetic method.

a measuring cell containing two windows for viewing the inner part of the cell. The cell is thermostated, either by electrical heating wires or by immersing it into a temperature-controlled liquid or air bath. The volume of the cell may be made variable by liquid mercury or by a movable piston inside the cell. The content of the cell is mixed by a stirrer, usually a magnetically moved object. The volume of the cell can be very small, typically in the range of several $cm^3$, since no samples are removed.

According to the procedure of the synthetic method, the mixture in the measuring cell is made up of accurately known quantities of the components, so that the total concentration of the system is very well established. For determining a point on the phase envelope, pressure is varied at constant temperature, or temperature is varied at constant pressure, until the appearance or the disappearance of a phase can be detected. Detection of a phase transition can be achieved directly by viewing the inner part of the cell and looking for a first or a last liquid drop or gaseous bubble, or indirectly by monitoring a physical property of the mixture in the cell which changes markedly, when a phase appears or disappears. Normally the pressure is varied, since this is the easier operation compared to a variation of temperature. Then it is conve-

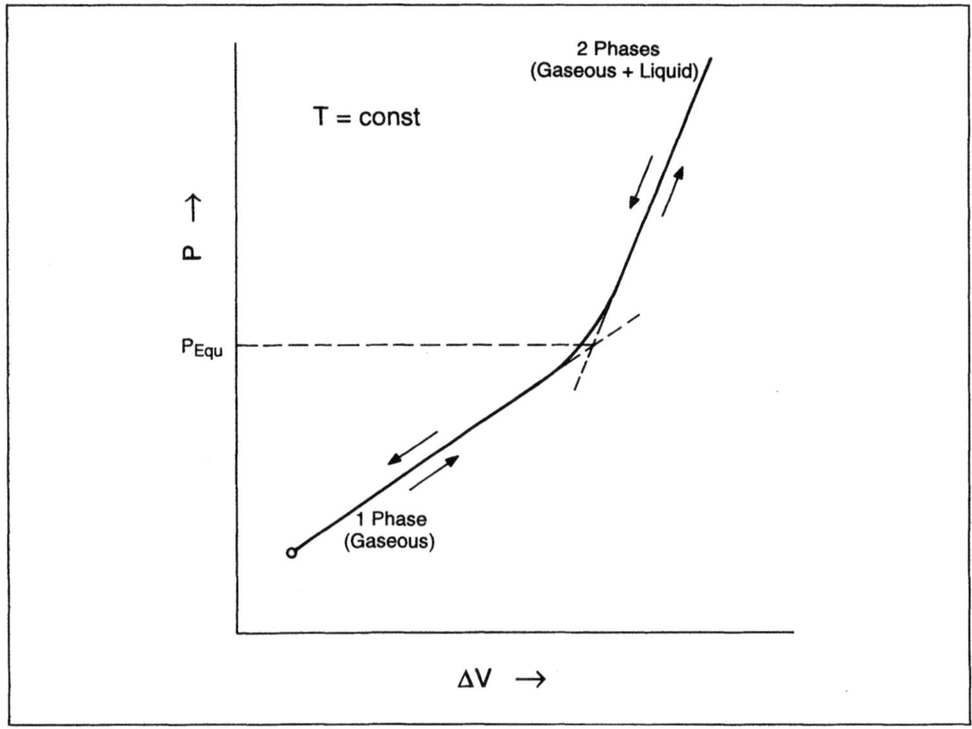



**Fig. 3.54.** Determination of a phase transition by variation of the volume of the equilibrium cell. $\Delta V$ is the volume change of the cell, induced by a piston, bellows or an inert liquid. $P_{Eq}$ is the equilibrium pressure, where the first part of a new phase emerges or the last part of a phase vanishes.

nient to monitor pressure and volume. At a phase boundary, compressibility of the mixture changes. Starting from a one-phase region, variation of pressure is linear with variation of volume over the very limited range applied in the experiment. If a second phase forms, the gradient of this line changes. In the two-phase region, pressure again changes linearly with volume. But often there is an extended transition range between the two lines, so that the phase boundary can only be determined by extrapolating both lines till they cross (see Fig. 3.54). If the volume of the equilibrium cell is known, information on density of the phase is obtained simultaneously.

Often, it is advisable to use both methods in parallel, e.g., the synthetic method to establish phase boundary lines and densities and the static-analytical method to determine equilibrium concentrations.

In binary systems the phase boundary determines the composition of an equilibrium phase which is the only one in the cell. For mixtures of more than two components, the phase envelope can be determined, i.e., it can be decided whether the system consists of one or two phases (if there may be three phases, direct viewing will help to detect the additional phase!).

### 3.8.4 The Dynamic Method

The third method is the dynamic method. In this method, the condensed phase remains in an equilibrium cell, while the gaseous phase flows through the equilibrium cell. During residence time of the gaseous phase in the cell, equilibrium concentra-

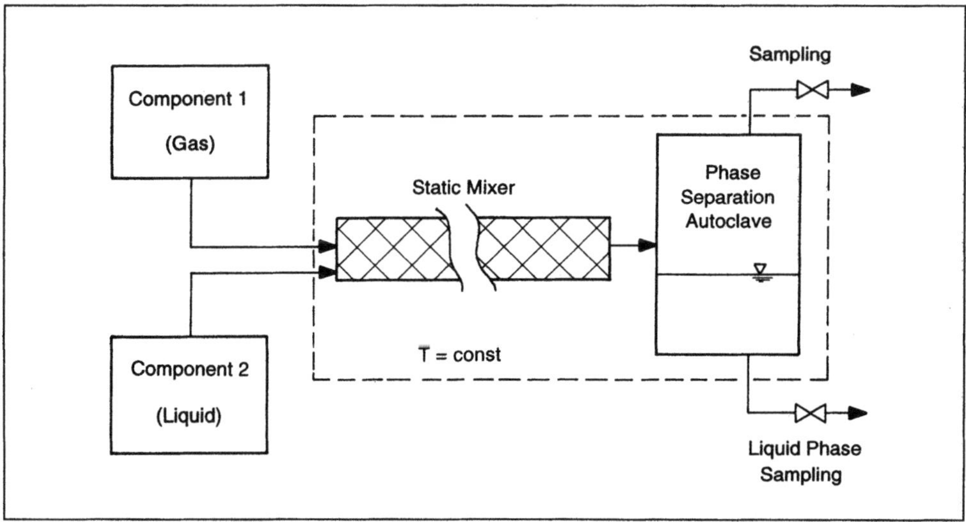


**Fig. 3.55.** Experimental equipment for measuring high pressure phase equilibrium by the dynamic method in systems with supercritical components (after Walter [65]).

tion is established in the gas, as well as in the condensed phase. The equilibrated gaseous phase is withdrawn from the cell and analyzed.

This method can be used for determining low concentrations in the gaseous phase, since the solute can be accumulated after the equilibrium cell and thus extremely low concentrations may be determined. It is also possible to determine several equilibrium points at one run by applying a transfer technique [51]. Determination of equilibrium data points is performed quickly, but establishment of equilibrium must be carefully observed during the experiments.

The dynamic method is best applicable to binary systems, since in multicomponent systems, due to different equilibrium solubility in the supercritical gas, the composition of the condensed phase changes. If applied for multicomponent systems, variation of the composition of the liquid phase must be taken into account. A flow scheme of the apparatus for determining phase equilibria by the dynamic method is shown in Fig. 3.55 [65].

### 3.8.5 What is the Best Method?

As one can imagine, there is no general answer. According to experience, the static-analytical method carried out with an equilibrium cell of around 1000 cm$^3$ is a method covering most of the needs of acquiring basic data for process design. But it has serious limitations in the critical region and at small concentrations if not carried out with chromatographic experimental techniques. In general, the method should be chosen according to the particular problem. Yet, often the question arises of whether a measurement can be performed with a certain set up or not. Therefore, static and synthetic method are discussed below.

The *synthetic method* has a relatively simple set-up. More than two phases can be detected quite easily. Because of the procedure for carrying out the method, all errors accumulate in the determination of total composition and equilibrium pressure. Total composition can be determined by carefully weighing the components and placing them totally into the cell. The equilibrium pressure is determined as the pressure at which a phase appears or disappears. This can be seen by emerging gas bubbles or a blackening of the field of vision due to tiny liquid droplets. By definition, the phase envelope is reached at zero quantity of a phase. Therefore, the equilibrium pressure cannot be detected very accurately, even if the phase envelope is exceeded repeatedly from both sides. Errors are greatest, where the equilibrium concentration of a phase envelope depends strongly on pressure. This is the typical case for gas extraction systems containing a supercritical solvent. If the phase envelope extends to very low concentrations, e.g., in the gaseous phase, no phase transition can be determined with the synthetic method, since the amount of the low-volatile compound is too small to detect a phase transition optically or by a change in compressibility.

The *static method* is more expensive. Beside the larger equilibrium cell, equipment for delivering compressed gas and a sampling device is needed. An experiment needs more time than is necessary for an experiment with the synthetic method.

For the static method, errors in the equilibrium concentration of a phase are characteristic. Errors in the equilibrium compositions are due to non-equilibrium at the moment of sampling, insufficient separation of the phases in the cell, and changes in composition during the sampling procedure. Insufficient separation of phases in the cell limits the application of the static method to the non-critical region of a system. Outside the critical region, equilibrium can be ensured by stirring long enough before sampling and by taking as small samples as possible in order to limit the pressure drop. Phase equilibria for gas extraction between compounds of very different volatility and at relatively low temperatures favor the static method, since composition of the phases is determined directly by sampling and the concentration dependence on pressure is relatively small. The main advantage of the static method is that by sampling it is possible to determine equilibrium data in systems containing an unlimited number of components.

Both methods may be used with similar accuracy and effort for binary systems of volatile components. The static-analytical method is more appropriate for systems with low volatile compounds with more than two components, and for the determination of the solubility of low volatile compounds in supercritical compressed gases, although at very low solubilities ($< 0.1$ %) a dynamic method may be useful. The static-analytical method is not applicable in the critical region. With skill and effort it may be used for three-phase systems. But for routine screening for two- or three-phase regions a synthetic method is much better. The synthetic method is appropriate for measurements of the phase envelope in the critical region.

## 3.8.6 Accuracy of the Measurements

Equilibrium measurements in connection with gas extraction are carried out to get basic data for process design. Therefore, the experiments should be carried out with the real substances used in the process.

In equilibrium experiments on the model system $CO_2$ – squalane [4] the accuracy of determining the composition of the equilibrium liquid phase was estimated to $\pm 0.0075$ in the mole fraction. The accuracy of the gas phase concentration, as determined with the static-analytical method was estimated to $\pm 0.002$ for squalane at low concentration. At higher equilibrium concentrations ($> 0.05$ mole/mole) uncertainty is about $\pm 0.005$ in the mole fraction. These estimates can be transferred to other systems of supercritical and low volatile compounds, especially also to the determination of the phase envelope in complex or multicomponent systems. The accuracy of determining equilibrium composition of a multicomponent mixture is further limited by the accuracy of the analytical method. In most cases chromatographic methods are applied. Accuracy in this case is at best 2 % of the actual value and only if an absolute calibrating method is applied. For complex mixtures accuracy may be much lower due to difficulties separating the sampled mixture into single compounds.

# 3.9 Correlation and Calculation of Phase Equilibria and Solubilities in Binary and Multicomponent Systems

## 3.9.1 Basic Equations

An overview on the needs, the compounds involved, the basic equations, the models and the methods for the correlation and calculation of phase equilibria and solubilities in binary and multicomponent systems is presented. A detailed description of a semi-empirical practical method for multicomponent systems with undefined components and components with unknown critical constants is discussed, the calculation of phase equilibria with pseudo-components and the correlations of component parameters for two equations of state is given.

Thermodynamic phase equilibrium determines the limits for mass transport between different phases, which are often involved in chemical engineering processes. Gas extraction in its various applications is such a process whose possibilities, to a large extent can be derived from information on equilibrium properties. Therefore, determination of phase equilibrium is important. Phase equilibria reveal the compositions of phases at equilibrium, including solubility of extracted compounds in the supercritical solvent and the solubility of the supercritical solvent in the extract phase, the quantities of the equilibrium phases, the distribution of individual compounds betweeen the equilibrium phases, and the variation of these quantities with pressure, temperature, and the concentration of various compounds.

The compounds involved in gas extraction comprise the supercritical solvent, the solute components and additional modifying compounds. These components are characterized by great differences in volatility, size, and chemical constitution. Information on the physico-chemical behavior is to be determined either by thermodynamic calculations or by experiments. Since experiments tend to be some orders of magnitude more expensive per data point, it may prove useful to have calculational methods at hand. At least these methods can be used for interpolation and thus reduce the necessary number of experiments.

The basic problem of phase equilibrium can easily be stated: A given quantity of a mixture of $m$ components is confined to a closed volume. The closed volume is held at constant temperature $T$ and pressure $P$. The mixture eventually divides itself into two homogeneous phases, say a liquid and a gaseous one. We now assign state variables to each phase, i.e., a temperature $T_v$, pressure $P_v$, and concentrations $y_i$ for the vapour phase, and $T_l$, $P_l$, $x_i$ for the liquid phase (Fig. 3.56). Since we are interested in equilibrium, we ask for the values of these variables at thermodynamic heterogeneous equilibrium (phase equilibrium). Equilibrium is established when there are no further changes of the variables with time. Since the closed volume in total is held at constant pressure and temperature, the equations for thermal and mechanical equilibrium are:

$$T_v = T_l \quad \text{and} \quad P_v = P_l. \tag{3.9}$$

Concentrations of the components in the two phases remain constant at equilibrium, but are in general not equal in both phases:

$$y_i = \text{const.}, \quad x_i = \text{const.}, \quad \text{or} \quad \partial y_i/\partial t = 0; \quad \partial x_i/\partial t = 0, \qquad 3.10$$

with $t$ for the time. To establish a relation between $y_i$ and $x_i$, general conditions for thermodynamic equilibrium are considered. An isolated system is in equilibrium, if the following equations are satisfied:

$$(\delta S)_U \leq 0 \qquad (\delta U)_S \geq 0, \qquad 3.11$$

with $S$ for the entropy and $U$ for the internal energy. The variation of the internal energy $U$ for different homongeneous regions (phases), denoted with superscripts 1, 2, ... $\pi$, can be written as:

$$T^1 \delta S^1 - P^1 \delta V^1 + \sum \mu_i^1 \delta n_i^1 + T^2 \delta S^2 - P^2 \delta V^2 + \sum \mu_i^2 \delta n_i^2 +$$

$$+ T^\pi \delta S^\pi - P^\pi \delta V^\pi + \sum \mu_i^\pi \delta n_i^\pi + \geq 0, \qquad 3.12$$

where $\mu_i$ are the differentials with respect to the number of moles $n_i$, the chemical potentials:

$$\mu_i = (\partial U/\partial n_i)_{S, V, n} = (\partial A/\partial n_i)_{T, V, n} = (\partial H/\partial n_i)_{S, P, n} = (\partial G/\partial n_i)_{T, P, n}. \qquad 3.13$$

Entropy, volume and mass of the system are constant since it is isolated, therefore

$$\delta S^1 + \delta S^2 + \ldots = 0; \quad \delta V^1 + \delta V^2 + \ldots = 0; \quad \delta n_i^1 + \delta n_i^2 + \ldots = 0, \qquad 3.14$$

for $i = 1, 2 \ldots m,$

with $m$ for the number of components. The equation for equilibrium is always satisfied, if the values for the temperature, the pressure and the chemical potentials are equal in the different phases:

$$T^1 = T^2 = \ldots = T^\pi;$$

$$P^1 = P^2 = \ldots = P^\pi;$$

$$\mu_1^1 = \mu_1^2 = \ldots = \mu_1^\pi; \qquad 3.15$$

$$\ldots\ldots\ldots$$

$$\ldots\ldots\ldots$$

$$\mu_m^1 = \mu_m^2 = \ldots = \mu_m^\pi.$$

This is easily verified, if the equation for equilibrium is rearranged and the equations for thermal, mechanical, and chemical equilibrium incorporated [49]:

$$T^1\,(\delta S^1 + \delta S^2 + \ldots) - P^1\,(\delta V^1 + \delta V^2 + \ldots) + \sum_i [\mu_i^1\,(\delta_{ni}^1 + \delta_{ni}^2 + \ldots)] \geq 0. \qquad 3.16$$

Equilibrium conditions may also be expressed in the form of other thermodynamic relations, like the Gibbs-function (free enthalpy) $G$:

$$(\delta G)_{T,\,P} \geq 0, \qquad 3.17$$

which means that the free enthalpy of a system at thermodynamic equilibrium is at a minimum. The variation of the free enthalpy of a system can be expressed by variations of pressure, temperature, and the number of moles of a component. These variables can be directly varied in a process by operational means and therefore the free enthalpy plays a central role in chemical engineering thermodynamics.

$$G(T,P,n): \quad dG = -\,SdT + VdP + \sum_1^m \mu_i dn_i. \qquad 3.18$$

The state of a homogeneous region (a phase) in thermodynamic equilibrium can be described by its temperature, pressure, and chemical potentials of the components, i.e., $m + 2$ variables. These variables are not all independent. The internal energy of a finite mass of a mixture representing a phase in our equilibrium system may be expressed as:

$$U = TS - PV + \sum_i \mu_i n_i. \qquad 3.19$$

Changes in the internal energy are possible by changes in any variable:

$$dU = TdS + SdT - PdV - VdP + \sum_i \mu_i dn_i + \sum_i n_i d\mu_i. \qquad 3.20$$

By comparison with the fundamental equation for the variation of the internal energy, the Gibbs-Duhem equation is obtained:

$$SdT - VdP + \sum_i n_i d\mu_i = 0. \qquad 3.21$$

This equation restricts the number of variables which can be independently varied in one phase to $m + 1$. If there are several phases in internal equilibrium, the number of variables for $\pi$ phases is $\pi\,(m + 1)$. If there is also equilibrium between these phases, the number of equilibrium conditions is $(\pi - 1)(m + 2)$, since for each variable there is one equilibrium condition. The number of variables, which can be specified in such an equilibrium system and the number of degrees of freedom $F$ can

130

be obtained from the total number of independent variables, $\pi(m + 1)$, minus the number of equilibrium conditions $(\pi - 1)(m + 2)$. The resulting equation is the phase rule for non-reacting systems: $F = m + 2 - \pi$ (Eq. 3.7).

In a binary two-phase system one concentration in one phase in addition to temperature *or* pressure can be specified. At given values of pressure and temperature the compositions of the two phases are fixed. In a ternary two-phase system in addition to temperature and pressure one concentration in one phase has to be specified in order to fix the problem for a unique solution.

The chemical potential is not easily understood as a physical quantity like pressure and temperature. Lewis therefore defined a function $f$, which he called *fugacity*. In the rigorous equation $(\partial \mu_i / \partial P)_T = V_i$, for $V_i$ the volume of an ideal gas $V_i = (RT)/P$ is incorporated and the equation is integrated at constant temperature. The resulting equation is the relation between the chemical potential of an ideal gas and pressure:

$$\mu_i - \mu_i^0 = RT\ln(P/P^0). \tag{3.22}$$

Real gases and substances will not follow this relation, since their volumetric behavior is not accurately represented by the ideal gas law. But at least for gases the real value will not be too far off and therefore a variable defined as *fugacity* (see below) instead of pressure will bear some resemblance to a corrected pressure.

The real behavior is represented by the *fugacity,* which is defined by the following equation:

$$\mu_i - \mu_i^0 = RT\ln(f_i/f_i^0). \tag{3.23}$$

For a pure ideal gas, fugacity is equal to pressure, for a component in an ideal gas mixture equal to the partial pressure. Since all substances approach the ideal gas at very low pressure, the definition of the fugacity is completed by the limit:

$$f_i/(y_iP) \rightarrow 1, \text{ for } P \rightarrow 0. \tag{3.24}$$

With fugacity the equilibrium condition for the chemical potential can be transformed. For a gaseous *(v)* and a liquid *(l)* phase the relation between chemical potential and fugacity is:

$$\mu_{i,v} - \mu_{i,v}^0 = RT\ln(f_{i,v}/f_{i,v}^0) \quad \text{and} \quad \mu_{i,l} - \mu_{i,l}^0 = RT\ln(f_{i,l}/f_{i,l}^0). \tag{3.25}$$

Substituting these equations into the equilibrium equation, we obtain:

$$\mu_{i,v}^0 + RT\ln(f_{i,v}/f_{i,v}^0) = \mu_{i,l}^0 + RT\ln(f_{i,l}/f_{i,l}^0). \tag{3.26}$$

If the standard states are equal, the equilibrium relation in terms of fugacity is readily obtained. But even if the standard states are not equal, they can be transformed by the relation between fugacity and chemical potential.

$$f_{i,v} = f_{i,l}. \tag{3.27}$$

The condition that for thermodynamic equilibrium of heterogeneous systems the chemical potentials of all components must be equal is thus transformed in general to the condition that the fugacities of all components in all the phases must be equal.

It is a matter of convenience whether chemical potential or fugacity is used. In international chemical engineering literature, fugacity is used, while in more fundamental approaches chemical potential is preferred.

If we define the fugacity coefficient $\varphi_i$:

$$\varphi_i = f_i/(y_i P), \tag{3.28}$$

equilibrium conditions for a gas-liquid system can be written:

$$\varphi_i^v y_i = \varphi_i^l x_i. \tag{3.29}$$

The fugacity coefficient can be calculated from volumetric properties:

$$RT\ln\varphi_i = \int_0^P [v_i - (RT)/P]\, dP, \tag{3.30}$$

where $v_i = (\partial V/\partial n_i)_{T, P, nj}$ is the partial molar volume of component $i$.

Since most equations of state, representing volume $V$, are explicit in $P$, a transformation for independent variables $V$ and $T$ may be more useful:

$$RT\ln\varphi_i = \int_V^\infty [(\partial P/\partial n_i)_{T,V,nj} - (RT)/V]\,dV - RT\ln z, \tag{3.31}$$

where $z = (PV)/(RT)$ is the compressibility factor. Beside the different independent variables, there is furthermore the difference between these two equations, that $v_i$ is a partial molar quantity, while $\partial P/\partial n_i$ is not.

The equations for the equilibrium phases can be combined and the fugacity coefficient eliminated. For a liquid and a gaseous phase in equilibrium we obtain:

$$RT\ln(x_i/y_i) = \int_{p+}^P [v_i(y_i) - v_i(x_i)]\, dP, \tag{3.32}$$

where $v_i(y_i)$ and $v_i(x_i)$ are the partial molar volumes of component $i$ in the vapor and the liquid phase, respectively, and $P^+$ is a reference pressure, low enough that ideal behavior can be assumed.

Now, the equations for phase equilibrium can be solved, provided a function $V = f(T,P,n_i)$ is available which allows to calculate integrals and derivatives with sufficient accuracy.

## 3.9.2 Equations of State for Mixtures

There are several possibilities for such functions, e.g. corresponding states principles, perturbation theories, and equations of state. For a more detailed discussion see [16, 49, 50] in combination with the literature cited. For purposes of calculating phase equilibria for gas extraction and related processes, equations of state are by far the most useful equations up to now. Therefore, within the scope of this book only equations of state are considered. Suitable equations of state are presented in Chapter 2.

For the calculation of phase equilibria it is now assumed that an equation of state is able to represent the properties of all the components of the mixture as well as the properties of the mixture as a pseudocomponent (one-fluid theory), which again is not the only possibility, but seems to suit our purposes best. The properties of the mixture are calculated by applying rules by which the properties of the mixture are related to the pure-component properties, the so called combining and mixing rules. Then, the parameters of the equation of state are mixture parameters. For the modified Redlich-Kwong equation of state:

$$P = \frac{RT}{V - b_m} - \frac{a_m(T)}{V(V + b_m)}, \qquad \qquad 3.33$$

where $a_m(T)$ and $b_m$ are the mixture parameters of the equation of state. It is now necessary to define the relation between the mixture parameters and the pure-component parameters, i.e., the dependence of mixture parameters on composition. It is common practice to define quadratic mixing rules which are for a binary mixture:

$$a_m(T) = x_1^2 a_{11}(T) + 2x_1 x_2 a_{12}(T) + x_2^2 a_{22}(T), \qquad \qquad 3.34$$

$$b_m \quad = x_1^2 b_{11} + 2x_1 x_2 b_{12} + x_2^2 b_{22}, \qquad \qquad 3.35$$

where $x$ stands for the concentration in one phase and $x_i$, $a_{ii}$, and $b_{ii}$ refer to pure component values, $b_{12}$ and $a_{12}(T)$ are binary parameters. The mixing rules for the other phases are analogue. The binary parameters are mostly written in the form:

$$a_{12}(T) = [a_{11}(T)a_{22}(T)]^{0.5}(1 - k_{12}), \qquad \qquad 3.36$$

$$b_{12} \quad = 0.5 \, (b_{11} + b_{22})(1 - l_{12}). \qquad \qquad 3.37$$

The absolute values $|k_{12}|$ and $|l_{12}|$ are small compared to 1. They are fitted to binary experimental data, usually concentrations of phases in heterogeneous equilibrium. In many cases, especially in systems with nonpolar or weak polar compounds one binary parameter in the attractive term ($k_{12}$) is sufficient. If there are no binary values for equilibrium phases known, the binary parameters can be set to zero. Then, the method is used in a predictive way, concluding from pure components directly to the behavior of the mixture, assuming a concentration dependence as given by the parameter mixing rules.

Although from the equations alone it cannot be concluded that fitting to one of the phases of a two-phase gas-liquid system should be preferred, fitting of $k_{12}$ to the liquid phase yields better results than fitting $k_{12}$ to the gaseous equilibrium phase. It may be argued that this is due to the higher density of the liquid phase resulting in more pronounced interactions, yet from a theoretical point of view also the gas phase should contain all the necessary information (which is the same reasoning we assumed for being able to calculate thermodynamic properties and phase equilibrium from volumetric properties alone). It is only the limitation in accuracy of our equations employed in practice which makes a difference. The binary values are not dependent on pressure but are functions of temperature (see examples).

For correlating phase equilibrium data, in order to employ experimental information in design calculations, the best possible fit to the experimental data is needed. In these cases several possibilities for fitting can be made use of: Fitting only to the liquid or gaseous phase (depending on in which phase greater accuracy is needed), fitting to both phases, e.g., $k_{12}$ to the liquid and $l_{12}$ to the gaseous phase, weighing the binary interaction parameters with some arbitrary function, and even using different $k_{12}$-values for the different phases in equilibrium. In general, these empirical methods will help somewhat with representing the experimental data, but will not change the character of the equations of state used. Therefore, do not expect too much increase in accuracy for representation of the phase-equilibrium data. Especially using two binary parameters in the mixing rules ($k_{12}$ and $l_{12}$) may lead to an indefinite number of paired values of these parameters, which represent the data with about the same deviations.

Binary systems of nonpolar and weakly polar compounds are correlated surprisingly well by any of the van der Waals-type equations of state, over a wide range of temperatures and pressures. Only in the critical region of the binary systems are there greater deviations from experimental values. Though experimental errors in this region are also larger, the deviations calculated with the equations of state tend to be greater than the experimental errors. In a mass transfer process the critical state of a mixture is of no use (the diffusion coefficient approaches zero and compositions of the phases approach one another). Therefore, calculation errors in the critical region are not important in practice. Even in calculations of the extent of the two-phase region, substantial errors can be tolerated, since operating conditions will be chosen well away from critical and limiting conditions.

The importance of binary systems is their relative simplicity and the deep insight into principal phase behavior, which can be extracted from binary systems. Yet any practical separation problem in gas extraction incorporates at least three components and many mixtures contain more than three components. For the calculation of multicomponent phase exquilibria it must be decided how many species are interacting. Common practice is to use binary interactions only and to calculate multicomponent properties with binary interaction coefficients. The mixing rules for a multicomponent system with $m$ components are then, if we use parameter $\alpha$ for the temperature dependence of the attractive parameter $a_{ii}$,

134

$$a_m(T) = \sum_{i=1}^{m} \sum_{j=1}^{m} (x_i x_j a_{ij}^+) \qquad\qquad 3.38$$

$$a_{ij}^+ = (a_{ii}a_{jj})^{0.5}\,(1 - k_{ij}) \text{ for } i \neq j \qquad\qquad 3.39$$

$$a_{ii} = a_i T^\alpha \qquad\qquad 3.40$$

$$b_m = \sum_{i=1}^{m} (x_i b_i), \qquad\qquad 3.41$$

if only one binary interaction parameter $k_{ij}$ is employed.

For practical correlation purposes this procedure sometimes must be modified. For ternary systems of type II (immiscibility in two binary systems), the phase equilibrium calculation will yield sufficiently accurate results using only interaction coefficients determined from binary systems. But for ternary type I-systems (immiscibility in one binary system), the results often are not adequate. Therefore, the binary interaction coefficients have to be fitted to ternary data. In general, for design purposes pseudo binary interaction coefficients should be calculated from multicomponent data, if experimental data are available.

The mixing rules, presented above, are simple, but proved to be inadequate for asymmetric systems and polar components. The introduction of composition dependent mixing rules is one way to overcome these problems. The most general one was proposed by Vidal, who introduced the excess free enthalpy $(G^E)$ into the mixing rule. For the excess free enthalpy any model, especially any activity coefficient model, can be employed. Recently, Wong and Sandler [66] proposed to use the excess free energy $(A^E)$ instead of the free excess enthalpy. The introduction of $G^E$ violates the quadratic composition dependence of the second virial coefficient. Using the excess free energy to introduce the composition dependence of the interactions, Wong and Sandler formulated mixing rules which satisfy the second virial composition dependence [41].

The Wong-Sandler mixing rule defines mixture parameters $a$ and $b$ by the following equations:

$$b - \frac{a}{RT} = \sum_i \sum_j \left( x_i x_j (b - \frac{a}{RT})_{ij} \right), \qquad\qquad 3.42$$

$$\frac{A_\infty^E}{CRT} = \frac{a}{bRT} - \sum_i \left( x_i \frac{a_i}{b_i RT} \right), \qquad\qquad 3.43$$

where C is a numerical constant characteristic of the cubic equation of state used.

For the $A^E$ term any model for liquid activity coefficients is suitable, since at low pressure excess free enthalpy and excess free energy are equal. For the binary interaction coefficient, the following term was proposed:

$$\left(b - \frac{a}{RT}\right)_{ij} = (b_i + b_j)/2 - \sqrt{a_i a_j}\,(1 - k_{ij})\,/\,(RT). \qquad 3.44$$

From the examples presented so far, it can be concluded, that this mixing rule has excellent capability for correlating phase equilibria.

### 3.9.3 Predictive Calculations

If the pure component parameters are known, the only unknown parameters are the binary interaction coefficients. If these can be correlated or determined independently beforehand, a real predictive calculation of phase equilibria is feasible. A first approximation of binary phase equilibrium can be achieved by assuming that the interaction coefficient $k_{12}$ is zero. For some systems, e.g., hydrocarbon systems, this is a reasonable approximation. But for other systems deviation from real behavior is unpredictable.

Correlations of binary interaction coefficients are another possibility. Since interaction coefficients in general are determined by fitting experimental data to the applied equation of state and mixing rules, experimental errors accumulate in the binary interaction coefficients, thus rendering useful correlations difficult. Hederer and Wolff [28] have proposed a correlation for the interaction coefficients of binary systems, which can be used in combination with the modification of the Redlich-Kwong EOS, proposed by Hederer, Peter and Wenzel (see Chapter 2). The correlation is

$$k_{12} = k_1 \frac{a_{11} a_{22}}{b_1 b_2}\, T^{(\alpha_1 + \alpha_2)} + k_2, \qquad 3.45$$

with
$k_1 = -0.49768 \cdot 10^{-6}\,\text{bar}^{-2}\text{l}^{-2}; \; k_2 = 0.10356.$

Correlation coefficients were calculated by linear regression from interaction coefficients determined by fitting to experimental data. Therefore, the interaction coefficients scatter around the regression function, Eq. 3.45. Since calculation results of phase equilibria are strongly dependent on the value of the interaction coefficient, results of phase equilibrium calculations, using $k_{12}$ from the correlation, are not necessarily better than using $k_{12} = 0$.

More promising is the combination of a predictive low-pressure method with phase equilibrium calculation by an equation of state, especially with the Wong-Sandler mixing rule. A group contribution method, like UNIFAC, with the parameters determined at normal pressure, may be combined with one of the equations of state. Results of such calculations are promising for many systems. For a more detailed discussion the reader is referred to the literature.

### 3.9.4 Examples for Phase Equilibrium Calculations

In the following figures, examples for calculation of phase equilibria with equations of state are presented. In the diagrams the EOS and the interaction parameters are indicated. Interaction parameters were determined by fitting to experimental phase equilibrium data.

In Fig. 3.56 the correlation of experimental data in the binary system $CO_2$ – oleic acid is shown. Actually, it is a quasi-binary system, since beside oleic acid there are appreciable amounts of other fatty acids present. But for our purposes, the system is treated as a binary system. Phase equilibrium can be represented using simple quad-

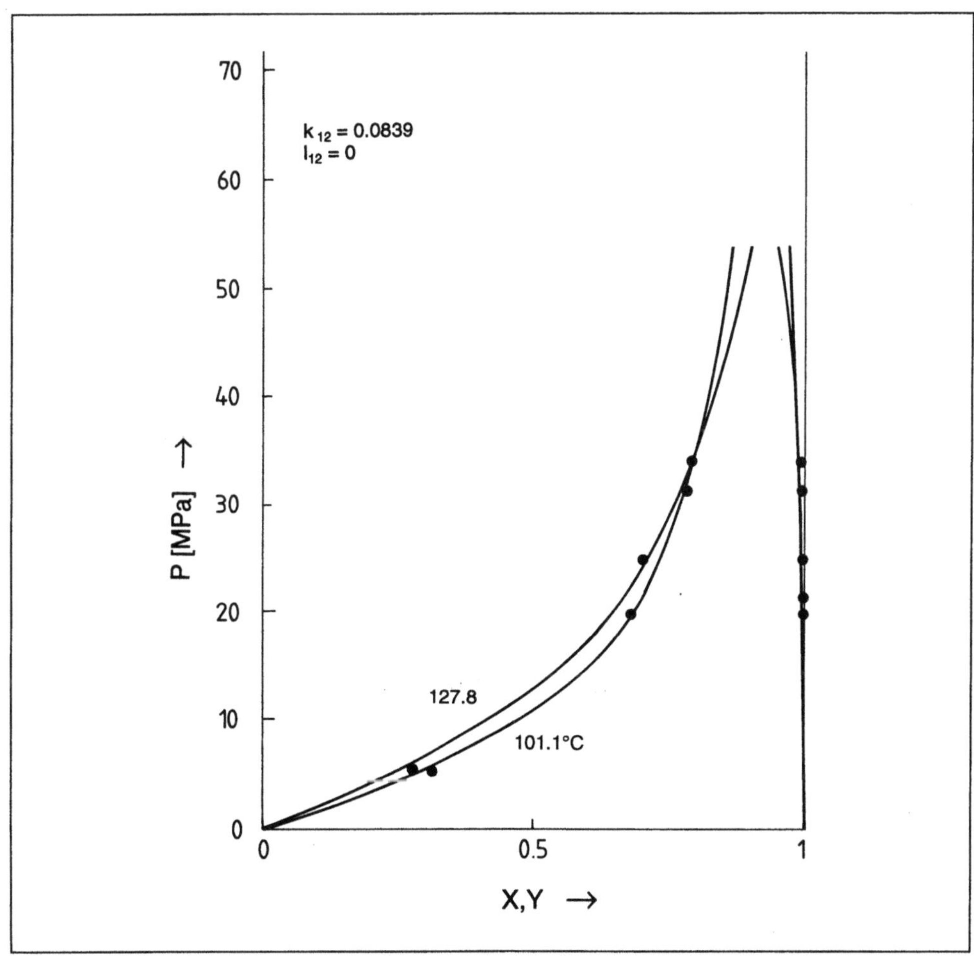


**Fig. 3.56.** *P,x*-isotherms of the binary system $CO_2$ – oleic acid [4]. Calculated with the HPW-modification of the Redlich-Kwong EOS.

ratic mixing rules and one binary interaction parameter $k_{12}$. If solubility in the gaseous phase is the most important quantity, then $k_{12}$ should be fitted to gas phase data. In the binary system $CO_2$ – benzene, Fig. 3.57, it is not possible to represent the phase equilibrium data with a simple quadratic mixing rule and a modified Redlich-Kwong EOS with only one interaction coefficient. It is necessary to use a second interaction coefficient, $l_{12}$, fitted to gas phase data, as illustrated in Fig. 3.57 in order to represent the data adequately.

The interaction coefficients are temperature dependent. In Fig. 3.58 the temperature dependence, as calculated with the HPW-modification of the Redlich-Kwong EOS, has been plotted for several binary systems. In most cases, $k_{12}$ is a relatively weak and smooth function of the inverse temperature.

The correlation of phase equilibrium data for complex mixtures is especially difficult, since there is no simple way to find the correct component parameters. Palm oil, for example consists of many triglycerides which behave similar with respect to carbon dioxide. Therefore, phase equilibrium is treated in a quasi-binary system. The liquid phase is well represented, as shown in Fig. 3.59, but solubility in the gaseous phase is not. This may be due to the effect that the composition of the solute, dissolved in carbon dioxide, is somewhat different from the composition of the palm oil in the liquid phase. If a true binary system is treated, the solubility in supercritical $CO_2$ can be satisfactorily calculated, as is illustrated in Fig. 3.60 on the system EPA-ethylester – $CO_2$. Another example is the solubility of $\alpha$-tocopherol in $CO_2$ (Fig. 3.61). In this system, as well as in the previous one, two interaction coefficients are necessary.

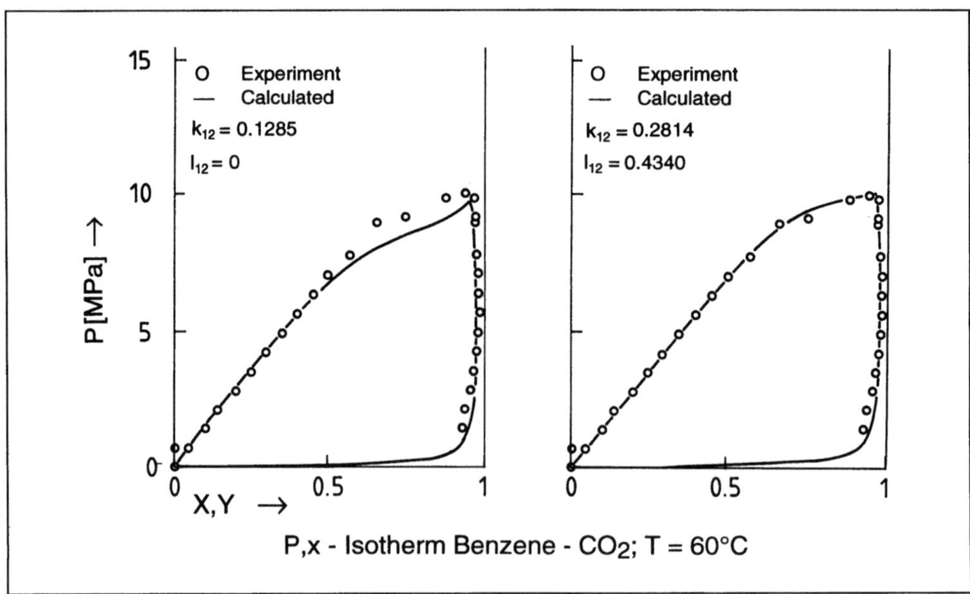



**Fig. 3.57.** $P,x$-isotherms of the binary system $CO_2$ – benzene [4]. Two binary interaction coefficients are necessary to represent the data adequately.

138

For quasi ternary systems, in Figs. 3.62 and 3.63 examples are presented in systems with an entrainer.

It is difficult to correlate the correct distribution coefficients and separation factors for low volatile substances in systems containing supercritical compounds. The properties of the supercritical component and the low-volatile substances are very different, while the properties of the low-volatile substances are quite similar. For fatty acid methyl esters, Fig. 3.64 shows an example of a good representation of the tie lines and the phase envelope.

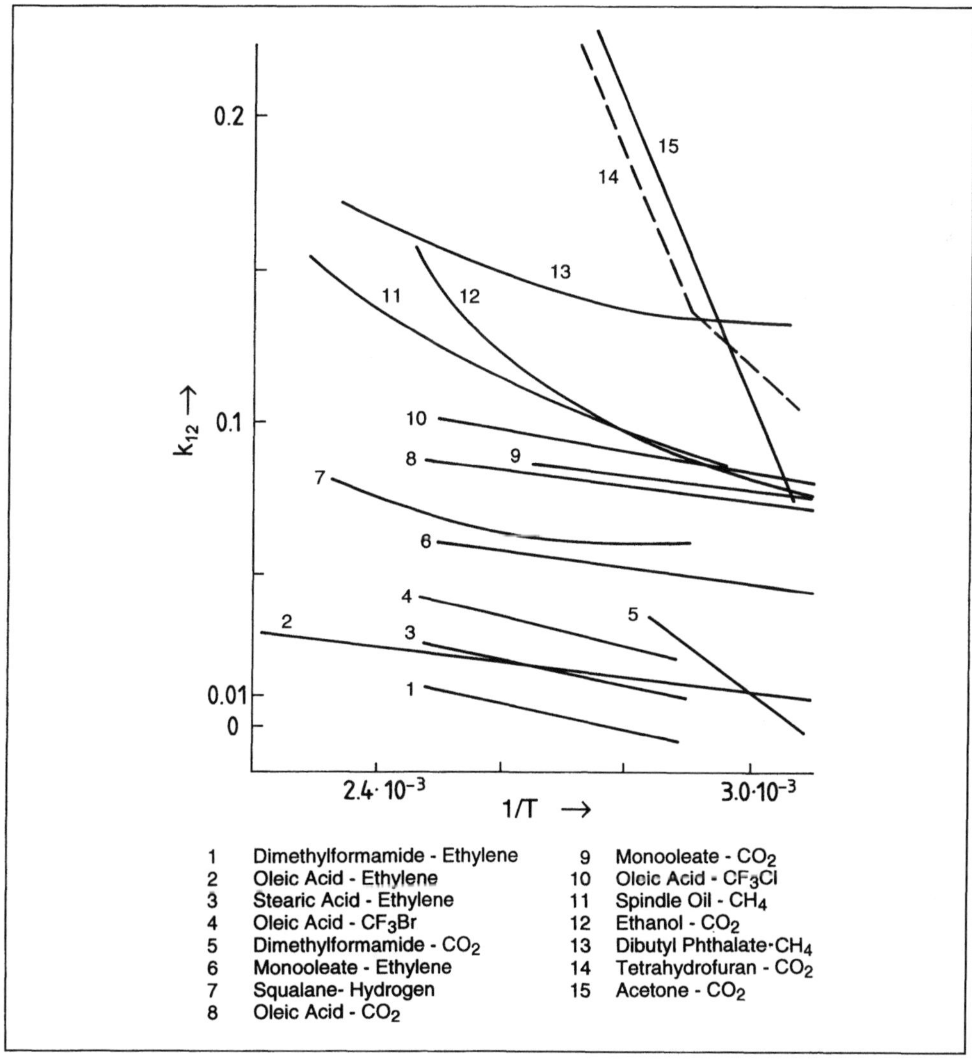


**Fig. 3.58.** Interaction coefficients for various binary systems. The coefficients may only be used with the HPW-modification of the Redlich-Kwong EOS, using the parameter listed in Table 2.5.

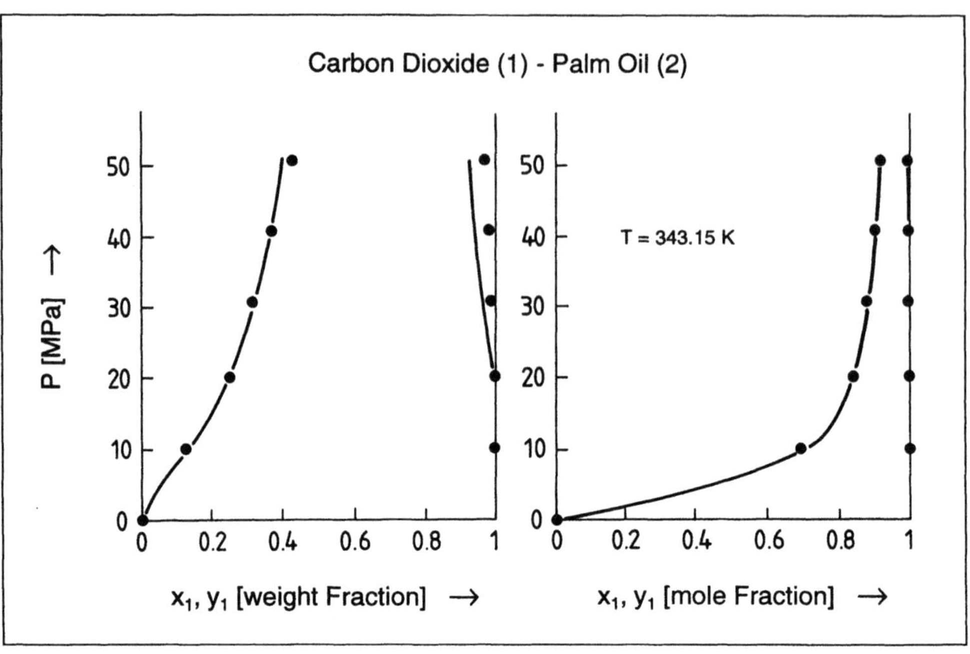


**Fig. 3.59.** Phase equilibrium in the quasi-binary system palm oil-carbon dioxide. $T = 343.15$ K; HPW modification of the Redlich-Kwong EOS. $k_{12} = 0.45$. Data from Brunner [4].

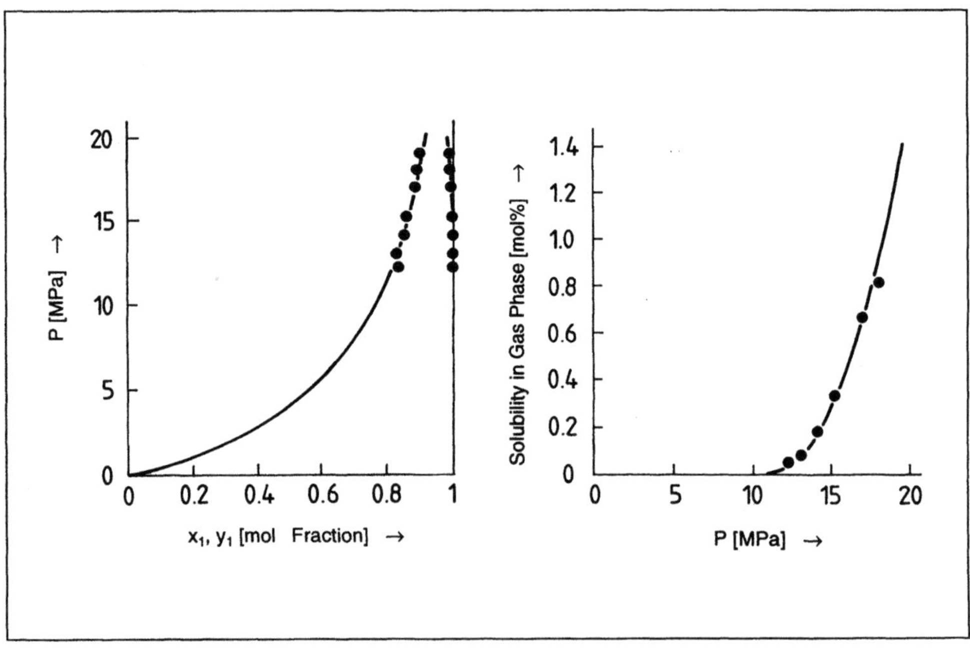


**Fig. 3.60.** Phase equilibrium in the binary system EPA-ethylester $- CO_2$ [9]. $T = 333.15$ K; Peng-Robinson EOS, two binary interaction coefficients: $k_{12} = 0.051$, $l_{12} = 0.041$. Data from Bharath et al. [2].

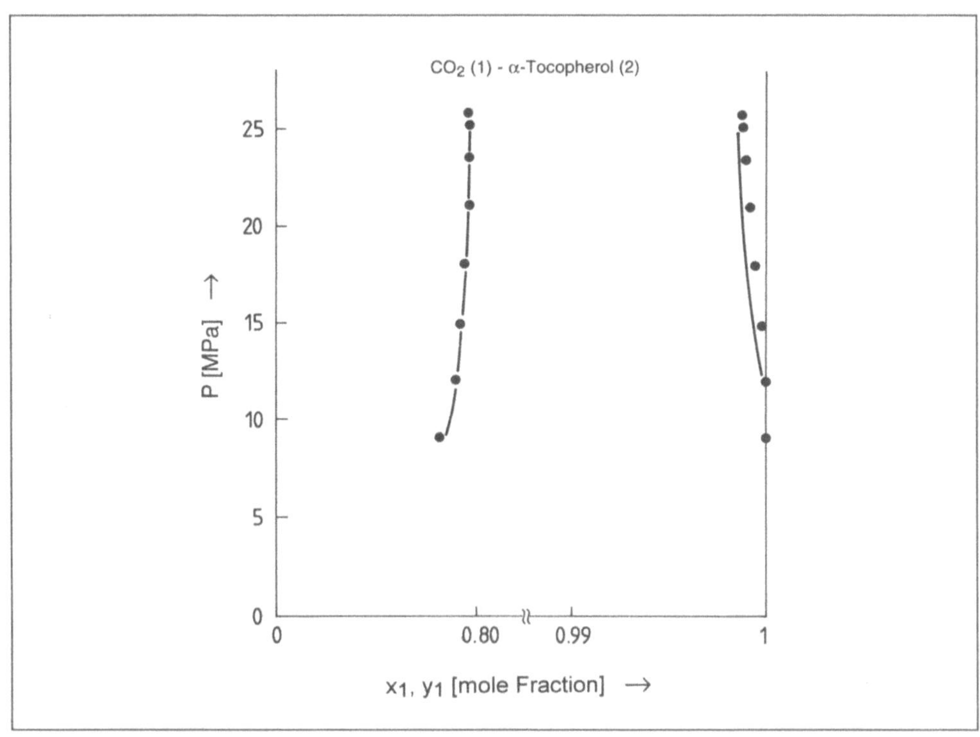



**Fig. 3.61.** Phase equilibrium in the binary system $\alpha$-tocopherol – $CO_2$ [38]. $T = 333.15$ K; HPW-modification of the Redlich-Kwong EOS. Two binary interaction coefficients: $k_{12} = 0.1195$, $l_{12} = 0.055$. Data from Pereira et al. [42].

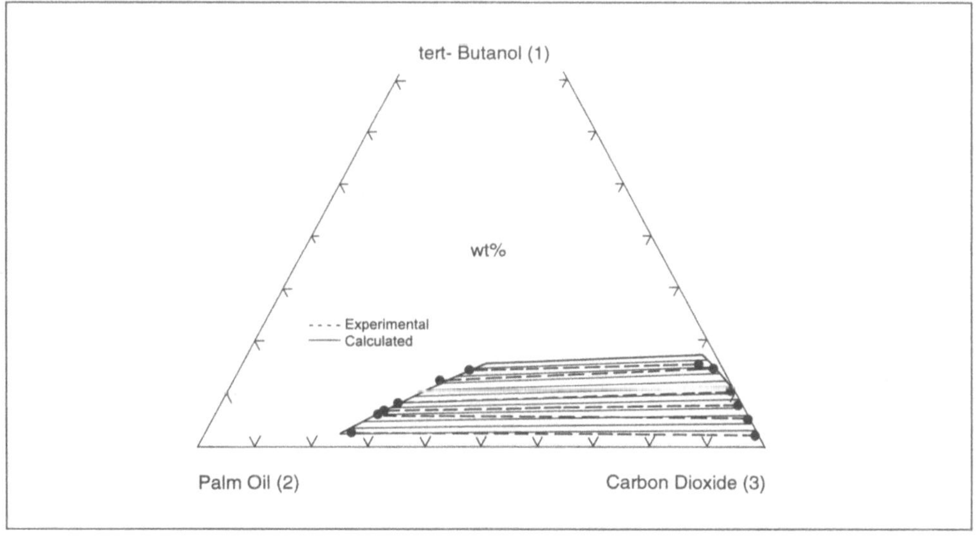



**Fig. 3.62.** Phase equilibrium in the quasi ternary system $CO_2$ – tert-butanol – palm oil [53]. $T = 343.1$ K, $P = 13.4$ MPa; HPW-modification of the Redlich-Kwong EOS. $k_{12} = -0.1049$; $k_{13} = 0.1749$; $k_{23} = 0.0379$. Data from Brunner [4].

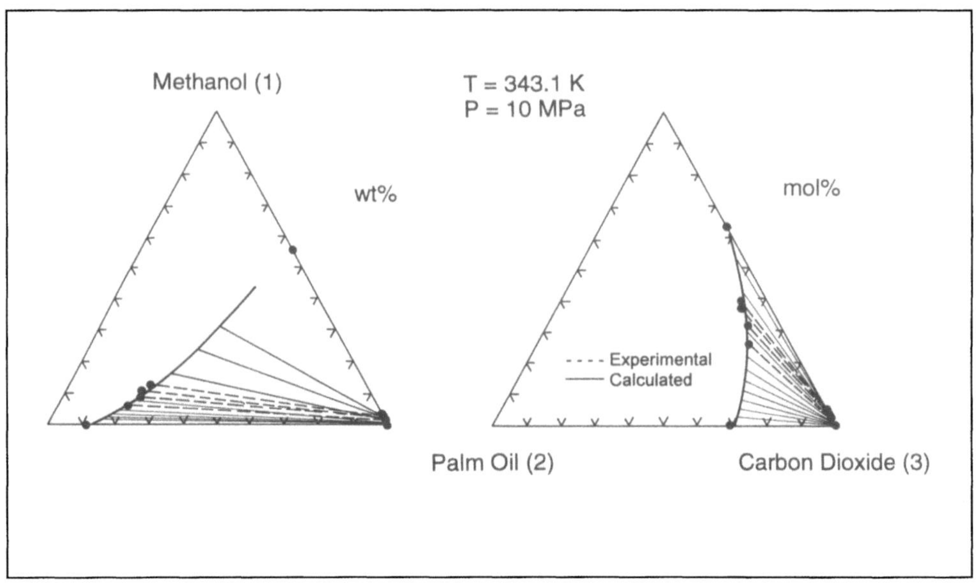



**Fig. 3.63.** Phase equilibrium in the quasi ternary system $CO_2$ – methanol – palm oil [53]. $T = 343.1$ K, $P = 10$ MPa; HPW-modification of the Redlich-Kwong EOS. $k_{12} = -0.2163$; $k_{13} = 0.0824$; $k_{23} = 0.0438$. Data from Brunner [4].

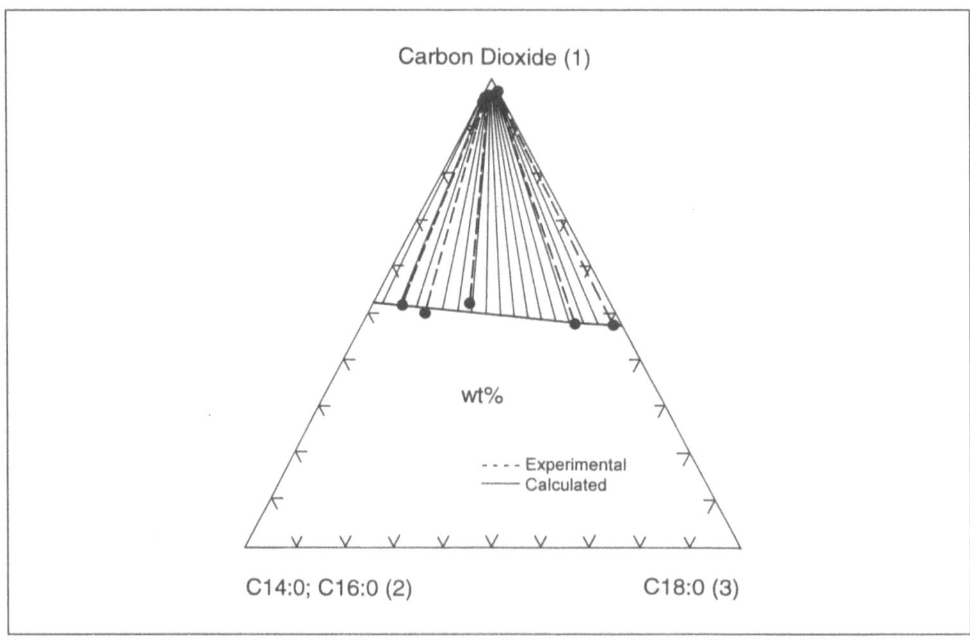



**Fig. 3.64.** Phase equilibrium in the quasi ternary system $CO_2$ – myristic acid/palmitic acid methylester – stearic acid methylester. $T = 333.15$ K, $P = 14.5$ MPa; Peng-Robinson EOS. Melhem-mixing rules. Two binary interaction coefficients. $k_{12} = 0.0832$; $k_{13} = 0.0446$; $k_{23} = 0.0015$. $L_{12} = 0.0200$; $L_{13} = -0.0150$; $L_{23} = 0.0$. Data from van Gaver [25].

142

The correlation of experimental phase equilibrium data is always the first step in calculating phase equilibria. But after that is achieved, a reasonable extrapolation may yield insight into the variation of phase equilibrium with pressure and temperature. As an example, phase equilibrium in the ternary system $CO_2 - H_2O - $ n-hexadecane calculated at a constant temperature of 200 °C for different pressures is shown in Fig. 3.65. The only experimental data used are that at 200 bar. Even in such a complicated system, phase equilibrium can be predicted, at least qualitatively.

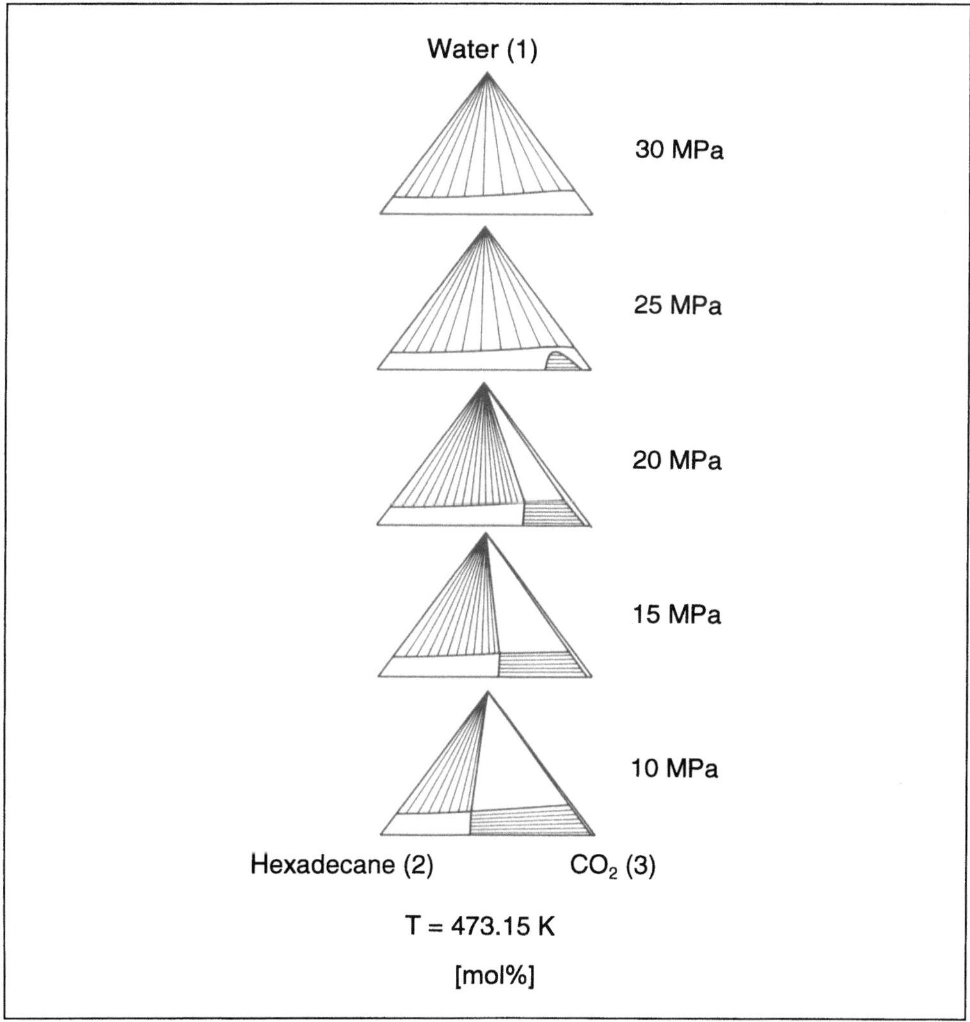


**Fig. 3.65.** Phase equilibrium in the ternary system $CO_2 - H_2O - $ n-hexadecane. $T = 473.15$ K; Peng-Robinson EOS (after Teich [61]). Data from Dohrn and Brunner [15].

## References

1. Andrews T (1869) On the continuity of the gaseous and liquid states of matter. Phil Trans 159: 575; Proc Roy Soc (London) 18: 42 – 45
2. Bharath R, Inomata H, Arai K, Shji K, Noguchi Y (1989) Vapor-liquid equilibria for binary mixtures of carbon dioxide and fatty acid ethyl esters. Fluid Phase Equilibria 50: 315 – 327
3. Brunner E (1990) Fluid mixtures at high pressures. IX. Phase separation and critical phenomena in 23 (n-alkane + water) mixtures. J Chem Thermodynamics 22: 335 – 353
4. Brunner G (1978) Phasengleichgewichte in Anwesenheit komprimierter Gase und ihre Bedeutung bei der Trennung schwerflüchtiger Stoffe. Habilitationsschrift, Universität Erlangen-Nürnberg
5. Brunner G, Peter S (1979) Die Trennung schwerflüchtiger Stoffe mit Hilfe komprimierter Gase in Anwesenheit von Schleppmitteln. Ber Bunsenges Phys Chem 83 : 1137
6. Brunner G, Peter S (1982) On the solubility of glycerides and fatty acids in compressed gases in the presence of an entrainer. Separation Science and Technology 17 (1): 199 – 214
7. Brunner G (1983) Selectivity of supercritical compounds and entrainers with respect to model substances. Fluid Phase Equilibria 10: 289 – 298
8. Brunner G (1986) Anwendungsmöglichkeiten der Gasextraktion im Bereich der Fette und Öle. Fette Seifen Anstrichmittel 88 (12): 464 – 474
9. Bünz A, Dohrn R (1992) Berechnung des Gas-flüssig-Phasengleichgewichtes im binären System $CO_2$-Eicosapentansäure-Ethylester. Arbeitsbericht, Technische Universität Hamburg-Harburg
10. Czech B (1991) Das Betriebsverhalten einer Gegenstromkolonne bei der Erzeugung von Diglyceriden mit Hilfe der nahekritischen Extraktion. Dissertation, Universität Erlangen-Nürnberg
11. Cotterman RL, Dimitrelis D, Prausnitz JM (1985) Design of Supercritical fluid extraction processes using continuous thermodynamics. In: Penninger JML, Radosz M, McHugh MA, Krukonis VJ (eds) Supercritical Fluid Technology. Elsevier, Amsterdam, pp 107 – 120
12. Cotterman RL, Prausnitz JM (1991) Continuous thermodynamic for phase-equilibrium calculations in chemical process design. In: Sandler SI, Astrita G (eds) Kinetic and Thermodynamic Lumping of Multicomponent Mixtures. Elsevier, Amsterdam, pp 229 – 275
13. Deiters (1993) private communication
14. De la Tour C (1822) Ann Chim (3) 21: 127 – 132
15. Dohrn R, Brunner G (1986) Phase equilibria in ternary and quaternary systems of hydrogen, water und hydrocarbons at elevated temperatures and pressures. Fluid Phase Equilibria 29: 535 – 544
16. Dohrn R (1994) Berechnung von Phasengleichgewichten mit Hilfe von Zustandsgleichungen. Habilitationsschrift, Technische Universität Hamburg-Harburg
17. Donelly HG, Katz DL (1954) Phase equilibria in the carbon dioxide-methane system. Ind Eng Chem 46: 511 – 517
18. Ebeling H, Franck EU (1984) Spectroscopic determination of caffeine solubility in supercritical carbon dioxide. Ber Bunsenges Phys Chem 88: 862 – 865
19. Ender U (1989) Der Einfluß der Druckpulsation auf die überkritische Fluidextraktion von Monoglyceriden. Dissertation, Universität Erlangen-Nürnberg
20. Ewald AH (1953) The solubility of solids in gases. Trans Faraday Soc 49: 1401
21. Ewald AH (1955) The solubility of solids in gases. Trans Faraday Soc 51: 347
22. Franck EU (1956) Zur Löslichkeit fester Stoffe in verdichteten Gasen. Z Phys Chem NF 6: 345 – 355
23. Franck EU (1971) Gas-Liquid and gas-solid equilibria at high pressures, critical curves and miscibility gaps. In: Jost W (ed) Physical Chemistry, vol I/Thermodynamics, New York
24. Gährs HJ (1984) Application of atmospheric gases in high pressure extraction. Ber Bunsenges Phys Chem 88: 894 – 897
25. Gaver D van (1992) Fractionatie van vetzuuresters met supercritische extractie. Dissertation, Universiteit Gent
26. Gutsche B (1986) Phase equilibria in oleochemical industry. Application of continuous thermodynamics. Fluid Phase Equilibria 30: 65
27. Hannay JB, Hogarth J (1879) On the solubility of solids in gases (preliminary notice). Chem News 40: 256, Proc Roy Soc (London) 29: 324
28. Hederer H, Wolff A (1991) Korrelation von binären Wechselwirkungsparametern mit den Reinstoffparametern einer kubischen Zustandsgleichung. Chem Ing Tech 63: 618 – 621


144

29. Hölscher IF, Spee M, Schneider GM (1989) Fluid phase equilibria of binary and ternary mixtures of $CO_2$ with hexadecane, 1-dodecanol, 1-hexadecanol and ethoxy-ethanol at 333.2 and 393.2 K and at pressures up to 33 MPa. Fluid Phase Equilibria 49: 103 – 113

30. Johannsen M, Brunner G (1994) Solubilities of the xanthines caffeine, theophylline and theobromine in supercritical carbon dioxide. Fluid Phase Equilibria, in print

31. Kay WB (1938) Liquid-vapor phase equilibrium. Relations in the ethane – n-heptane system. Ind Eng Chem 30: 459 – 465

32. Kay WB (1968) The critical locus curve and the phase behavior of mixtures. Accounts of chemical research 1: 344 – 351

33. Kehlen H, Rätzsch MT (1983) Liquid-liquid phase separation in polymer systems and polymer compatibility by continuous thermodynamics. Z phys Chem Leipzig 264: 1153

34. Kennedy GC (1950) A portion of the system silica-water, Econ Geol 45: 629 – 653

35. King MB (1969) Phase Equilibria in Mixtures. Pergamon Press

36. Koningsveld R, Diepen GAM (1983) Supercritical phase equilibria involving solids. Fluid Phase Equilibria 10: 159 – 172

37. Konynenberg PH van, Scott RL (1980) Critical lines and phase equilibria in binary van der Waals mixtures. Phil Trans Roy Soc 298: 495 – 540

38. Lehmann R (1992) Korrelation von Meßdaten für Phasengleichgewichte des Systems $CO_2$-$\alpha$-Tocopherol im Bereich hoher Drücke. Studienarbeit, Technische Universität Hamburg-Harburg

39. Lentz H, Franck EU (1978) Phase equilibria and critical curves of binary ammonia-hydrocarbon mixtures. Angew Chem Int Ed Engl 17: 728 – 730

40. Mitra S, Wilson NK (1991) An Empirical Method to Predict Solubility in Supercritical Fluids. J Chromatographic Science 29, July: 305 – 309

41. Orbey H (1994) Mixing rules for the estimation of vapor-liquid equilibrium of highly non-ideal mixtures using cubic equations of state. In: Kiran E, Levelt-Sengers JMH (eds) Supercritical Fluids – Fundamentals for Application. NATO ASI, Kluwer

42. Pereira PJ, Goncalves M, Coto B, Azevedo EG de, Ponte MN da (1993) Fluid Phase Equilibria 91: 133

43. Peter S, Brunner G, Riha R (1973) Zur Trennung schwerflüchtiger Stoffe mit Hilfe fluider Phasen. DECHEMA-Monographie 73: 197 – 206

44. Peter S, Brunner G, Riha R (1976) Zur Abtrennung des Monoglycerids der Ölsäure aus einem Glyceridgemisch mit Hilfe von komprimiertem Kohlendioxid in Gegenstromkolonnen. Fette Seifen Anstrichmittel 78: 45 – 60

45. Peter S, Schneider M, Weidner E, Ziegelitz R (1986) Die Trennung von Lecithin und Sojaöl in einer Gegenstromkolonne mit Hilfe eines überkritischen Extraktionsmittels. VDI-Berichte 607: 851 – 868

46. Poynting JH (1881) Change of State: Solid-Liquid. The London, Endinburgh and Dublin Phil Mag 4, 12: 32

47. Prausnitz JM, Joshi DK (1984) Supercritical fluid extraction with mixed solvents. AIChE J 30: 522 – 525

48. Rowlinson JS, Swinton FL (1982) Liquids and liquid mixtures, 3rd edn. Butterworths, London

49. Sadus RJ (1992) High pressure phase behaviour of multcomponent fluid mixtures. Elsevier, Amsterdam – London – New York – Tokyo

50. Sandler S, (1994) Equations of state for phase equilibrium computations. In: Kiran E, Levelt-Sengers JMH (eds) Supercritical Fluids – Fundamentals for Application. NATO ASI, Kluwer

51. Schlichting H (1991) Experimentelle Bestimmung und Korrelierung der Löslichkeit verschiedener Lösungsmittel in Hochdruckgasen. Dissertation, Technische Universität Berlin

52. Schmitt WJ (1984) Doctoral Thesis, Dept Chem Eng Mass Inst Tech, Cambridge MA

53. Schmoll A (1993) Berechnung von Phasengleichgewichten von Palmöl und Schleppmitteln mit überkritischem Kohlendioxid. Studienarbeit, Technische Universität Hamburg-Harburg

54. Schneider GM (1978) Physicochemical principles of extraction with supercritical gases. Angew Chem Int Ed Engl 17: 716 – 727

55. Schneider GM (1983) Physicochemical aspects of fluid extraction. Fluid Phase Equilibria 10: 141 – 157

56. Schneider GM (1988) Thermodynamics of fluid mixtures at high pressures. Basis of supercritical fluid technology. In: Perrut M (ed) Proc Int Symp Supercritical Fluids, Nice, vol1: 1 – 17

57. Smits A (1905) Über Erscheinungen, welche auftreten, wenn die Faltenpunktskurve der Löslichkeitskurve begegnet. Z phys Chem 51: 193 – 221.

58. Smits A (1905) Über Erscheinungen, welche auftreten, wenn bei binären Gemischen die Faltenpunktskurve der Löslichkeitskurve begegnet. Z phys Chem 52: 587 – 601

59. Stahl E, Glatz A (1984) Extraction of natural substances with supercritical gases, 10. Communication: Qualitative and quantitative determination of solubilities of steroids in supercritical carbon dioxide. Fette, Seifen Anstrichm 86: 346 – 348

60. Sunol AK, Hagh B, Chen S (1985) Entrainer selection in supercritical extraction. In: Penninger JML, Radosz M, McHugh MA, Krukonis VJ (eds) Supercritical Fluid Technology. Elsevier, Amsterdam, pp. 451 – 464

61. Teich J (1992) Phasengleichgewichte bei erhöhten Drücken und Temperaturen in ternären Systemen aus Wasserstoff, Kohlendioxid, Wasser und Kohlenwasserstoffen. Dissertation, Technische Universität Hamburg-Harburg

62. Todd DB, Elgin JC (1955) Phase equilibria in systems with ethylene above its critical temperature. AIChE J 1: 20 – 27

63. Tsekhanskaya Yu (1964) Solubility of naphthalene in ethylene and carbon dioxide under pressure. Russ J Phys Chem (Engl Transl) 38: 1173

64. Villard MP (1896) Dissolution des liquides et des solides dans les gaz. Journ de Phys theor et appl 5: 453 – 561

65. Walter D (1992) Messung und Korrelation von Hochdruck-Dampf-Flüssigkeits-Gleichgewichten in binären Mischungen aus Kohlendioxid und Benzolderivaten bei Temperaturen von 313 K bis 393 K und Drücken bis 22 MPa. Dissertation, Universität Kaiserslautern

66. Wong DSH, Sandler S, (1992) A theoretically correct mixing rule for cubic equations of state. AIChE J 38: 671 – 680

# 4 Heat and Mass Transfer

In this chapter an introduction is given to heat and mass transfer relevant to gas extraction. Heat and mass transfer data are best presented as dimensionless-group correlations for heat and mass transfer coefficients. Some are presented and, as far as possible, compared to experimental data in gas extraction. A special section is dedicated to the treatment of heat transfer during boiling under elevated pressures.

Kinetic processes, like heat and mass transfer, not only depend on conditions of state and the variation of component properties with location and time, but, in general, also on hydrodynamics and the construction of the mass or heat transferring equipment. Only in case of non-convective transfer, in conductive transport of heat and in diffusive transport of matter, is the resulting flow determined by conditions of state, material properties, and the field potential of the flow- inducing variable.

## 4.1 Heat Transfer

In gas extraction heat transfer may not be considered as being predominant, since heat transfer with media under pressure is not at all new. Therefore, only basic principles of heat transfer in combination with the effect of elevated pressures and under near critical conditions are discussed.

### 4.1.1 Conductive Heat Transfer

Heat transfer by conduction is determined by a temperature gradient and the heat conductivity of the material through which heat is transferred. Conductive heat transfer in general is three dimensional and heat conductivity varies with conditions of state and the properties of the material.

For a comprehensive treatment of equations and phenomena of conductive heat transfer the reader is referred to textbooks on heat transfer. In gas extraction, fluids in most cases can be considered as being isotropic with respect to heat conductivity. Solids, where heat conductivity is more likely to depend on direction and location, may also be considered as being isotropic. If not, the reader again is referred to textbooks on heat transfer, since heat transfer to or from solids is not a major problem in gas extraction (contrary to mass transfer, as will be discussed later in Section 4.2

and Chapter 7). If heat transfer by conduction can be considered as being prevalent in one direction, a one-dimensional relation between the amount of heat $q$ transferred through area $F$ and the temperature gradient can represent heat transfer:

$$q = \lambda\, F \frac{\mathrm{d}T}{\mathrm{d}x}, \qquad\qquad 4.1$$

with
$q$ = amount of heat transferred per unit of time;
$\lambda$ = heat conductivity;
$F$ = area through which heat is transferred vertical to coordinate $x$;
$x$ = coordinate in direction of temperature gradient;
$T$ = temperature.

The effect of elevated pressure can be taken into account by the dependence of heat conductivity on pressure, $\lambda = \lambda(P)$, as discussed in Section 2.4.2 and presented in Figs. 2.26 to 2.30.


## 4.1.2  Heat Transfer by Convection

Heat transfer by convection is the transfer of heat from a medium through a surface, along which a fluid flow takes up or delivers the amount of heat transferred. A linear relation between the amount of heat transferred and the temperature gradient is sufficient for representing the associated phenomena. It is assumed for the purpose of treating heat transfer in gas extraction that cross-effects, as can be derived from an analysis by non-equilibrium thermodynamics, are neglected. For a more detailed discussion, the reader is referred to textbooks on heat transfer and on irreversible thermodynamics.

The amount of heat $q$ transferred by convective heat transfer per unit of time is represented by the following linear relation between heat flow and temperature gradient, also known as Newton's law for heat transfer:

$$q = \alpha\, F\,(T_1 - T_2), \qquad\qquad 4.2$$

with
$\alpha$ = heat transfer coefficient, defined by the above equation;
$F$ = area through which heat is transferred vertical to the temperature gradient;
$T_1, T_2$ = temperatures at different locations between which $q$ is transferred.

The heat transfer coefficient $\alpha$ depends on conditions of state, on material properties, and on impulse transport, which again is determined by hydrodynamics and the construction of the heat transfer equipment or by the shape of the heat transferring surface. The heat transfer coefficient is, in principle, also dependent on simultaneous mass transfer. But as stated above, this interconnection will be neglected. Heat transfer coefficients are calculated from semi-empirical equations, containing dimensionless groups of parameters, which reduce the number of independent variables. Experimental data, as far as possible, are presented in such dimensionless equations to enable transfer to other applications. Examples for basic equations for heat transfer at forced convection and free convection are given in Eqs. 4.3 and 4.4. These equations may be modified, i.a., due to different flow regimes, geometry, and type of the components of a mixture.

**Forced convection:**

$$Nu = \frac{\alpha\, d}{\lambda} = C\, Re^m\, Pr^n. \qquad\qquad 4.3$$

**Free convection:**

$$Nu = \frac{\alpha\, d}{\lambda} = D\, (Gr\, Pr)^o, \qquad\qquad 4.4$$

with
$Nu$    = Nusselt number;
$Re$    = Reynolds number, $Re = (ud)/\nu$;
$Pr$    = Prandtl number, $Pr = \nu/D$;
$Gr$    = Grashoff number, $Gr = (\beta_T \Delta Tgl^3)/\nu^2$;
$d$    = characteristic length, e.g., diameter of a heat transferring pipe;
$l$    = characteristic length, e.g., length in direction of flow;
$u$    = flow velocity;
$\nu$    = viscosity;
$D$    = diffusion coefficient;
$\beta_T$    = thermal expansion coefficient;
$g$    = accelaration due to gravity;
$m, n, o$ = exponents fitted to experimental data.

The dimensionless correlations are determined from experimental data on different mixtures, at different conditions of state, and in differently constructed devices. They represent experimental data within reasonable limits of error, and may only be applied in certain regions of the parameters, which usually are specified. The accuracy is lower for transport coefficients than for thermodynamic equilibrium properties; for heat transfer coefficients, as a rule of thumb, the accuracy is about $5-10\,\%$. If dimensionless correlations for heat transfer coefficients are applicable under conditions of gas extraction, then heat transfer coefficients can be determined from these correlations and the influence of parameter variations may be derived also from these correlations. While the influence of pressure on heat transfer in gases or during boiling of liquids has been investigated, heat transfer coefficients for mixtures, typical for gas extraction, are rare.

Colburn [7] has reported experimental results on heat transfer coefficients in the system hydrogen – nitrogen (3:1) at pressures between 3 and 90 MPa and at Reynolds-numbers $4 \cdot 10^4 < Re < 4.4 \cdot 10^5$. The results can be represented by one line in a log-log diagram. Results of measurements of heat transfer coefficients in air at 0.7 MPa can be also represented by this line, if they are reduced with heat capacity $c_p$. This is shown in Fig. 4.1.

Free convection is fluid flow, induced by density gradients, e.g., due to temperature gradients. In gas extraction the supercritical solvent is subject to density variations with only slight changes in pressure and temperature. Furthermore, flow velocities within the processing equipment are low, so that flow due to free convection may be important. Therefore, conditions for free convective flow must be considered in such type of systems. According to Kraussold [15], the criterion for free convection

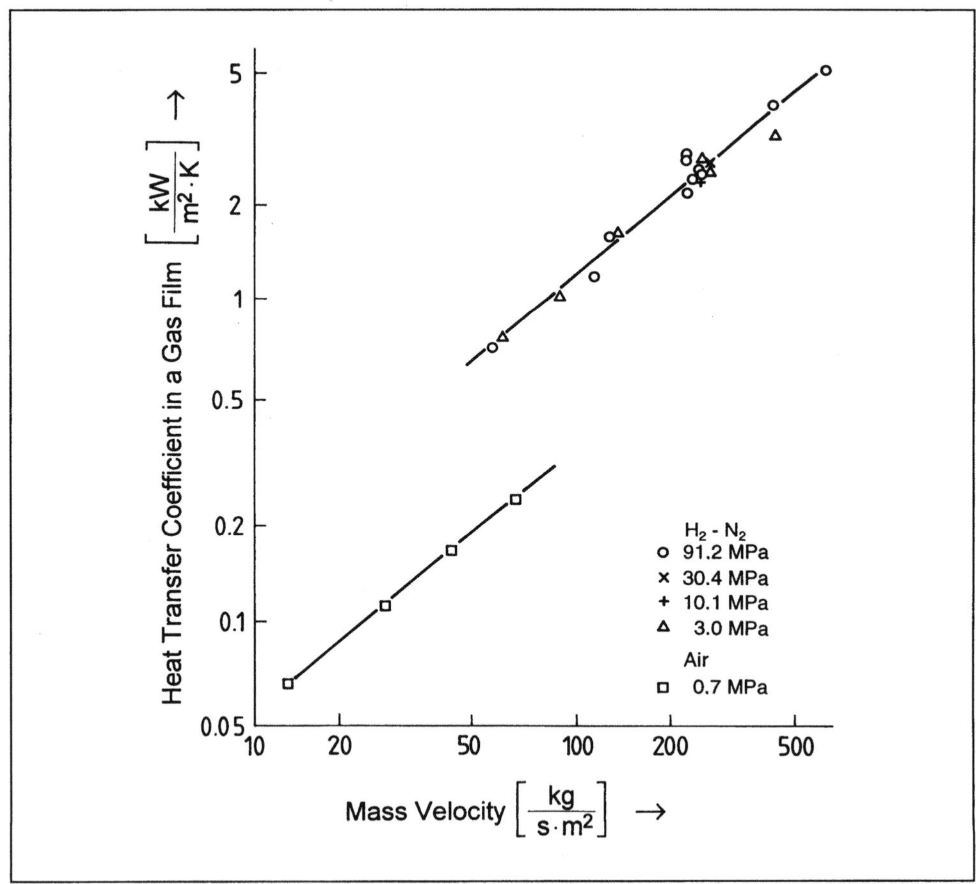



**Fig. 4.1.** Heat transfer coefficients in gases at elevated pressures [7].

150

is that the product of the Grashoff-number and the Prandtl-number must be greater than a certain limit, which was found to be 600:

$$Gr \cdot Pr > 600. \qquad\qquad 4.5$$

The limiting condition for free convective flow was confirmed by the experimental results of various authors. At higher pressures, free convective flow is more easily established than at low pressures. Lenoir and Comings [16] found that for carbon dioxide, free convective flow can be induced at 38 °C with a temperature difference of only 3 K. Figure 4.2 shows data on $(Gr \cdot Pr)$ for $CO_2$ and $N_2$ at 38 °C. The tendency for free convective flow is clearly more pronounced for carbon dioxide, which is near its critical point. Further information on heat transfer during natural convection can be found in common textbooks.

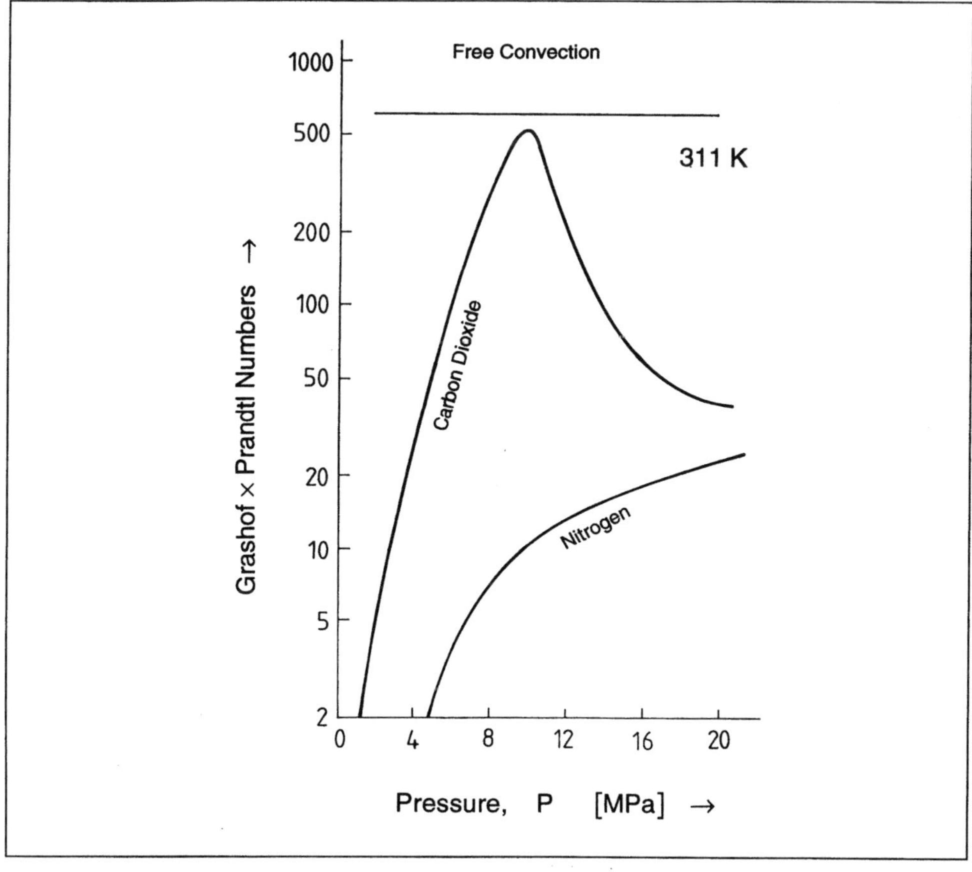


**Fig. 4.2.** The term *(GrPr)* during heat transfer at elevated pressures for $CO_2$ and $N_2$ (after Lenoir and Comings [16]).

## 4.1.3 Boiling

During boiling of a liquid, heat flux, as a function of the temperature difference between heat transferring surface and heat absorbing liquid, varies in a characteristic way, as schematically plotted in Fig. 4.3. The first region, at low temperature differences, is subcooled boiling. Free convection, due to the density differences in the liquid, combines with local evaporation. Gas bubbles formed on the heat transferring surface condense again as they travel through the subcooled bulk liquid (i.e., liquid below the boiling temperature). At higher temperature differences, bubbles are formed and leave the liquid phase to form the gaseous phase. This region is the "normal" boiling range. With increasing temperature difference, the frequency of bubble formation increases and distances between bubbles on the heat transferring surface get smaller and smaller. Heat flux increases with increasing temperature difference up to a maximum. At even higher temperature differences, heat flux decreases, due to the effect that emerging bubbles increasingly combine at the heat transferring surface, forming areas of gaseous phase, where heat transfer is lower due to reduced density of the gaseous phase. Heat flux therefore decreases. Eventually a coherent

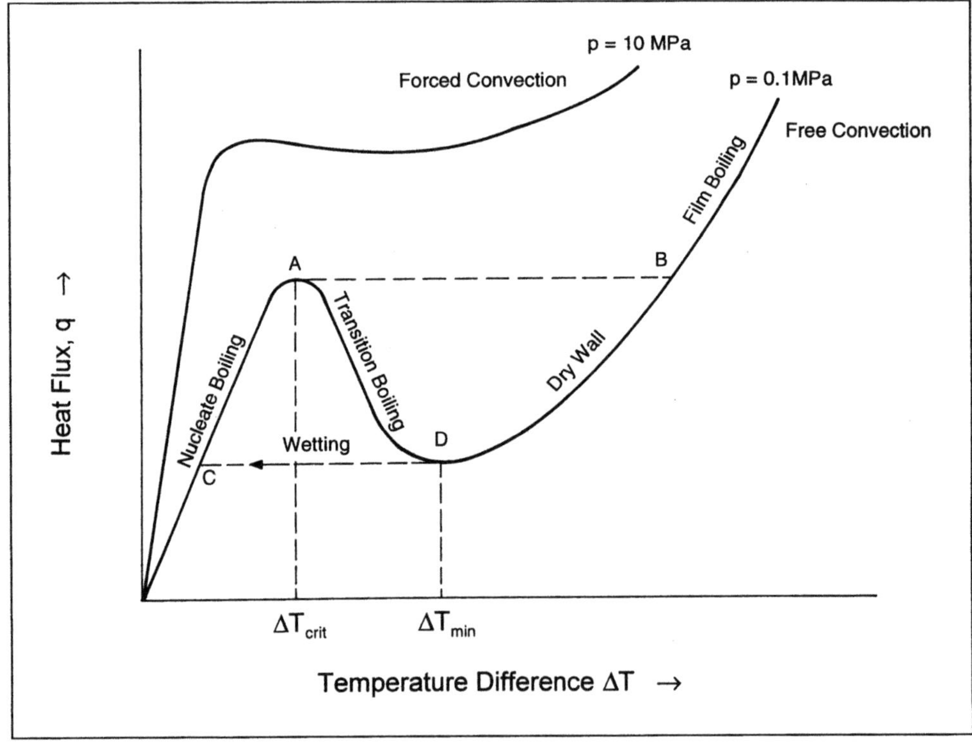


**Fig. 4.3.** Heat flux during boiling. $\Delta T$: temperature difference between heat transferring surface and heat absorbing liquid [19].

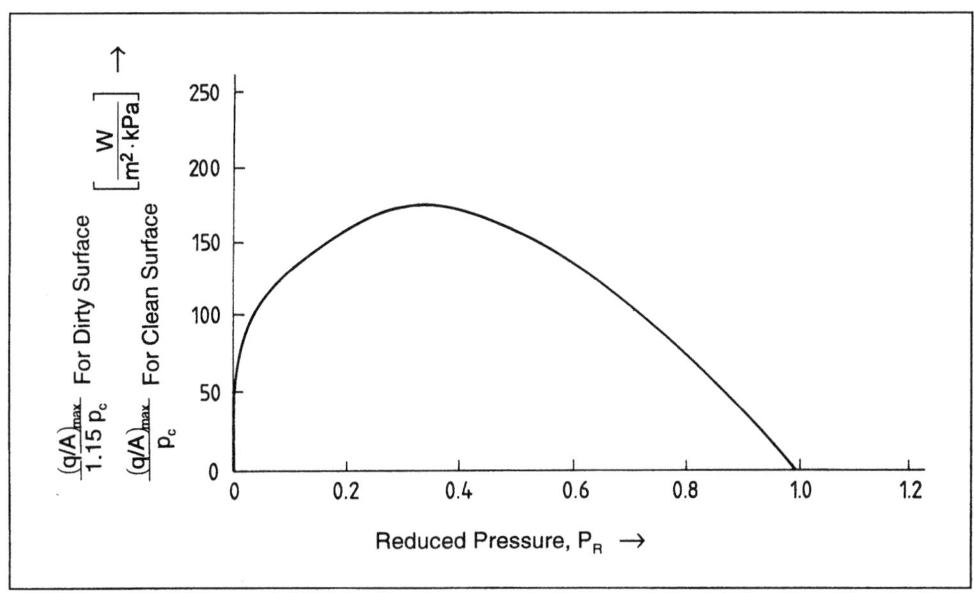

**Fig. 4.4.** Maximum heat transfer during boiling [6].

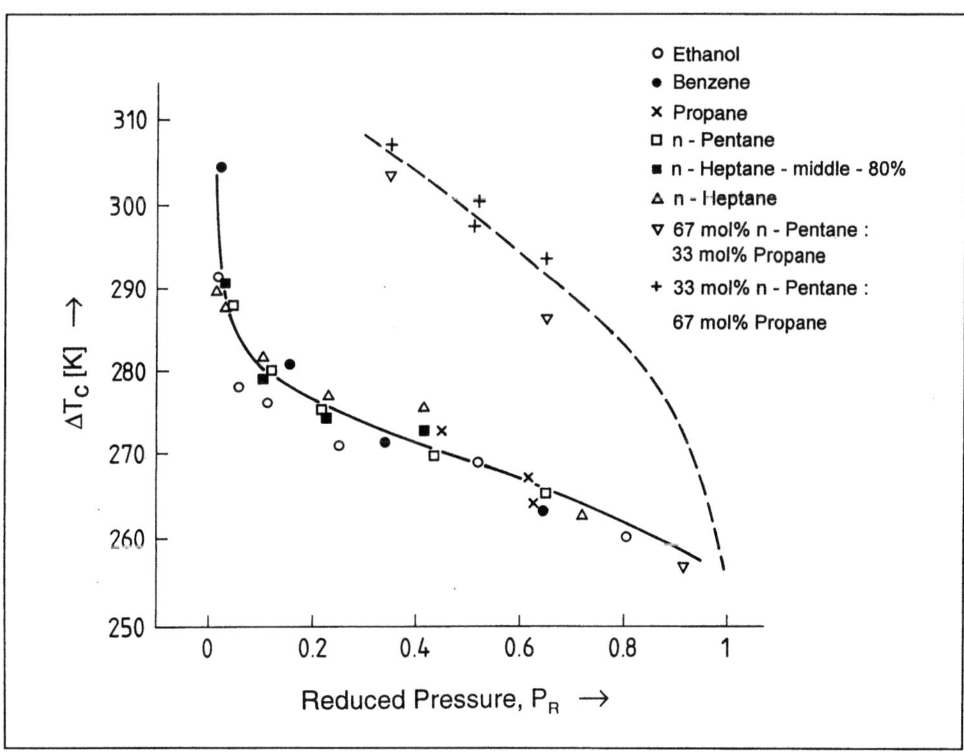



**Fig. 4.5.** Critical temperature difference $\Delta T_c$ at maximal heat flux for liquids and binary mixtures [6].

153

gaseous phase is formed on the heat transferring surface. Boiling under these conditions is called "film boiling". Since heat transfer in film boiling is considerably lower than in bubble boiling, the maximum of heat transfer is of some importance. This maximum is pressure dependent and is shifted with increasing pressure to smaller temperature differences between heating surface and liquid. Maximum heat flux has its highest values at a reduced pressure of about 0.35 (Fig. 4.4). In binary systems the temperature difference at maximal heat flux is much higher (Fig. 4.5).

The influence of pressure on heat transfer during boiling has been investigated and correlations have been established. For bubble boiling the following relations have been found:

$$\alpha/\alpha_{0.3} = 0.175 + 2.75\,(P/P_c), \text{ after Danilova [9];} \tag{4.6}$$

$$\alpha/\alpha_{0.3} = 0.175 + (2.02 + 0.51)/(1 - (P/P_c))(P/P_c), \text{ after Haffner [11],} \tag{4.7}$$

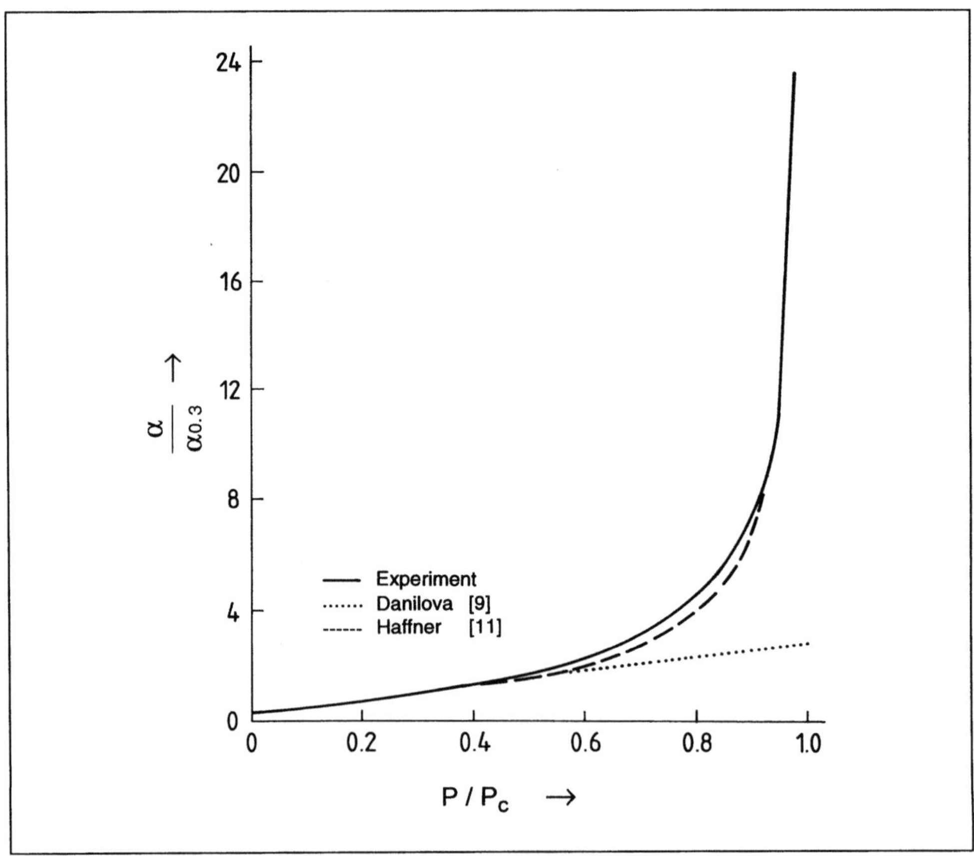


Fig. 4.6. Reduced heat transfer coefficient as a function of reduced pressure [2].

154

with:

$\alpha$ = heat transfer coefficient;

$\alpha_{0.3}$ = $\alpha$ at $P/P_c = 0.3$;

$P$ = saturation pressure at boiling conditions;

$P_c$ = critical pressure.

The reduced heat transfer coefficient for some refrigerants during pool boiling at saturation pressures is plotted in Fig. 4.6 as a function of reduced pressure [2]. Experimental results for refrigerants can be reproduced up to a reduced pressure of about 0.5 by the correlation of Danilova, and up to a reduced pressure of about 0.9 by the correlation of Haffner. Heat transfer coefficients increase with pressure. For $P_r >$ 0.8, influence of pressure is strong. Maximum heat flux on the other hand decreases rapidly with pressure, therefore, the high heat transfer coefficients near the critical point of the boiling liquid are of limited practical value.

## 4.1.4 Dimensionless Correlations for Heat Transfer Including Boiling

For heat transfer, including boiling, various dimensionless correlations have been published [12]. Some of them are listed below and compared to experimental results.

For **natural convection:**

$$Nu_d = \frac{\alpha_m\, d}{\lambda} = 0.5 \sqrt[4]{Pr\, Gr} \sqrt[4]{Pr/Pr_w}, \text{ after Michejew [20],} \qquad 4.8$$

with:

$\alpha_m$ = mean heat transfer coefficient;

$Nu$ = Nusselt number;

$Pr$ = Prandtl number, $Pr = v/D$;

$Pr_w$ = Prandtl number at the heat transferring surface;

$Gr$ = Grashoff number, $Gr = (\beta_T\, \Delta T g l^3)/v^2$;

$d$ = characteristic length, e.g., diameter of a tube;

$l$ = characteristic length, e.g., length in direction of flow;

$v$ = viscosity;

$\lambda$ = heat conductivity;

$D$ = Diffusion coefficient;

$\beta_T$ = thermal expansion coefficient;

$g$ = accelaration due to gravity;

$\Delta T$ = temperature difference between heat transferring surface and bulk of the fluid.

Experimental values for $CO_2$ compare well with the results of the correlation for $0.5 < GrPr < 5 \cdot 10^3$.

For **bubble boiling**, after Vaihinger and Kaufmann [27], and Müller [21]:

$$Nu = \frac{\alpha\, d_A}{\lambda_L} = 0.078\, K_1^{0.62}\, K_2^{0.133}\, K_3^{0.234}\, Re^{0.078}\, Pr^{1.032},$$

(4.9)

with:
$K_1, K_2, K_3$ = dimensionless coefficients;
$d_A$ = bubble diameter.

Experimental results for $CO_2$ scatter appreciably around the correlation.

For **bubble boiling**, after Stephan [26] and Fedders [10], taking into account bubble formation frequency and the condition of the surface:

$$Nu = \frac{\alpha\, d_A}{\lambda_L} = 10^{-5} \left(\frac{q\, d_A}{T_s\, \lambda_L}\right)^{0.7} \left(\frac{d_A T_s \lambda_L}{\nu_L\, \sigma}\right)^{0.8} \left(\frac{R_p\, \varrho_V\, \Delta h_V}{d_A\, \varrho_L\, (f\, d_A)^2}\right),$$

(4.10)

with:

$$f\, d_A = 0.59 \left(\frac{g\, \sigma\, (\varrho_L - \varrho_V)}{\varrho_L^2}\right)^{1/4},$$

(4.11)

$f$ = frequency of bubble formation;
$q$ = heat flux density;
$R_p$ = surface roughness ($d = 0.3$ mm; $R_p \approx 0.3$ μm);
Index $L$ = liquid phase;
Index $V$ = gaseous phase;
$d_A$ = bubble diameter;
$T_s$ = boiling temperature;
$\sigma$ = surface tension;
$\varrho$ = density;
$\Delta h_V$ = enthalpy of vaporization;
$g$ = acceleration due to gravity.

For **film boiling**, including **radiation** (after Pitschmann and Grigull [23]):

$$Nu_c = \frac{\alpha\, d}{\lambda_V}\quad C = 0.9\, Ra_c^{0.08} + 0.8\, Ra_c^{0.2} + 0.02\, Ra_c^{0.4},$$

(4.12)

with:
$Ra_c = C\, Ra = C\, (\varrho_L - \varrho_V)g(h_V - h_L)d^3/(\lambda_V\, \eta_V\, \Delta T);$

(4.13)

$$C = 1 - (q_{rad}/q);$$ 4.14

$q_{rad}$ = heat flux according to radiation;

$q$ = $Q/(d \pi L)$: average density of heat flux.

Good agreement was obtained between experimental and calculated results for $10^{-5} < CRa < 10^{10}$.

For **film boiling** after Nishikawa [22]:

$$Y = 0.22 + 0.15\, X + 0.0058\, X^2;$$ 4.15

$$Y = \log Nu_c, \text{ with } Nu_c = (\alpha dC)/\lambda_V;$$ 4.16

$$X = (\log Gr)\, [Pr_N^2/(Pr_N + 1.33)]$$ 4.17

with:

$$Gr = \frac{\varrho_V(\varrho_L - \varrho_V)\, g d^3}{\eta_V^2};$$ 4.18

$$Pr_N = (\nu_V/a_V)\, (1 + 3.33\, \Delta h_V)/(c_{pV}\, \Delta T);$$ 4.19

$C$ = as in Eq. 4.14;

$a_V$ = thermal diffusivity;

$\eta$ = dynamic viscosity ($\eta = \nu \cdot \varrho$).

These correlations represent experimental values for $CO_2$ with sufficient accuracy.

For the maximum heat flux and the minimum heat flux correlations have also been established. For the **maximum heat flux** after Addoms [1]:

$$q_{max} = 2.4\, \Delta h_V \varrho_V \left(\frac{\varrho_L - \varrho_V}{\varrho_V}\right)^{1/2} \left(\frac{g\, \lambda_L}{\varrho_L\, c_{pL}}\right)^{1/3},$$ 4.20

and for the **minimum heat flux** $q_{min}$, according to Zuber [30]:

$$q_{min} = \frac{\pi^2}{60} \left(\frac{4}{3}\right)^{1/4} \Delta h_V \varrho_V \left(\frac{\sigma g\, (\varrho_L - \varrho_V)}{(\varrho_L + \varrho_V)^2}\right)^{1/4}.$$ 4.21

Heat transfer coefficients during boiling of a binary mixture depend on concentration. Deviations increase from linearly interpolated values with increasing pressure. The effect of a second substance on heat transfer coefficients is likely to increase with difference in boiling points of the substances. For mixtures of propane and carbon dioxide Böhm et al. [3] have published some data which seem to indicate that experimentally obtained values differ appreciably from that obtained from known correlations.

### 4.1.5 Reductive Parameters for Dimensionless Correlations

Dimensionless groups are characteristic for a certain phenomenon. The Reynolds-number, $Re = (ul)/\nu$, given by the product of flow velocity $u$ and a characteristic length $l$ (e.g., the diameter of the pipe in which the mixture flows), divided by the viscosity $\nu$, characterizes fluid flow phenomena. Fluid flow of two different mixtures is physically similar, with respect to hydrodynamics, if $Re$-numbers are equal:

$$Re_I = Re_{II} \quad \text{(Condition for similarity of fluid flow).} \qquad 4.22$$

Using this relation for physical similarity, scale up, or comparison of flow, for two different mixtures is possible.

Thermodynamic similarity, i.e., similarity of conditions of state for a compound or a mixture, can be represented by the theory of corresponding states (Eqs. 2.2 and 2.3). This thermodynamic relation may also be applied to transport properties, especially if the transport properties only depend on the nature of the substance and conditions of state. In other cases (e.g., non-Newtonian fluids) this can only be a first approximation and applicability has to be established for each case. As reducing variable all state variables can be applied, but do not have the same effect. The relation between volumes of the gaseous and the liquid phase may be correlated with reduced pressure, but not with reduced temperature [19].

Viscosity $\nu$, heat conductivity $\lambda$, surface tension $\sigma$ and enthalpy of evaporation $\Delta h_V$ are material properties which determine certain processes. It may be useful to transfer data from one mixture to another mixture. Mayinger [19] has shown that, although absolute values for two different components differ considerably, the values can be correlated by a factor which must be individually determined for the compounds under consideration, but is constant for the whole region of conditions of state. For scale up or transfer to other media, this factor must be taken into account.

Scaling factors for different phases are different. Therefore, multiphase processes, in principle, require more than one scaling factor. This leads to different scaling factors for each phase. Obviously, for a two-phase flow this is not practicable and therefore a decision must be made about which scaling factor to use. For fluid flow the scaling factor for the phase is chosen which determines fluid dynamics; the influence of the other phase is neglected. Therefore, in complex systems, primary and secondary parameters for scale up have to be defined.

# 4.2 Mass Transfer

## 4.2.1 Influence of Pressure and Gravity

Mass transfer, as heat transfer, depends on conditions of state. For gas extraction, especially the influence of pressure is important. From absorption processes it is known that the product of mass transfer coefficient $k$ and pressure at constant mass flow of gas is constant.

$$(k \cdot P)_G = \text{const.} \tag{4.23}$$

For conditions of forced convection a variety of correlations has been proposed. Yet for Re > 10, the form given by Eq. 4.24 which is analogous to heat transfer, seems to be adequate [17].

$$Sh = C_1 \, Re^{1/2} \, Sc^{1/3}, \tag{4.24}$$

with
$Sh$ = Sherwood number, defined as $Sh = (\beta l)/D$;
$Re$ = Reynolds number, $Re = (ul)/v$;
$Sc$ = Schmidt number, $Sc = v/D$;
$\beta$ = mass transfer coefficient;
$C_1$ = coefficient depending on geometry and configuration of flow;
$l$ = characteristic length;
$u$ = flow velocity;
$v$ = viscosity;
$D$ = diffusion coefficient.

Natural convection can occur easily in gas extraction systems since linear flow velocity is low (in the range of 1 to 20 mm/s), viscosity is low, and direction of flow may be against the force of gravity. For natural convection the influence of $Re$-number disappears, and the general expression for the dimensionless mass transfer coefficient is given by Eqs. 4.25 and 4.26:

$$Sh = C_2 \, (Sc \, Gr)^{1/4} \quad \text{for laminar natural convection} \tag{4.25}$$

$$Sh = C_3 \, (Sc \, Gr)^{1/3} \quad \text{for turbulent natural convection,} \tag{4.26}$$

with
$Sh$ = Sherwood number, defined as $Sh = (\beta l)/D$;
$Gr$ = Grashoff number, $Gr = (\beta_T \Delta T g l^3)/v^2$;
$Sc$ = Schmidt number, $Sc = v/D$;
$C_2$ = coefficient depending on geometry and configuration of flow;
$C_3$ = coefficient depending on geometry and configuration of flow;
$l$ = characteristic length, e.g., length in direction of flow;
$u$ = flow velocity;
$v$ = viscosity;
$D$ = diffusion coefficient;
$\beta_T$ = thermal expansion coefficient;
$g$ = acceleration due to gravity.

Values for the coefficients $C_2$ and $C_3$ have been reported in the literature; $C_2 \approx 0.5$, and $C_3 \approx 0.1$, but have to be determined from individual experimental data. If natural convection is dominant, then the mass transfer coefficients may be calculated from Eqs. 4.25 and 4.26.

In an intermediate region, forced and natural convection both are of influence on the mass transfer process. Then, Grashof number and Reynolds number must be taken into consideration. The resulting Sherwood number for the intermediate region (index $I$) is calculated from Sherwood number for forced convection (index $F$) and Sherwood number for natural convection (index $N$) [17] for assisting flow:

$$Sh_I^3 = Sh_N^3 + Sh_F^3, \tag{4.27}$$

and for opposing flow

$$Sh_I^3 = Sh_N^3 - Sh_F^3. \tag{4.28}$$

Equations 4.24 to 4.28 cannot represent all experimental data with sufficient accuracy. Therefore, individual correlations are presented, some of which will be presented in the following sections.


## 4.2.2 Solid-gas Systems

In gas-extraction processes solid-gas systems occur during dissolution and precipitation of solids, and in extraction processes involving solid materials. Mass transfer in extraction systems always includes the interaction with the solid matrix from which substances are extracted. This process is discussed in Chapter 7 in context with the extraction from solid materials.

Mass transfer for pure substances has been determined experimentally in a few investigations. Preferably naphthalene and caffeine have been used as model substances for determining mass transfer coefficients. For naphthalene – $CO_2$ Lim et al. [17] obtained the following correlation for opposing flows:

$$\frac{Sh_I}{(Sc\,Gr)^{1/4}} = 0.1813 \left( \frac{Re^2\,Sc^{1/3}}{Gr} \right)^{1/4} (Re^{1/2}\,Sc^{1/3})^{3/4}$$

$$+ 1.2149 \left[ \left( \frac{Re^2\,Sc^{1/3}}{Gr} \right)^{3/4} - 0.01649 \right]^{1/3}, \tag{4.29}$$

for a packed bed and laminar flow: $4 < Re < 135$. Catchpole et al. [5] reported measurements on mass transfer of benzoic acid from a packed bed into near-critical $CO_2$. Their data could be represented by the equation:

$$Sh = 0.839\,Re^{0.667}\,Sc^{1/3}, \tag{4.30}$$

in the range of: $2 < Re < 55$, and $4 < Sc < 16$.

160

Knaff and Schlünder [14] reported the following correlation for mass transfer in a cylinder:

$$Sh = Sh_\infty + (1 + 0.14(d_i/d_o)^{1/2}) \frac{0.19 \ (Re \ Sc \ d_h/L)^{0.8}}{1 + 0.117 \ (Re \ Sc \ d_h/L)^{0.467}}, \qquad 4.31$$

with:
$Sh_\infty$ = $Sh$ at infinite dilution;
$Sh_\infty$ = $3.66 + 1.2 \ (d_i/d_o)^{-0.8}$;
$d_i$   = diameter of sample;
$d_o$   = outer diameter of annular duct;
$d_h$   = $d_o - d_i$: hydraulic diameter of annular duct;
$L$   = length of sample.

Experimental results for the dissolution of caffeine in $CO_2$ could be represented in the approximate range of $90 < (Re \ Sc \ d_h)/L < 260$.

For the extraction of theobromine from cocoa seed shells, Brunner and Zwiefel-hofer [4] fitted kinetic parameters to a model for the extraction process (see Chapter 7). The result for the mass transfer coefficient $\beta$ for external mass transfer solid-gas is shown in Fig. 4.7, and is compared to the dimensionless correlation for mass transfer from flat particles as proposed by Comiti and Renaud [8]:

$$Sh = 1.5 \ Re^{0.43} \ Sc^{1/3}, \text{ for } 5.6 < Re < 28.1, \qquad 4.32$$

with
$Sh$  = Sherwood number, defined as $Sh = (\beta d_p)/(XD)$;
$Re$  = Reynolds number, $Re = (u d_p)/v$;
$Sc$  = Schmidt number, $Sc = v/D$;
$\beta$   = mass transfer coefficient;
$d_p$  = characteristic length, defined as $d_p = 6/(X a_s)$;
$X$   = $a_d/a_s$, with:
$a_d$  = solid surface in contact with extracting supercritical fluid;
$a_s$  = surface area of solid material;
$u$   = flow velocity;
$v$   = viscosity;
$D$   = diffusion coefficient.

The values for $\beta$ obtained by parameter-fitting deviate from those obtained from the correlation for the mass transfer coefficient. This is attributed to radial dispersion which was not taken into account. A similar observation was reported by Martin [18], who was able, using a by-pass model, to show that this type of deviation is due to the different flows in the central and peripheral region of a fixed bed. A sensitivity analysis for the theobromine extraction showed that the diffusion coefficient in the solid material does not influence overall mass transport essentially. Mass transfer rates from the surface of the solid material to the bulk of the extracting gas are dominant.

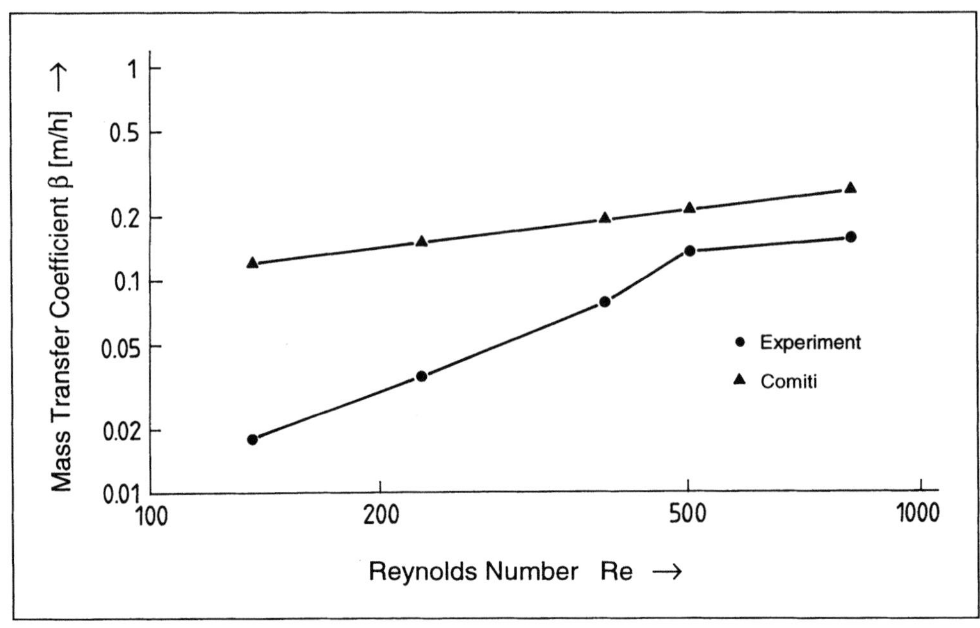


**Fig. 4.7.** Mass transfer from a fixed bed of cocoa seed shells to supercritical $CO_2$ [4].


## 4.2.3 Liquid-gas Systems

Mass transfer coefficients for liquid-gas systems in gas extraction have been reported in a few investigations. Still, a systematic treatment seems not to be possible since data are not sufficient. Therefore, results of Zehnder and Trepp are reported as an example. Zehnder and Trepp [29] measured mass transfer at laminar flow for non-volatile substances like $\alpha$-tocopherol and $CO_2$ in tubes and ducts, and correlated their results with dimensionless mass transfer equations. Such mass transfer equations usually are of the form:

$$Sh = C_4\, Re^{1/2}\, Sc^{1/3}\, (d_h/l)^{1/3},\qquad\qquad 4.33$$

with
$Sh$ = Sherwood number, $Sh = (\beta l)/D$;
$Re$ = Reynolds number, $Re = (ul)/v$;
$Sc$ = Schmidt number, $Sc = v/D$;
$C_4$ = coefficient depending on geometry and configuration of flow;
$Re$ = Reynolds number, $Re = (ul)/v$;
$u$ = flow velocity;
$l$ = characteristic length;
$D$ = diffusion coefficient;
$v$ = viscosity;
$d_h$ = hydraulic length, equivalent to diameter of a tube.

Zehnder and Trepp [29] were only able to represent their measurements by introducing a concentration-dependent diffusion-hindering term, defining an effective diffusion coefficient $D_{eff} = D\, x_E$, with:

$D$ = diffusion coefficient according to the equation of Sassiat [25], a modified Wilke-Chang equation [28];

$x_E$ = equilibrium concentration of the transferred substance (mass fraction).

Their resulting equation is given by:

$$Sh\, x_E = 0.13\, Re^{C_5}\, (Sc\, x_E)^{0.632}\, (d_h/l)^{0.33}, \qquad\qquad 4.34$$

with
$C_5$ = exponent depending on flow conditions; $0.33 < C_5 < 0.5$.

Clearly, more experimental information is needed before generally applicable correlations can be derived.

## References


1. Addoms JN (1948) Heat transfer at high rates to water boiling outside cylinders. Doctoral dissertation, Massachusetts Institute of Technology, Cambridge
2. Bier K, Gorenflo D, Wickenhäuser G (1977) Pool boiling heat transfer at saturation pressures up to critical. In: Hahne E, Grigull U (eds) Heat Transfer in Boiling. Academic Press, New York, San Francisco London; Hemisphere, Washington, London, pp 137 – 158
3. Böhm F, Heinisch R, Peter S, Weidner E (1989) Design, construction and operation of a multi-purpose plant for commercial supercritical gas extraction. In: Johnston KP, Penninger JML (eds) Supercritical Fluid Science and Technology. ACS Symposium Series No. 406, pp 499 – 510
4. Brunner G, Zwiefelhofer U (1993) Mass transfer and scale up of the extraction from solids with supercritical carbon dioxide. AIChE Annual Meeting, St. Louis, 1993.
5. Catchpole O, Simoes P, King MB, Bott TR (1990) Film mass transfer coefficients for separation processes using near-critical $CO_2$. In: DECHEMA, GVC (eds) 2nd International Symposium High Pressure Chemical Engineering, Abstract Handbook, pp 153 – 158, Erlangen
6. Cichelli MT, Bonilla CF (1945) Heat transfer to liquids boiling under pressure. Trans AIChE 41: 755 – 787
7. Colburn AP, Drew TB, Worthington H (1947) Ind Eng Chem 39: 958
8. Comiti J, Renaud M (1991) Liquid-solid mass transfer in packed beds of parallelepipedal particles: Energetic correlation. Chem Eng Sci 46: 143 – 154
9. Danilova GN (1965) Influence of pressure and temperature on heat exchange at boiling of Freons. Kholod Tekh 42/H2: 36 – 42
10. Fedders H (1971) Messung des Wärmeuberganges beim Blasensieden von Wasser an metallischen Rohren. KFA Jülich GmbH Jül-740-RB
11. Haffner H (1970) Dissertation, Technische Universität München. Research report BMWB K 70-24
12. Hahne E, Feurstein G (1977) Heat transfer in pool boiling in the thermodynamic critical region: Effect of pressure and geometry. In: Hahne E, Grigull U (eds) Heat Transfer in Boiling. Academic Press, New York, San Francisco London; Hemisphere, Washington, London
13. Happel O (1977) Heat transfer during boiling of binary mixtures in the nucleate and film boiling ranges. In: Hahne E, Grigull U (eds) Heat Transfer in Boiling. Academic Press, New York, San Francisco London; Hemisphere, Washington, London

14. Knaff G, Schlünder EU (1987) Mass transfer for dissolving solids in supercritical carbon dioxide. Part I: Resistance of the boundary layer. Chem Eng Process 21: 151 – 162
15. Kraussold H (1934) Forsch Gebiete Ingenieurw 5B: 186
16. Lenoir JM, Comings EW (1951) Chem Eng Progr 47: 223
17. Lim GB, Holder GD, Shah YT (1990) Mass transfer in gas-solid systems at supercritical conditions. J Supercritical Fluids 3: 186 – 197
18. Martin H (1978) Chem Eng Sci 33: 913 – 919
19. Mayinger F (1982) Strömung und Wärmeübergang in Gas-Flüssigkeits-Gemischen. Springer, Wien New York, pp 198 – 219
20. Micheyev MA (1962) Grundlagen der Wärmeübertragung. VEB Technik, Berlin, pp 74 – 81
21. Müller F (1966) Wärmeübergang bei der Verdampfung unter hohen Drücken. Dissertation, Technische Universität Berlin
22. Nishikawa K et al (1971) Pool film boiling heat transfer from a horizontal cylinder to saturated liquids. Int J Heat Mass Transfer 14: 19 – 26
23. Pitschmann P, Grigull U (1970) Filmverdampfung an waagerechten Zylindern. Wärme Stoffübertrag 3: 75 – 84
24. Rance RW, Cussler EL (1974) AIChE J 20: 353
25. Sassiat PR, Mourier P, Caude MH, Rosset RH (1987) Anal Chem 59: 1164
26. Stephan K (1964) Beitrag zur Thermodynamik des Wärmeübergangs beim Sieden. In: Abhandlungen der Deutschen Kälte- u Klimatech Ver; no 18. CF Müller, Karlsruhe
27. Vaihinger D, Kaufmann D (1972) Zum Druckeinfluß auf den Wärmeübergang bei ausgebildeter Blasenverdampfung. Chem Ing Tech 15: 921 – 927
28. Wilke CR, Chang P (1955) AIChE J 1: 264
29. Zehnder B, Trepp C (1993) Mass-transfer coefficients and equilibrium solubilities for fluid-super-critical-solvent systems by online near-IR spectroscopy. J Supercritical Fluids 6: 131 – 137
30. Zuber N (1956) Hydrodynamic Aspects of Boiling Heat Transfer. Thesis AECV-4439

# 5 Solvent Cycle

In this chapter the circuit of the supercritical solvent during a gas extraction process and the operations for recycling the solvent are discussed. The supercritical solvent and modifiers, which have been added to the solvent, in general must be recycled. A flow sheet of a solvent cycle is shown in Fig. 5.1.

The extract, dissolved in the supercritical solvent, must be separated from the solvent (1, Fig. 5.1), the solvent must be cleaned for reuse in the extraction process (2), and the solvent must be removed from the remaining substrate or raffinate (3). In case a solvent mixture is employed, composition of the mixture has to be adjusted (4) before it is again employed in the gas extraction process step.

The necessary operations for changing conditions of state and composition in the solvent circuit can be carried out in different ways which depend on the nature of the substances involved, the scale of the process unit, and the operating conditions of the processing unit. The main difference in solvent cycles is whether the solvent is cycled in the supercritical (gaseous) or subcritical (liquid) state. In both cycles the solvent can be driven by a pump or by a compressor [1, 2].

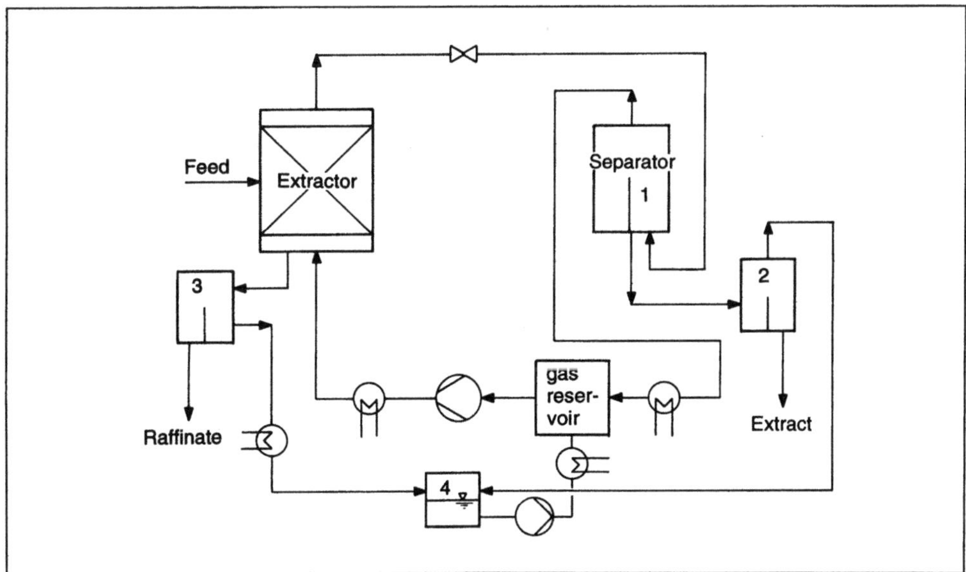


**Fig. 5.1.** Flow scheme of solvent cycle in a gas extraction process.

# 5.1 Gas Circuit in the Compressor Mode

First, a compressor-driven solvent circuit is described, in which the solvent is liquefied for recycling.

Figure 5.2 illustrates the process cycle in a temperature-entropy ($T,S$) diagram. The process for which the supercritical solvent is employed (e.g., fixed bed extraction process, countercurrent multistage separation, chemical reaction, chromatograhic separation or other processes) is carried out under conditions, marked 3 in Fig. 5.2. The gaseous phase leaves the processing unit containing the dissolved substances which are then separated from the supercritical solvent. In a compressor mode cycle this is achieved (Fig. 5.2) by isenthalpic throttling into the subcritical region ($3 \rightarrow 4$). The throttling process and the following evaporation ($4 \rightarrow 5/1$) separates the extracted substances from the solvent and regenerates the solvent for reuse. The gaseous solvent is then compressed to the pressure of the extraction, in Fig. 5.2 shown as an idealized, isentropic compression ($1 \rightarrow 2$). Temperature of the supercrit-

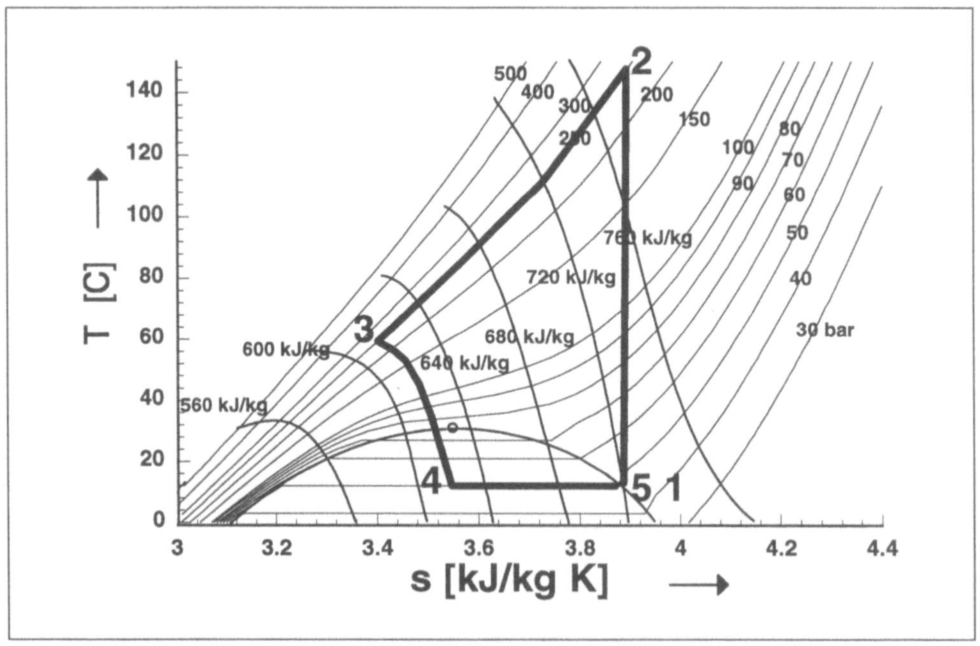


**Fig. 5.2.** Compressor process of a solvent circuit of a gas extraction process in a $T,S$-diagram.

| | |
|---|---|
| 1 | = precipitation of extract and regeneration of solvent; |
| $2 \rightarrow 3$ | = isobaric cooling; |
| 3 | = extraction; |
| $3 \rightarrow 4$ | = isenthalpic throttling; |
| $4 \rightarrow 5/1$ | = evaporation; |
| $1 \rightarrow 2$ | = isentropic compression. |

ical solvent increases during compression, therefore, the solvent is cooled to the temperature of the extraction ($2 \rightarrow 3$). Now the cycle is closed and can be run through again.

Recycling of the solvent which has been removed from the substrate or raffinate and from the extract, has not been considered separately. In principle, the same process steps are necessary to recycle these parts of the solvent. The main difference is that the substrate or raffinate and extract are eventually expanded to ambient conditions and an economic analysis must show whether recycling of the solvent from such low pressure is economically profitable.

Other possibilities for separating the extract from the solvent are:

1) Changing temperature in such a way that the solvent power of the supercritical solvent is reduced substantially, maintaining supercritical conditions or cooling down to subcritical conditions,

2) employing an additional mass separating agent (e.g., water as absorbing medium or active charcoal as adsorbens) while maintaining supercritical conditions for the solvent.

Separation of extract and solvent by changing temperature depends on the variation of equilibrium solubility of the extract in the solvent. The effectiveness can be concluded from phase equilibrium as discussed in Chapter 3. In general, a change of temperature will not be effective enough to clean the solvent sufficiently for reuse in the extracting process. But change in temperature can well be applied in addition, or if total regeneration of the solvent is not necessary.

Separation of extract and solvent by an absorbens or adsorbens makes feasible a solvent cycle at nearly constant pressure. Considering energy consumption, this is the most effective operating mode. But it is only advantageous if the downstream following separation of extract and absorbens or adsorbens is feasible with less effort than the compression (equal efficiency of both process steps with respect to separation of extract and solvent assumed).

In case of an absorption or adsorption for regenerating the solvent, conditions of state of the solvent remain supercritical for the solvent. These conditions of state can be near the two-phase region and only slightly in the supercritical region, if changing of pressure and temperature are combined, or near the operating conditions of the extraction, if only pressure drop over the process units accounts for changes in conditions of state.

The area enclosed by the solvent circuit in the $T,S$-diagram represents the thermodynamic work needed for the process of cycling the supercritical solvent. The lower the pressure drop needed for separating the dissolved components from the solvent, the lower the necessary work. From this point of view, a mass separating agent is advantageous, since only hydrodynamic pressure drop must be compensated. But removing the components from the mass separating agent may be difficult, if not impossible, so the individual case must be considered.

If a mixed solvent is used, the composition of the solvent in general will change during the individual process steps. The initial composition of the solvent must be established before it is used in the process again.

In Fig. 5.3 a compressor process with throttling to the two phase area of the solvent is shown. The necessary heat and electric energy have been calculated, assuming that

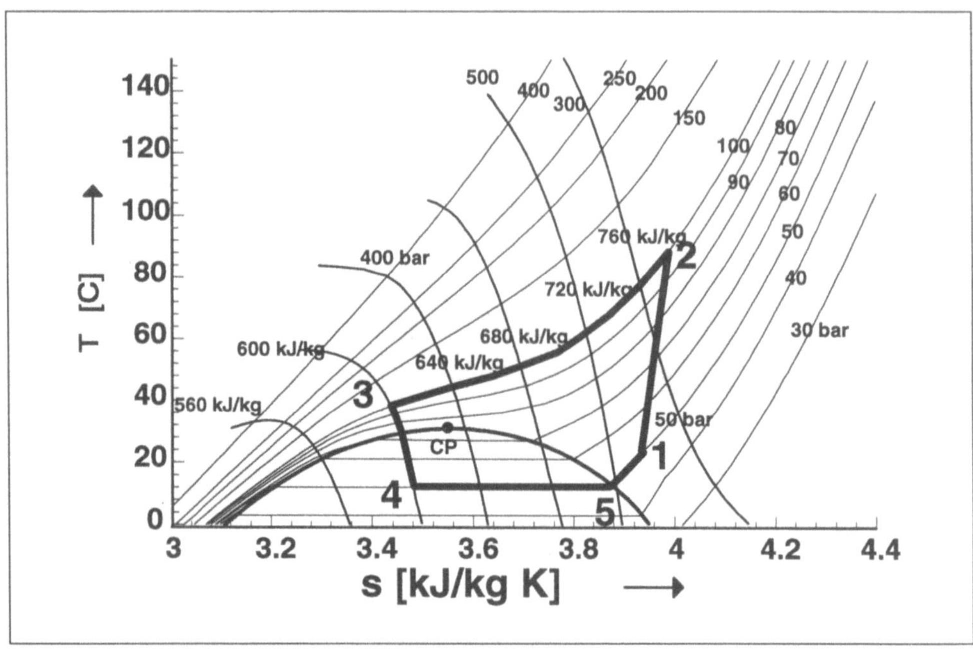



**Fig. 5.3.** Compressor process for extraction conditions of 40 MPa, 313 K. Regeneration of the solvent at a pressure of 5 MPa.

the cooling of the gas ($2 \rightarrow 3$) is achieved by evaporating the condensated liquid solvent ($4 \rightarrow 5 \rightarrow 1$) and cooling by an air-cooler. The amounts of energy are listed in Table 5.1.

**Table 5.1.** Energy consumption of a compressor process. Throttling to the subcritical state. Supercritical solvent: $CO_2$. Conditions of the extraction: 40 MPa, 313 K; Conditions of the regeneration: 5 MPa, 299 K.

| Process step | Energy transferred [kJ/kg] | Heat rejected [kJ/kg] | Heat removed by cooling [kJ/kg] | Electrical energy [kJ/kg] |
|---|---|---|---|---|
| $4 \rightarrow 5/1$ | 168.2 | 0 | 0 | 0 |
| $1 \rightarrow 2$ | 38.7 | 0 | 0 | 38.7 |
| $2 \rightarrow 3$ | − 206.1 | 0 | 36.0 | 0.3 |
| | 0 | 0 | 36.0 | 39.0 |

In Fig. 5.4 a compressor process with throttling in the supercritical state of the solvent is shown. The necessary heat and electric energy have been calculated for cases of energy recovery and without energy recovery. The amounts of energy are listed in Table 5.2.

168

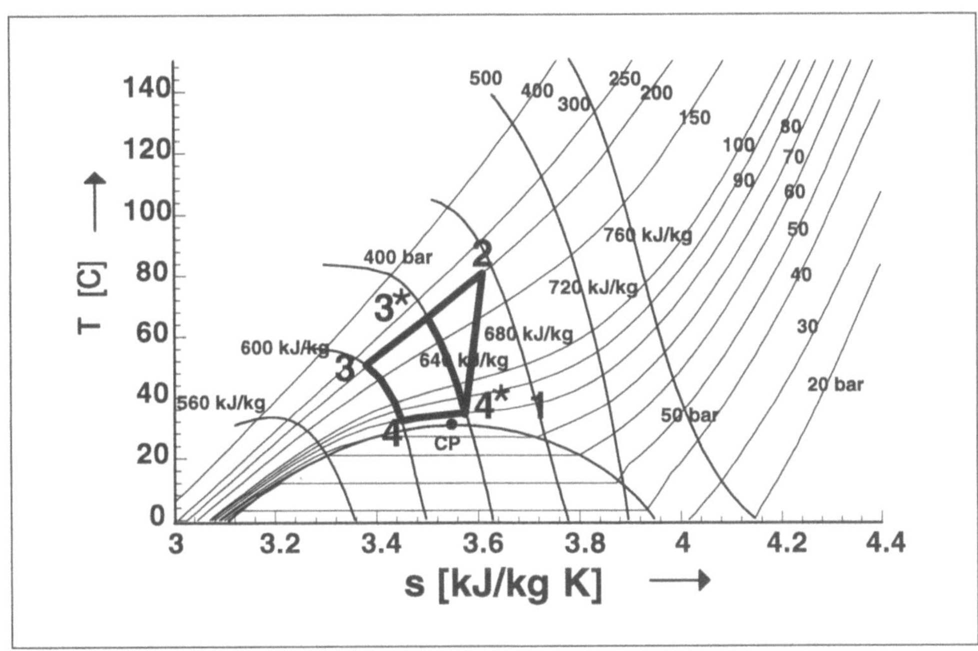


**Fig. 5.4.** Compressor process for extraction conditions of 40 MPa, regeneration of the solvent at a pressure of 8 MPa. See Table 5.2.

**Table 5.2.** Energy consumption of a compressor process. Throttling in the supercritical state. Supercritical solvent: $CO_2$.

1) Solvent cycle with heat recovery:
   Conditions of the extraction: 40 MPa, 323 K; Conditions of the regeneration: 8 MPa, 309 K:

| Process step | Energy transferred [kJ/kg] | Heat rejected [kJ/kg] | Heat removed by cooling [kJ/kg] | Electrical energy [kJ/kg] |
|---|---|---|---|---|
| $4 \rightarrow 1$ | 38.0 | 0 | 0 | 0 |
| $1 \rightarrow 2$ | 30.7 | 0 | 0 | 30.7 |
| $2 \rightarrow 3$ | − 68.7 | 0 | 30.5 | 0.3 |
| | 0 | 0 | 30.5 | 31.0 |

2) Solvent cycle without heat recovery:
   Conditions of the extraction: 40 MPa, 339 K; Conditions of the regeneration: 8 MPa, 309 K.

| Process step | Energy transferred [kJ/kg] | Heat rejected [kJ/kg] | Heat removed by cooling [kJ/kg] | Electrical energy [kJ/kg] |
|---|---|---|---|---|
| $4^* \rightarrow 1$ | 0 | 0 | 0 | 0 |
| $1 \rightarrow 2$ | 31.0 | 0 | 0 | 31.0 |
| $2 \rightarrow 3^*$ | − 31.0 | 0 | 31.0 | 4.1 |
| | 0 | 0 | 31.0 | 35.1 |

# 5.2 Solvent Circuit in the Pumping Mode

The main difference of this solvent circuit to the compressor cycle is that process steps are carried out in a clockwise direction in a $T,S$- diagram.

Figure 5.5 illustrates the cycle of the supercritical solvent in a $T,S$-diagram. The process for which the supercritical solvent is employed (fixed bed extraction process, countercurrent multistage separation, chemical reaction, chromatograhic separation or other processes) is carried out at conditions marked 3 in Fig. 5.5. The gaseous phase leaves the processing unit. The dissolved substances are subsequently separated from the solvent. In a pump driven solvent cycle this is achieved, as in a compressor driven cycle, by changing pressure and temperature to reduce the solvent power of the supercritical solvent. In Fig. 5.5, isenthalpic throttling is shown ($3 \rightarrow 4$). The substances, dissolved in the solvent precipitate and form a separate phase which can be removed, or, if soluble in the liquid solvent phase, are recovered by evaporating the liquid solvent phase ($4 \rightarrow 5$). Conditions of state for the solvent after throttling are in the two-phase region (4). Part of the solvent is gaseous (5). This part is

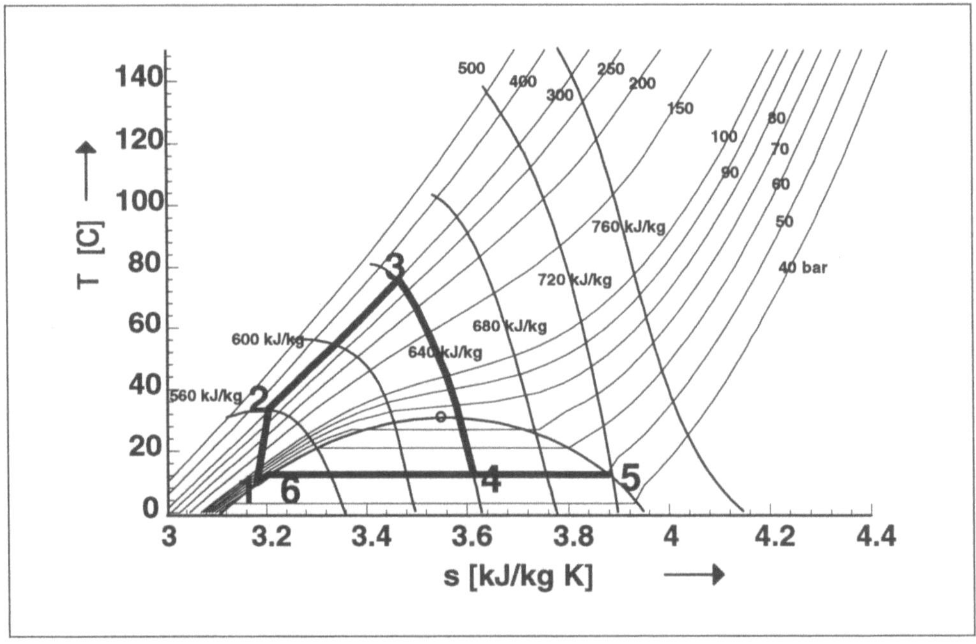



**Fig. 5.5.** Pump process for a solvent circuit of a gas extraction process in a $T,S$-diagram.

| | | | |
|---|---|---|---|
| 1 | = subcooled liquid solvent; | $6 \rightarrow 5$ | = evaporation; |
| $1 \rightarrow 2$ | = pumping to extraction pressure; | 5 | = precipitation; |
| $2 \rightarrow 3$ | = isobaric heating; | 6 | = separation of condensed extract phase; |
| 3 | = extraction; | $5 \rightarrow 6$ | = condensation; |
| $3 \rightarrow 4$ | = isenthalpic throttling; | $6 \rightarrow 1$ | = subcooling. |
| $4 \rightarrow 5$ | = evaporation; | | |

cooled, condensed to the liquid state (6) and subcooled to (1). The other part of the solvent after throttling is in the liquid state. For removing substances still dissolved, the liquid solvent can be evaporated (6 → 5). From this gaseous state the solvent is liquefied (5 → 6) and subcooled (1). The subcooled liquid is pumped to processing pressure (1 → 2) and temperature is adjusted to the processing temperature (2 → 3).

The area enclosed by the solvent circuit in the $T,S$-diagram represents the thermodynamic work needed for the process of cycling the supercritical solvent. Latent heats for evaporation and condensation have to be taken into account. If a mixed solvent is used the composition of the solvent must be adjusted before it is used in the process again. Solvent removed from extract and raffinate may also be reintroduced to the solvent circuit.

In Fig. 5.6 a pump process with throttling to the two-phase area of the solvent is shown. The necessary heat and electric energy have been calculated, assuming that the liquid is subcooled. The amounts of energy are listed in Table 5.3.

In Fig. 5.7 a pump process with throttling in the supercritical state of the solvent is shown. The necessary heat and electric energy have been calculated and the amounts are listed in Table 5.4.

**Table 5.3.** Energy consumption of a pump process. Throttling to the subcritical state. Isobaric heating after the pump. Supercritical solvent: $CO_2$. Conditions of the extraction: 40 MPa, 339 K; Conditions of the regeneration: 6 MPa, 301 K.

| Process step | Energy transferred [kJ/kg] | Heat rejected [kJ/kg] | Heat removed by cooling [kJ/kg] | Electrical energy [kJ/kg] |
|---|---|---|---|---|
| 4 → 1 | − 175.5 | 0 | 175.5 | 10.9 |
| 1 → 2 | 50.0 | 0 | 0 | 50.0 |
| 2 → 3 | 7.3 | 7.3 | 0 | 0 |
| 3 → 4 | 0 | 0 | 0 | 0 |
| 4 → 5 | 119.5 | 0 | 0 | 0 |
|  |  | 7.3 | 175.5 | 60.9 |

**Table 5.4.** Energy consumption of a pump process. Throttling in the supercritical state, heating after throttling. Supercritical solvent: $CO_2$. Conditions of the extraction: 40 MPa, 328 K; Conditions of the regeneration: 8 MPa, 309 K.

| Process step | Energy transferred [kJ/kg] | Heat rejected [kJ/kg] | Heat removed by cooling [kJ/kg] | Electrical energy [kJ/kg] |
|---|---|---|---|---|
| 4 → 1 | − 98.0 | 0 | 98.0 | 5.9 |
| 1 → 2/3 | 65.2 | 0 | 0 | 65.2 |
| 3 → 4 | 0 | 0 | 0 | 0 |
| 5 | 34.0 | 0 | 0 | 0 |
|  | 0 | | 98.0 | 71.1 |

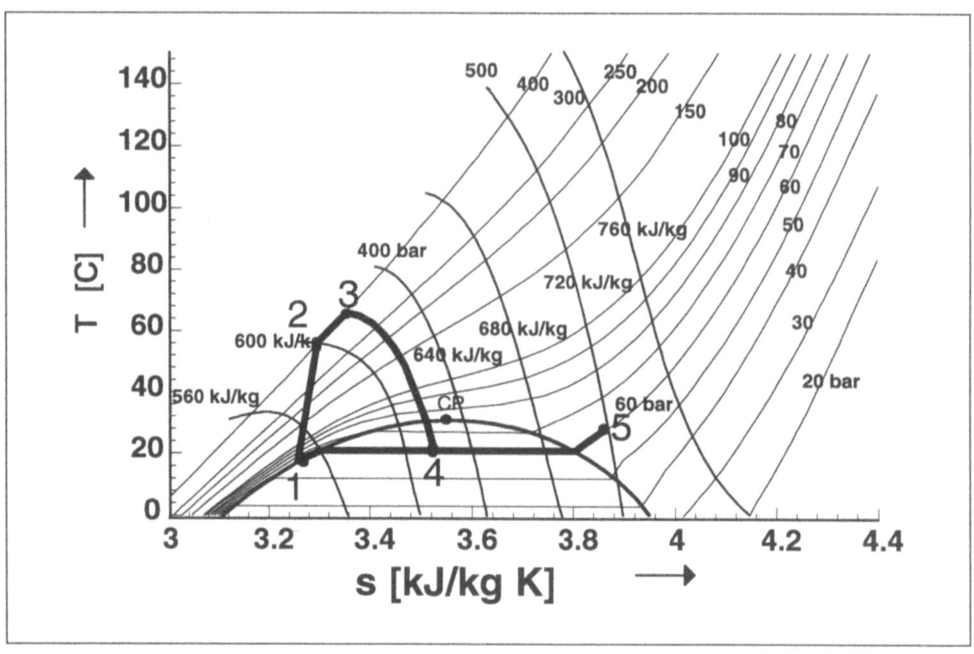

**Fig. 5.6.** Pump process with throttling to the two-phase area. Subcooling of liquid, isobaric heating after the pump.

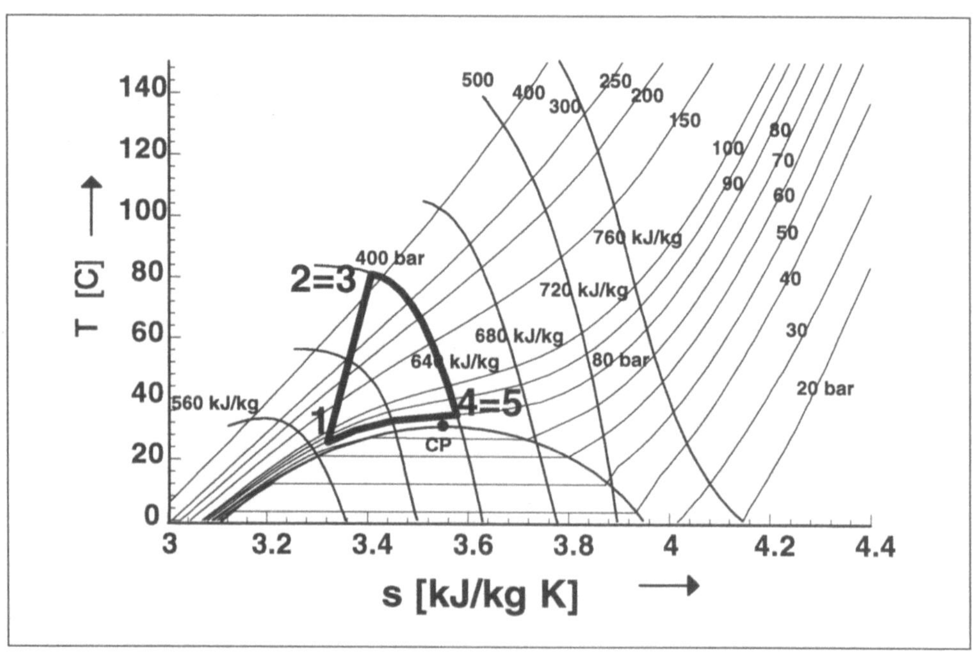

**Fig. 5.7.** Pump process with throttling in the supercritical state of the solvent.

172

# 5.3 Comparison of Compressor and Pump Cycle

A comparison of both cycle processes is presented in Fig. 5.8 and in Table 5.5. Energy consumption of both cycles increases with extraction pressure, without much difference between either if precipitation and regeneration is carried out in the super-critical state. Differences are small even for subcritical precipitation and regeneration for an extraction pressure of around 30 MPa. At higher pressures the compressor process needs more energy, at lower extraction pressures the pump process needs more energy.

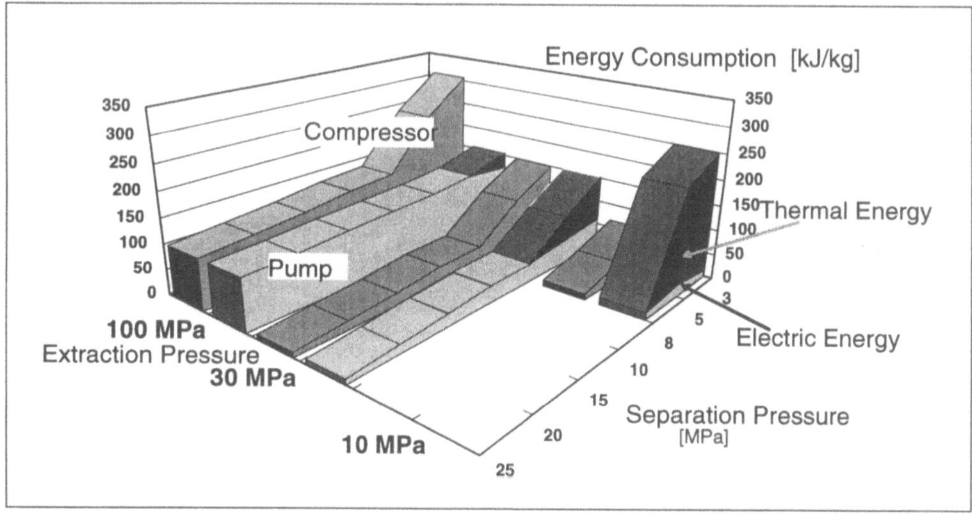


**Fig. 5.8.** Comparison of compressor and pump solvent cycle. Extraction temperature 313 K.

**Table 5.5.** Comparison of electrical energy [$kJ/kg_{CO_2}$] needed for the gas circuit in dependence on pressure in the extraction and precipitation step. Supercritical solvent: $CO_2$.

| Precipitation pressure [MPa] | Pump circuit | | | Compressor circuit | | |
|---|---|---|---|---|---|---|
| | Extraction pressure [MPa] | | | | | |
| | 10 | 30 | 100 | 10 | 30 | 100 |
| 3 | 12.9 | 51.4 | 127.9 | 65.2 | 150 | 307 |
| 5 | 24.2 | 47.4 | 131.3 | 34.5 | 103.8 | 240 |
| 8 | 13.7 | 36 | 122.3 | 11 | 38.5 | 138.1 |
| 10 | | 33.2 | 118.2 | | 35.9 | 132.2 |
| 15 | | 24.7 | 114 | | 24.8 | 120 |
| 20 | | 17.1 | 106.9 | | 17.2 | 110.2 |
| 25 | | 9.9 | 101.5 | | 9.6 | 101.5 |

Some advantages of the pump process are: Lower investment cost for a pump than for a compressor, good control of solvent mass flow, less energy consumption than the compressor process at pressures higher than about 30 MPa.

Disadvantages of the pump process are: Heat exchangers and condensator necessary, additional heat energy needed at low extraction pressures.

Advantages of the compressor cycle are: Only one heat exchanger necessary, low heat energy consumption.

Disadvantages of the compressor cycle are: Mass flow control more difficult than in the pump process, higher consumption of electrical energy, higher investment cost for the compressor, higher energy consumption at pressures higher than about 30 MPa.

## References


1. Birtigh A (1993) Gaskreisläufe aus thermodynamischer Sicht. Private communication. Technische Universität Hamburg-Harburg
2. Lack EA (1985) Kriterien zur Auslegung von Anlagen für die Hochdruckextraktion von Naturstoffen. Dissertation, Technische Universität Graz, Austria

# 6 The Separation of Solvent and Dissolved Substances: Methods for Precipitation

The separation of dissolved substances from the solvent in gas extraction is a necessary operation. Any suitable separation process can be employed, but processes are preferred which introduce no additional substances with only small changes in conditions of state. Additional fractionation of the extract during removal of the solvent from the extract, e.g., by stepwise reducing solvent power of the solvent, is a further aspect.

Substances dissolved in the supercritical solvent can be removed by reducing solvent power of the solvent or by applying a mass separating agent. In the first case, a condensed phase is formed, which is then separated from the remaining gaseous phase. Reduction of solvent power results from changes in conditions of state which result in reduced density, or from adding substances with low solvent power.

## 6.1 Separation by Reduced Solvent Power

### 6.1.1 Reducing Pressure and Increasing Temperature

Solvent power of supercritical solvents depends on pressure and temperature. Therefore, different solubilities at different conditions of state can be used for separating dissolved substances from the solvent. In general, solvent power of the supercritical component increases with increasing density and vice versa. Lower density can be achieved by reducing pressure or increasing temperature.

Reduced **pressure** leads to lower concentrations of the dissolved substances, because of lower density. The effect of pressure on the solute, the Poynting-effect, is low in case of a solid solute. In case of a liquid solute phase, pressure determines the concentration of the supercritical solvent in this liquid phase. Concentrations of the supercritical solvent can be high, and the effect on the structure of the liquid contributes to the concentration of the solute in the gaseous phase. Concentration of the supercritical solvent in the liquid phase increases with pressure and leads to a wider packing of the solute molecules, contributing to the tendency to adopt the gaseous state.

Increasing **temperature** leads to a decrease in density (at constant pressure) and a lower solvent power of the supercritical solvent. Increasing temperature also increases vapour pressure of the solute. At low pressures, where density decreases strongly with temperature, concentrations of the solute components in the supercritical solvent decrease with temperature. At high pressures, where density decreases moderately with increasing temperature, the increase of vapour pressure dominates. The result is an increasing concentration of solute with increasing temperature (at constant pressure).

Contrary to this statement, which refers to the "normal" behavior, there are conditions of state, where a reverse behavior can occur and, e.g., solubility decreases with pressure, even at high pressures. This behavior depends on the binary phase behavior and has been discussed in Chapter 3.

After establishing thermodynamic conditions of reduced solubility, a condensed phase is formed which is separated from the remaining gaseous phase in one or more separators in series, operated at identical or different conditions. An additional fractionation can be achieved if the extract consists of more than one compound, by either operating two separators at different conditions or collecting fractions from one separator at different time intervals. An example is presented in Chapter 7 for the separation of extract from palm fruits from supercritical carbon dioxide.

It can be assumed that phase equilibrium is approximately achieved in the separators as long as mass flow is not too high and residence time is sufficiently long. In the separators used in experimental installations, residence time is in the range of about 80 to 200 s.

### 6.1.2 Separation by Expansion into the Two-Phase Region of the Supercritical Solvent

The extraction phase is expanded to conditions where the solvent is subcritical. A two-phase mixture is formed, consisting of the liquefied solvent containing the extract and the remaining gaseous phase of the solvent. A three-phase mixture occurs if the extract is not soluble in the liquid phase of the solvent. Then, the extract phase can be separately removed from the separator. The remaining liquid solvent phase is evaporated and liquefied again for recycling. During evaporation of the liquid solvent, extract compounds precipitate and are collected. This sequence of operations is very effective with respect to recovery of extract and regeneration of solvent, but costly in energy requirement.

## 6.2 Separation of Phases: Improved Separation by Cyclones

With increasing solvent mass flow, increasing amounts of the condensed phase are entrained. Separation of the phases is incomplete and product losses and backmixing are the consequences. Several devices for phase separation, used in low pressure processes, can be employed to reduce entrainment in gas extraction. Demisters, consisting of wire mesh packing, deflectors and filters of different kinds, especially sinter-metal filters are useful. These devices may be used as additional measures for phase separation at the outlet of a precipitator. At the inlet of a precipitator, an improved separation of the phases can be achieved by means of cyclones which are placed downstream to the expansion. Cyclones have been effectively used for the separation of gaseous, liquid and solid phases. Therefore, the application in gas extraction for dense gases is obvious. In a high pressure pyrolysis plant, excellent separation of fine mineral particles from the pyrolysis gases was achieved [2]. Perrut [3] also reported successful use of a cyclone as phase separator.

## 6.3 Separation of Extract and Solvent by a Mass Separating Agent

### 6.3.1 Separation by Absorption

In certain cases the solute downstream to a gas extraction processing unit may be advantageously separated from the supercritical solvent by absorption. Then the solvent circuit may be operated at nearly constant pressure. The absorbing liquid must dissolve the solute, but should not be soluble in the supercritical solvent. The miscibility gap between supercritical solvent and absorbens must be wide.

Furthermore, the absorbing liquid must not interfere with the quality of the products. Therefore, the ideal absorbing liquid consists of components which are part of the feed material. Organic solid substrates normally contain water. Therefore, water is a good absorbens if solubility for the extracted compounds is good. An ideal example is the decaffeination of coffee beans. Caffeine is hydrophilic and forms hydrates with water. In addition, carbon dioxide and water are only slightly mutual soluble at the temperatures applied ($< 373$ K). Solubility of caffeine in water is strongly dependent on temperature, so that the dissolved caffeine can be separated from the absorbens by crystallization at lower temperatures.

### 6.3.2 Separation by Adsorption

Adsorption, like absorption, allows to operate the circuit of the supercritical solvent at constant pressure (neglecting pressure drop). Furthermore, adsorption is very effective with respect to removing the extract from the gas and regenerating the gaseous solvent. On the other hand, it may be difficult to remove the extract from the adsorbens.

Adsorption has been effectively used in decaffeination for adsorbing caffeine, without recovering caffeine from the employed active charcoal.

Adsorption may be employed as additional operation for removing residual quantities of extract from the gaseous solvent, before the solvent is recycled to the gas extraction process. Due to the high selectivity, which can be obtained in an adsorption process, adsorption may be the advantageous operation for separating extract from solvent if concentration of extract in the supercritical solvent is low, provided that the separation of the extract from the adsorbens can be achieved effectively. Research is needed to make available suitable adsorbing materials.

## 6.3.3 Separation by Membranes

Filtering devices can be used for separating extract compounds after condensation. From a homogeneous gaseous phase, separation of the extract components is possible with membranes, since the difference in molecular weight between supercritical solvent and the extract compounds is sufficiently high. Typically, the molecular weight of the solvent is about 50 kg/mol, while the molecular weight of the solute compounds is in the range of 200 to 800 kg/mol. Membranes allow the operation of the solvent circuit at low pressure differences of some MPa. Membranes for separation of extract and solvent in gas extraction have been proposed by Gehrig [1] and are under experimental investigation on laboratory scale [4].

## 6.3.4 Separation by Adding a Substance of Low Solvent Power

Solvent power of a supercritical gas can be reduced by adding a substance of low solvent power. An example is the solubility of caffeine in carbon dioxide, which is reduced remarkably by adding nitrogen (Fig. 3.37). Before applying the solvent again in the process, the added compound must be removed, e.g., by membrane separation.

A similar effect can be achieved if an entrainer is employed in the process and the entrainer is removed. The solubility of a compound in a supercritical solvent can be enhanced if a relatively small amount of entrainer is employed. For separating extract and solvent, the entrainer can be removed by adsorption. It is likely that the extract compounds are also adsorbed, but since extract and entrainer molecules compete for adsorption sites, the mixed adsorbate may be removed more easily from the adsorbens than the extract compounds alone. Again, the advantage of such a circuit would be the operation at nearly constant pressure.


## References

1. Gehrig M (1986) Verfahren zur Gewinnung von Coffein aus verflüssigten oder überkritischen Gasen. Offenlegungsschrift DE 3443390 A1
2. Hoffmann R, Künstle K, Brunner G (1984) Production of liquid fuels and electricity from oil shale by hydrogen retorting. Preprints Int Symp High Pressure Chemical Engineering, Erlangen, pp 155 – 161
3. Perrut M (1989) High performance extract solvent-separators. In: Perrut M (ed) Proceedings of the International Symposium on Supercritical Fluids, Vol 2, pp 627 – 632
4. Sarrade S, Rios GM, Veyre R, Soria R (1994) Supercritical $CO_2$-extraction coupled with nanofiltration separation, interest and preliminary studies. In: Kiran E, Levelt-Sengers JMH (eds) Supercritical Fluids – Fundamentals for Application. NATO ASI, Kluwer

# 7 Extraction of Substances with Supercritical Fluids from Solid Substrates

## 7.1 Introduction

The extraction of valuable materials from solid substrates by means of supercritical gases has been carried out on a commercial scale for more than a decade. Large scale processes are related to the food industry like the decaffeination of coffee beans and black tea leaves and the production of hops extracts. Smaller scale processes comprise extractions of spices, flavoring compounds and other highly valued compounds. The extraction of high quality edible oils is under investigation. It would be a large scale process, but as for all commodity products, the value added is not high, so the economy of the process is the main problem.

Gas extraction from solids is carried out by continuously contacting the solid substrate with the supercritical solvent. The solid substrate in most cases forms a fixed bed. The supercritical gas flows through the fixed bed and extracts the product components until the substrate is depleted.

It is interesting to know what components can be extracted with supercritical solvents. This was treated in Chapter 3 in context with the properties of supercritical solvents. Modeling the extraction allows to determine the time of extraction, which is important for an optimal utilization of the plant, and to influence the course of extraction by adjusting parameters according to the results of modeling the process. In order to achieve this, mechanisms of mass transport in the extraction process have to be considered and models evaluated with respect to their applicability.

Contacting a solid in a fixed bed is a one-stage process. There may be some advantages in contacting the solid substrate several times and under different conditions. Even countercurrent contacting can be taken into consideration.

Normally, extractions are first carried out in laboratory scale plants. Therefore some installations will be presented and the mode of operation and the evaluation of the results will be discussed. The ultimate goal is commercial large scale operation of gas extraction.

# 7.2 The Extraction Process

## 7.2.1 General Description of the Extraction Process and an Extraction Plant

Gas extraction from solids consist of two process steps: 1) the extraction, and 2) the separation of the extract from the solvent (Fig. 7.1). In the extraction, the supercritical solvent flows through a fixed bed of solid particles and dissolves the extractable components of the solid. The solvent is fed to the extractor and evenly distributed to the inlet of the fixed bed. The loaded solvent is removed from the extractor and fed to the precipitator. The direction of flow of the supercritical solvent through the fixed bed can be upwards or downwards. Advantages and disadvantages are discussed below. At high solvent ratios (ratio of flow of supercritical solvent to the amount of solid) the influence of gravity is negligible (compare also Chapter 4). The shape of the fixed bed can also be a matter of design consideration. Height to diameter for cylindrical fixed beds, cylindrical shaped layers of solid material, and combination of extractor and precipitator are some possible variations.

The solid material will be depleted from the extractable material in the direction of flow. Concentration of extract components increases in the direction of flow in the supercritical solvent and in the solid material. The shape of the concentration curve depends on operating conditions and the kinetic extraction properties of the solid material and the solvent power of the supercritical solvent, as discussed below.

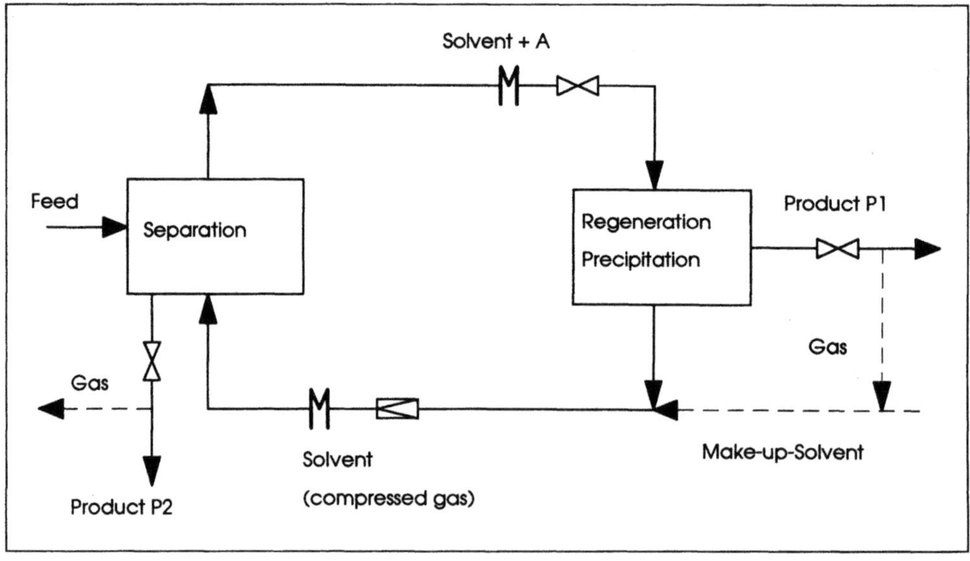


**Fig. 7.1.** Flow scheme of gas extraction from solids.

The properties of the supercritical solvent have been treated extensively in Chapters 2 and 3. In gas extraction from solid material, the properties of the solid are of utmost importance and will be discussed below.

## 7.2.2 Mechanisms of Transport in the Solid Phase

Transport of substances can occur within a solid and across the borders of a structure (the plant material's structure), which differs from substrate to substrate. Even material from the same type of plant is different from harvesting period to harvesting period, according to conditions and treatmenat after harvesting, according to its age and the treatment prior to extraction. These influences refer only to the primary structure. Furthermore, the solid material can consist of particles of different size and form and the size distribution of the particles may vary as shown in Fig. 7.2. The bed of particles may form different geometries and even change its geometry during the process. The solid may form a fixed bed of various geometry, it may be stirred or even fluidized. A wide variety of parameters may be relevant for modeling of mass transport of substances in solid substrates.

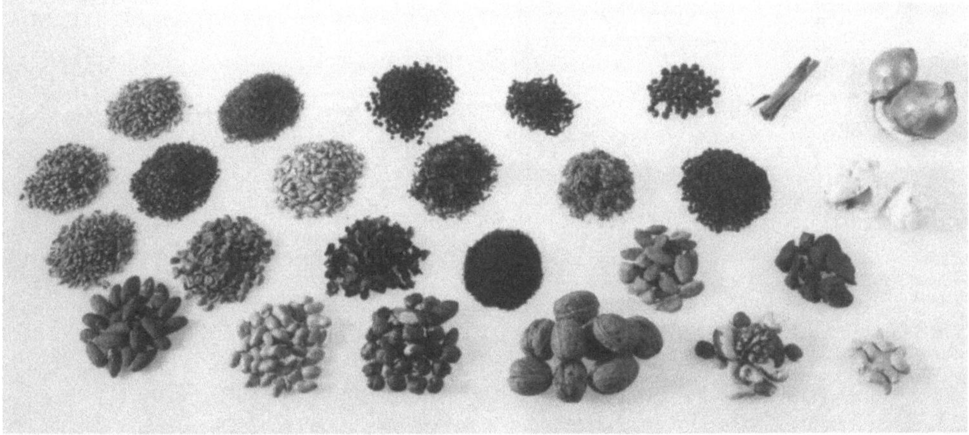

**Fig. 7.2.** Some examples of solid materials which are of interest for gas extraction.

A further question is the initial distribution of the extract substances within the solid substrate. The substances may be adsorbed on the outer surface, on the surface of pores, evenly distributed within the solid or within the plant cells. Each of these different distributions has some influence on the course of the extraction. On the other hand, the integral extraction curve is of relative simple form (see Fig. 7.7). It is not possible to calculate many parameters with significance from such a curve.

The extraction of soluble compounds from solid plant material proceeds in several parallel and consecutive steps:

1) The plant matrix absorbs the supercritical solvent and other fluids which are deliberately added to influence the extraction process. The cell structure swells, i.e., the cell membranes and the intercellular channels are widened. Mass transport resistance is lowered by these measures.

2)  In parallel, the extract compounds are dissolved by the solvent. A chemical reaction may occur previous to solvation.

3) The dissolved compounds are transported to the outer surface of the solid. Diffusion is the most important transport mechanism.

4) The dissolved compounds pass through the outer surface. A phase change may occur at that place.

5) The compounds are transported from the surface layer into the bulk of the supercritical solvent and are subsequently removed with the solvent from the bulk of the solid material.

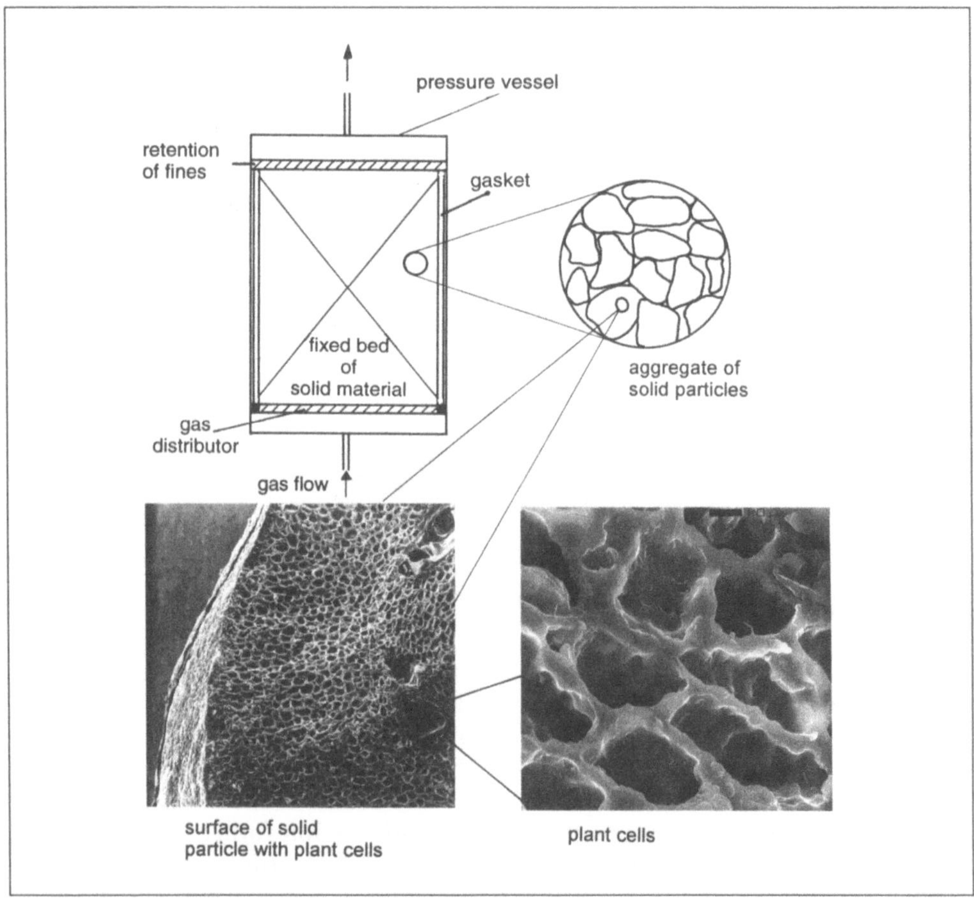


**Fig. 7.3.** Schematic drawing of a solid substrate in a fixed bed.

A solid substrate is schematically shown in Fig. 7.3. A plant cell is part of a cell structure which, in turn, is part of a particle and an aggregate of particles in a fixed bed of plant material placed in an extraction vessel. A plant cell is shown in Fig. 7.4 [9, 40]. The sub-structures of the cell, like the cytoplasm with the vacuoles, the cell wall, the intercellular cavities and the plasma-membrane, the tonoplast and the pits, are determining the mass transfer in the extraction process, if the extract substances are part of the cell. The extract substances are dissolved in the cytoplasm or the vacuoles.

In the cell sap crystallized substances like caffeine are present, beside carbohydrates, fats, proteins, and soluble salts. Transport of these substances can take place across the various membrane systems. Part of these membrane systems is the elementary membranes. An elementary membrane, Fig. 7.5, consists of three layers of proteins and lipoids. A model of such an elementary membrane according to Danielli [9, 40] is represented in Fig. 7.6. The lamelles are made of a bimolecular lipid layer. In this layer the hydrophobic ends of the molecules take adjacent positions, while the

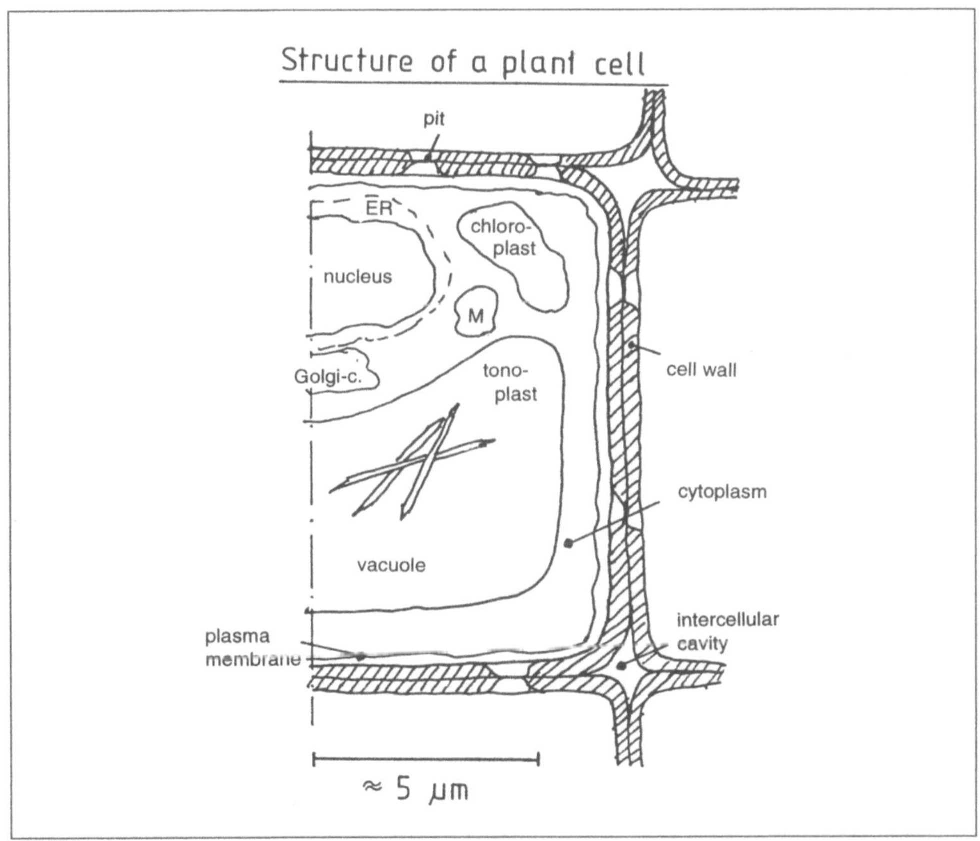


**Fig. 7.4.** Cross-section of a plant cell [40].

hydrophilic ends point to the hydrated proteins. The elementary membrane has pores which can be passed by water and by lipids. A dynamic model of such an elementary membrane may explain the role of water during an extraction process with plant materials. According to Kavanau [9, 40], the lipid pillars of an elementary membrane change with water content. If there is not enough water in the system, the pil-

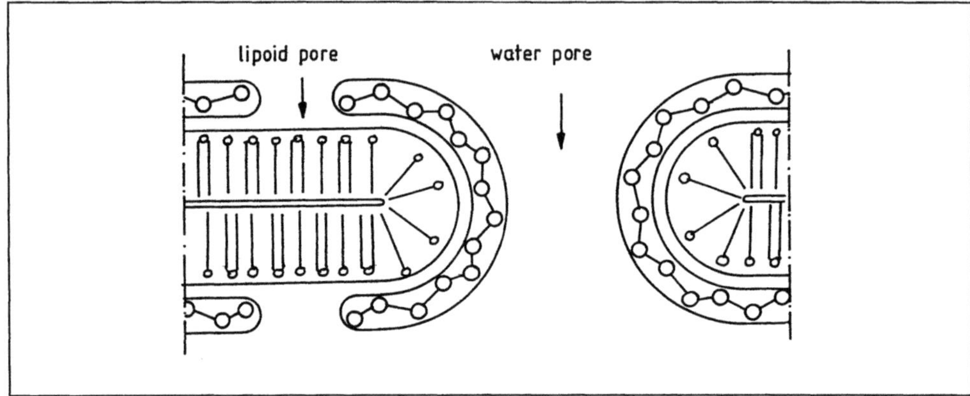


**Fig. 7.5.** Schematic drawing of an elementary membrane [40].

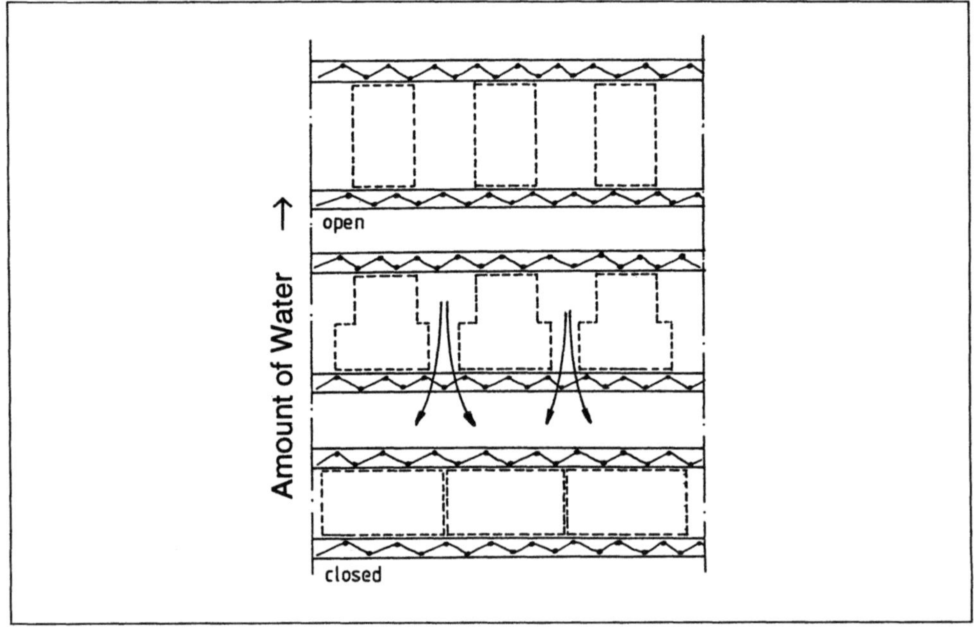


**Fig. 7.6.** Model of an elementary membrane (after Danielli [40]).

184

lars close the membrane, making it impermeable. If enough water is available, the membrane becomes permeable.

The elementary membrane of a living cell is semipermeable. It can be passed by the solvent, but not by the solute. The cell wall consists of several layers providing the necessary structural stability. At certain points, the pits, the cell wall is reduced to the middle layer, mainly elementary membranes, through which mass transport can occur. The intercellular cavities are formed by adjacent cells. The middle lamella of the cell wall parts and cavities emerge, and form long hollow structures suitable for transport of matter. The cells form the tissue, which is the storage place for starch, proteins, sugars, and fats.

In special cases it may be necessary that the cell structure and the cells remain intact during the extraction process. This is the case for the decaffeination of green coffee beans. They are roasted after the extraction of caffeine. During the roasting process, the typical substances for the coffee aroma are formed. The reactions depend i.a. on the pressure, which can be reached in the cells during the roasting, which may be in the range of 1 to 1.5 MPa. In damaged cells and cell structures the resulting pressure is most probably much lower, being therefore detrimental to the coffee aroma.

On the other hand, it is obvious that the transport paths are shorter and mass transfer is enhanced if the plant material is crushed and cells and cell structures are destroyed. Then, the transport resistance across membranes will be of minor importance. Other transport mechanisms like diffusion in the solid, desorption from a solid surface, and diffusion in a laminar flowing layer of solvent are prevailing.

## 7.2.3 Course of the Extraction of Substances from Solids with Supercritical Solvents

The extraction of components from solid material is carried out by contacting the solid substrate with a continuous flow of the supercritical solvent. The solid substrate in most cases forms a fixed bed, through which the supercritical gas flows and extracts the product components until the substrate is depleted. For the solid as well as for the solvent, this is an unsteady process. The course of the extraction process can be followed by determining the amount of extract against time of extraction. From these data, more information on the process can be deduced, which is discussed below.

### 7.2.3.1 Total Amount of Extract

The amount of extract accumulating during the course of the extraction will in principle follow the schematical curve of Fig. 7.7. The first part of the curve may be a straight line, corresponding to a constant extraction rate (type 1 extraction rate curve, see Fig. 7.9). The second part forms a graph, approaching a limiting value which is given by the total amount of extractible substances (type 2 extraction rate curve, Fig. 7.9).

The gradient of the first part of the graph can be given by the equilibrium solubility. Then, from this gradient the equilibrium solubility may be determined. On the other hand, a straight line can be caused by constant mass transfer resistance and is no proof that equilibrium solubility is obtained during the extraction.

The curve for the total amount of extract is a response curve to the flow of supercritical solvent entering the extractor. The response curve depends on process parameters and all the phenomena occurring during the extraction in the fixed bed. Some of these phenomena are: radial distribution of the solvent at the inlet, backmixing of the supercritical solvent during the flow through the fixed bed caused by the uneven size, surface, and distribution of the solids, the self-diffusion of the solvent, and radial distribution of the solvent. Due to the kinetic of mass transfer, concentration of extracted substances in the supercritical solvent has an axial concentration profile. The axial concentration profile in the gaseous phase corresponds to an axial concentration profile in the solid. A radial concentration profile in the solid and in the gaseous phase overlays the axial concentration profile. In Fig. 7.8 the course of an extraction from solid material is schematically illustrated by a sequence of concentration profiles.

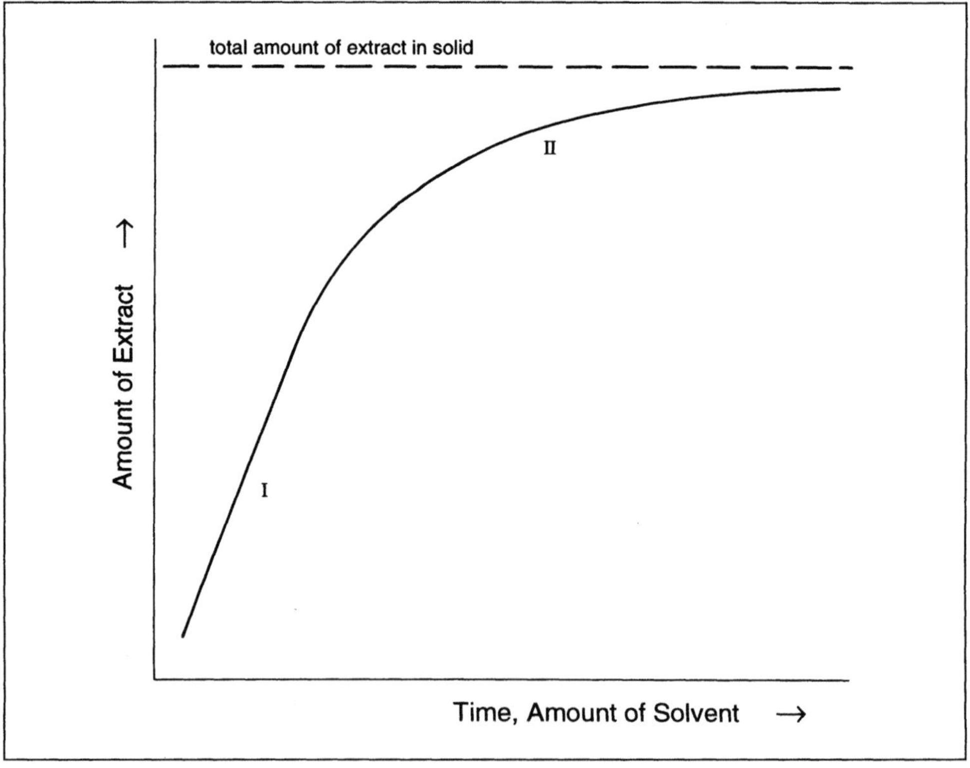


**Fig. 7.7.** Integral extraction curve. Total amount of extract against time of extraction. Part I is linear; Part II is nonlinear and may approach a limiting value, given by the amount of extractible substances.

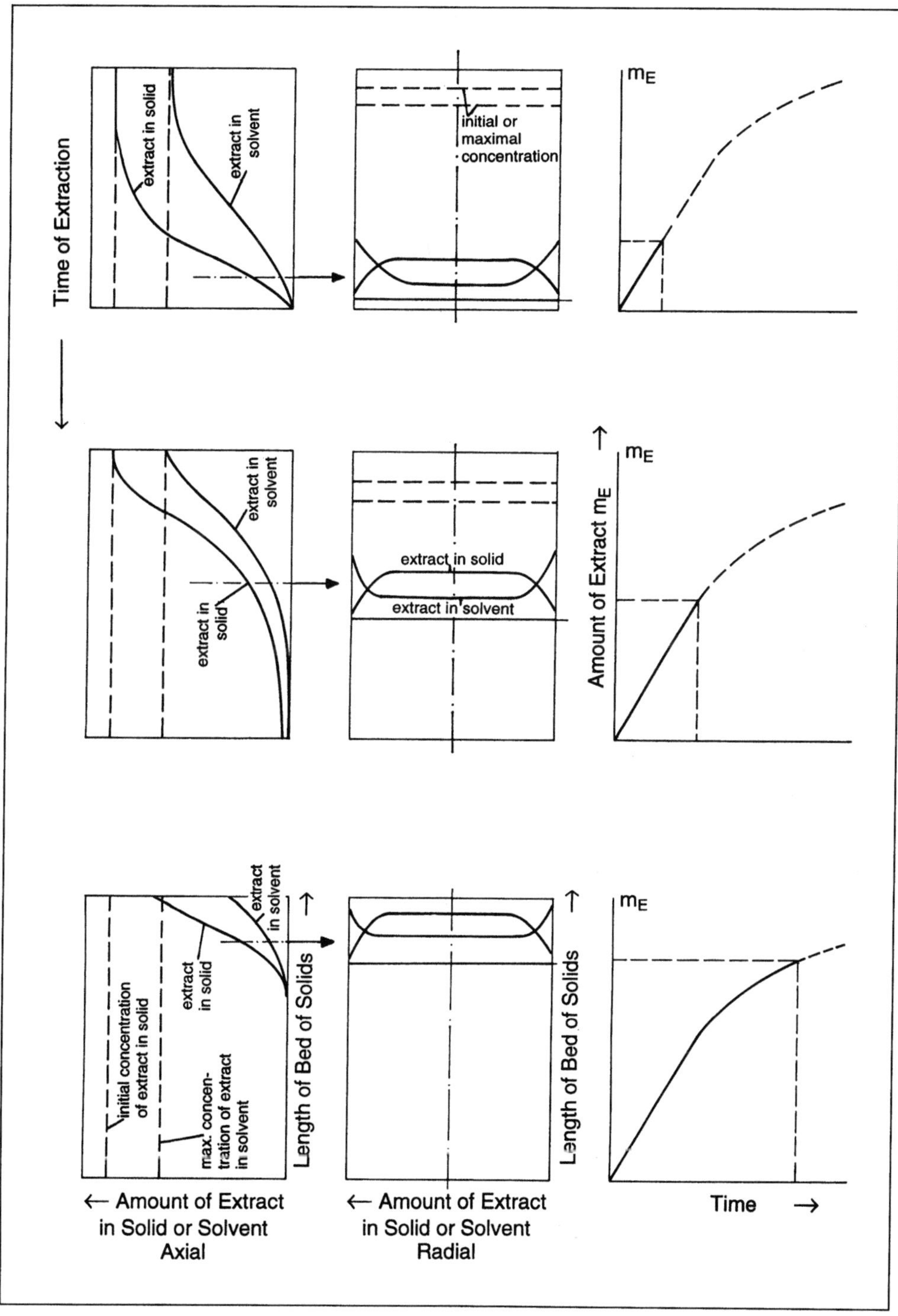



**Fig. 7.8.** Course of an extraction from solid material, illustrated in sequences of concentration profiles for the solid substrate and the supercritical solvent.

Due to the various influences on the curve of total amount of extract, the meaning of this curve for comparing extractions on different materials and in different extractors is very limited. But the information from this curve and the information which can be derived from it is useful for comparing extraction results within a series of experiments on the same substrate and in the same extracting equipment.

The extraction process, as discussed below, can be viewed from the point of a single particle, from the point of the fixed bed, and from the solvent. Modeling, which takes into account the effects as far as necessary to reproduce experimental data, is discussed in Section 7.3. Subsequently, the information available from experimental data is discussed.


### 7.2.3.2 Extraction Rate

The course of the extraction from solid substrates follows two types of extraction curves for the extraction rate, as shown in Fig. 7.9.

Curve 1 represents the extracted quantity per unit of time, the extraction rate, in case of a high initial concentration of extract in the solid substrate, and for an extract

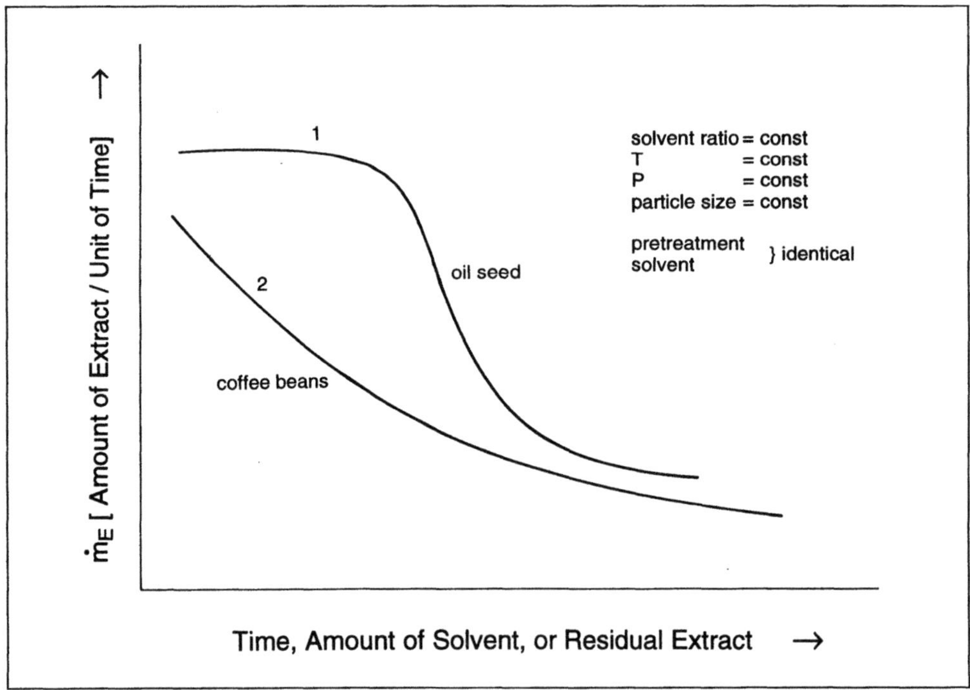


**Fig. 7.9.** Extraction rate curve. Amount of extract in a certain time interval, plotted against time of extraction, amount of solvent, remaining content of extract in the solid or the fraction of the extractable amount of substances.

which is readily accessible for the solvent [10, 20]. During the first part of the extraction, mass transfer is then constant (at constant operating conditions) and determined by the transition at the interface between the solid and the fluid. The part of the fixed bed of solids, for which constant mass transfer can be assumed, travels with increasing time of extraction through the fixed bed.

In the second part of the extraction, two effects may add and cause a declining medium concentration of extract in the outflowing solvent:

1) The extract in the solid substrate near the interface solid/gas is depleted for most of the solid substrate. Transport of the extract within the solid to the interface then adds an additional transport resistance.

2) The length of the fixed bed containing the initial content of extract is not long enough to enable the maximum loading of the solvent.

Curve 2 represents the extraction rate in cases of a low initial concentration of extract in the solid substrate, or an extract not readily available for the solvent, so that transport within the solid to the interface solid substrate-fluid solvent is dominating mass transport from the beginning. Curve 2 also corresponds to the second part of curve 1, since a depletion phase always follows the first extraction phase of constant concentration at the outlet of the supercritical solvent.

In addition, the extraction curves may exhibit an initial region, where the extraction rate increases with time. This is due to starting up the extraction equipment. The process begins with an extract free solvent. This is enriched with extract during the initial part of the extraction. Therefore, at the outlet of the vessel, an increasing concentration of extract or an increasing amount of extract per unit of time is observed. The extent of this initial part depends on the residence time of the solvent in the extraction vessel. For solvent ratios (see below) relevant for technical purposes, and for a small void volume within the extraction vessel, the initial part is small, if not negligible. But for low solvent ratios, the initial part may extend over a substantial part of the extraction. Then, an analysis of the results has to take into consideration the residence time of the solvent in the extraction vessel. The initial part of an extraction from solids cannot be avoided, since for the solid substrate the process is always unsteady.

### 7.2.3.3 Remaining Amount of Extract in the Solid

The course of extraction can also be followed by the remaining amount of extractible components in the solid. The resulting graph is shown in Fig. 7.10. The extract is depleted monotonously in the solid substrate with increasing time of extraction or amount of solvent. The graph is a straight line, in case of the first part of a type 1 extraction rate curve. Mass transfer resistance may be attributed to the fluid phase only. The gradient of the graph represents the extraction rate.

In case of a type 2 extraction rate curve, the course of extraction follows an exponential function. There may be a total depletion of the substrate of extractible components, if the concentration of these compounds in the solvent is zero and there are no irreversible reactions of the extractible compounds with the substrate. Otherwise the extraction curve approaches an asymptote, given by the distribution coefficient corresponding to the initial concentration of extract in the solvent. Experimental determination of the extraction curves is discussed in Section 7.4.

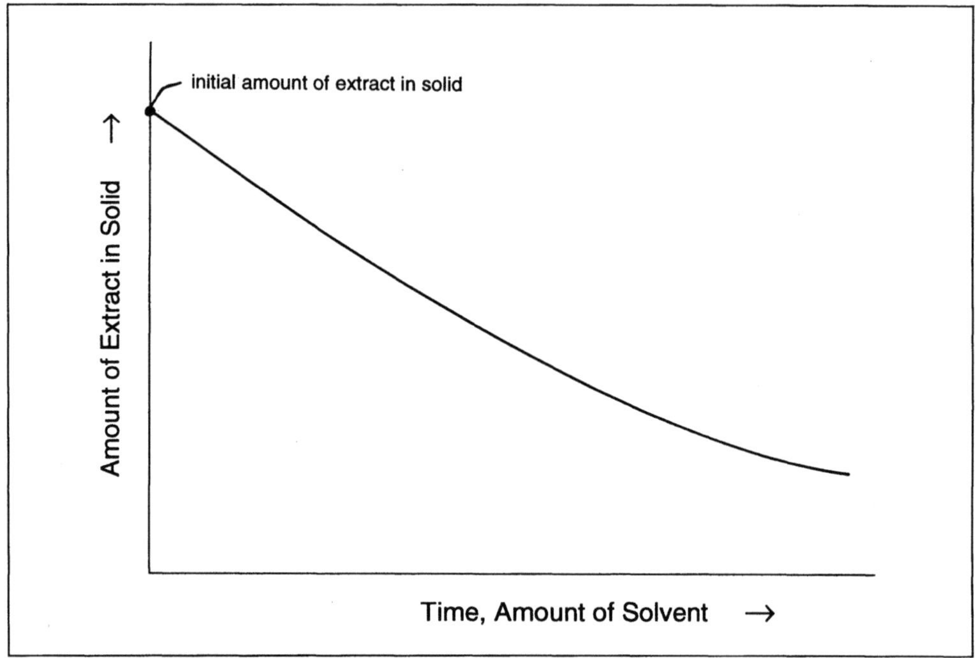


**Fig. 7.10.** Remaining amount of extract in the solid.

### 7.2.3.4 Concentration of the Extract in the Supercritical Solvent

Another parameter which characterizes the course of the extraction is the concentration of extract in the supercritical solvent. In Fig. 7.11 ideal and real curves for the extract content in the supercritical solvent are shown for the cases of extraction rate curves of types 1 and 2. The ideal curve for type 1 is constant for the first part of extraction. In the second part of the extraction, the concentration decreases with ongoing extraction. In the first part, mass transfer resistance in the fluid phase dominates. The extract compounds are readily available at the interface solid/fluid. Therefore, constant process conditions assumed, a constant amount of extract is transferred to the bulk of the supercritical solvent phase, resulting in constant concentration at constant solvent flow rate. The value of this concentration depends on mass transfer and on solvent ratio (Reynolds-number, see Chapter 4).

190

The maximum value of extract concentration in the supercritical solvent is given by the equilibrium solubility of extractible components in the supercritical solvent. Due to different transport resistances and equilibrium distribution coefficients for different compounds, these may be successively extracted at different rates. The maximum concentration of extract in the supercritical solvent may then be a function of extraction time.

Carrying out the extraction at equilibrium concentration of extract in the supercritical solvent results in the minimum quantity of solvent needed for a specified extraction. But this is not equivalent to the minimum time of extraction. An increased solvent rate and a somewhat lower concentration in the solvent may enhance the extracted quantity per unit of time to far higher values than achievable with equilibrium loading (compare the discussion on solvent ratio in Section 7.2.4).

In the second part of an extraction proceeding according to a type 1 extraction rate curve, the concentration of the extract decreases with ongoing extraction, due to increasing mass transfer resistances and depletion of extract in the solid phase. The level of the concentration is lower than the equilibrium concentration.

In many cases, concentration of extract components will even be far lower than the equilibrium solubility of the isolated extract, since within the solid substrate additio-

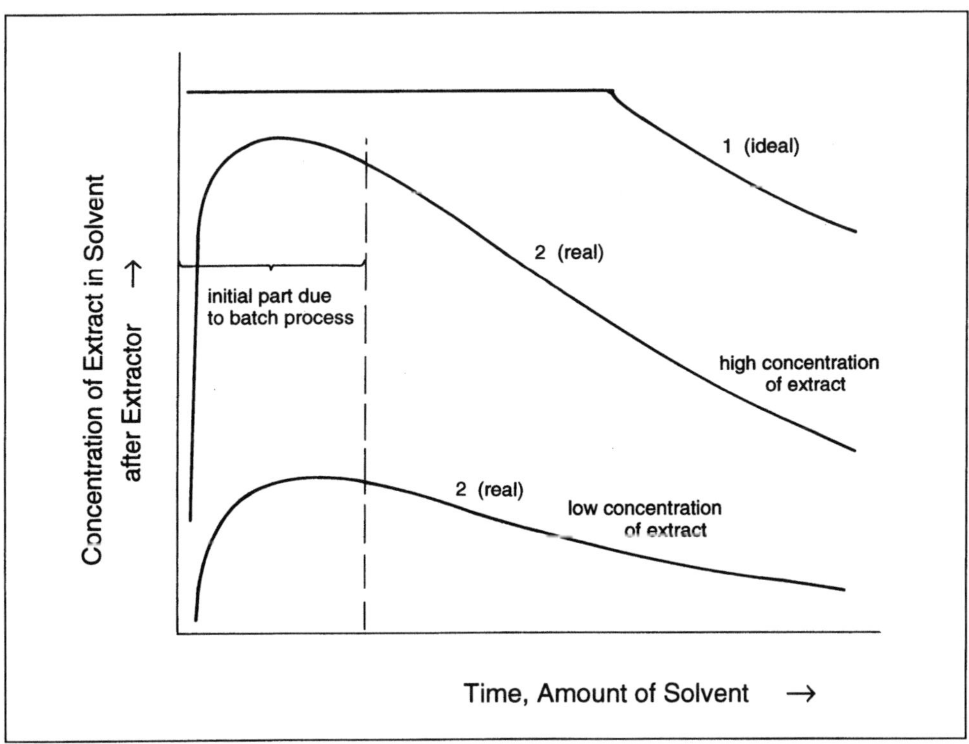


**Fig. 7.11.** Concentration of the extract in the supercritical solvent.

nal interactive forces influence the equilibrium value. These equilibria often are not known. Therefore, it is common practice to refer to the equilibrium solubility of the isolated components in the supercritical solvent.

Equilibrium values, including the solid substrate (adsorption equilibria), depend also on the concentration of the extractible compounds in the solid substrate. It may be easier and sufficient to use the pure component equilibria for modeling the extraction process, since concentrations during an extraction are far from equilibrium values.

An example is the extraction of caffeine from coffee beans. The equilibrium solubility of pure caffeine in carbon dioxide for the conditions of the extraction is about 0.4 wt.-% (4000 ppm) at 350K and 30 MPa. Equilibrium concentration of caffeine in coffee beans in carbon dioxide is about 200 ppm at the same conditions, and depends on the residual caffeine content in the beans. Concentration of caffeine in the supercritical solvent during most time of the extraction is lower than 100 ppm.

A reverse example is the extraction of theobromine from cocoa seed shells. Theobromine is about a factor of 100 less soluble in carbon dioxide than caffeine. The concentration of theobromine in supercritical carbon dioxide during the extraction is always near the equilibrium value. In this case, it is necessary to use the equilibrium value as determined with the cocoa seed shells.

### 7.2.4 Influence of Process Parameters and the Condition of the Solid Substrate on the Extracting Process

The influence of pressure, temperature, and solvent ratio is discussed in this chapter. The role of the supercritical solvent was treated in Chapter 3.

**Pressure**

At process conditions of gas extraction, the solvent capacity in general increases with pressure at constant temperature. Therefore, the remaining content of extract in the solid substrate after a certain time of extraction will decrease with pressure, as illustrated in Fig. 7.12 for the decaffeination of coffee beans. For the coffee beans, the amount of extracted caffeine increases with pressure.

To characterize the extraction result, the remaining content of extract, the time of extraction, or the solvent quantity needed for a certain degree of extraction can be used and plotted on the ordinate.

**Temperature**

A higher temperature often causes a higher extraction rate, if pressure is not low. One reason is the dependence of solvent power on temperature, which has been discussed in Chapter 3. At relatively low pressures, decrease of density and solvent power with increasing temperature prevails, while at relatively high pressures, increase in vapor-pressure with temperature prevails. The other reason for a higher amount of extract per unit of time is increasing mass transfer rates with temperature. So extraction temperatures with supercritical carbon dioxide can be in the neighborhood of the critical temperature, but a temperature increase to about 100 °C (if possible with respect to the extract) can enhance extraction rate. An example is shown in Fig. 7.13 for the extraction of caffeine from coffee beans with $N_2O$.

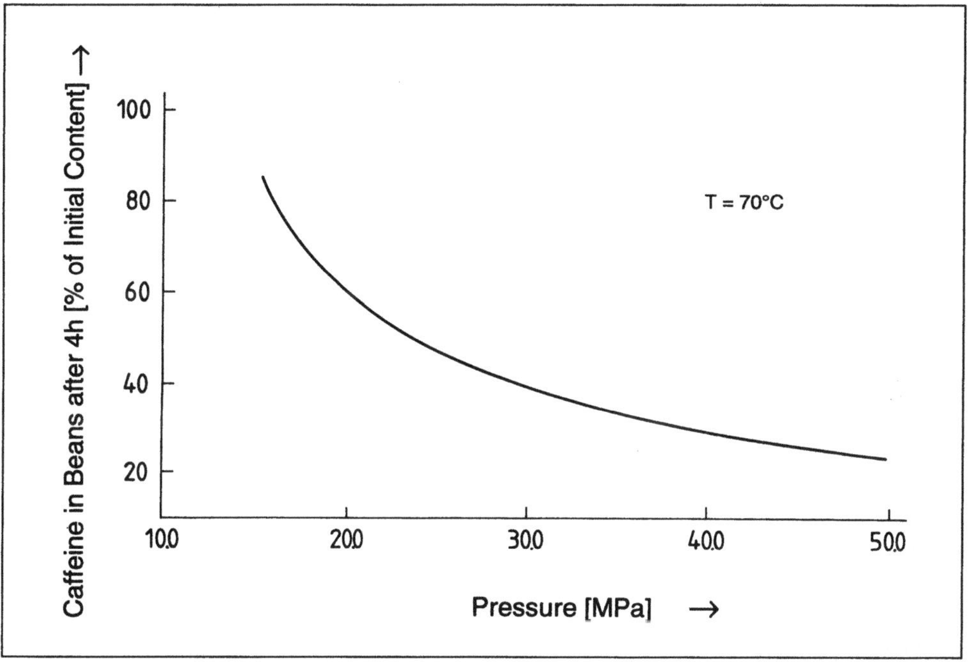


**Fig. 7.12.** Dependence of amount of extract on pressure. Extraction of caffeine from raw coffee beans with supercritical nitrous oxide ($N_2O$) [12].

## Density

With increasing density, the extraction rate increases at constant temperature. Density is responsible for the capacity of a solvent, since, as discussed in Chapter 3, the solubility of a compound rises with increasing density. In the extraction process mass transfer is also of importance. Therefore, the extraction results will be different for the same density at different temperatures. This is illustrated in Fig. 7.14.

## Solvent ratio

Solvent ratio is the most important parameter for gas extraction, once approximate values of pressure and temperature are selected. With increasing solvent ratio

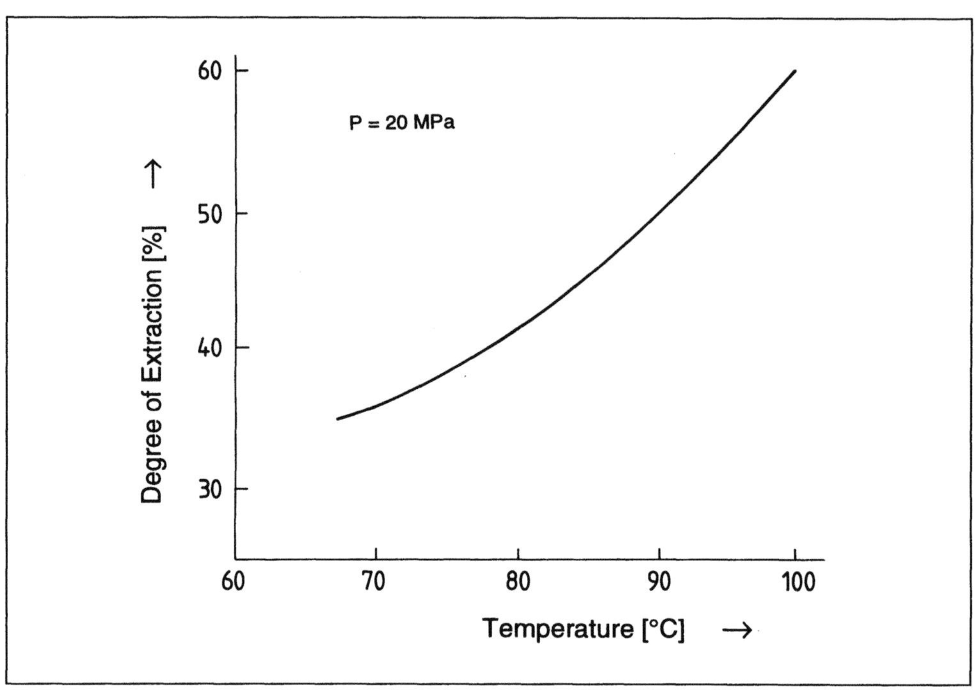



**Fig. 7.13.** Dependence of amount of extraction rate on temperature. Extraction of caffeine from raw coffee beans with supercritical nitrous oxide ($N_2O$) [12].

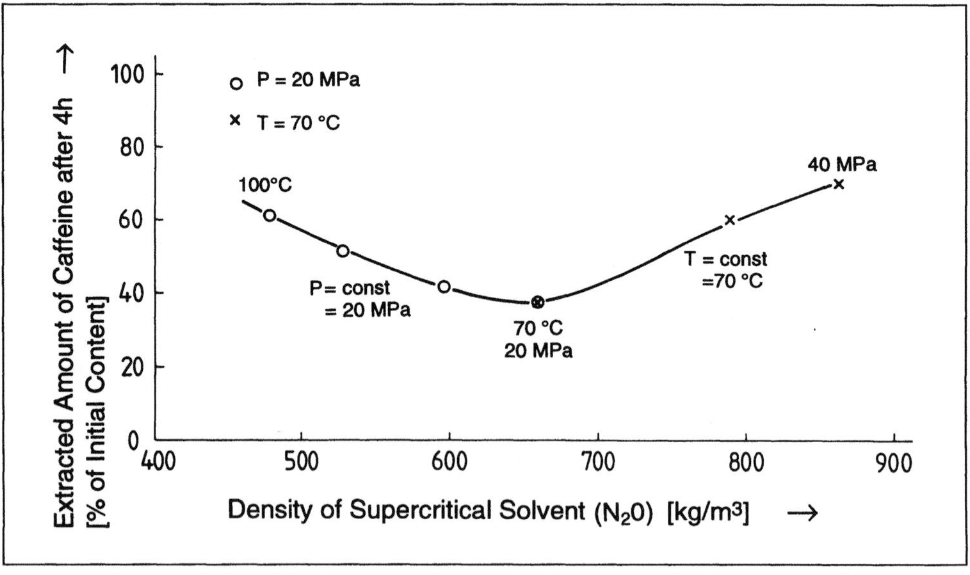



**Fig. 7.14.** Dependence of extraction on density. Extraction of caffeine from raw coffee beans with supercritical nitrous oxide ($N_2O$).

194

the extraction rate can be enhanced more than with changing process parameters in a relatively narrow limit. In Fig. 7.15, dependence of the extracted amount on solvent ratio is shown for constant process parameters for the example of the extraction of caffeine from coffee beans.

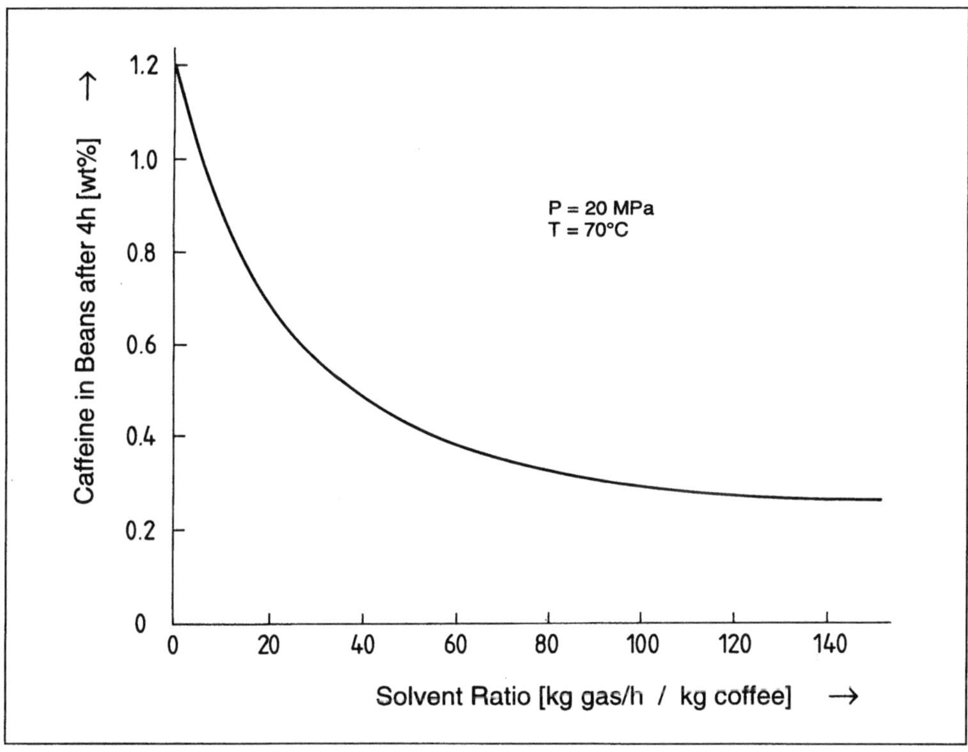


**Fig. 7.15.** Dependence of the amount of extract on solvent ratio. Extraction of caffeine from raw coffee beans with supercritical nitrous oxide ($N_2O$) [12].

At low solvent ratios, the remaining amount of extract in the solid substrate is high after a certain time of extraction. In a medium range for the solvent ratio, its influence on the extraction result is greatest. At very high solvent ratios, the remaining extract content seems to approach a lower limit.

The influence of the solvent ratio cannot be discussed without considering economic consequences. Solvent ratio influences production costs in two ways:

1) A high solvent ratio, resulting in short extraction times, causes enhanced operating costs per unity of product quantity, if the loading of the solvent decreases with increasing solvent ratio. In addition, capital costs are higher for high solvent ratios, because the equipment for the supercritical solvent cycle is more expensive due to its larger size.

2) A high solvent ratio on the other hand increases the amount of extract and throughput of solid material, thus decreasing production cost per unit quantity of product. Since gas extraction is a process with high monetary costs, this effect in general is dominant, but there may be limitations, if cycling the solvent is costly.

If monetary costs are predominant the solvent ratio should be chosen according to the maximum extraction rate achievable. With increasing solvent ratio, the extraction rate increases. But loading of the supercritical solvent decreases at high solvent ratios, due to short residence times of the solvent. At some solvent ratio the extraction rate will be maximal. On the example of the extraction of caffeine from coffee beans, this is shown in Fig. 7.16. At high solvent ratios, beyond about 150 $kgN_2O/(kg_{solid}\,h)$, the total amount of extract per unit of time declines.

If costs for cycling the supercritical solvent per unit quantity of product are dominant, then the solvent ratio will be more determined by the minimum quantity of solvent. The minimum quantity of solvent corresponds to a flow rate of the solvent with which equilibrium loading of the solvent at the exit of the extractor is just achieved. The actual optimum solvent ratio depends on the individual conditions and must be determined for each process and location. In this optimization process the possibilities of the regeneration of the solvent must be included, especially considering the residual concentration downstream to the regeneration. In addition to economical considerations, there may be further restrictions, e.g., a negative influence of extraction time on the quality of the product.

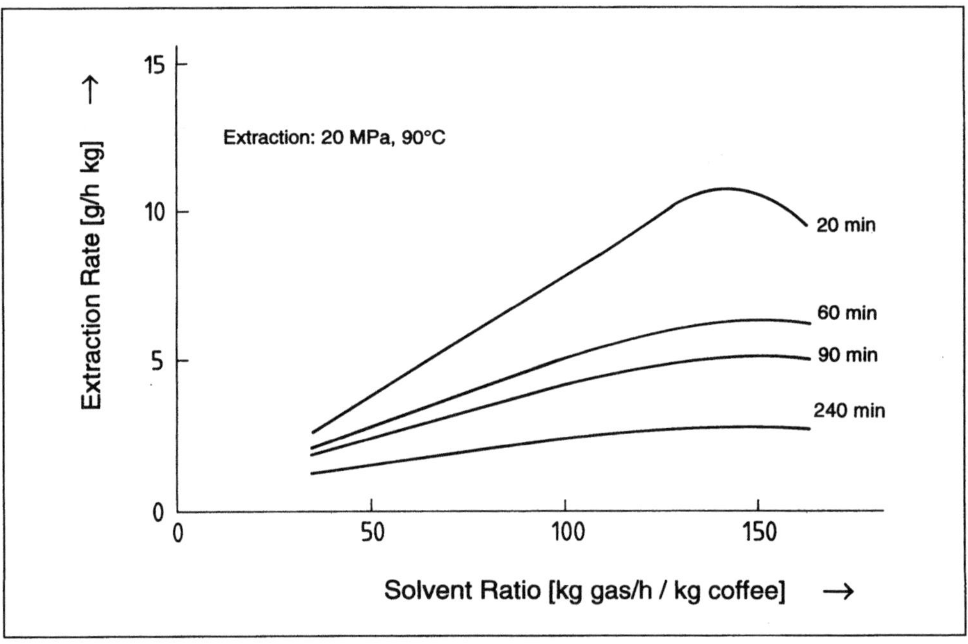


**Fig. 7.16.** Dependence of extraction rate on solvent ratio. At high solvent ratios a maximum is obtained.

**Size of solid particles**

Mass transfer in gas extraction from solid substrates in most cases depends heavily on the transport rate in the solid phase. The length of the transport path determines mass transport in the solid phase. In general the extraction rate increases with decreasing particle size. An example is given in Fig. 7.17 for the decaffeination of coffee beans which have been ground to different particle sizes. In this case, the amount of extract in a certain time (4 h) increases with decreasing particle size.

On the other hand, mass transfer has to be achieved into the fluid phase. If the smaller particles hinder fluid flow in the fixed bed, then mass transfer rate decreases with smaller particles. This can be shown by the extraction results of theobromine from cocoa seed shells, as shown in Fig. 7.18. For this solid substrate, mass transfer rate is lowered if smaller particles are applied for extraction. The reason is the geometry of the shell particles. Larger particles retain the ellipsoid shape of a bean, providing space for fluid flow between different layers of particles. Smaller particles are nearly flat sheets, forming dense layers which leave no channels for fluid flow.

Particle size can only be a process parameter in cases where further use of the particles is compatible with size reduction or other pre-processing methods. Green coffee beans and tea leaves cannot be changed in size, but for oil seeds size reduction and other pretreatment methods which open the plant cells and increase the surface of the solid are of advantage. In Fig. 7.19 the extraction curves for differently pretreated oil seeds are shown.

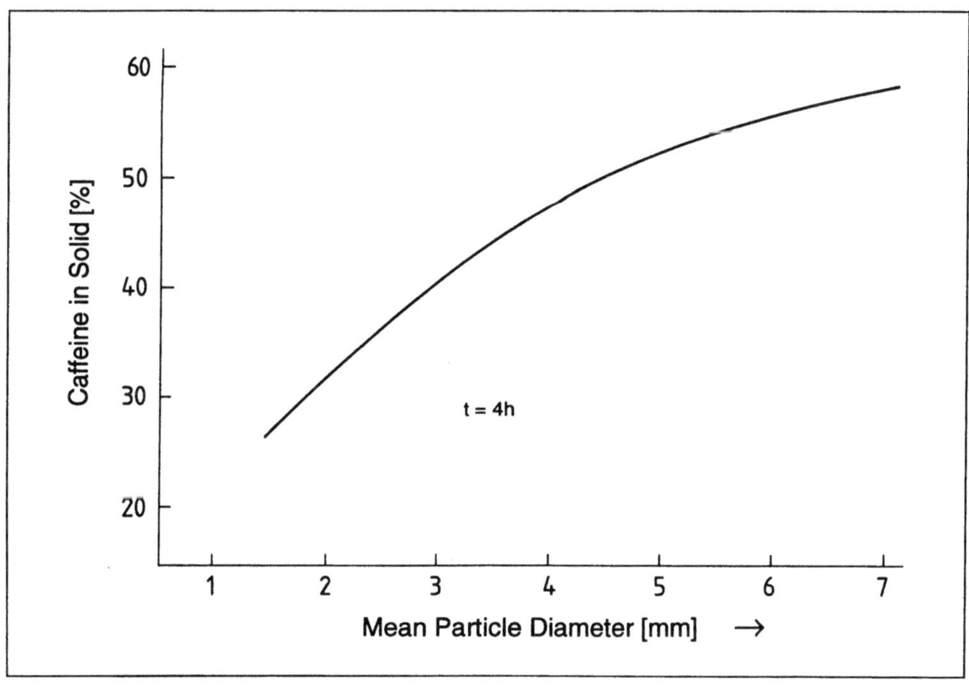


**Fig. 7.17.** Dependence of extraction rate on the size of the particles of the solid substrate [12].

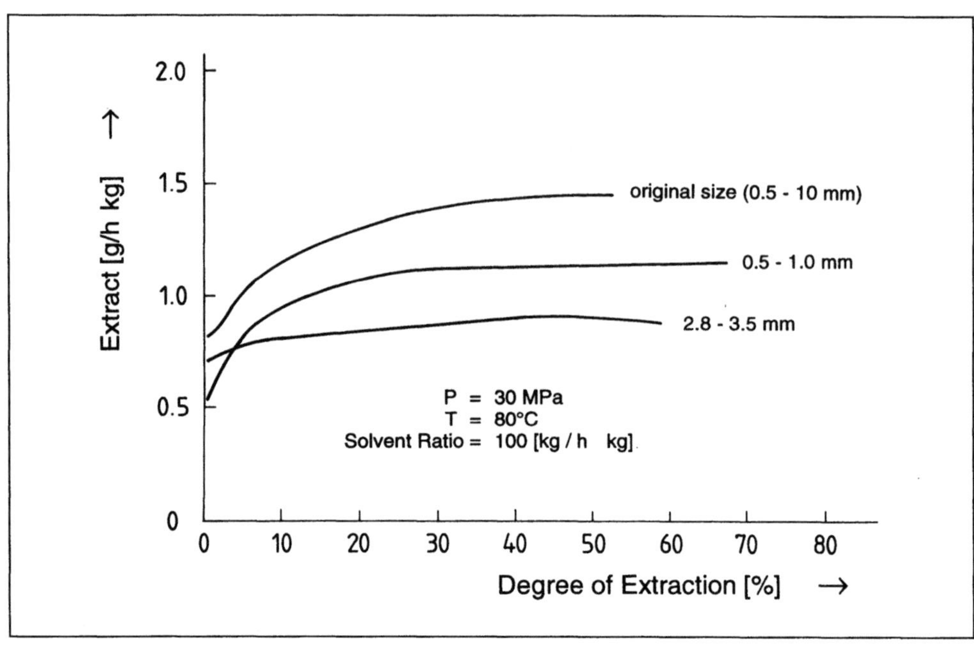



**Fig. 7.18.** Limiting mass transfer by small solid particles. Extraction of theobromine from cocoa seed shells [11, 38].

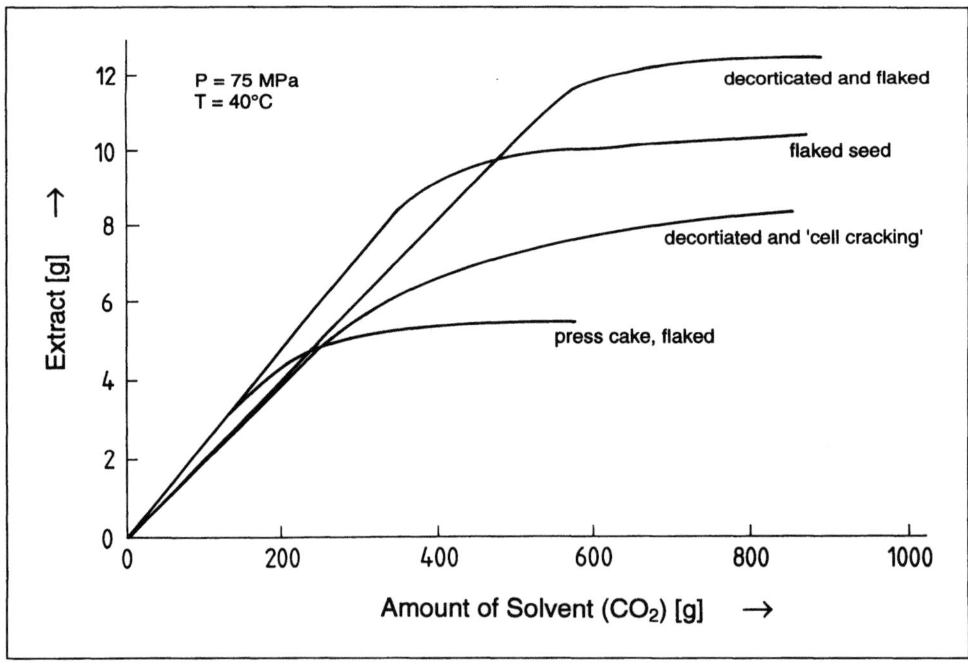



**Fig. 7.19.** Extraction of differently pretreated rape seeds (after Eggers et al. [18]).

198

# 7.3 Modeling the Extraction

The extraction of substances from solid substrates with supercritical solvents can be analyzed and modeled in a simple way by considering only medium values and determination of unknown coefficients by fitting to the extraction curve and a mass balance. The results are simple equations which can represent parts of the extraction curve sufficiently, but fail for others, especially during the first part of the extraction. If the process is to be modeled more accurately, the analysis is far more complex. In the following, the simple way is discussed first. Then, an overview is given on several modeling equations and a somewhat simplified solution is presented and compared to experimental results. Only a fixed bed of solids is considered.

**Table 7.1.** Parameter of an extraction from solids with supercritical gases.

**Fluid phase (extract phase)**

Concentration of the extract downstream of the extraction vessel:
- accumulated quantity of extract;
- quantity of extract per unit of time;
- composition of the extract in dependence of time.

Concentration of the extract in the extraction vessel:
- medium concentration over the total volume;
- concentration throughout the extraction vessel for plug flow;
- local concentrations considering radial distribution (no backmixing, but no plug flow);
- local concentrations considering radial and axial distributions (backmixing):

**Solid phase (raffinate phase)**

Concentration of the extractible substances in the bulk solid:
- accumulated depletion of the solid (mean value for the extraction vessel);
- depletion of the solid related to the remaining content of extractible substances (mean value for the extraction vessel);
- remaining concentration of extractible substances: radial and axial distribution.

Concentration of the extractible substances in single particles:
- mass transport by diffusion;
- mass transport resistance by chemical reactions and/or phase transitions;
- simple geometric particles, complex shape of particles;
- monodispersity of the solid particles (size), multidispersity of the solid particles (size distribution).

**Operating parameters**

- Pressure, temperature, density of the fluid, quantity of solvent per unit of time and mass of solid (solvent ratio);
- chemical composition of the extracting solvent.

**Pretreatment of the solid**

- size reduction and enlargement of surface;
- destruction of the plant cells;
- adjustment of the water content;
- chemical reactions for setting free the extract compounds.

### 7.3.1 Basic Considerations

The extraction system involves a bulk solid phase and a fluid phase. The fluid phase comprises the supercritical solvent and the dissolved extract. In many cases the fluid can be assumed to be identical with the supercritical solvent. But this has to be examined for each case. The solid phase remains within the extraction vessel, the fluid phase is passed through the extraction vessel (semi-continuous operation). Mass transport occurs between the two phases. Substances are transported from the solid to the fluid phase. In both phases the following parameters may be of interest (Table 7.1).

Mass transfer in the solid and fluid phases is coupled by a mass balance for the transferred substances and by the equilibrium relations for phase transitions. Operating parameters and pretreatment of the solid control the extraction process.

The extraction process can be modeled using various relationships between the parameters. These relations contain models with specific coefficients, e.g., mass transport models. The coefficients of these models may be obtained by independent methods, but can be calculated by fitting the result of the extraction to the model. The goal of the modeling procedure is to get a sufficient quantitative representation of the process with a simple system of equations and few and physically meaningful parameters.

### 7.3.2 Steady State Approximation of the Extraction from Solids

The extraction of substances from solid material can be approximately modeled by assuming a steady state process. This is a reasonable first approximation since the extraction processes are slow, residence time of the fluid is in the order of minutes, and extraction time for the solid is in the range from about 10 min to several hours.

The amount of substances, $\overset{*}{m}$, extracted per unit of time $t$, is given by the following:

$$\frac{dm}{dt} = \overset{*}{m} = -m_s \frac{dc_m}{dt},$$
7.1

with
$m$ = mass of extract components;
$m_s$ = mass of the solid substrate;
$c_m$ = mean concentration of extractible components in the solid.

The extract is transported within the solid to the interface solid-fluid and from there into the bulk of the fluid:

$$\overset{*}{m} = \beta_s A (c_m - c_0);$$
7.2

$$\overset{*}{m} = \beta_F A (c_0 - c_\infty).$$
7.3

200

Total mass transfer resistance is given by Eq. 7.4, if no other transport resistances must be taken into account and there is no phase transition at the interface solid-fluid (or the equilibrium distribution coefficient is set to 1).

$$\frac{1}{k} = \frac{1}{\beta_s} + \frac{1}{\beta_F},$$

$\qquad$ 7.4

with

$\beta_s$ = mass transfer coefficient in the solid phase;
$\beta_F$ = mass transfer coefficient in the fluid phase;
$k$ = total mass transfer coefficient;
$c_0$ = initial mean concentration of extractible components in the solid;
$c_I$ = concentration of extractible components at the interface solid-fluid;
$c_\infty$ = concentration of extracted components in the bulk of the fluid;
$A$ = mass transfer area.

If a phase transition occurs, the equilibrium at the interface has to be taken into account in addition.

The differential equation can be integrated if the total mass transfer coefficient $k$ is assumed to be constant. For $k = $ const., Eq. 7.5 for the mean concentration of the extract components in the solid can be obtained:

$$\frac{c_m - c_\infty}{c_0 - c_\infty} \approx \exp\left(-\frac{k A}{m_s c_0} t\right).$$

$\qquad$ 7.5

If it can be assumed, that mass transport resistance is dominant in the solid, then the result is:

$$\frac{1}{\beta_F} \ll \frac{1}{\beta_s} \quad \text{and} \quad k \approx \beta_s.$$

$\qquad$ 7.6

Equation 7.5 is able to represent the extraction process after the initial extraction period (long time solution). The first part of the extraction (short time solution), which is important for the total extraction, cannot be modeled with this simple model.

It would be necessary to include in the model one or two terms which consider mass transfer resistances by chemical reaction ocurring in the solid during the dissolution of extract components and the afore-mentioned phase transition term which, in principle, is always necessary, since extract components are transferred from a condensed state into the gaseous state in the supercritical solvent. Using the simple model without these terms means to include them into the mass transfer coefficients.

The effective mass transport coefficients $\beta_s$, $\beta_F$ or $k$ are generally unknown. They may be calculated from a total mass balance or by means of a transport model. Then,

they can only be used in connection with this model or the method with which they
have been obtained.

Since simple models of this kind are not able to quantitatively represent the extraction, they are not further discussed.

If there is a first part of the extraction with a constant rate of extraction (compare
Fig. 7.9, curve 1 and the related discussion), then mass transfer in this part of the
extraction can be calculated by dimensionless correlations, e.g., the correlation for a
fixed bed, determined by Wakao and Kaguei [42] from the analysis of many data (but
not including gas extraction systems):

$$Sh = 2 + 1.1\, Re^{0.6}\, Sc^{0.33} \quad \text{for} \quad 3 < Re < 3000, \qquad 7.7$$

with

$$Sh = \frac{\beta_F\, d}{D_G}, \qquad Re = \frac{u\, d}{v}, \qquad Sc = \frac{v}{D_G}, \qquad 7.8$$

$\beta_F$ = mass transfer coefficient solid-fluid;
$d$ = diameter of a volume equivalent sphere;
$D_G$ = self diffusion coefficient of the fluid;
$v$ = viscosity of fluid;
$u$ = linear flow velocity of fluid.

According to this correlation, mass transport is proportional to the diffusion coefficient of the gas and the volumetric flow $V$ of the gas with an exponent of 3/5, ($V^{3/5}$).


### 7.3.3 More Complex Models for the Extraction from Solids: The One-Dimensional Dispersion – Single Particle Model

The extraction from solids involves at least two phases. At the interface, the concentrations are unsteady functions. Therefore, the phases have to be modeled separately. If the variation of a variable, mostly concentrations, is considered only in one
dimension, normally along the direction of flow, a one dimensional dispersion model
is obtained. In such a model plug flow is assumed, which is superimposed by axial
mass transport, represented in the model by an effective axial dispersion coefficient.

For modeling the transport processes in the solid, the important mass transport
mechanisms or transport resistances must be represented in the model. In many, diffusion is one of the dominant, if not the only transport mechanism. A concentration
gradient across the solid particle is established during extraction. This can be represented by the second law of Fick. Then, a single particle model can be obtained. The
equations can be simplified, if there are no concentration gradients, or if the concentration curve is known beforehand. Such models have been used for modeling heat

transfer in a fixed bed, e.g., by Anzelius [1], Schumann [34], Nusselt [26], and Hausen [22]. In the fluid phase no axial dispersion and in the solid a linear transport kinetic was assumed. In 1952, Rosen [29] modeled the adsorption in a fixed bed of particles applying a particle model for the solid phase, but with no axial dispersion in the fluid phase.

The one-dimensional dispersion – single particle model consists of equations which consider axial dispersion in the fluid phase and transport resistances in the solid particles. The mass balance for the extract compounds or an individual component in differential form, i.e., for an infinitesimal length of the extraction vessel, can be written as follows:

$$c = c_F + c_{sm}, \qquad\qquad\qquad 7.9.a$$

$$\frac{\partial c}{\partial t} = -\nabla (\vec{u}\, c) + D\, \nabla^2 c, \qquad\qquad\qquad 7.9.b$$

which reduces with only axial transport in the fluid phase and a summation of all effects in the solid phase by the parameter $c_{sm}$ to:

$$\varepsilon \frac{\partial c_F}{\partial t} + (1 - \varepsilon) \frac{\partial c_{sm}}{\partial t} = \varepsilon\, D_{ax} \frac{\partial^2 c_F}{\partial z^2} - u_0 \frac{\partial c_F}{\partial z}, \qquad\qquad 7.9.c$$

with the following initial and boundary conditions:

$$c_F = c_{F0} \quad \text{for} \quad t = 0, z > 0, \qquad\qquad\qquad 7.10$$

$$u_0 c_{F0} = u_0 c_F - \varepsilon D_{ax} \frac{\partial c_F}{\partial z} \quad \text{for} \quad z = 0, t > 0, \qquad\qquad 7.11$$

$$\frac{\partial c_F}{\partial z} = 0 \text{ for } z = L, t > 0, \qquad\qquad\qquad 7.12$$

$c$    = concentration of extract components in the extraction volume;  
$c_F$    = concentration of extract components in the fluid phase;  
$c_{sm}$    = mean concentration of extract components in the solid particle;  
$D_{ax}$    = axial dispersion coefficient;  
$u_0$    = free volume velocity of the fluid;  
$z$    = coordinate of axial length;  
$L$    = length of the fixed bed;  
$\varepsilon$    = void volume fraction (porosity) of the fixed bed;  
$t$    = time of extraction.

Equation 7.9.c connects the concentration of the extract components in the fluid to the concentration of the extract components in the solid as functions of axial length and time of extraction $t$. The terms on the lefthand side of the equation represent the accumulation or depletion of the extract components in the fluid and solid phase. The terms on the righthand side of the equation represent mass transport by convection and dispersion in the fluid phase. The boundary conditions, Eq. 7.11 and Eq. 7.12, satisfy the condition of plug flow at the lower and upper ends of the fixed bed.

Equation 7.9.c contains the assumption that the density of the fluid is constant. It is necessary to think about the applicability of this assumption. The density of the fluid may be changed by the extracted components and by variations of temperature and pressure. In the case of gas extraction, the fluid solvent is a supercritical gas whose density is strongly dependent on temperature and pressure. In general, the extraction vessels are operated at constant temperature. Pressure drop across the fixed bed of solids is low, typically in the range of 0.1 MPa at a total pressure of 20 to 30 MPa. This is due to the low viscosity of the supercritical gas and the relatively high porosity of the fixed bed of $\varepsilon > 0.5$. The loading of the fluid solvent in the case of a supercritical gas is normally in the range of 1 to 3 wt.-%. Therefore, the change in density is small and negligible compared to the density. But there may be cases where the assumption of constant density is not applicable, especially if this model is applied to packed column chromatography.

### 7.3.4 Particle Model

The nonsteady concentration distribution of a substance during the process of extraction from a solid substrate can be expressed by the second law of Fick. For spherical particles the following equation is obtained:

$$\frac{\partial c_s}{\partial t} = D_{es} \left( \frac{\partial^2 c_s}{\partial r^2} + \frac{2}{r} \frac{\partial c_s}{\partial r} \right), \qquad 7.13$$

with
$D_{es}$ = diffusion coefficient of extract components in the solid;
$R$ = radius of the particle;
$r$ = radial coordinate in the particle;
$\beta_F$ = mass transfer coefficient for the fluid phase;
$c_s$ = concentration of extract components in the solid.

Boundary conditions are given by:

$$c_s = c_{s0} \qquad \text{for } t = 0, 0 \leq r \leq R, \qquad 7.14$$

$$\frac{\partial c_s}{\partial r} = 0 \qquad\qquad \text{for } r = 0,\, t > 0, \qquad\qquad 7.15$$

$$\beta_F\,(c_{sR} - c_F) = -\,D_{es}\,\frac{\partial c_s}{\partial r} \qquad \text{for } r = R,\, t > 0. \qquad\qquad 7.16$$

The mean concentration of the extracted substance in the particle is obtained by integrating the concentration profile over the particle volume $v$:

$$c_{sm} = \frac{3}{c_{s0}\,\pi\,R^3} \int\limits_0^R [4\pi r^2\, c_s(r,t)]\,\mathrm{d}v. \qquad\qquad 7.17$$

In this approach the solid substrate is treated as an isotropic medium. Real solid substrates in most cases are anisotropic in a complicated way. The details of mass transport within the particle normally are not of interest, given that the extraction result can be obtained by simpler methods. Therefore mass transport in such solid substrates for our purposes is treated for the isotropic case. The transport coefficient is then interpreted as an effective diffusion coefficient or an effective transport coefficient.

## 7.3.5 Coupling Condition between Fluid Phase Model and Particle Model

At the interface solid-fluid phase the models describing the two phases must be coupled. In case of a homogeneous extraction (an extraction wherein the solvent does not change its phase at the interface), the coupling equation is simple: At the interface solid-fluid the concentration of the extracted substances must be equal.

In case of a heterogeneous extraction (an extraction wherein the solvent changes its phase at the interface) a partition coefficient for the transported components must be taken into account, in most cases given by the thermodynamic equilibrium partition coefficient.

$$K = \frac{c_F}{c_s} \qquad \text{for } r = R. \qquad\qquad 7.18$$

Equations 7.9.c, 7.13, 7.17, 7.18 and boundary conditions given by Eqs. 7.10, 7.11, 7.12 provide a system of equations for modeling an extration from a solid substrate.

## 7.3.6 Dimensionless Groups

The solution of this system of equations is simplified by combining parameters to dimensionless groups which are commonly used in the analysis of technical problems. Thus, the number of variables is reduced and the comparison to other results for other substrates and process conditions as well as scale up is made more feasible than with the original parameters. These dimensionless groups are:

$$Fo = \frac{D_{es}\,t}{R^2} \qquad \text{Fourier-number} \qquad 7.19$$

$$Bi = \frac{\beta_F\,R\,u_0}{D_{es}} \qquad \text{Biot-number} \qquad 7.20$$

$$Bo = \frac{u_0\,L}{D_{ax}} = Pe\,\frac{L}{d} \qquad \text{Bodenstein-number} \qquad 7.21$$

$$Pe = \frac{u_0\,d}{D_{ax}} \qquad \text{Peclet-number} \qquad 7.22$$

The solution of the resulting system of equations, comprising the model for the extraction of substances from solid substrates, encounters problems with the mathematical solution and with determination of the parameters. The dimensionless groups require knowledge on the three kinetic parameters $\beta_F$, $D_{es}$, and $D_{ax}$. At least for $D_{es}$ and $D_{ax}$ the existing experimental data base is too small for a generalization. Therefore, these parameters have to be determined experimentally under process conditions by fitting them with the extraction process model to the experimentally measured course of extraction. With increasing amount of data, correlations for these kinetic coefficients will become available.

In the following, some aspects of modeling and some simplifications are discussed in more detail.

## 7.3.7 Some Aspects for Modeling the Fluid Phase

### 7.3.7.1 Hydrodynamic Behavior and Axial Dispersion

In supercritical solvent extraction from solid substrates an axial dispersion of fluid flow can be observed (often indirectly by the result of the extraction). In a fixed bed,

in most cases plug flow is assumed. Since hydrodynamic flow across the fixed bed is not uniform, an axial dispersion is the result. The reasons are:
- nonuniform distribution of the solvent at the entrance of the extractor;
- radial distribution of the porosity;
- radial distribution of the viscosity of the solvent (due to concentration and temperature gradients);
- non steady or non stabile hydrodynamic flow.

An axial dispersion is taken into consideration by the Peclet-number. The higher the axial dispersion (axial backmixing), the lower the Peclet-number.

Nonuniform distribution of the solvent can be avoided by appropriate construction of the solvent distributor. For a gaseous solvent a perforated plate or a sinter metal plate with a pressure drop of around 0.1 MPa is sufficient.

A radial distribution of viscosity is mainly encountered during heat transfer from the wall of the vessel. Viscosity is strongly dependent on temperature. Therefore, temperature gradients near the wall of the vessel change viscosity and affect hydrodynamic flow. For supercritical gases, this effect can be important, since in the range of conditions for gas extraction properties of the solvent strongly depend on temperature (and pressure). Especially for fixed beds of large diameters it may be difficult to ensure an even temperature throughout the bed. On the other hand, temperatures in most cases are relatively low in gas extraction and heat transfer mainly occurs by the solvent. Therefore, a radial temperature distribution can be neglected in most cases. If not, a model taking radial distribution into consideration may be employed (see below).

Radial distribution of porosity is a more severe reason for axial dispersion. As is well known, porosity $\varepsilon$ in a fixed bed of particles increases in the neighborhood of the wall of a vessel up to the value of $\varepsilon = 1$ at the wall (Fig. 7.20). The radial distribution of the flow velocity of a gas in a tubular reactor is shown in Fig. 7.21 for the isothermal and non-isothermal case.

The influence of porosity decreases with increasing ratio of vessel diameter $d_V$ to particle diameter $d_p$. For values of $d_V/d_p > 10$, the influence is negligible [17]. Note that this is true only for fixed beds of monodisperse particles and without channeling.

Instabilities in hydrodynamic flow are a further reason for axial dispersion. Such instabilities occur because of gradients in density, if density increases against gravity, or because of gradients in viscosity, if viscosity increases in direction of the flow and a stability limit for the flow regime is passed. Such gradients are caused by the solvation process of the solvent.

If the solvent flows against gravity, the critical flow velocity can be calculated [17]:

$$u_{crit} = K'g \frac{d\varrho}{d\mu}, \qquad\qquad 7.23$$

$K'$ = permeability of the fixed bed, e.g., [m$^2$];
$g$ = acceleration due to gravity, e.g., [m/s$^2$];
$\mu$ = viscosity.

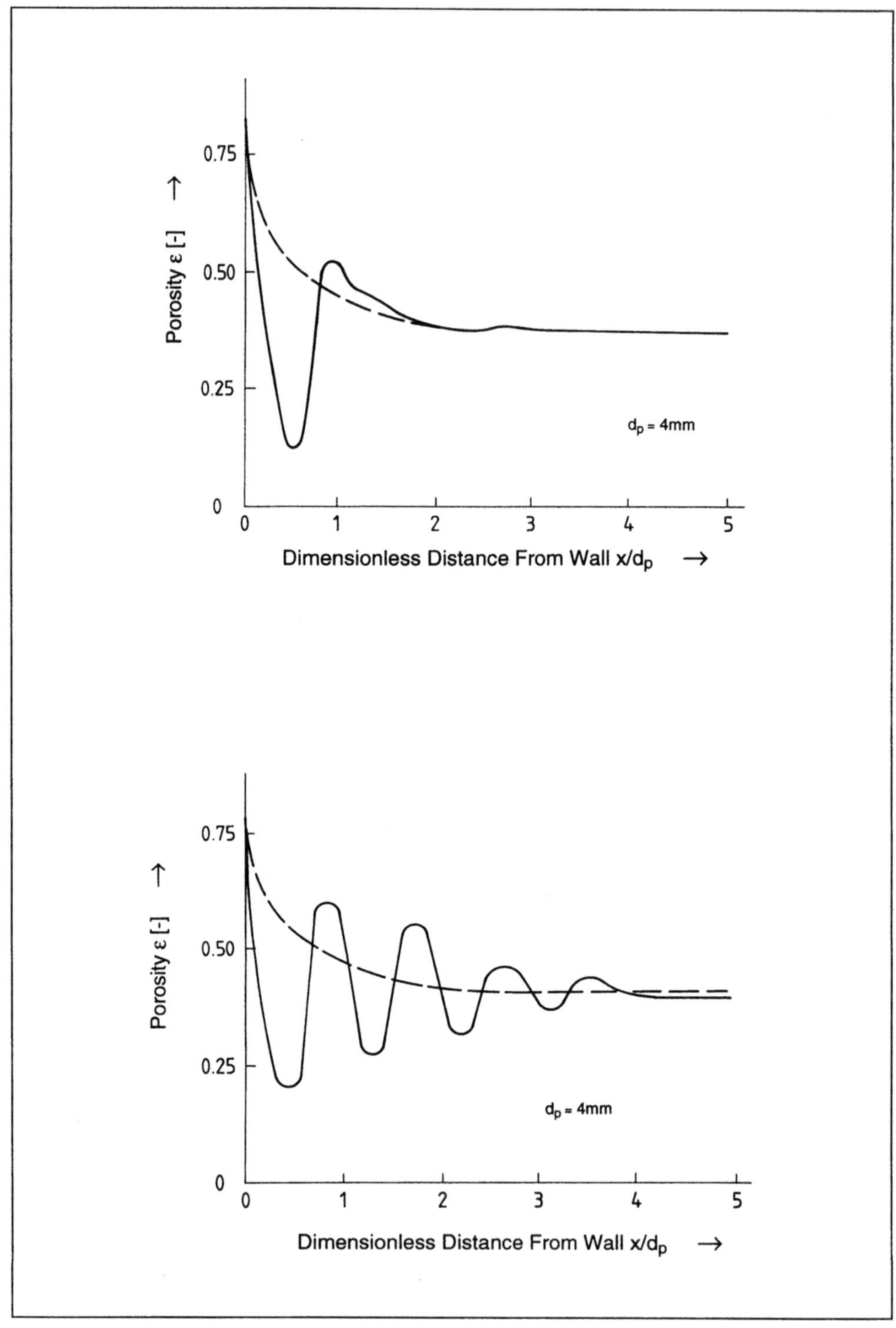

**Fig. 7.20.** Radial distribution of porosity in a fixed bed of particles (after Bennati et al. [3]).

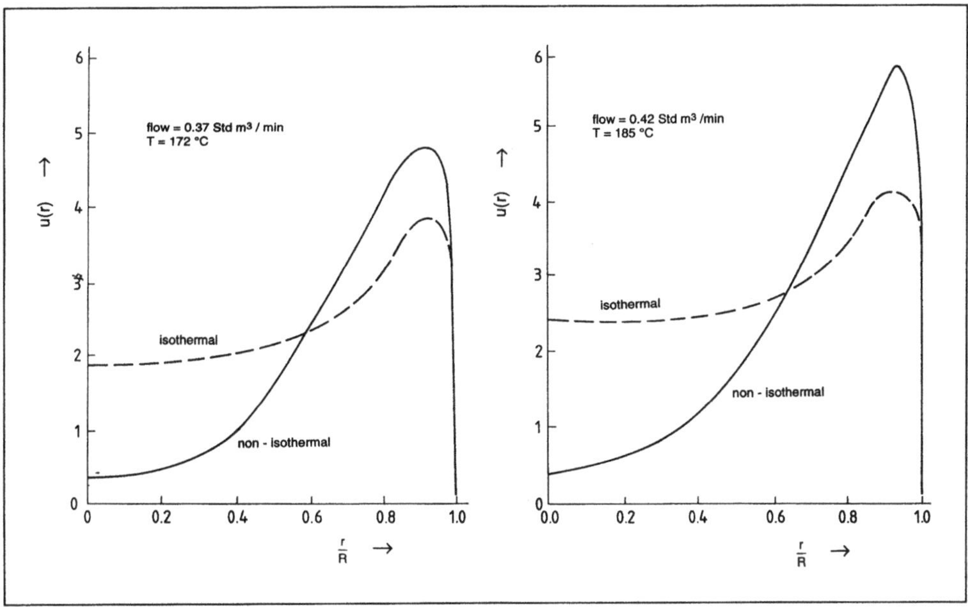


**Fig. 7.21.** Radial distribution of flow velocity in a tubular reactor (after Schertz et al. [33]).

Instabilities may occur if the critical velocity $u_{crit}$ is exceeded. Decreasing the velocity will stabilize the flow regime. For the gas extraction process it follows that significant differences should occur if the fixed bed is contacted by a gas flow against gravity or in direction of gravity. Such an influence was experimentally found by various authors, e.g., Dams [15, 16]. Some of his results are shown in Fig. 7.22. Naphthalin was extracted by carbon dioxide from a fixed bed of porous spheres. The results show that the process cannot be modeled without considering axial dispersion, if the flow direction is against gravity, and linear flow velocity is low. Beutler et al. [4, 5] found that the rate of extraction from pepper and draff was enhanced if the supercritical carbon dioxide flowed from upside down in the direction of gravity.

This argumentation is only valid for relatively small solvent ratios resulting in low flow velocities of the solvent. At higher solvent ratios these effects are negligible, as is shown in Section 7.4 for the example of the extraction of theobromine from cocoa seed shells, and has been experienced by the author during the development of decaffeination processes. High solvent ratios provide a more efficient utilization of the expensive high pressure volume of the extractor and are therefore preferred in large scale applications.

The process of dissolution of extracted compounds in the solvent during the extraction process decreases the Peclet-number and therefore increases axial dispersion. If the value for the axial dispersion coefficient is determined in a fixed bed of inert particles, higher values for the Peclet-number are obtained than are valid during a real extraction process. These differences may be very significant. Schwartzberg et al.

[36] measured the extraction of sugar from ground, roasted coffee. With mass transfer of sugar, the resulting Peclet-number was 0.15; without mass transfer it amounted to 0.25.

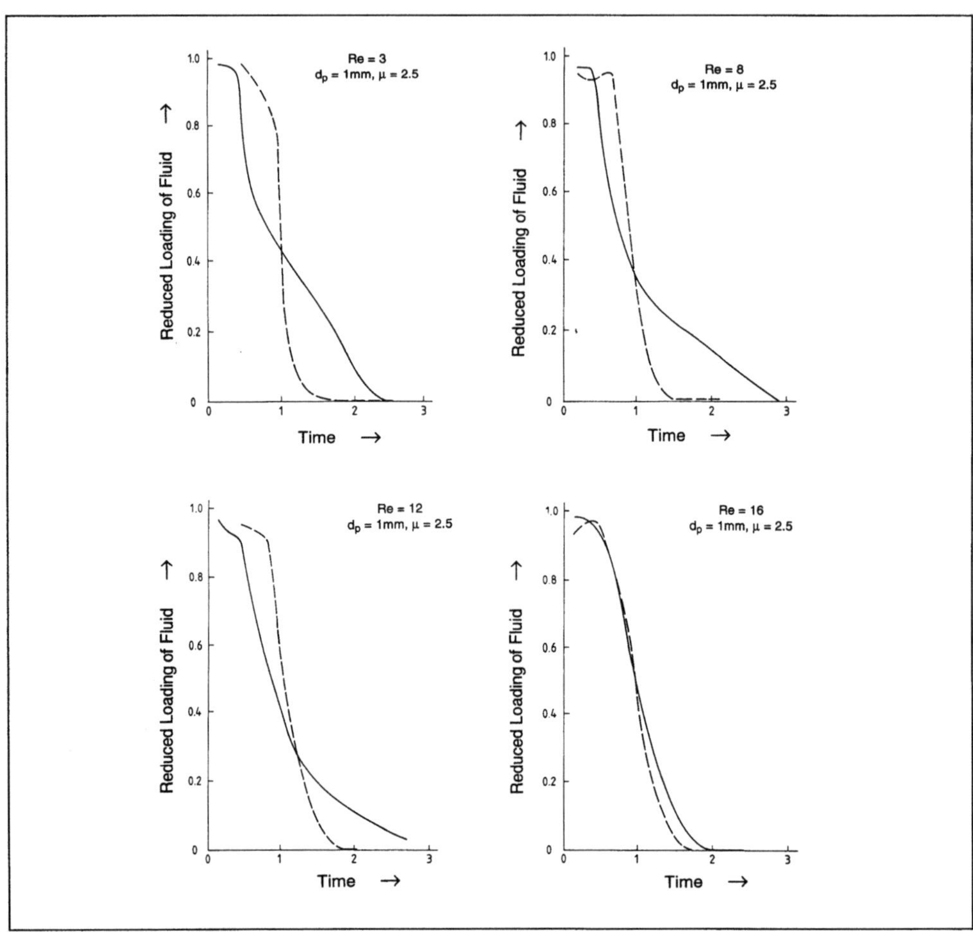


**Fig. 7.22.** Influence of direction of flow on the extraction result. Naphthalene-$CO_2$ (after Dams [16]).

## 7.3.7.2 Model without Axial Dispersion

If axial dispersion is considered negligible against convective flow, the first term on the right side of Eq. 7.9.c can be omitted, as well as Eq. 7.12. The boundary condition, Eq. 7.11, can also be simplified. For this system of equations, Rosen [29, 30] has given analytical and numerical solutions. In Fig. 7.23 the numerical solution is shown for breakthrough curves in an adsorption. The rate parameters for mass transfer and pore diffusion serve as parameters in the diagram. The parameters in the diagram have the following meaning:

$$Da_1 = (\varrho_s k_1 S_m L)/u_0,$$  7.24

$\varrho_s$ = bulk density of solid;
$k_1$ = adsorption rate coefficient;
$S_m$ = specific surface;
$L$ = length of adsorption bed;
$u_0$ = free volume velocity.

$$\psi = R_P \sqrt{(k_1 S_m \varrho_P)/D_e} ,$$  7.25

$R_P$ = radius of particle;
$\varrho_P$ = density of particle;
$D_e$ = effective diffusion coefficient.

$f_1$ = $C_f/C_{fo}$ remaining part in the void volume of the fixed bed.

$$\varrho = (\varepsilon_S \varrho_P) / [\varrho_S (K_L \varrho_P + \varepsilon_P)] \quad \text{(capacity factor)},$$  7.26

$\varepsilon_S$ = porosity of the fixed bed;
$\varepsilon_P$ = porosity of the particles;
$K_L$ = equilibrium constant.

If the remaining part is defined appropriately, this solution can be used for modeling the extraction process. For negligible mass transfer resistance ($\psi^2/(3 Bi Da_1) = 0$), the breakthrough curves are steepest. Steep curves are also the result of high effective diffusion coefficients and small particles. For $Fo \leq 50$, Rosen [30] found an approximative solution which is sufficiently accurate in many cases for the calculation of the breakthrough curves:

$$\frac{C_{fout} - K_L c_{s0}}{C_{fin} - K_L c_{s0}} = 0.5 \{1 + \mathrm{erf}\,[F(C \varepsilon / (1 - \varepsilon))]\},$$  7.27

$C$ = constant, specific for the system;
$F$ = f($Bi$, $Fo$) for $Bi < \infty$;
$F$ = f ($Fo$) for $Bi = \infty$;
$c_{s0}$ = initial content in solid.

Resistance to mass transport during the extraction from solid substrates in many cases lies in the solid phase. This means that the Biot-number is high. For Biot-numbers $Bi > 200$, the resulting error is smaller than 1 % if resistance to mass transfer at the interface is neglected. The mass transfer coefficient then approaches infinity [37].

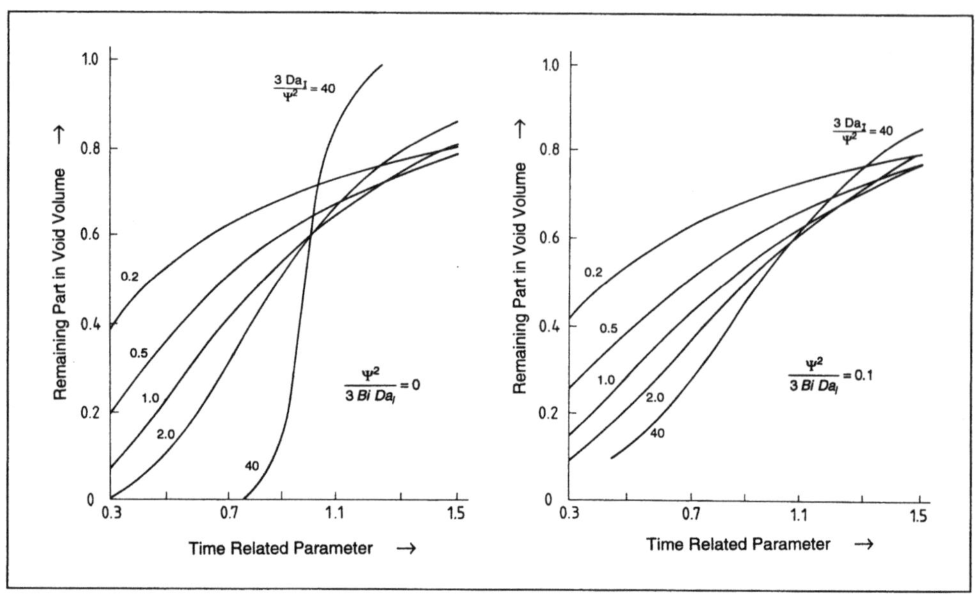


**Fig. 7.23.** Numerical solution for breakthrough curves in adsorption (after Rosen [29, 30]).

In certain cases this simplification is not applicable, especially for
– very small particles;
– low equilibrium partition coefficients;
– heterogeneous extractions, i.e., some gas extractions.


## 7.3.8 Some Aspects for Modeling the Solid Phase


### 7.3.8.1 Concentration Gradients in Fixed Bed of Particles

In a fixed bed of particles the amount of substances which can be extracted is given by the initial content. Any extraction process will therefore lead to an unsteady process with respect to concentration in the solid substrate. Moreover, there must be a concentration profile in the direction of the flow of the solvent. The front of the concentration profile may be steep or flat, depending on the specific process. Of course, we prefer steep profiles, since then most of the fixed bed has the same concentration of remaining extractible components. Radial concentration profiles result from inhomogeneities in flow and are not a necessary consequence of the extraction process, like the axial profiles.

Several questions arise with respect to a practical operation of the fixed bed process:

212

– Is the geometry of the fixed bed (e.g., height to diameter) influencing the overall extraction result?
– To what extent do axial dispersion effects influence the axial concentration profile?
– Are there radial concentration profiles?

The first question was investigated in a bench scale equipment. The solid substrate consisted of seeds (green coffee beans) of about 4 to 7 mm size, with a concentration of extractible substances in the range of 1 wt.-%. The solvent was nitrous oxide and the solvent ratio about 30 [kg/(kgh)], (kg gas per h and kg of dry solid).

In relatively small beds with amounts of material of 0.4 kg, where only the mean value of the total fixed bed could be determined, no influence of the geometry of the bed on the mean value of extracted quantity could be found [2]. Height $h$ of the fixed bed was varied between 50 mm and 175 mm, the diameter $d$ of the bed between 50 mm and 250 mm, resulting in a variation of $h/d$ of: $0.6 < h/d < 3$. The overall extraction result was not influenced by this variation of geometry.

If there exist axial and radial distribution effects due to flow inhomogeneities during the extraction process, then the degree of extraction should depend on the position of a sample within the fixed bed. For this determination in several experiments the fixed bed was partitioned vertically and horizontally. The different sections were analyzed after the experiment for the remaining amount of extracted compounds. Radial and axial concentration profiles were found, which are shown in Fig. 7.24. Similar results werde obtained for cocoa seed shells with a totally different particle geometry (see Section 7.4).

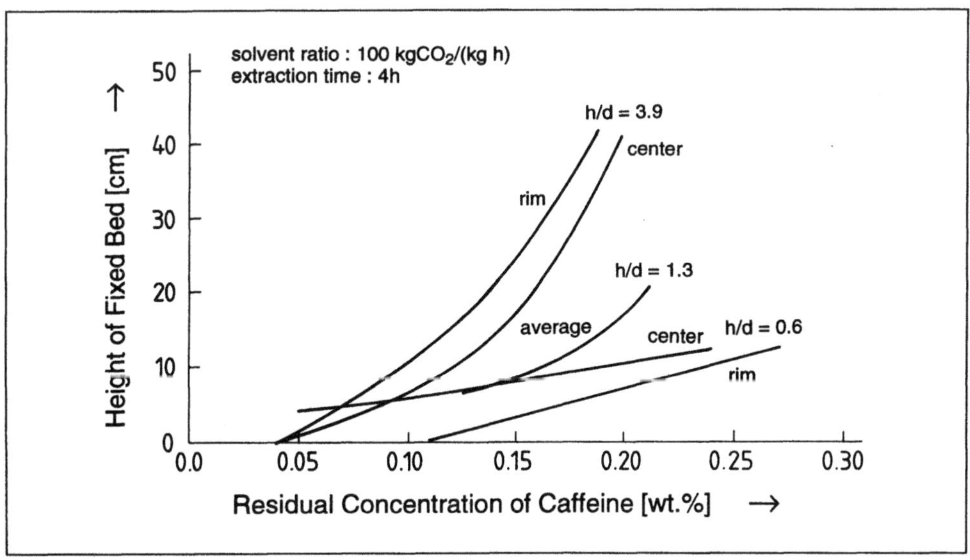


**Fig. 7.24.** Axial and radial concentration profiles in gas extraction from solids. Extraction of caffeine from coffee beans.

## 7.3.8.2 Determination of the Effective Diffusion Coefficient

All considerations on the extraction of substances from a solid, even with a super-critical solvent, assume that diffusion of the extracted components in the solid substrate is an important, if not the only decisive effect for the extraction process. All the models therefore employ an effective diffusion coefficient which must be determined experimentally in combination with a mass transport model. The effective diffusion coefficient therefore is a model-dependent transport coefficient. In general, Fick's second law is applied. The experimental methods are selected in such a way that the boundary conditions of Eq. 7.9 to 7.12 are cancelled and mass transport resistance at the interface is negligible, thus simplifying Eq. 7.16. Such conditions can be verified

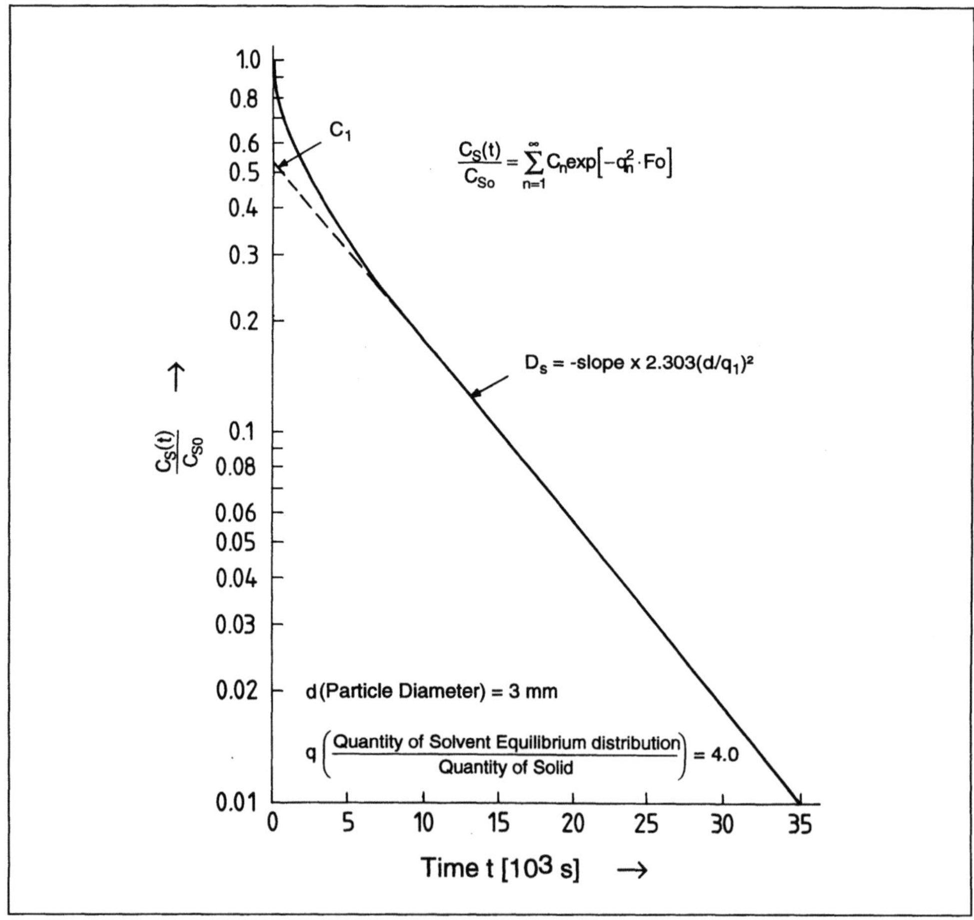


**Fig. 7.25.** Effective diffusion coefficient for the extraction of sugar from coffee-grounds (after Schwartzberg and Chao [37]).

in an ideally stirred vessel or a differential cycle reactor. Solutions of Fick's second law can be found in textbooks [14].

A special solution for the concentration of the extract in the solid, $c_s$, with respect to the initial concentration, $c_{s0}$, is presented in Eq. 7.28 for spherical particles, constant concentration at the interface solid-fluid, and long extraction times:

$$c_s(t) / c_{s0} = 6 / \Pi^2 \exp(-\Pi^2 Fo), \qquad\qquad 7.28$$

$Fo$ = Fourier-number.

If the second law of Fick is valid for the mass transport in the solid phase, Eq. 7.28 represents a straight line in a semi-logarithmic plot. From its gradient the effective diffusion coefficient can then be determined. Figure 7.25 shows this for the extraction of sugar from coffee-grounds [37].

Some natural substances behave differently during extraction. If the extraction of oil from soja flakes is plotted according to Eq. 7.28, no single straight line is obtained. Furthermore, two different regions can be seen, meaning that different mass transport mechanisms are effective in the different regions. At first, a rapid extraction takes place, which is even faster if smaller flakes are extracted. For this region the oil is easily accessible for the extracting solvent. This is due to the destruction of the cells by the flaking process and the relatively high oil content of the flakes. In the second region, extraction is significantly slower. Oil has to be transported in the solid phase, e.g., from intact cells to the interface solid-fluid. Mass transfer resistance is higher than during the first extraction period. This can be due to a structured solid substrate consisting of two different layers with different extent and transport characteristics. Osborne and Katz [27] have applied two different effective diffusion coefficients and thus were able to model the extraction (see Fig. 7.26).

### 7.3.8.3 Modeling the Extraction by Means of Special Solutions of Fick's Second Law

The extraction of caffeine or theobromine from natural substrates is an unsteady process. The main mass transfer resistance usually is attributed to diffusion of the extract substances in the solid material. Experimental results show, however, that in addition there must be a mass transfer resistance at the interface solid-fluid depending on the Reynolds-number of the solvent. In the following it will be proved that there is a mass transfer resistance for the transport of the extract substances into the bulk of the fluid which is important enough to influence the extraction result. If there would be only a mass transfer resistance in the solid substrate, solutions of Fick's second law should reveal a constant effective diffusion coefficient. If the effective diffusion coefficient is not constant, in addition, other resistances than diffusional mass transport resistances are of influence.

The depletion of extract components by gas extraction causes concentration curves in the particle which are similar to unsteady cooling curves, Fig. 7.27. At the starting point of the process, the distribution is uniform. After some time, the ranges near the surface of the particle have begun to adjust their conditions to those outside the particle, while at the middle of the particle conditions have practically not changed. Some time later, the ranges at the surface of the particle have reached the condition corresponding to the surrounding, e.g., equal temperature or a concentration corresponding to the equilibrium distribution coefficient. In the center of the particle conditions have now begun to change. Even later, conditions throughout the particle correspond to conditions of the surrounding.

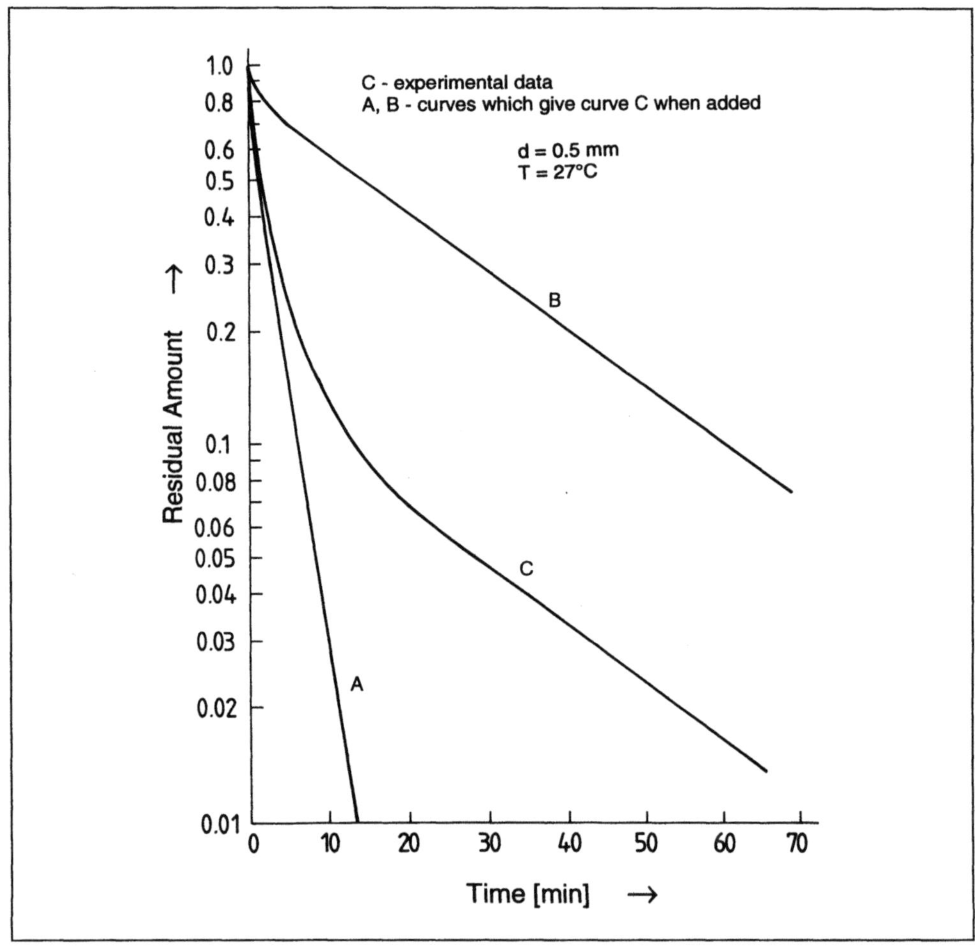


Fig. 7.26. Extraction of oil from soja bean flakes. Modeling the extraction by two different effective diffusion coefficients (after Osborne and Katz [27]).

216

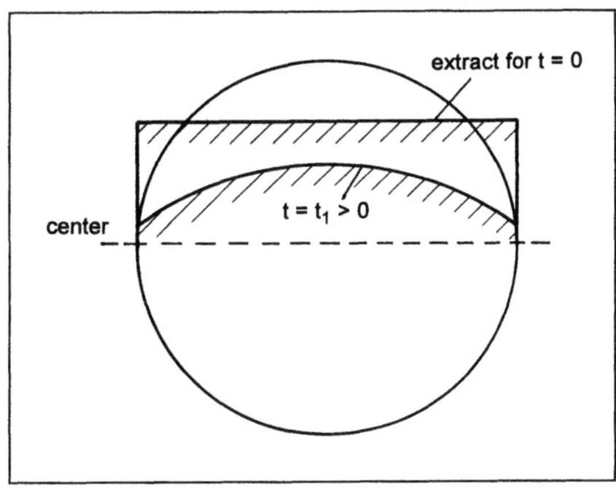



**Fig. 7.27.** Depletion of extract components in a spherical particle.

In heat transfer these processes are modeled with Fourier's law, while in mass transport processes Fick's second law applies [7, 8, 35]:

$$\frac{\partial c_s}{\partial t} = D_{es} \frac{\partial^2 c_s}{\partial x^2}.$$

7.29

The fixed bed can be assumed to consist of particles of the same size and geometry, or can itself be considered as quasi-homogeneous. Solutions for spheres, plates, and cylinders have been published in textbooks [14].

For simple boundary conditions, Eqs. 7.35 to 7.38, and even plates, the solution is given by:

$$1 - E(t) = C_n \exp(- q_n^2 Fo),$$

7.30

$n$ = variable index, $n = 1, 2, 3, \ldots$;
$m_{et}$ = mass of extract in the solid at time $t$;
$m_{e0}$ = mass of extract in the solid at time $t = 0$;
$E(t)$ = degree of extraction, defined by Eq. 7.31:

$$1 - E(t) = m_{et}/m_{e0},$$

7.31

$$C_n = \frac{8}{\pi^2 (2n + 1)},$$

7.32

$$q_n = \left( \frac{(2n + 1)\,\pi}{2} \right)^2,$$

7.33

$$Fo = \frac{D_{es}\,t}{l^2}.$$ 7.34

Boundary conditions are given by the following equations:

$$c_s = c_{s0}, \text{ for } t = 0,\ -l \le x \le l,$$ 7.35

$$c_s = 0, \text{ for } x = l,\ t \ge 0,$$ 7.36

$$\frac{\partial c_s}{\partial x} = 0, \text{ for } x = 0,\ t \ge 0,$$ 7.37

$$D_{es} = \text{const.}, \text{ for } Bi \to \infty.$$ 7.38

An effective diffusion coefficient can be obtained for any time of the extraction process, if course and result of the extraction are known. An example is presented in Fig. 7.28.

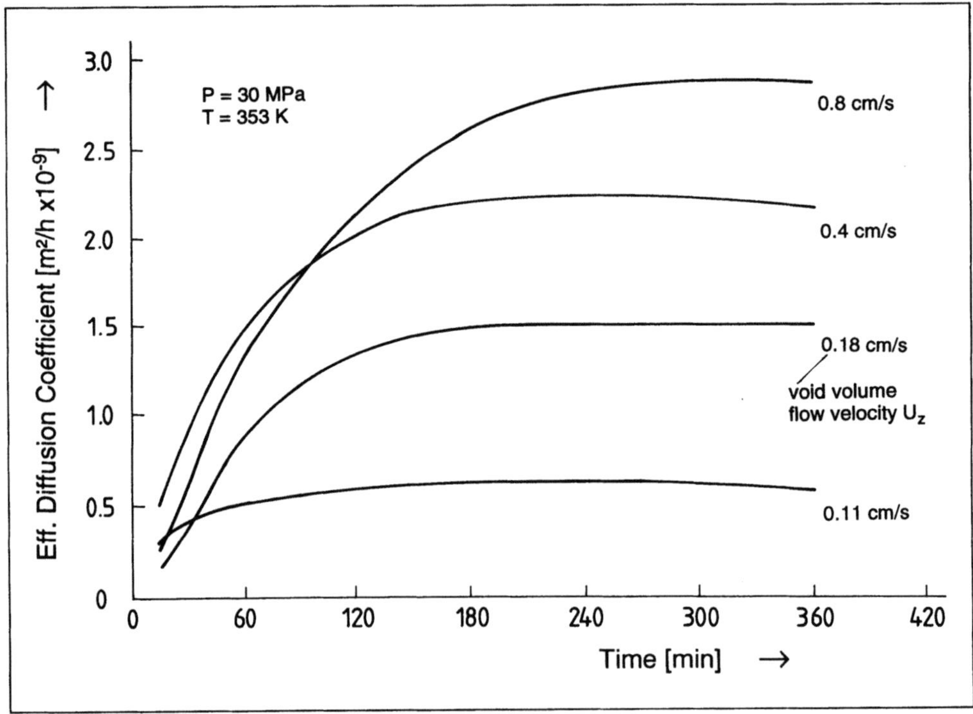


**Fig. 7.28.** Effective diffusion coefficient during an extraction of theobromine from cocoa seed shells with supercritical carbon dioxide [35, 43].

The effective diffusion coefficient is not constant during the extraction. In the first part of the extraction, there is obviously an influence of the mass transfer at the interface solid/fluid. The influence of convective mass transfer has been calculated from several experiments with different solvent ratio and plotted in Fig. 7.29 as the effective diffusion coefficient for long extraction times, against the void volume velocity of the solvent. The effective diffusion coefficient increases with increasing velocity. At high velocities, corresponding to high solvent ratios, the effective diffusion coefficent approaches a limiting value.

That the transport coefficient is dependent on extraction time is of substantial practical importance. Therefore, the first part of the extraction can be influenced by the solvent ratio. A high solvent ratio at the beginning of the extraction can substantially reduce the necessary total extraction time.

Only at very long extraction times does the effective diffusion coefficient reach a constant value, and mass transport resistance in the solid substrate dominates. The solvent ratio is of no importance for the rate of extraction. With the effective diffusion coefficient, determined from long extraction times, the end part of the extraction, or the total amount extracted can be represented, since the diffusion coefficient has been calculated from the amount of extract (Fig. 7.30).

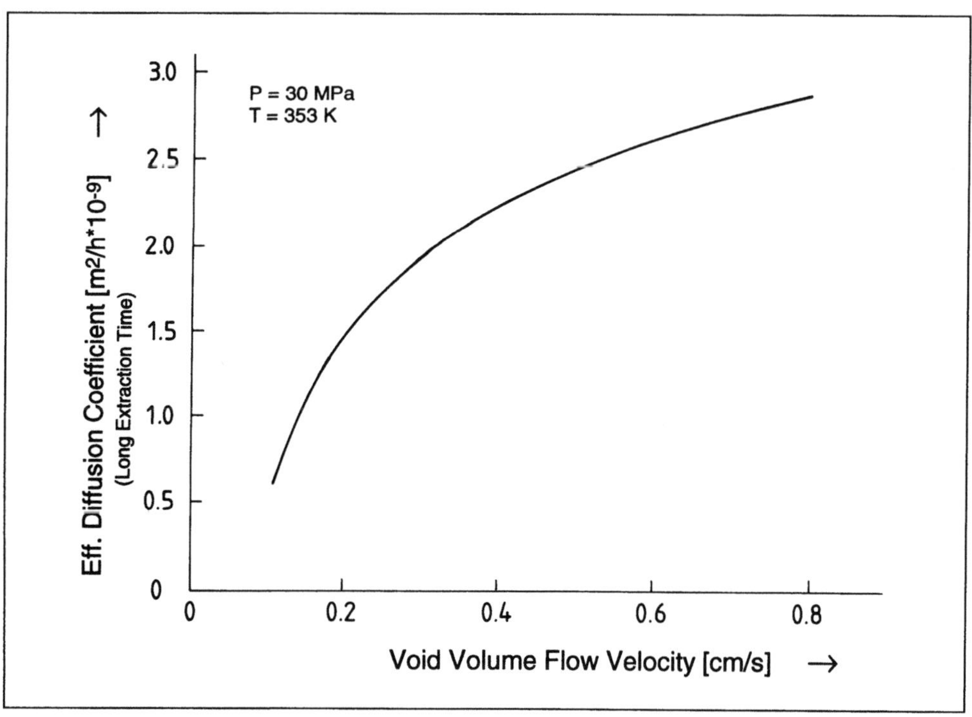


**Fig. 7.29.** Influence of flow velocity of the solvent on the effective diffusion coefficient.

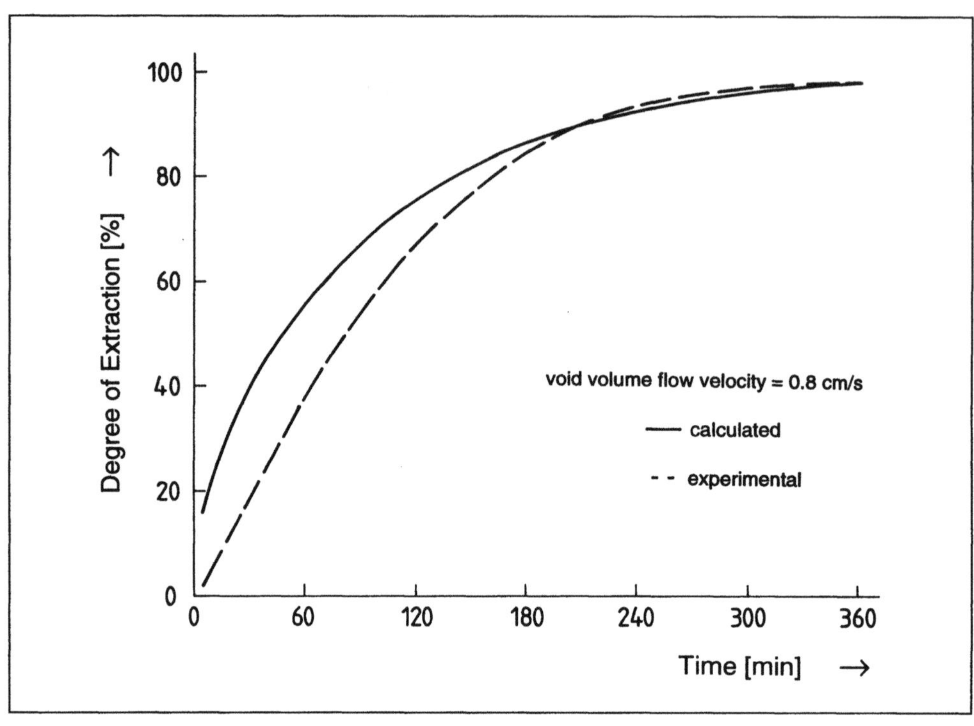


**Fig. 7.30.** Modeling the extraction of theobromine from cocoa seed shells with a constant effective diffusion coefficient, calculated from the total amount of extract.

Deviations of calculated and measured extraction curves for the first part of the extraction may be due to several effects, the major ones probably being an effective diffusion coefficient depending on concentration and a transport resistance at the interface solid-fluid and in the surface layer. In order to take into account these effects, the boundary conditions have to be modified. Equation 7.16 is replaced by the following:

$$D_{es} \frac{\partial c_s}{\partial t} = \beta_F (c_{F,l} - c_{F,\infty}),$$

$$\text{7.39}$$

with
$c_{F,l}$ = concentration of extract in the fluid phase at the interface, $x = l$;
$c_{F,\infty}$ = concentration of extract in the bulk of the fluid phase.

The Biot-numbers assume finite values and the constants $C_n$ in Eq. 7.32 and $q_n$ in Eq. 7.33 then depend on the Biot-number:

220

$$C_n = \frac{2 \sin^2 q_n}{(q_n + \sin q_{n*} \cos q_n)\, q_n}; \qquad \cot q_n = \frac{q_n}{Bi}; \qquad\qquad 7.40$$

It is now possible to model the extraction curve with a constant effective diffusion coefficient for different Biot-numbers, representing different transport resistances in the surface layer due to different solvent velocities. The result for a model calculation for an extraction carried out in a bench scale extraction (25 l extraction volume, extraction of theobromine from cocoa seed shells) is presented in Fig. 7.31. For comparison the curves of Fig. 7.30 have also been included (curves 1 and 0). The curves for Biot-numbers of 5 and 2 enclose the experimental extraction curve. Therefore, the existence of a finite Biot-number can be derived, proving that there is a transport resistance in the fluid.

Still, the model is not fully appropriate. This may be due to several reasons:

1) The fixed bed was treated as a single homogeneous body. The influence of the packing has been neglected (no axial or radial dispersion). This assumption is

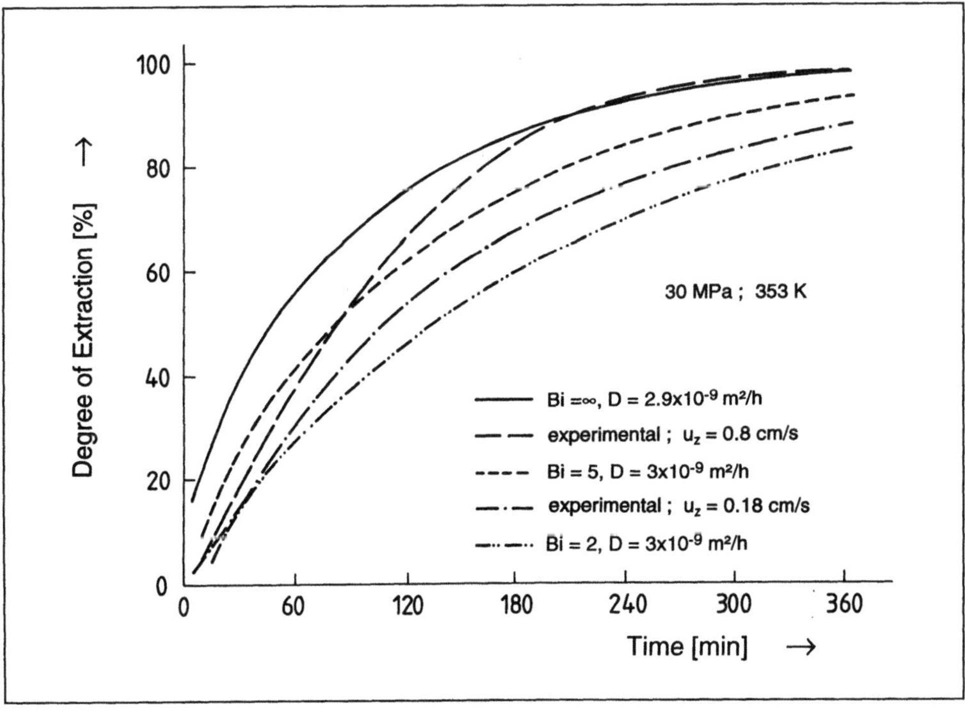



**Fig. 7.31.** Modeling the extraction of theobromine from cocoa seed shells with a constant effective diffusion coefficient, including mass transfer resistance to the bulk of the fluid.

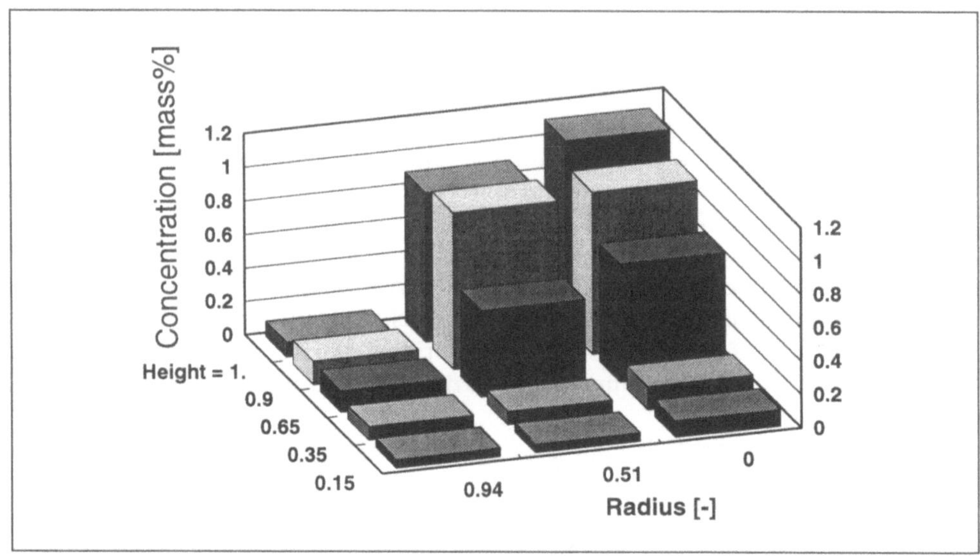


**Fig. 7.32.** Experimental results of the theobromine extraction from cocoa seed shells, showing radial distribution effects.

only justified in case of low concentrations in the fluid phase and an equal distribution of the compounds between the phases and may not be applicable for the modeled extraction of theobromine from cocoa seed shells.

2) Extract components other than theobromine were not considered. Caffeine and fat compounds are extracted beside theobromine, and water is exchanged between the phases.

3) Radial dispersion: There is some radial dispersion during the extraction, as can be seen from the experimental results in Fig. 7.32, while in the model no radial dispersion has been assumed.

### 7.3.8.4 Simplification of the Mass Balance for the Solid Phase

Particle models have been used for modeling gas extraction [8, 23, 39]. The main problem is to know the effective diffusion coefficient in the solid phase. This problem can be avoided if mass transfer resistances in the solid and in the fluid phase are combined to one mass transfer coefficient which is used in combination with a linear concentration gradient. The total solid substrate is then characterized by a mean concentration for the extractible substances. The corresponding mass balance equations for the fluid and solid phase can then be written as follows:

222

Fluid phase:

$$\varepsilon \, \frac{\partial c_F}{\partial t} = D_{ax} \, \frac{\partial^2 c_F}{\partial z^2} - u_z \, \frac{\partial c_F}{\partial z} + k_{oG}a(c_F^* - c_F), \qquad 7.41$$

solid phase:

$$(1 - \varepsilon) \, \frac{\partial \bar{c}_s}{\partial t} = - k_{oG}a \, (c_F^* - c_F), \qquad 7.42$$

with:

$k_{oG}$ = overall mass transfer coefficient related to the gaseous (fluid) phase;

$c_F^*$ = equilibrium concentration in the fluid phase, corresponding to the mean concentration of extract $\bar{c}_s$ in the solid phase;

$a$   = specific surface of the solid serving as mass transfer area;

$\varepsilon$   = porosity of the fixed bed (void volume fraction);

$D_{ax}$ = axial dispersion coefficient (diffusion coefficient of the fluid phase).

For this model two kinetic parameters, $k_{oG}$ and $D_{ax}$, must be known. They can be calculated from extraction curves. Some difficulties may arise to determine the effective specific surface area. Therefore, the product $(k_{oG} \, a)$ in many cases is the more useful parameter.

This model is applicable if concentration gradients are small in the solid particles, diffusion in the solid substrate is not an important transport mechanism, or the concentration profile can be specified beforehand. Goto et al. [21] assumed a parabolic concentration profile for the extraction of lignin from wood particles, using supercritical tert-butylalcohol. The authors obtained the following expression for the mass transfer coefficient $k_{oG}$:

$$k_{oG} = \frac{5 \, k_F}{5 + Bi}, \qquad 7.43$$

$k_F$ = mass transfer coefficient for the fluid phase.

The above described model has been used for modeling various gas extraction experiments: The extraction of naphthalene-phenanthrene mixtures with supercritical carbon dioxide from a bed of porous spheres [16], for the extraction of oil from rape seeds [13, 25] and for the extraction of alkaloids [31].

### 7.3.9 The VTII-Model for the Extraction from Solids Using Supercritical Solvents

Extraction from solids with supercritical gases can be a complicated process if the two phase flow, the properties of the solid material, and the inhomogeneities of the fluid and the solid phases have to be considered. For most cases, a far simpler model is sufficient. In this model, the following parameters are considered sufficient to calculate course and result of an extraction [28, 43]:
- equilibrium distribution between solid and supercritical solvent (adsorption isotherm);
- diffusion in the solid (effective diffusion coefficient or effective transport coefficient as defined by the transport model);
- mass transfer from the surface of the solid to the bulk of the fluid phase (supercritical solvent);
- axial dispersion (effective dispersion coefficient, taking into account inhomogeneities of the fixed bed, the solvent distribution and the influence of gravity).

The equations of the model are given in Eqs. 7.44 to 7.47:

Mass balance for the fluid phase:

$$\frac{\partial c_F(z)}{\partial t} = D_{ax} \frac{\partial^2 c_F(z)}{\partial z^2} - \frac{u_z}{\varepsilon} \frac{\partial z_F(z)}{\partial z} - \frac{1-\varepsilon}{\varepsilon} \frac{\partial \bar{c}_s(z)}{\partial t}. \tag{7.44}$$

Mass balance for the solid phase:

$$\frac{\partial \bar{c}_s(z)}{\partial t} = a k_{oG} \left( c_F(z) - \bar{c}_s(z) \frac{K(\bar{c}_s)}{\varrho_s} \right). \tag{7.45}$$

Equilibrium between fluid phase and solid phase:

$$K(\bar{c}_s) = k_1 \bar{c}_s \exp^{-k_2}. \tag{7.46}$$

Overall mass transfer coefficient:

$$\frac{\beta_F}{k_{oG}} = 1 + \frac{Bi\, K(\bar{c}_s)}{6}, \tag{7.47}$$

$$Bi = \frac{\beta_F R K}{D_{es}} \qquad \text{Biot-number,} \tag{7.20}$$

224

with

$\bar{c}_s$ = mean concentration of extract components in the solid phase;
$c_F$ = concentration of extract in the fluid (supercritical solvent);
$D_{ax}$ = axial dispersion coefficient;
$u_z$ = void volume linear velocity of supercritical solvent;
$K(\bar{c}_s)$ = equilibrium distribution coefficient between solid and fluid phase
$\quad$ $K(\bar{c}_s) = c_F/\bar{c}_s$;
$D_{es}$ = effective diffusion coefficient in the solid phase;
$k_{oG}$ = overall mass transfer coefficient related to the fluid (gaseous) phase;
$z$ = coordinate in axial direction;
$\varepsilon$ = void volume fraction (porosity of the fixed bed);
$t$ = time of extraction;
$a$ = specific surface of solid phase (mass transferring surface area);
$\varrho_s$ = density of solid;
$k_1, k_2$ = coefficients of the sorption isotherm (Freundlich-isotherm);
$\beta_F$ = mass transfer coefficient for the fluid phase.

For the balance for the fluid phase some assumptions have been made: Gradients of any kind are neglected in radial direction. It is assumed that gradients in radial direction can be taken into account by applying the model to several shells in axial direction, for which parameters can be considered constant. In the fluid phase convection and axial dispersion cause dispersed plug flow. The extraction process is isothermal. The loading of the supercritical solvent is relatively low. Sorption enthalpy can therefore be neglected.

In the solid phase the following assumptions have been made: Transport in the solid can be considered as one-dimensional. The solid is uniform and the extract is evenly distributed over the solid material. Real transport phenomena, like membrane transition, pore diffusion, diffusion in the solid, etc. are summarized in an effective transport coefficient (effective diffusion coefficient). At the interface solid/fluid, phase equilibrium is assumed.

The equations have been written for the extract mixture. They may be used for individual components. Interactions between the individual components of the extract and the influence on the extraction process are neglected.

Experimental extraction curves have been represented by this model by fitting the kinetic coefficients ($\beta_F$, $D_{es}$, $D_{ax}$) to the experimental curves, obtained from a laboratory installation. An example is presented in Fig. 7.33. With the optimized parameters it is possible to model the whole extraction curve with reasonable accuracy. These parameters have been used to model the extraction curve obtained from a pilot plant. In Fig. 7.34 experimental and calculated extraction curves are compared; the coincidence is sufficient. Therefore, the model can be used to determine the kinetic parameters from a laboratory experiment and they can be used for scaling up the extraction.

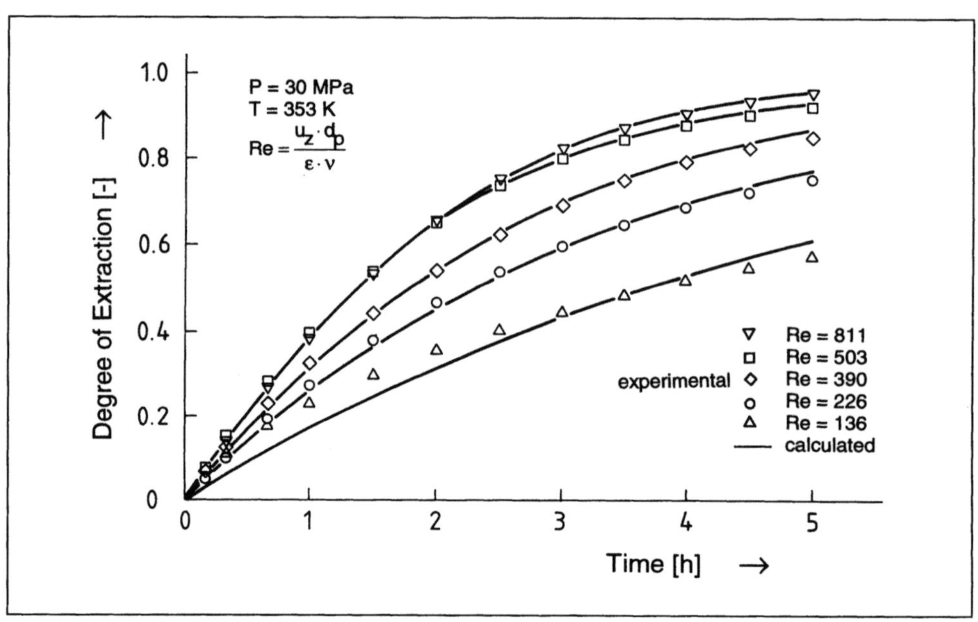

**Fig. 7.33.** Representation of a laboratory extraction of theobromine from cocoa seed shells with $CO_2$ by fitted parameters according to the VTII-model [43].

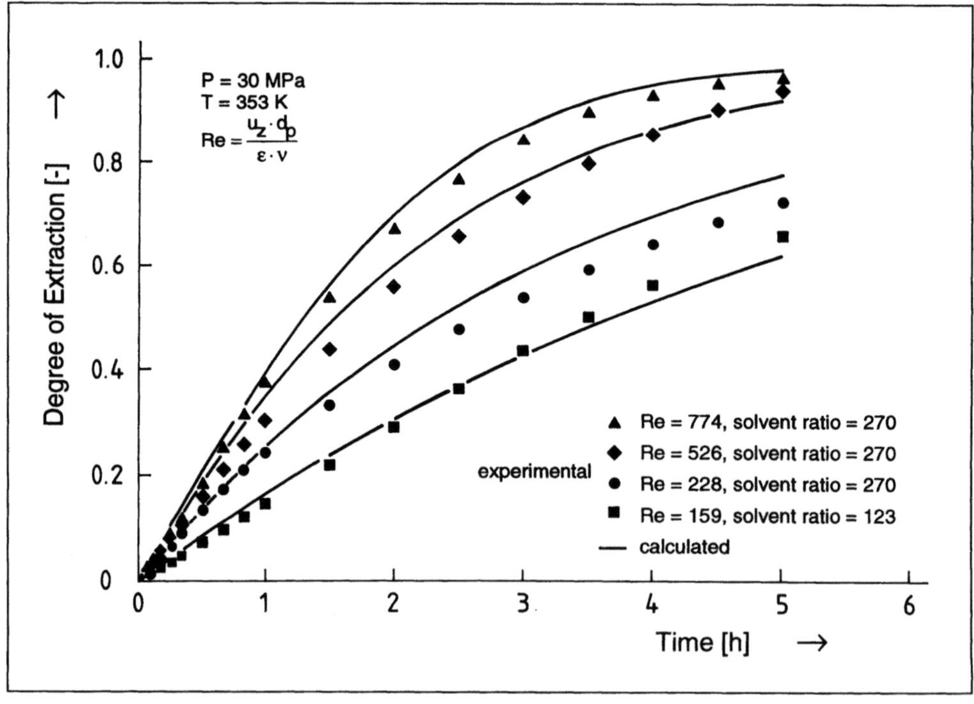

**Fig. 7.34.** Calculation of the extraction of theobromine in a pilot plant (scale up factor $\approx 40$) using the optimized parameters from the laboratory extraction [43].

# 7.4 Examples

## 7.4.1 Extraction of Theobromine from Cocoa Seed Shells

The extraction ot theobromine from cocoa seed shells has been mentioned several times already. Some additional results for this example are presented below.

Cocoa seed shell is a waste product from the processing of cocoa beans. In Germany, in 1992, about 40 000 t of shells were produced. Most of them are burned. Cocoa seed shells are of interest due to their theobromine content of about 1 to 2 wt.-%. For the chemical structure of theobromine, see Fig. 3.14. Theobromine has similar properties as caffeine and is used for pharmaceutical purposes. The shells are separated from the production process in dome-shaped particles of 0.2 mm thickness and a size distribution of 0.1 to 8 mm.

The equilibrium distribution of theobromine between the cocoa seed shells and supercritical carbon dioxide is shown in Figs. 7.35 and 7.36, which show the influence of pressure, temperature and water content on the equilibrium distribution.

Extraction experiments were carried out with a laboratory and a pilot plant, which are explained in detail in Section 7.5. Some extraction curves for the laboratory extraction are presented in Fig. 7.33. Water content is an important parameter in the extraction of plant material, which can be confirmed for cocoa seed shells, as shown

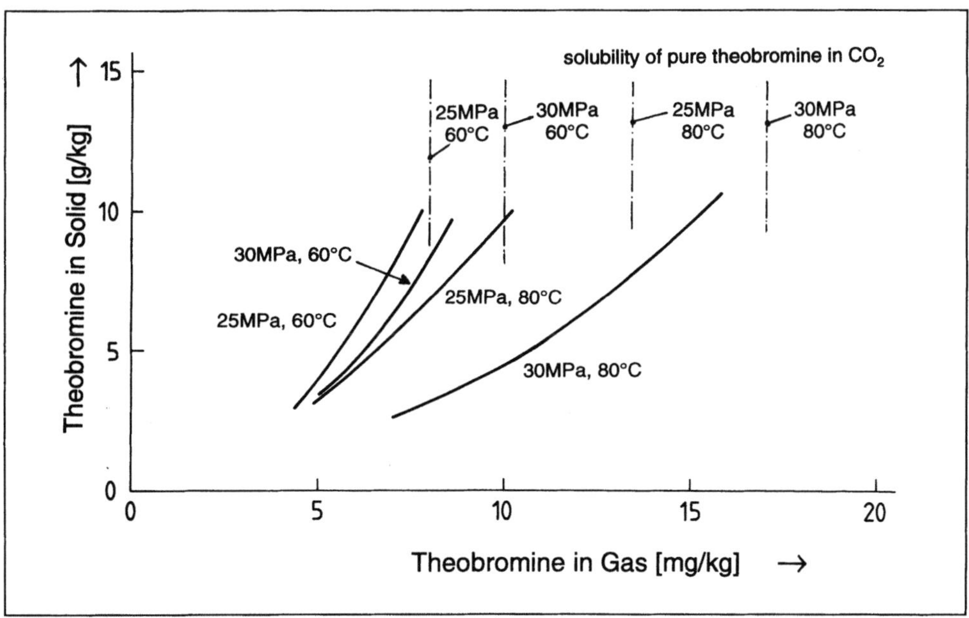


**Fig. 7.35.** Equilibrium solubility of theobromine in carbon dioxide. Solubilities of pure theobromine [24], and of theobromine contained in cocoa seed shells [43].

in Fig. 7.37. The influence of fluid flow velocity, or solvent ratio, is important for the extraction process, as can be seen from Fig. 7.38, where the degree of extraction has been shown as a function of Reynolds-number. In addition, results from the laboratory and pilot plant are compared. They coincide within experimental error. Experimental uncertainty is influenced i.a. by the error of the analytical procedure for the theobromine content in the solid, which is guessed to be about 5 to 10 %, depending on the concentration range (higher error at lower concentrations). Bodenstein-numbers, calculated from the fitted kinetic coefficients, are in the range from 2 to 12, indicating a substantial backmixing effect. The aspect of mass transfer is also treated in Section 4.2.2, where mass transfer coefficients, calculated from optimized fitted kinetic coefficients, are compared with correlations for dimensionless groups. The values calculated from the experiment are lower than the values from the correlations. Therefore, the correlations cannot be used for simulating gas extraction from solids, within the range of parameters investigated.

Concentration profiles for the pilot plant extraction, shown in Fig. 7.39 reveal that the assumption of dispersed plug flow is only an approximation. It may be used for the high solvent ratio of the experiment, shown in Fig. 7.39, where radial concentration differences are relatively small. If the solvent ratio is reduced by a factor of 2, the extraction at the rim of the fixed bed is far better than in the center, due to uneven fluid flow.

The course of the extraction as simulated with the VTII-model is plotted in Fig. 7.40. The simulation result, representing a medium value across the fixed bed (bro-

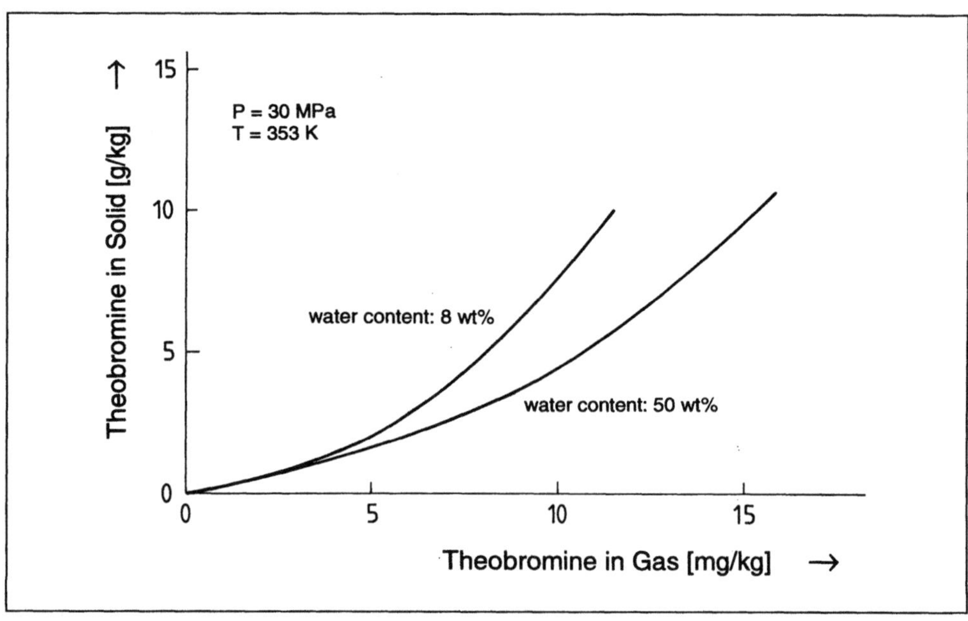


**Fig. 7.36.** Equilibrium solubility of theobromine in carbon dioxide. Influence of water content of the cocoa seed shells [43].

228

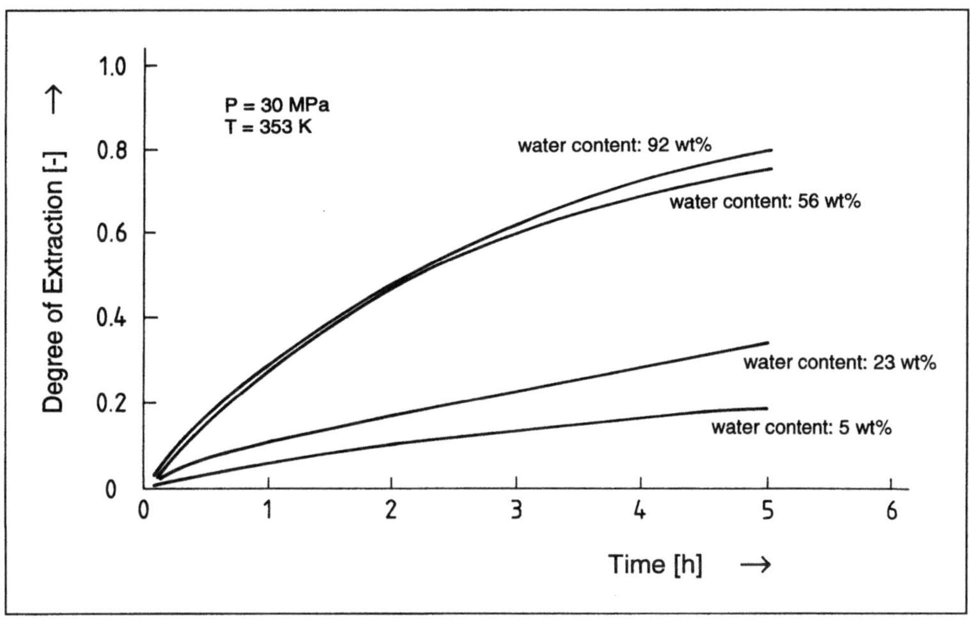

**Fig. 7.37.** Extraction curves for the extraction of theobromine from cocoa seed shells. Influence of water content [43].

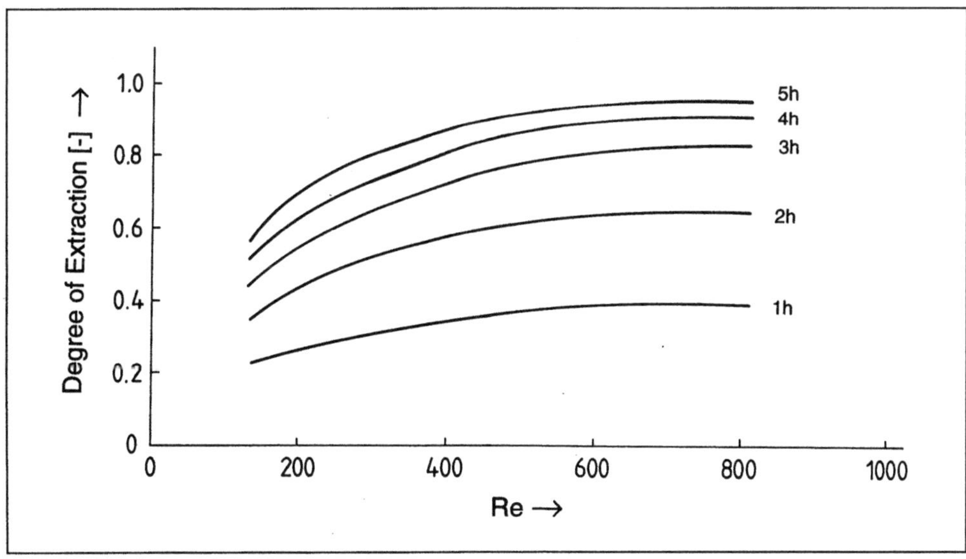

**Fig. 7.38.** Extraction curves for the extraction of theobromine from cocoa seed shells. Influence of fluid flow velocity or solvent ratio [43].

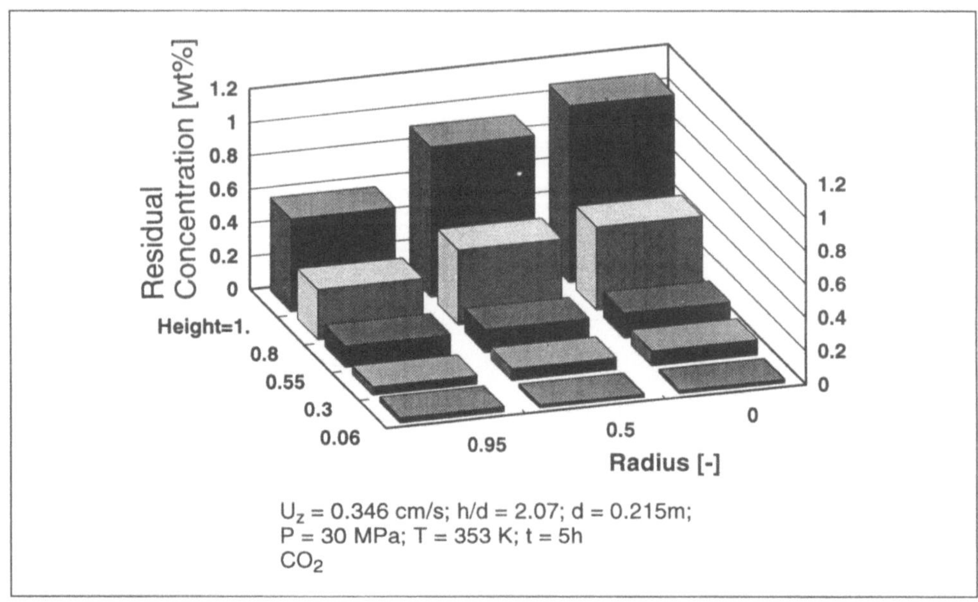



**Fig. 7.39.** Concentration profiles for the extraction of theobromine from cocoa seed shells. Pilot plant experiment [43].

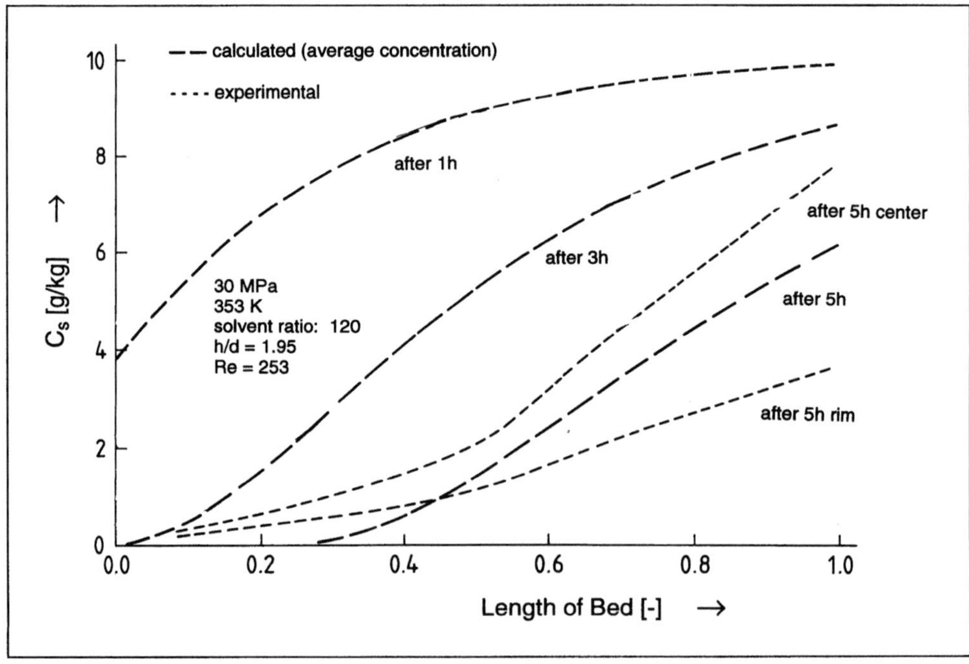



**Fig. 7.40.** Simulation of concentration profiles with the VTII-model for the extraction of theobromine from cocoa seed shells with supercritical carbon dioxide [43].

ken lines) is between the experimental values for the rim (squares) and the center (circles). Therefore, the model can be used for simulating the medium concentration as a function of bed length and extraction time. For the simulation of the radial distribution, the VTII-model can be used separately for the center and the rim of the fixed bed. Then, the amount of solvent for each section is an additional parameter.

## 7.4.2 Extraction of Oil from Oil Seeds

The extraction of oil from oil seeds would be one of the primary targets of gas extraction, if the value added by extraction for most of the oils was not too low. There may remain special oils or valuable components extracted with the oils for commercial application of extraction with supercritical solvents. The following examples are meant to show the possibilities of gas extraction of oils.

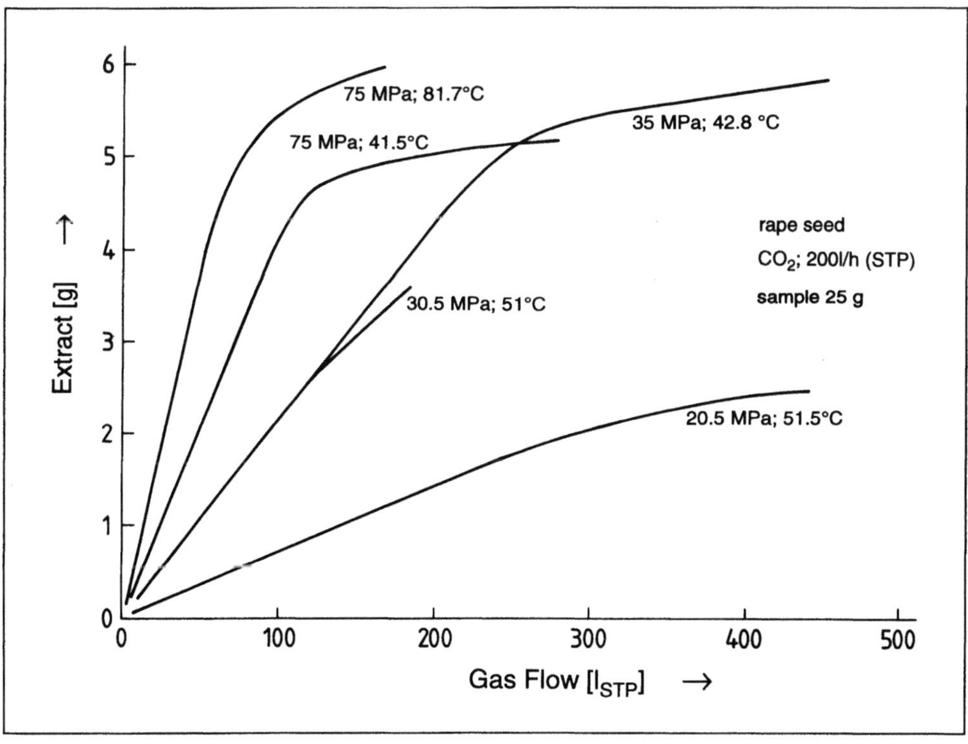


**Fig. 7.41.** Extraction of oil (triglycerides) from pretreated rape seeds. Influence of different conditions for the extraction.

Oils from oil seeds mainly consist of triglycerides of C16- to C20-fatty acids. Triglycerides are fairly soluble in supercritical carbon dioxide, but far more so in short chain paraffines, like propane (compare Chapter 3).

The extraction rate clearly depends on conditions of state for the extraction, which determines the solvent power of the supercritical solvent, as can be seen from Fig. 7.41 for the deoiling of pretreated rape seed. The total amount of extract for conditions of reasonable solubility is in general the same, but at relatively low pressures, the amount of extract is lower than at higher pressures. The mean concentration of extract compounds in the supercritical fluid may be strongly dependent on solvent ratio, as shown in Fig. 7.42 for the extraction of edible oil. In this case, equilibrium solubility of extract compounds in the solvent is not achieved at the outlet of the fixed bed. Kinetic and hydrodynamic behavior of the extraction system determine the actual concentration at the outlet. A further increase in flow rate (solvent ratio) decreases the extract concentration. The total amount of extract may even decrease with increasing solvent ratio, depending on the value of the product of flow rate times mean concentration in the gas. Note that Fig. 7.42 illustrates a special case of a single-stage laboratory extraction. Comparison to other experimental results and transfer

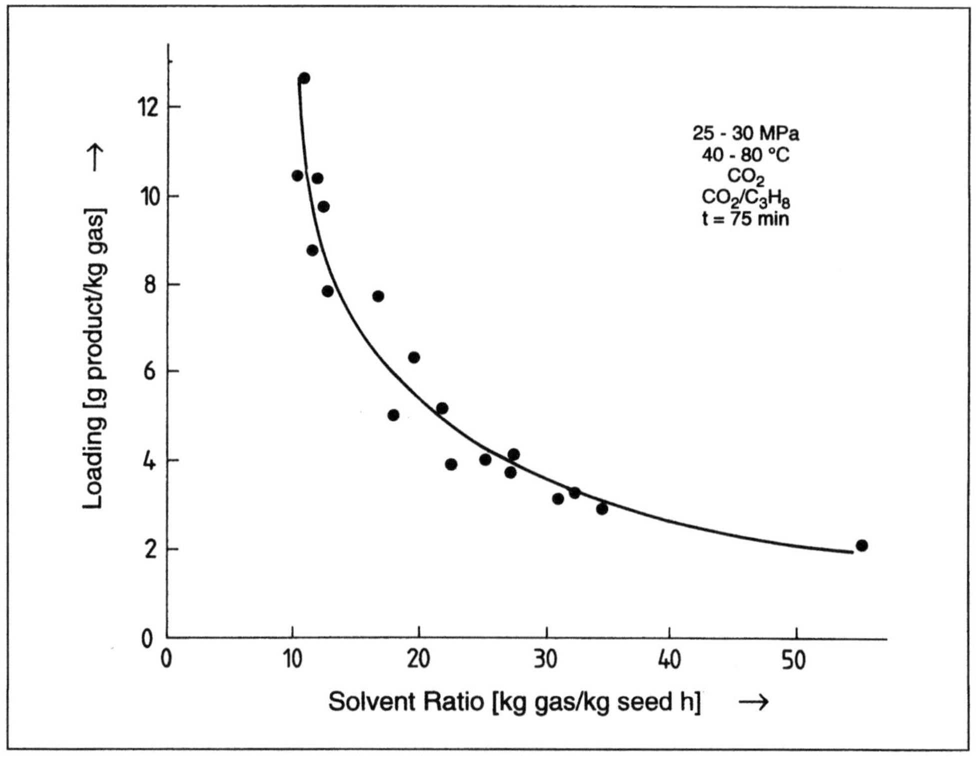


**Fig. 7.42.** Extraction of oil (triglycerides) from pretreated rape seeds. Influence of solvent ratio.

of results to other extraction systems (still using the same feed and solvent) are only possible if all important parameters listed in Table 7.1 are taken into account. A higher load of the solvent can be achieved by different means, e.g., increased mass transfer area, a packing of higher performance, recycling of the extract phase, and passing the extract phase through a sequence of extractors (see Section 7.4.4). Pretreatment of the oil bearing material is of major influence. In Fig. 7.19 differently pretreated rape seeds are compared in their extraction behavior, and in Fig. 7.43 the different extraction curves due to different pretreatment of palm fruits are shown. Here the amount of extract definitely depends on pretreatment. Pressure and temperature dependence of the extraction of palm oil from palm fruits is shown in Fig. 7.44. The rate of extraction increases with temperature and pressure.

An example for the extraction of oil from a wild palm from the Amazon region is shown in Fig. 7.45.

If solvents are used which are better solvents for triglycerides, the rate of extraction can be enhanced at moderate conditions. In Fig. 7.46 this is shown for the extraction of soy bean oil from flakes with different solvents. According to those results, $CO_2 + C_3H_8$/ethanol and $CO_2$ + ethanol as solvent mixtures would be very effective.

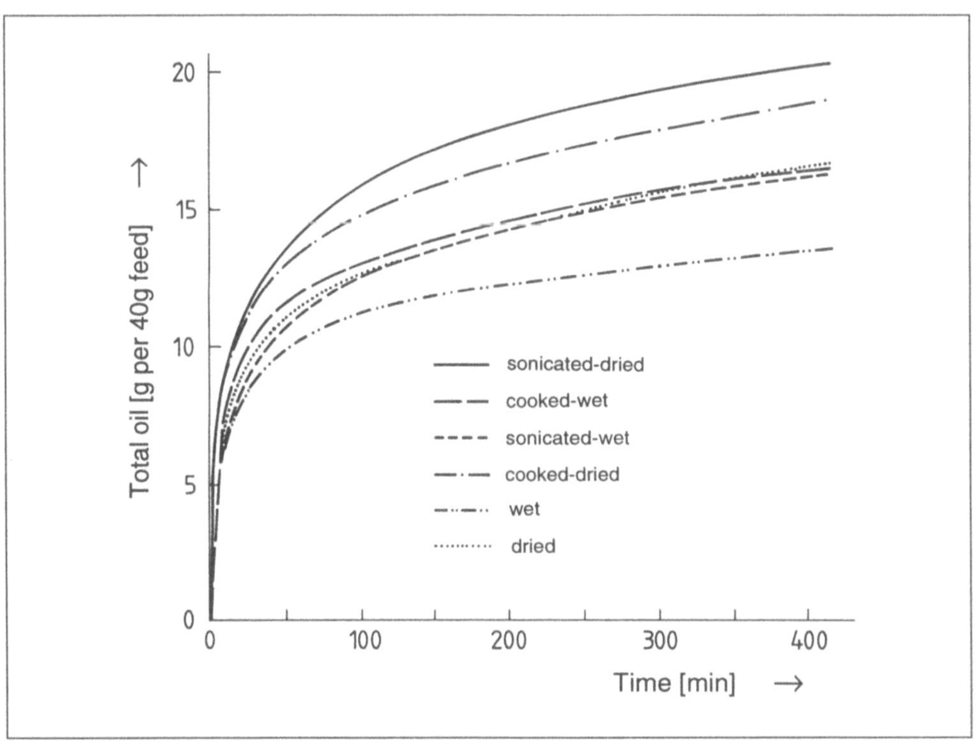


**Fig. 7.43.** Extraction of oil (triglycerides) from pretreated palm fruits. Influence of pretreatment method [6].

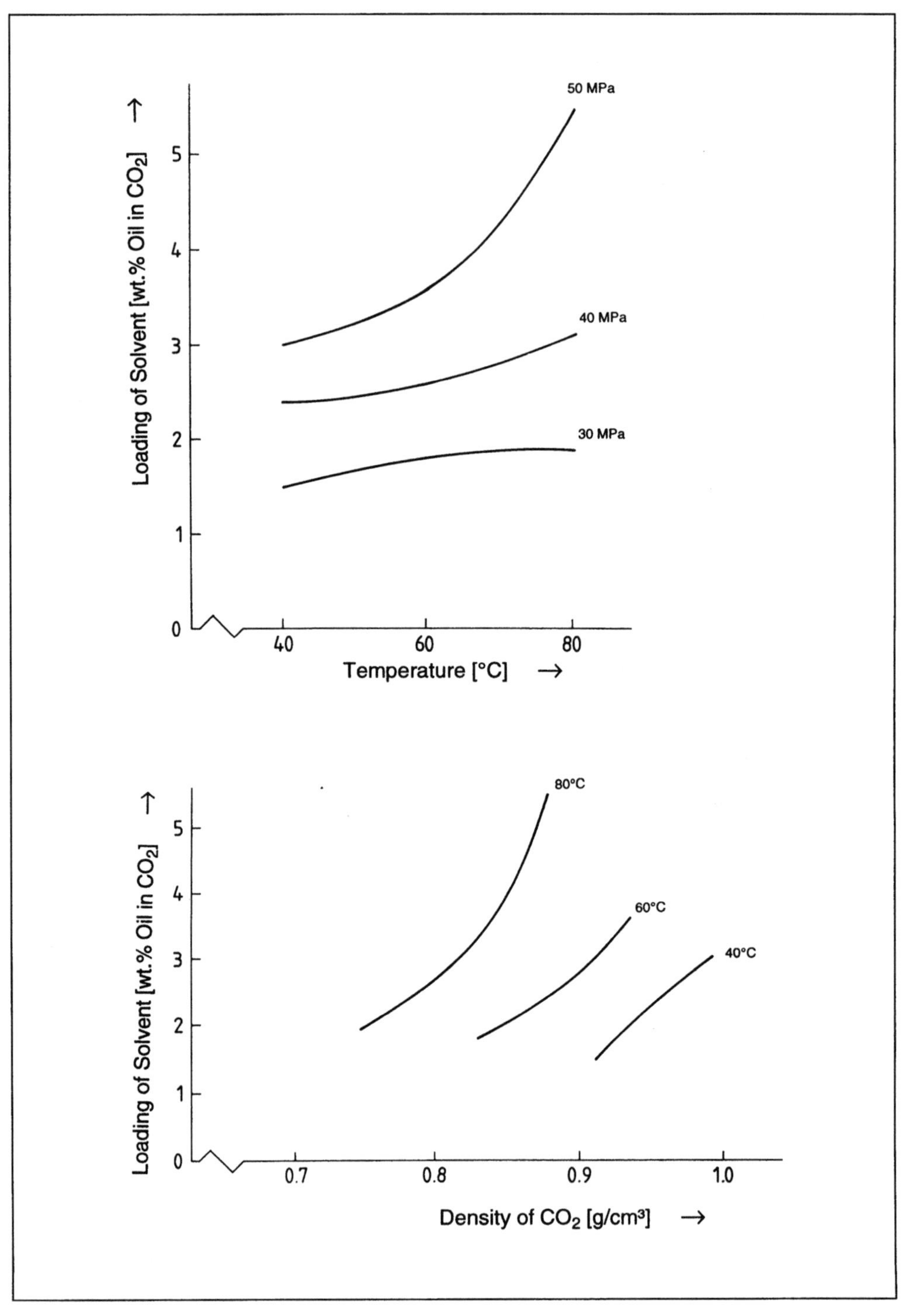


**Fig. 7.44.** Effect of pressure and temperature on the extraction of palm oil from palm fruits (Loading of solvent during period of constant extraction rate) [6].

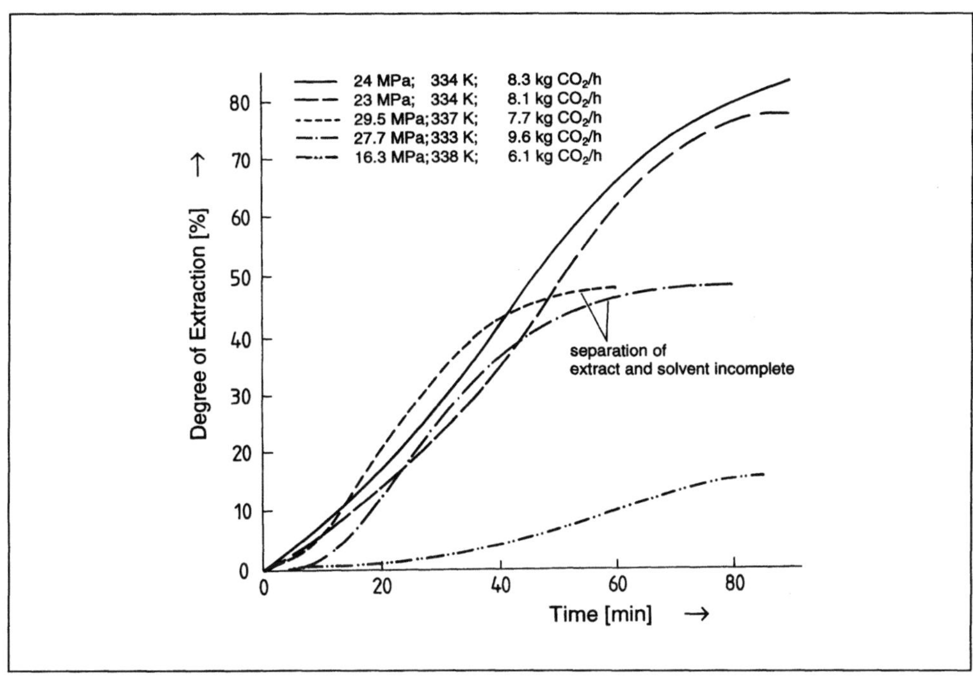



**Fig. 7.45.** Extraction of oil from a wild species of a tropical oil palm [19].

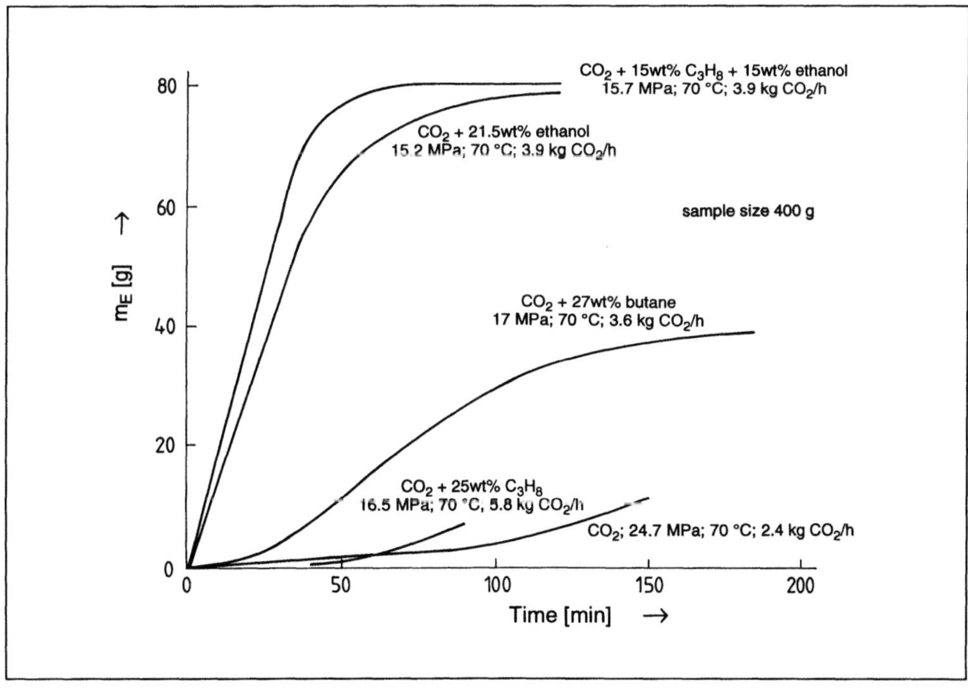



**Fig. 7.46.** Extraction of soya bean oil from soya bean flakes by different supercritical solvents and different entrainers.

The same is true for monochlorodifluoromethane (refigerant 22), and other entrainers in combination with carbon dioxide.

The dissolved oil must be separated from the supercritical solvent. During the precipitation process, kinetic and equilibrium effects cause substances to precipitate differently. The consequence is an enrichment in substances which cannot be achieved by extraction since extraction is determined by the solid substrate, the extract, and the solvent while precipitation is determined by the extract and the solvent only. As an example, the first fraction of the precipitation of palm fruit extract is shown in Fig. 7.47. The content of $\beta$-carotene in this fraction is enhanced by a factor of 10 against the content in the extract.

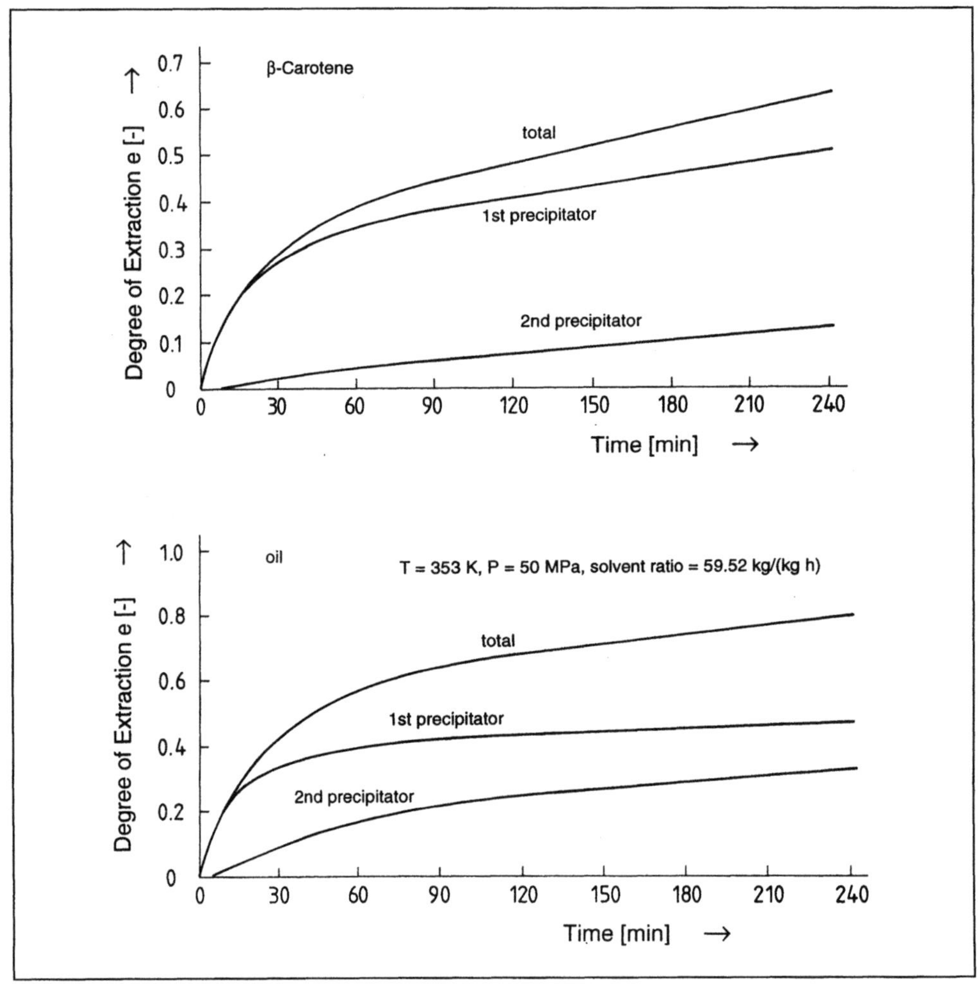


**Fig. 7.47.** Enriching of a substance during the precipitation from the fluid phase after extraction. Example: $\beta$-carotene from a palm fruit extract [2].

236

### 7.4.3 Comparison of Extracts from Liquid Extraction and Extraction with Supercritical Carbon Dioxide

One question is whether the extract from gas extraction is different from the extract of liquid solvents. At first glance, this seems trivial, since different liquid solvents produce different extracts. Therefore, the extract from a supercritical solvent could also be different; yet little is known about the difference. As an illustrative example, in Fig. 7.48 chromatograms of extracts from extractions with different solvents are compared.

Liquid solvents like n-hexane, methylene chloride, and ethanol produce different extracts as can be seen from the different peaks in the chromatograms. n-Hexane extracts unpolar compounds, ethanol extracts polar compounds. The extraction with supercritical carbon dioxide was carried out under conditions whereby $CO_2$ was saturated with water. After expansion of the solvent, the precipitated extract divided into two parts: a liquid one, containing water and dissolved substances and a solid one. The corresponding chromatograms of the two precipitates are also shown in Fig. 7.48. From a comparison with the chromatograms of the extracts of liquid solvents, it can be concluded that water-saturated carbon dioxide extracts polar and unpolar compounds. But there are also some other peaks in the $CO_2$-extract which cannot be found in the liquid extracts.

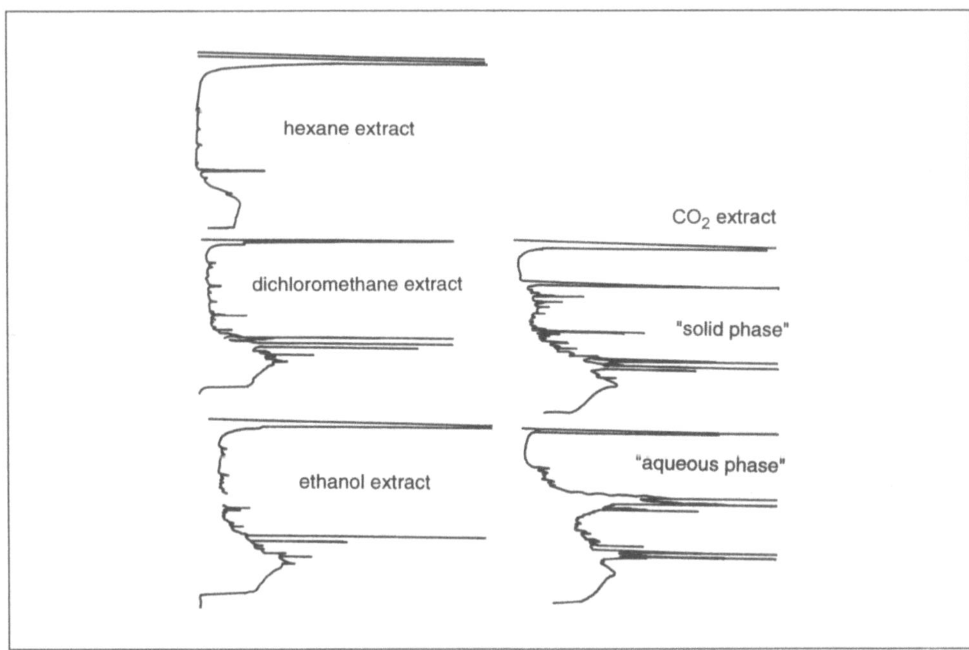


**Fig. 7.48.** Extracts obtained with different solvents. Solid material: Aparisthmium cordatum.

### 7.4.4 Solids in Multiple Stages and Countercurrent Operation in Gas Extraction from Solids

The simplest mode of operation for a fixed bed extraction consists of contacting it until a certain amount of extract has accumulated or a certain mean residual concentration in the solid raffinate is achieved. Even for an extract consisting of a pure component this is not the best way to carrying out the process. During the extraction process, extraction kinetics change due to the depletion of the solid substrate and therefore optimum process conditions change. In addition, loading of the solvent may be enhanced by carrying out the extraction in several stages.

For an extract mixture of different compounds, an extraction in several stages can yield different extract products, as has been illustrated in Fig. 7.47. Each extraction stage may be designed for other process conditions, even the solvent can be different.

The concentration of extract in the supercritical solvent at the outlet of an extraction vessel depends on the extract concentration in the solid material at the outlet. The highest concentration achievable is the equilibrium concentration corresponding to the initial concentration of extract in the solid. If concentration profiles are steep, equilibrium concentration may be achieved for most of the extraction. But if concentration profiles are flat, equilibrium concentration will not be achieved during the greatest part of the extraction. Then, and if the loading of the supercritical solvent is far from equilibrium, it may be useful to try to achieve a higher loading by the operating mode.

Multistage countercurrent contacting is the most effective mode. It reduces the amount of solvent and makes possible continuous production of extract. Real countercurrent contact is not easily established for solids, since special effort is necessary for moving the solid, with increased difficulties at elevated pressure. Therefore, it is easier to not move the solid material and to achieve countercurrent contact by other measures. Either several fixed beds can be applied and contacted in such a way that the bed with the highest extract concentration is contacted with the gas with the highest loading of extract and vice versa, or one fixed bed in a column is used and individual sections are formed by inlets or outlets for the solvent streams.

This may be advantageous, if many contacting steps are necessary, for relatively small beds and for a short contacting time per cycle. The arrangement is somewhat complicated if the solid has to be removed after contacting. It is more appropriate for applications where the solid acts as an additional separating agent, like in adsorption or chromatography, and is not removed. Then, a semi-continuous countercurrent contact can be achieved by only switching of valves. This has been used for separation of isomers by adsorption and for continuous chromatography (see Section 9.2).

For the purposes of extraction from solids with supercritical solvents, several fixed beds in countercurrent contact with the solvent are the best configuration. The process is explained, using Fig. 7.49 as reference.

The process of extraction is carried out in four extractors, each containing a fixed bed of solids. Three extractors are simultaneously contacted by the supercritical solvent. One extractor is shut off, emptied and reloaded with fresh solid material. The regenerated solvent first enters extractor 2, which contains the most depleted material, then flows through extractor 3 and extractor 4. Extractor 4 contains fresh

solid material at the beginning of the extraction period. The concentration of extract
in the solvent increases with each contacted fixed bed. After leaving extractor
number 4, the solvent is pumped to the regenerator and then recycled to bed no. 2.
The gas cycle is operated in this configuration for a certain time. It is changed after
the mean concentration in bed no. 2 has reached a lower concentration value, preset
by the specifications of the process. Then, bed no. 2 is removed and bed no. 1 is added
to the solvent cycle. The solvent now enters the battery of extractors at bed no. 3, it
being now the most depleted one, and leaves the battery at bed no. 1. In the next tact,
the solvent enters at bed no. 4 and leaves at bed no. 2, and so on.

By this operation, the extraction is split into three parts which are carried out
simultaneously. Therefore, (ideally) only one-third of processing time and also one-
third of solvent quantity is needed as compared to a process with one fixed bed. This
mode of operation results in an appreciable reduction of production costs, since the
costs of the two necessary additional extractors is outweighed by their benefit of additional capacity.

The flow scheme of Fig. 7.49 can be taken as an example for a decaffeination plant
for green coffee beans [12]. In this case, it is useful to apply a high solvent ratio at the
beginning of the extraction, while towards the end of the extraction a lower solvent
ratio is appropriate. Therefore, in addition to countercurrent contacting, the solvent
flow of two cycle compressors is combined for the fixed bed with fresh material and
then divided for two lines of extractors.

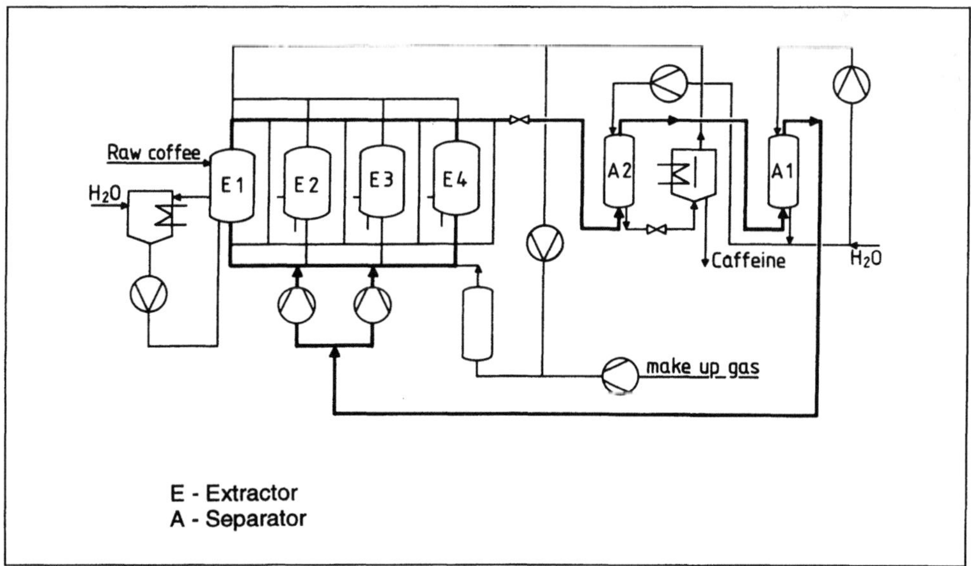


**Fig. 7.49.** Flow scheme of a multistage extraction. E1, E2, E3, E4: Extractors. A1, A2: Precipitators
(Absorber) [12].

# 7.5 Laboratory and Pilot Plant Equipment

## 7.5.1 General Considerations

Laboratory experiments are carried out to prove the extractability of substances and to obtain extract material that can be analyzed for its components. A laboratory plant for the extraction of solids with supercritical solvents is described and discussed.

A laboratory plant for the investigation of the extraction of substances from solid substrates with supercritical gases consists of various typical parts. Some types of equipment are available for extractions with liquid solvents also, beside those of the high pressure capacity, others are specific for gas extraction. In Fig. 7.50 a generalized scheme of an experimental apparatus is shown. The laboratory plant consists of the gas supply (1), including storage vessel and supply pump, the extraction vessel (2), and the precipitator (3) for separating extract and supercritical solvent. In some cases the supercritical solvent is mixed with a modifying compound (entrainer, modifier) in a separate vessel or by a separate pump (4). For experiments with somewhat higher gas flows and for simulation of process conditions, a gas cycle will be needed, comprised of cycle pump or compressor (5) with an upstream (6) and a downstream conditioning of the gas (7).

Instrumentation depends on the information needed from the investigation. For a minimum outfit it is necessary to install measuring equipment for pressure and temperature in the extractor, precipitator, before and after the pump or compressor and

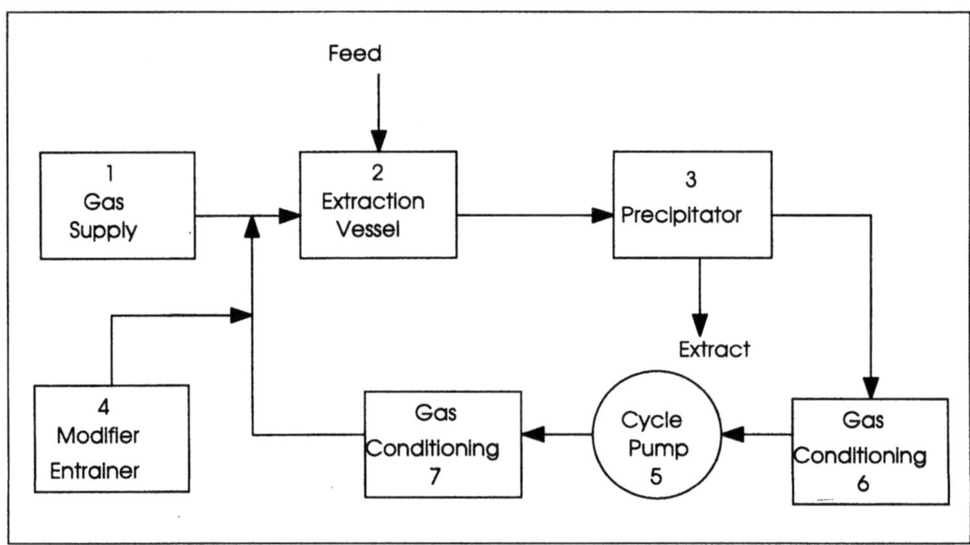


**Fig. 7.50.** Generalized flow scheme of a laboratory gas extraction plant for the extraction of solids.

in the gas supply. Furthermore, gas flow, concentration of extract and modifier in the supercritical gas, downstream of the extractor and the precipitator must be determined. It is useful to install temperature measurement devices at different locations in the extractor and precipitator and downstream to any heat transfer equipment.

Automation is mainly a question of the type of the investigation program. The greater the number of experiments to be carried out with the same configuration of the laboratory plant, the more automated the equipment can be. But at the minimum, the automatic control of the operating pressure should be possible. Temperatures and mass flow can be adjusted manually, but it is inconvenient to constantly adjust pressure, which normally changes due to sampling or introduction of modifier or absorption liquid into the process gas cycle.

The equipment actually needed for the different parts of a laboratory extraction depends on size, maximum operating pressure, and temperature of the extraction vessel and on the existence and type of a gas cycle. In the following three laboratory plants of different size and one pilot plant will be described. The plants have been designed or essentially modified to serve our purposes at the institute for Thermische Verfahrenstechnik at the Technische Universität Hamburg- Harburg. The plants have been in operation for at least several years. They have advantages (hopefully) and disadvantages (definitely) and they need improvements. Automation is kept to a minimum. The selection of equipment does not necessarily mean that it is the best. Equipment of other suppliers may be as good or better. This is simply the equipment we use because of various reasons. There are commercial type of laboratory extraction plants available; they tend to be more expensive than the sum of the individual components.

The course and the result of the extraction must be measured in order to make use of the experimental information for further considerations. The remaining extract in the solid can be verified by taking samples from the solid phase at different times and analyzing them for the extract compounds. Normally, this will not be possible, since it is not easy to take solid samples from a pressurized vessel. Alternatively, the extraction curve can be established by carrying out a series of extractions under identical conditions with different extraction times. This is tedious, expensive, and not necessarily accurate, since identical process conditions may pose a problem. In most cases it is easier to follow the course of the concentration of the extracted compounds in the solvent and calculate the concentration in the solid phase by means of a mass balance. Unfortunately, in practice this method also has its serious constraints, since accurate determination of the extract concentration in the gaseous solvent is also not easy. On-line analysis is of great help, but since analysis methods also have their accuracy limits, it is not simple to establish mass balances with an accuracy better than 10 %.

Gas extraction from solid substrates is investigated with the intention to acquire different type and amount of information:

I) Basic data on the extractability of a compound and the composition of the extract,

II) data on the extraction and precipitation process in dependence on pressure, temperature, solvent ratio, type and concentration of modifying compounds and on the pretreatment of the substrate,

III) data on scale up and demonstration of the whole process scheme.

For case I), a screening unit of high flexibility and availability is needed. The extraction volume can be small. For case II) the amount of extract should be apt to determine the course of the extraction with time. It has proved that about 100 to 500 g of solid substrate is appropriate, meaning that the extraction vessel volume should be about 1000 cm$^3$. If a gas cycle is added, a parameter study can be carried out on extraction and precipitation. For case III) the equipment should be able to carry out the total process or to simulate all the process steps in sequence in the same manner as intended in a production process. This means that for purposes of screening and a parameter study the precipitation of caffeine in a decaffeination process can be carried out by adsorption on active charcoal or by absorption in water. But for demonstration of process principles, the individual steps have to be carried out by the same unit operation and with the same type of mass transfer equipment as intended for use in a large scale process. Otherwise, no indication of the joint operation can be obtained. For information on scale up, the extractor volume must be at least one order of mangnitude larger than in the laboratory type of equipment. In the extraction from solids, scale up from 20 to 50 l vessels to extractors of several m$^3$ seems possible (scale up factor of 100 to 1000), if the experiments are carried out in a pilot plant unit.

## 7.5.2 Screening Unit

In Fig. 7.51 a simple screening unit for gas extraction from solids is schematically shown. It consists of the gas supply, an extractor, an expansion valve, a sampling unit, and a gas flow meter.

The gas supply for $CO_2$ is a gas compressor (Hofer, Mülheim, FRG), heated to temperatures higher than the critical temperature. For other gases, gas bottles in combination with a cooling unit and a liquid pump were used. Furthermore, a heat exchanger is necessary for heating the liquefied solvent to supercritical temperatures.

As extractor a commercial piece of 1″ (25 mm) pipe (Autoclave Engineers, Erie, PA, USA) of 250 mm length, material 316 SS, was employed. It was heated with electric heating jackets to operating temperature. Back pressure was regulated by a regulating valve in line with a shut-off valve (316 SS, 1/8″, Autoclave Engineers, Erie, PA, USA).

The expansion was carried out down to ambient pressure. The expansion valve was heated with an electric heating coil or with a hot air fan. The expanded gas and the extract passed a cooled glass sampling flask. Gas flow was determined by a wet gas meter (Elster, Darmstadt, FRG). Fittings and connecting pipes were of 1/8-inch-type (316 SS, Autoclave Engineers, Erie, PA, USA).

The screening plant can be operated up to 75 MPa and about 420 K, but mostly, runs were carried out up to 35 MPa and 370 K. Typical results from this unit comprise

242

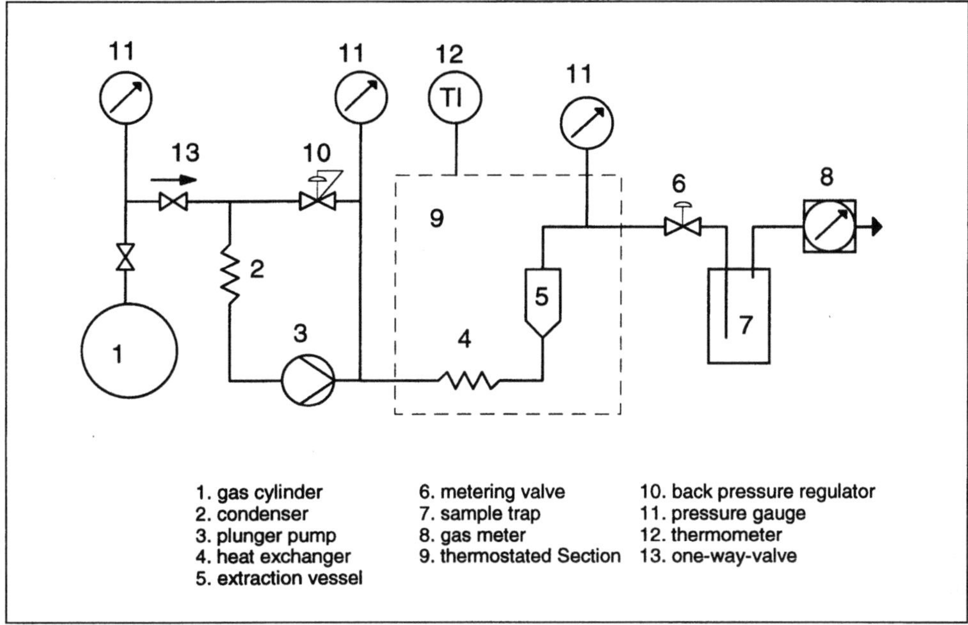


**Fig. 7.51.** Flow scheme of a screening unit for the extraction from solids.

the total amount of extract, the composition of the extract, the loss of weight of the solid substrate, and the total amount of solvent needed at operating conditions. For extract compounds of high concentration, as for oil seeds, even the course of the extraction can be measured from the samples taken during the extraction. An example of such results obtained with this type of equipment is the extraction curves shown in Fig. 7.41 for the extraction of rape seed oil.

## 7.5.3 Laboratory Extraction Plant, Small Extraction Volume

The flow sheet of the small laboratory extraction plant is presented in Fig. 7.52 [2, 6, 43]. This unit can be used for preliminary studies and parameter investigations. It is located at the Universiti Sains Malaysia in Penang, Malaysia, and is shown in Fig. 7.53 (right) together with a separation column (left) for countercurrent separations. It consists of two extractors (2000 ml and 200 ml), two 100 ml precipitators, and a 100 ml saturator (all from Haage, Mühlheim, FRG). The saturator is employed for saturating the supercritical solvent with water or with modifying compounds.

All the vessels are submersed in thermostated liquid-baths. The baths can be moved vertically, so that the vessels are easily accessible. Pressure can be reduced in two steps upstream to the precipitators. After the second precipitator the gas is expanded to ambient pressure. Gas flow is determined by a wet gas flow meter. Then the gas is discharged to the atmosphere. Gas supply is achieved by pumping (Burdosa, Butzbach, FRG) the liquefied gas to operating pressure and heating it to supercritical temperatures. Maximum operating pressure is 75 MPa. The maximum design operating temperature is 420 K and is determined by the heating liquid. Valves, fittings, and pipes are of the 1/8-inch-type (316 SS, Autoclave Engineers, Erie, PA, USA).

Results from this unit comprise the ones obtainable by the screening unit and, in addition, the amount of extract precipitated in two steps. Such a result is shown in Figs. 7.44 and 7.47 for the extraction of palm oil from palm fruits [2, 6].

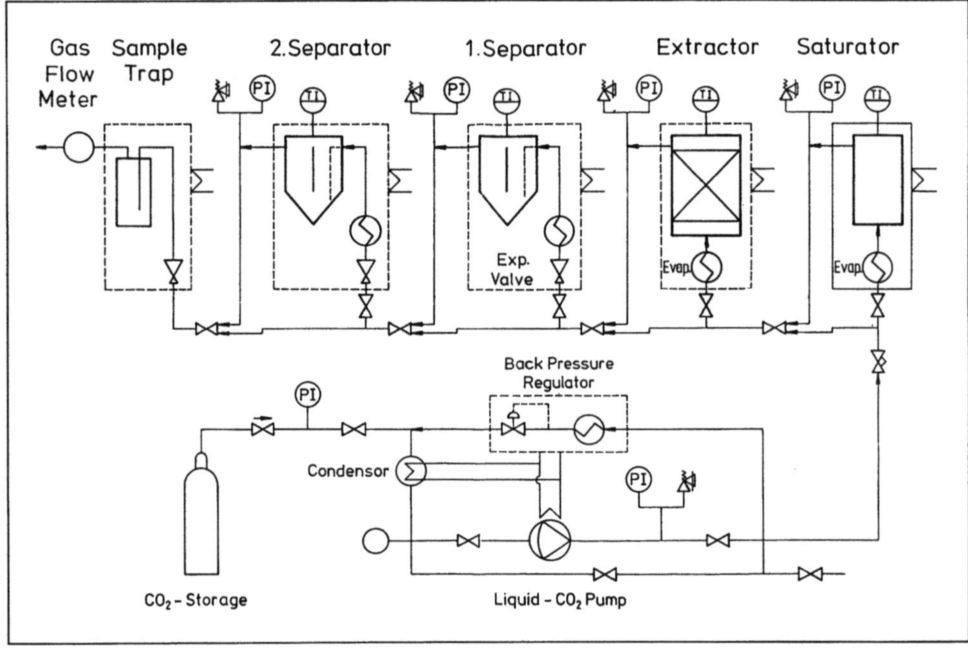


**Fig. 7.52.** Flow scheme of a small laboratory plant for extractions from solids with supercritical gases [2, 6, 43].

## 7.5.4 Laboratory Extraction Plant, Large Extraction Volume

The flow scheme of a laboratory extraction plant with a larger extraction volume (4 l) is shown in Fig. 7.54. Its initial design and construction stems from NOVA-

**Fig. 7.53.** Photograph of the small laboratory unit, located at USM, Penang, Malaysia, together with a separation column for countercurrent contacting [2, 43].

Swiss, but many modifications have been carried out due to the different purposes of the investigations carried out since its installation.

The plant consists of four main parts: 1) gas supply and gas cycling, 2) the part for adding a modifying compound (saturator), 3) the extraction vessel, and 4) the precipitation of the extract and simultaneous regeneration of the supercritical gas.

1) The gas (carbon dioxide) is taken from a storage vessel under vapor pressure at room conditions, is heated to supercritical conditions and compressed by one (P2) of two membrane compressors (NOVA-Swiss), with a maximum delivery pressure of 100 MPa, a maximum suction pressure of 20 MPa, and a maximum delivery rate of about 20 kg $CO_2$/h, to experimental conditions.

During the extraction experiment the gas is cycled by the second membrane compressor (P1) through extractor and precipitator, while the first compressor is used only intermittently to make up for the pressure drop caused by taking samples. The flow of gas is adjusted by a compressor bypass. On the high pressure side of the compressor, amplitudes of the pressure pulsation are relatively high, but since the saturation vessel (B1) for adding the modifying compound dampens the pulsation, pressure variations at the entrance of the extractor (B2) are negligible.

Compression of the gas enhances the temperature of the gas, which has to be adjusted to extraction temperature before entering the extractor. In this set-up, temperature is adjusted in the saturation vessel simultaneously to the saturation of the extraction gas by the compound in the saturator.

2) The saturator is a 2.5 l thermostated autoclave. It contains the modifying compound and some mass transfer equipment like rings, spirals or static mixing elements. The gas flows upwards through the saturator. At the exit of the saturator, the extraction gas is approximately saturated with modifier according to the temperature at the exit. Depending on the nature of the compounds, mass transfer equipment must be adjusted to achieve saturation. At high flow rates it is advisable to insert a liquid/gas separator at the top end, inside the saturator vessel, in order to minimize carryover of modifier liquid to the extraction vessel. This mode of adding modifier is convenient, since the amount of modifier can be adjusted by temperature. But it is limited to the two phase region of modifier and supercritical gas. At extraction pressures of 20 to 30 MPa and higher most modifiers are completely miscible with carbon dioxide. Water

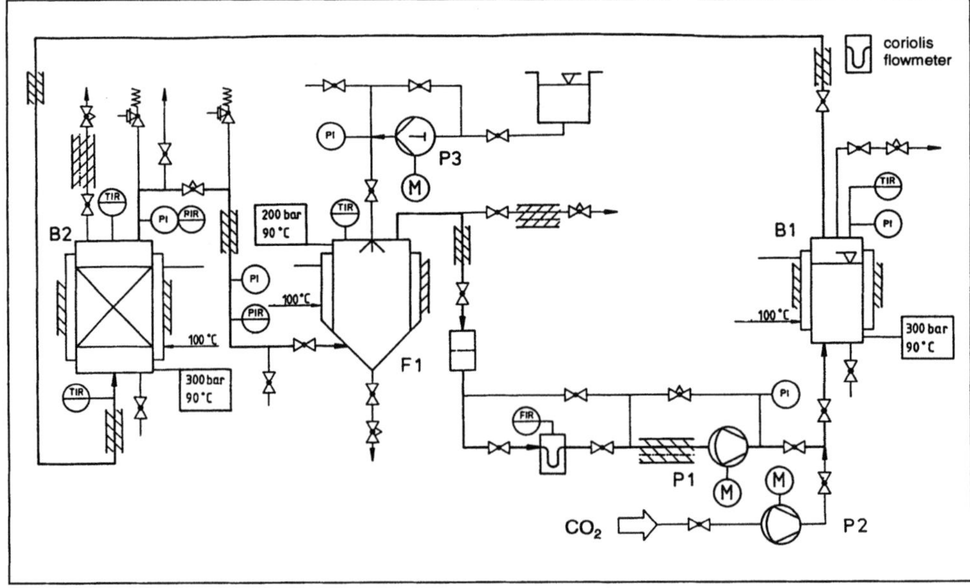


**Fig. 7.54.** Flow scheme of a laboratory plant with larger extraction volume.

has a two-phase region with supercritical compounds in this pressure range. In principle, it is possible to adjust the amount of modifier by mass transfer kinetics. But in practice this is not simple to achieve under experimental conditions. Therefore, in cases of total miscibility of modifier and supercritical solvent, the modifier is added by a metering pump.

3) The extraction vessel is a 4 l autoclave (NOVA-Swiss, max. pressure 100 MPa), thermostated by a liquid in heating jackets. The solid material is inserted in the autoclave by means of an extraction gasket made of a stainless steel pipe of somewhat smaller diameter than the inner diameter of the autoclave. The gasket is closed at its ends by sinter metal plates (pore diameter 5 – 20 μm, pressure drop across the plate about 0.1 MPa) which distribute the gas at the gas entrance and withhold solid particles at the gas exit. One of the sinter metal plates is removable for refilling the gasket. The gasket is sealed gas tight against the autoclave wall, in order to prevent gas flow around it.

The gas flows through the fixed bed of solids and is enriched with the extractible components from the solid material. At both ends of the extractor, sample valves are installed for taking samples from the gas flow.

4) Separation of the extract from the supercritical solvent consists of the precipitation vessel F1 (2 l, NOVA-Swiss, max. pressure 100 MPa) and a pressure relief valve. In some cases a heat exchanger is employed between pressure reduction and inlet to the precipitator.

Precipitation of the extract is achieved by pressure reduction and/or absorption of the extracted compounds in an absorbing liquid. For the extraction of xanthines (caffeine, theobromine and theophylline) the absorbing liquid is water. The supercritical gas, loaded with the extract, enters the precipitator at the lower end and flows upwards against the absorbing water, which is sprayed from the top of the vessel.

The regenerated gas leaves the vessel at the top after passing a liquid/gas separator for removing droplets. Due to the maximum suction pressure of 20 MPa of the cycle compressor, pressure must be reduced in the precipitator to less than 20 MPa. This pressure reduction improves the separation of supercritical gas and extract.

Results of extractions carried out with this plant are shown in Fig. 7.38 [32, 38, 43].

### 7.5.5 Pilot Plant

While a laboratory size plant has to be as flexible as possible, in order to be able to experimentally demonstrate the feasibility of a process step under a broad range of processing conditions, a pilot plant is designed to demonstrate the technical feasibility of certain process steps on a larger scale. Ideally, a pilot plant is a small version of a processing plant, but normally only parts of the technical process can be simulated on this scale. The variability of the plant is much more limited than that of a laboratory scale plant. The pilot plant described below was designed for a constant pressure gas cycle with absorptive removal of the extract from the gas.

The plant consists of the same groups of equipment as the laboratory scale plant; 1) gas supply and gas cycling, 2) the part for adding a modifying compound (saturator),

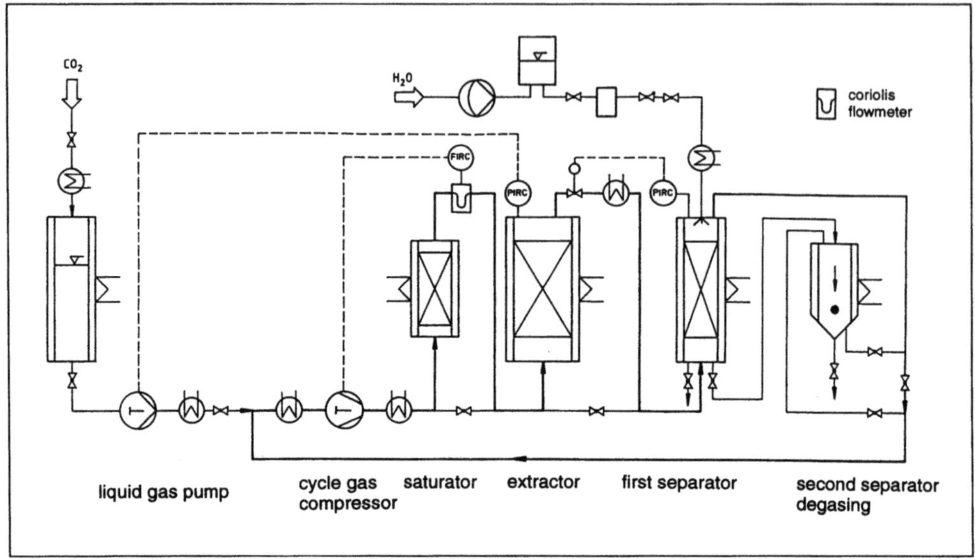


**Fig. 7.55.** Flow scheme of pilot plant.

3) the extraction vessel, and 4) the precipitation of the extract and simultaneous regeneration of the supercritical gas. The flow scheme of the plant is shown in Fig. 7.55. Size of the extractor is $2 \times 25\,000\ cm^3$, of the absorber: $8\,000\ cm^3$. Capacity of the pressure-enhancement pump 200 kg $CO_2$/h, of the cycle pump 300 kg $CO_2$/h. Since the bench scale plant is operated in nearly the same way as the laboratory plant, the description is omitted. The main difference, besides size, is that the pressure drop across the whole plant is in the range of 2 MPa. Examples for extraction results obtained with this plant are shown in Fig. 7.39 [41, 43].

Some dozens of medium-size and large-scale plants for extracting valuable substances from solids have been erected and are in operation. They are designed and operated to the principles discussed in this book, but will not be discussed, since information on production-scale gas extraction plants is available from a variety of companies which design and construct high-pressure plants.

# References


1. Anzelius A (1926) Über Erwärmung vermittels durchströmender Medien. ZAMM 6: 291–294
2. Bednarski B (1992) Untersuchungen zur Extraktion und Raffination von Palmöl mittels überkritischer Gase. Diplomarbeit, Fachhochschule Hamburg
3. Bennati RF, Brosilow CB (1962) Void fraction distribution in beds of spheres. AIChE J 8/3: 359–361


248

4. Beutler HJ, Gährs HJ, Lenhard U, Lürken F (1988) Einfluß der Lösungsmittelführung auf den Hochdruck-Extraktions-Prozeß. Chemie-Ing-Tech, 60: 773 – 776

5. Beutler HJ, Lenhard U, Lürken F (1988) Fat Sci Technol 90: 550 – 554

6. Bisunadan MM (1993) Extraction of oil from oil palm fruits using supercritical carbon dioxide. MSc thesis, Universiti Sains Malaisia, Penang

7. Brunner G (1983) Mass transfer from solid material in gas extraction. Proc. ISEC 83 (International Solvent Extraction Conference) Denver, p 521

8. Brunner G (1984) Mass transfer from solid material in gas-extraction. Ber Bunsenges Phys Chem 88: 887 – 891

9. Brunner G (1984) Gas extraction from solid material. Preprints Int. Symp. High Pressure Chemical Engineering, Erlangen, pp 311 – 316

10. Brunner G (1986) Anwendungsmöglichkeiten der Gasextraktion im Bereich der Fette und Öle. Fette Seifen Anstrichmittel 88: 464 – 474

11. Brunner G, Zwiefelhofer U, Simon A (1992) Extraction of xanthines from plant materials with carbon dioxide. In: Sekine T (ed) Solvent Extraction 1990. Proc of the International Solvent Extraction Conference (ISEC '90), Elsevier, Amsterdam, Part B, pp 1661 – 1670

12. Brunner G (1987) Decaffeination of raw coffee by means of compressed nitrous oxide. ASIC Proc 12 Internationales Wissenschaftliches Kolloq über Kaffee, Montreux, pp 294 – 305; also: Perrut M (ed) Proc International Symposium on Supercritical Fluids, Vol 2. Nice, pp 691 – 698

13. Bulley NR, Fattori M, Meisen A (1984) Supercritical fluid extraction of vegetable oil seeds. JAOCS 61/8: 1362 – 1365

14. Crank J (1989) The mathematics of diffusion. Oxford University Press, Oxford

15. Dams A (1989) Chemie-Ing-Tech 61/9: 712 – 715

16. Dams A (1990) Stoffübertragung bei der überkritischen Extraktion zweier schwerflüchtiger Aromaten und ihren Gemischen aus porösen Feststoffen. Dissertation, Universität Karlsruhe

17. Dumore JM (1964) Soc Pet Eng AIME Pap 4/4: 356 – 362

18. Eggers R, Sievers U, Stein W (1984) Preprints Int Symp High Pressure Chemical Engineering, Erlangen, p 57 – 61

19. Franca LF de, Correa NCF (1991) Extracao com gas supercritico-perspectivas para produtos naturais. Rev Technol Belem 3/1: 113 – 117

20. Gottschau T, Brunner G (1989) Extraktion von Ölsaaten mit verdichteten Gasen. GIT Fachz Lab 33: 1133 – 1139

21. Goto M, Smith JM, McCoy BJ (1990) Kinetics and mass transfer for supercritical fluid extraction of wood. Ind Eng Chem Res 29/2: 282 – 289

22. Hausen H (1950) Wärmeübertragung im Gegenstrom, Gleichstrom und Kreuzstrom. Berlin

23. Ikushima Y, Saito N, Hatakeda K, Ito S, Asano T, Goto T (1986) A supercritical carbon dioxide extraction from mackerel (Scomber Japonicus) powder: Experiment and modeling. Bull Chem Soc Jpn 59: 3709 – 3713

24. Johannsen M, Brunner G (1994) Solubilities of xanthines in supercritical carbon dioxide. Fluid phase equilibria 95: 215–226

25. Lee AKK, Bulley NR, Fattori M, Meisen A (1986) Modeling of supercritical carbon dioxide extraction of Canola oilseed in fixed beds. JAOCS 63/7: 921 – 925

26. Nusselt W (1927) Die Theorie des Winderhitzers. VDI Z 7: 85 – 91

27. Osborne JO, Katz DL (1944) Trans Am Inst Chem Eng 40: 511 – 531

28. Reese D (1992) Erstellung eines Programmsystems zur Modellierung von Festbettextraktionen und zur Schätzung der Modellparameter. Diplomarbeit, Thermische Verfahrenstechnik, Technische Universität Hamburg-Harburg

29. Rosen JB (1952) Kinetics of a fixed bed system for solid diffusion into spherical particles. J of Chem Phys 20: 387 – 394

30. Rosen JB (1954) General numeric solution for solid diffusion in fixed beds. Ind Eng Chem 46/8: 1590 – 1594

31. Schaeffer ST, Zalkow LH, Teja AS (1989) Modeling of the supercritical fluid extraction of monocrotaline from Crotalaria Spectabilis. J Supercritical Fluids 2/2: 15 – 21

32. Scheefer M (1991) Diplomarbeit, Fachhochschule Hamburg

33. Schertz WW, Bischoff KB (1969) Thermal and material transport in nonisothermal packed beds. AIChE J 15/4: 597 – 603
34. Schumann TEW (1929) Heat transfer: A liquid flowing through a porous prism. J Franklin Inst 208: 405 – 416
35. Schütt M (1992) Untersuchung und Modellierung der Extraktion von Theobromin aus Kakaoschalen mit überkritischem Kohlendioxid als Extraktionsmittel. Studienarbeit Arbeitsbereich Verfahrenstechnik II, Technische Universität Hamburg-Harburg
36. Schwartzberg HG, Torres A, Zaman S (1982) AIChE – Symposium Series No 218, Vol 78, "Food Process Engineering"
37. Schwartzberg HG, Chao RY (1982) Solute diffusivities in leaching processes. Food Technology Feb 1982, pp 73 – 86
38. Simon A (1990) Extraktion von Theobromin und Coffein aus Kakaoschalen mittels überkritischem Kohlendioxid. Diplomarbeit Fachhochschule Hamburg
39. Triday J, Smith JM (1988) Dynamic behavior of supercritical extraction of kerogen from shale. AIChE J 34/4: 658 – 668
40. Vogel G, Angermann H ( 1982) dtv-Atlas zur Biologie. 17th edn, Vol 1, pp 8 – 11
41. Wagener P (1992) Extraktion von Theobromin aus Kakaoschalen mit überkritischem Kohlendioxid im Technikumsmaßstab. Diplomarbeit, Fachhochschule Hamburg
42. Wakao N, Kaguei S (1982) Heat and Mass Transfer in Packed Beds. Gordon and Breach, New York London Paris
43. Zwiefelhofer U (1994) Stoffübertragung in Festbetten bei der Extraktion mit überkritischem Kohlendioxid am Beispiel der Extraktion von Theobromin aus Kakaoschalen. Dissertation, Technische Universität Hamburg-Harburg

# 8 Countercurrent Multistage Extraction

## 8.1 Basic Considerations

### 8.1.1 Countercurrent Processing

"Countercurrent extraction" in this context refers to multistage countercurrent separation of fluid mixtures with supercritical gases as solvents. Multistage separations become necessary when the separation factor between components is in the order of 1 to 10.

In such a countercurrent separation process, the components distribute between the solvent (extract phase) and the liquid (raffinate phase) which countercurrently flow through the separation equipment, in most cases a separation column.

Countercurrent operation of a separation device reduces the amount of solvent necessary, increases throughput, and enables higher extract concentrations in the solvent and lower residual concentrations in the raffinate than does single-stage or multistage cross current operation. Countercurrent operation is therefore useful for separations with high separation factors, as, for example, the extractions from solid substrates described in Chapter 7. But countercurrent operation is absolutely necessary for achieving a reasonable separation between two substances with a relatively low separation factor. The advantages of countercurrent multistage contacting are discussed in more detail in the literature on separation processes [23, 24, 40].

### 8.1.2 The Extraction Process

In Fig. 8.1 the process scheme of a countercurrent gas extraction for the separation of two components into practically pure substances is shown. Process equipment consists of the separation column (1) where gaseous and liquid phases are contacted countercurrently, a separator at the top for separating solvent and extract (2), devices for feeding reflux to the column (3), for recovering top product (4), for delivering feed to the column (5), for recovering product at the lower end of the column (6), and for recycling the solvent (7).

The separation column consists of two separation cascades. In the upper one (enriching section) the bottom product compounds are separated from the top product compounds and rejected to the lower section (stripping section). In the stripping section the top product compounds are separated from the bottom product compounds and transported to the enriching section.

At the top of the column the separator removes the extract from the solvent. From the extract a specified part is separated and introduced at the top of the column as reflux. The remaining part of the extract is the top product. The solvent is reconditioned (filtered, sometimes liquefied and again evaporated for removing trace substances, pressure and temperature are adjusted) and recycled by the cycle pump as the supercritical solvent at the bottom of the column. The feed is introduced at an intermediate location in the column by the feed pump; for a mixture of two components of about the same concentration, this location is at about the middle of the column. More details on the equipment are presented in Section 8.6 on laboratory plants.

The separation of two compounds into practically pure substances is possible with multistage countercurrent gas extraction. Two separation cascades are needed for this separation. A separation cascade is a sequence of separation steps. Gaseous or

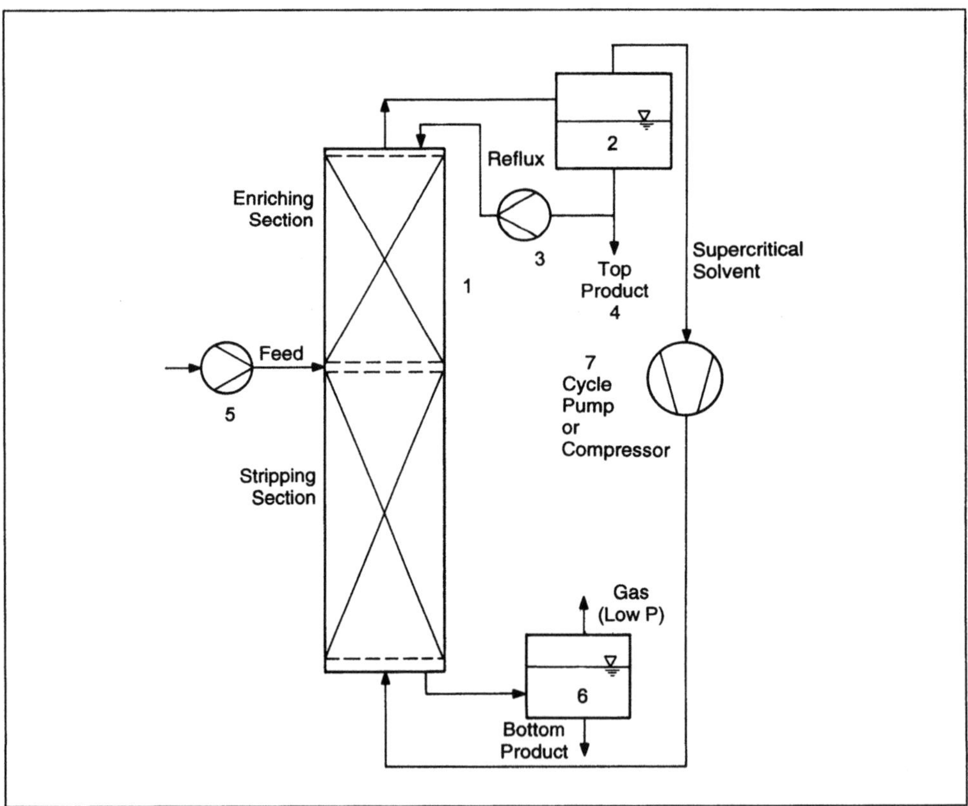


**Fig. 8.1.** Process scheme of a countercurrent gas extraction for the separation of two components.

liquid streams enter or leave the cascade only at the ends. No sidestreams or additional feedstreams are allowed. A separation of two components is specified by two values, e.g., the concentration of a compound at one end of the cascade and the amount recovered of this compound. In order to achieve this specified separation, two separation cascades are needed. The upper one serves to separate the bottom product compound from the top product, and the lower one serves to separate the top product compound from the bottom product. In multistage distillation the separation cascades are known as rectifying and stripping sections. In most cases the two cascades for one separation are unified in one separation column.

The separation of two components is the basic case. This case is not very common in practice, since multicomponent mixtures prevail. But the main features of a multicomponent separation can be clearly shown by a two component separation. Therefore, it is common practice to try to reduce multicomponent separations to two-component separations.

For the separation of more than two components into pure substances two more cascades are needed for each additional compound. For the separation of $n$ components $2(n - 1)$ separation cascades are needed. If two cascades are unified in a column, for the separation of $n$ components, $n - 1$ columns are necessary. The same reasoning holds for cases where components are actually fractions of a multicomponent or complex mixture.

## 8.1.3 Design Considerations and Basic Equations

The separation process of countercurrent gas extraction can be modeled using the common basic equations: Mass balance, energy balance, equilibrium distribution coefficients, and rate equations for mass transfer, as listed in Eqs. 8.1 to 8.5. The resulting system of differential equations can be solved, but the solution is not easily accessible in the case of gas extraction systems, since pure component properties, properties of mixtures, equilibrium relations, and mass transfer correlations are so far not available in most cases, especially not for multicomponent mixtures.

Multicomponent separations can be approximated by two component separations, assuming that it is sufficient to consider the separation between two selected main components, which are called key components. Examples are presented below.

The ultimate goal of analyzing a separation problem is to determine either the possible separation in an existing column (operating problem) or to determine the necessary size of a column with respect to separation efficiency and throughput for a given separation (design problem). The solution of these problems is split into three parts:

I) Determination of number of theoretical stages (or number of transfer units, see below). This part is related to phase equilibrium and is independent from any realization in separation equipment.

II) Size of a separation device with respect to separation performance, often the height of a separation column. This part is related to mass transfer and depends strongly on the type of mass transfer equipment.

III) Capacity of a separation device with respect to throughput, often determined by the diameter of a separation column. This part is related to hydrodynamic behavior in the separation equipment and strongly depends on the type of the internal equipment of the separation column and on the amounts of the counter-currently flowing phases.

Separation performance, expressed as number of theoretical stages or number of transfer units, is separately determined from the efficiency of mass transfer, related to a certain type of equipment, expressed as tray efficiency, height of a theoretical stage or height of a transfer unit. Then, the capacity of the mass transfer equipment for a separation is treated to determine the necessary diameter of a column. The upper capacity limit of a countercurrently operated column is given by conditions, where countercurrent flow can no longer be achieved. The lower capacity limit is determined by the minimum amount of flow for operating the separation equipment with reasonable efficiency.

Separation performance and flow of the phases are connected. Separation performance is enhanced by increased reflux, being best at total reflux with no top product. For practical application, an appropriate value for the reflux must be chosen. This value depends also on non-technical boundary conditions.

The foregoing aspects are common to countercurrent separation processes in general and are discussed in this chapter with respect to countercurrent gas extraction. Basic equations are given as Eqs. 8.1 to 8.5. A guideline for solving a countercurrent gas extraction separation problem is summarized in Table 8.1.

**Table 8.1.** Steps for solving a multicomponent separation problem in countercurrent gas extraction.

| |
| --- |
| Define the mixture by components or pseudo-components |
| Define the separation: Identify key components |
| Specify the separation: purity and recovery rate of one of the key compounds |
| Determine separation performance (as a function of reflux ratio):<br>      number of theoretical stages ($n$) or<br>      number of transfer units ($NTU$) |
| Determine efficiency of mass transfer equipment:<br>      tray efficiency or height equivalent to a theoretical stage or<br>      height of a transfer unit |
| Determine limits for mass flow of countercurrent streams:<br>      maximum flow (entrainment, flooding) and<br>      minimum flow (for effective mass transfer) |
| Decide for a certain ratio between countercurrent flows (reflux ratio) |
| Calculate separation performance or size of a column for the chosen equipment and operating conditions |

Mass balances:

$$\frac{\mathrm{d}L_i}{\mathrm{d}z} - \frac{\mathrm{d}V_i}{\mathrm{d}z} = 0, \qquad\qquad\qquad 8.1$$

$$\sum L_i = L \qquad \sum V_i = V. \qquad\qquad\qquad 8.2$$

Enthalpy balances:

$$\frac{\mathrm{d}(H_L L)}{\mathrm{d}z} - \frac{\mathrm{d}(H_V V)}{\mathrm{d}z} - q = 0. \qquad\qquad\qquad 8.3$$

Equilibrium relations:

$$V_i^* = \frac{K_i V}{L}\, L_i. \qquad\qquad\qquad 8.4$$

Rate equations for mass transfer:

$$\frac{\mathrm{d}V_i}{\mathrm{d}z} = \frac{k_{Gi} a P}{V}\, (V_i - V_i^*), \qquad\qquad\qquad 8.5$$

with:

$z$ = axial coordinate in the separation device;
$L_i, V_i$ = flow of component $i$ in the liquid and gaseous phase;
$L, V$ = total flow of liquid and gaseous phase;
$H_V, H_L$ = enthalpy of gaseous and liquid phase;
$k_{Gi}$ = mass transfer coefficient of component $i$, related to the gaseous phase;
$a$ = mass transfer area per volume of transfer device;
$P$ = total pressure;
$K_i$ = equilibrium partition coefficient of coefficient $i$ between gaseous and liquid phase;
$V_i^*$ = equilibrium concentration of component $i$ in the gaseous phase;
$q$ = heat energy.

## 8.1.4 Some Additional Considerations

From the point of view of separation processes, gas extraction in its various realizations is an equilibrium-determined separation process with a supercritical solvent as mass separating agent. Common chemical engineering methods can be applied to analyze the gas extraction process.

Countercurrent gas extraction is carried out near phase equilibrium. This is easy to visualize for a plate column where one or two plates may be considered as the volume in which equilibrium is established. But even in a packed column, countercurrent flows are not far from phase equilibrium. In a packed column, concentrations change continuously with column length. Some distance to equilibrium is always necessary to enable mass transfer. But concentrations of the components in the phases are not far from equilibrium, due to long residence time (low linear flow velocities), low viscosity, and high transport coefficients. Therefore, changes in concentration are slow and smooth.

For characterizing the composition of the phases, especially the gaseous phase, it is useful to distinguish between the distribution of the supercritical solvent, determining the extent of the two-phase area and the distribution of the solute compounds. This distinction is even useful for calculations, since common phase equilibrium calculation methods with equations of state give good results for the extent of the two phase area, but not as good results for the separation factor of similar compounds.

In order to establish two-phase flow, the components must not be miscible. This condition poses a problem in liquid-liquid extraction, but can easily be overcome in gas extraction, since the supercritical solvent always makes it possible to establish a region with two coexistent phases. Therefore, the first problem to solve in countercurrent gas extraction is to determine the extent of the two-phase area. This may be determined by well known methods, discussed in Chapter 3.

For the total separation of two components in a countercurrent column, a two-phase flow up to the end of the column is necessary. If pure component $A$ is to be removed at the top of the column in a separation of components $A$ and $B$, pure $A$ has to be reintroduced at the top as reflux. Operating conditions have to be chosen such that at the top end of a separation column still a two-phase flow is possible for pure $A$ dissolved in the supercritical solvent. The ternary system of components $A$ and $B$ and the supercritical solvent $S$ must be of ternary type 2, with mixing gaps in both binary subsystems $A$-$S$ and $B$-$S$ (see Chapter 3 on phase equilibria).

If pure component $A$ is miscible with the supercritical solvent under extraction conditions, the maximum concentration of $A$ in the top product is limited by the extent of the two-phase area, as illustrated in Fig. 8.2.

The streams introduced into the column are not in equilibrium with the countercurrent streams at the point of introduction. Equilibration takes place in the column. This has to be taken into account in the solution of the equations for analyzing the separation problem. The introduced streams may be presaturated, but that causes the equilibration process to take place at another location.

The separation column for countercurrent gas extraction is operated similar to other countercurrent separation devices. The main differences stem from the high pressure, which poses problems, for example, with handling mass flows or with observing hydrodynamic behavior. Other difficulties arise from the properties of typical mixtures which can be treated by gas extraction, e.g., multicomponent systems with unidentified compounds, a high viscosity or a high melting point temperature.

In general, the most effective mode of operating a countercurrent column is continuously and in a steady state. But it is possible to carry out countercurrent gas extraction batchwise, analogous to batch multistage distillation or rectification. It can also be analyzed with the same methods.

A considerable number of investigations has been carried out since the early 1970s. Yet the types of mixtures treated in a countercurrent separation are only a few, comprising e.g. mono-, di-, and triglycerides, free fatty acids, fatty acid esters with lower alcohols, compounds which occur in plant oils, like lecithin, sterols and tocopherols,

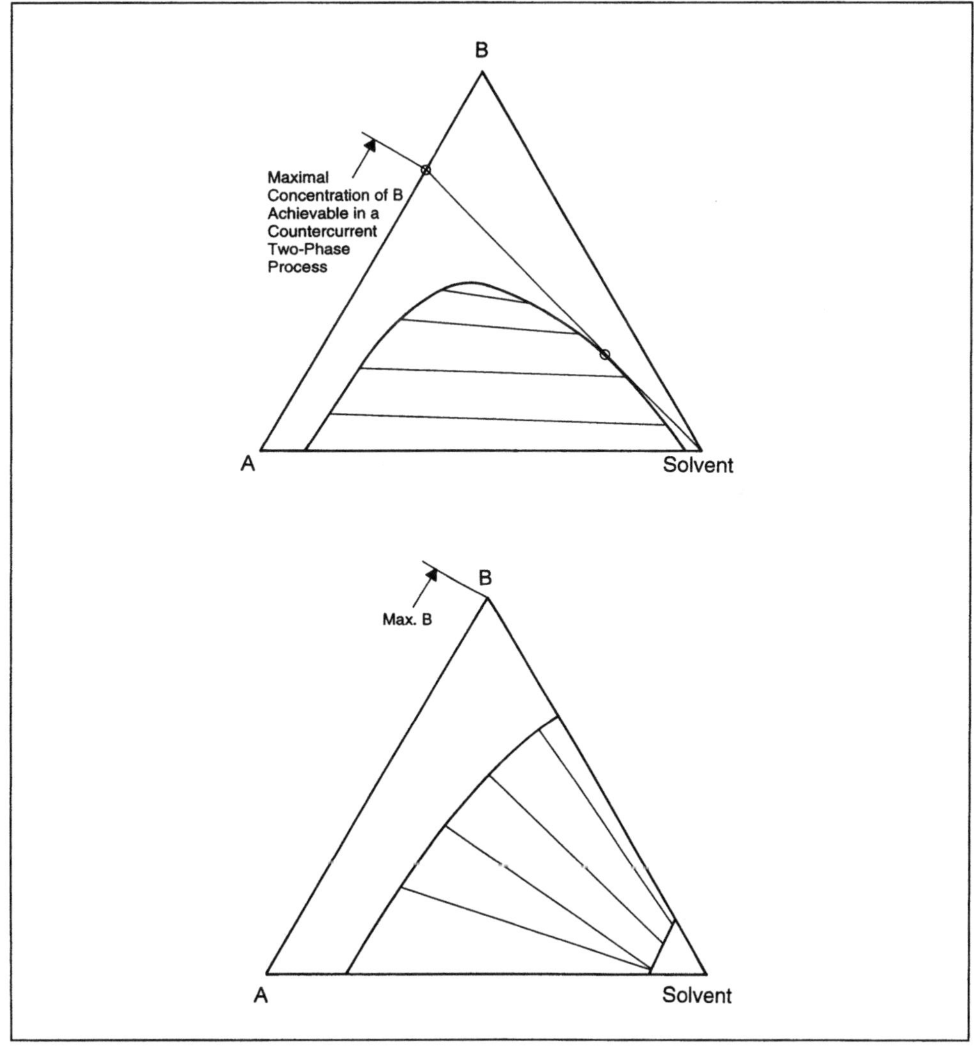


**Fig. 8.2.** Maximum concentration of a component which can be achieved in a separation with countercurrently flowing phases.

mixtures of water and alcohols, and mineral oil fractions or similar complex mixtures [2 – 14, 16 – 21, 25 – 35, 37 – 39, 42, 43].

In the following, various aspects of countercurrent gas extraction are treated in a generalized way, as far as possible, and are illustrated by two separation problems: the separation of tocopherols from water vapor distillation condensates and the concentration of unsaturated fatty acid esters.

# 8.2 Modeling Countercurrent Gas Extraction

Modeling countercurrent gas extraction aims to the formulation of a system of equations which describes the process of gas extraction. For gas extraction the basic equations on phase equilibria, mass and energy balance and the kinetic equations for mass transfer are available (Eqs. 8.1 to 8.5) and must be solved. As mentioned above, the problem is further split into a pure separation problem and a problem of the effectivity and capacity of the separation equipment which enables us to study the separation without taking into consideration the many possible types of mass transfer equipment.

From these equations the number of theoretical stages or number of transfer units should be extractable, as well as the concentration of individual components along the separation device.

## 8.2.1 The Concept of Equilibrium or Theoretical Stages

As methods for analyzing the separation problem, two well known methods are available: The method using equilibrium or theoretical stages, and the method of **Number of Transfer Units** (*NTU*). Both methods will be discussed subsequently, but since the equilibrium stage method is more expressive it will be used in the following for a detailed discussion. Transfer units are treated in Section 8.2.9.

The method of equilibrium stages cuts the countercurrent separation into a sequence of discrete sections, which may be parts of a column. The streams leaving an individual section are supposed to be in thermodynamic heterogeneous equilibrium (phase equilibrium), while the streams entering the section are not in equilibrium. Such a section is defined as an "equilibrium stage," or "theoretical stage".

In each theoretical stage, concentrations of the components are changed from the concentration of the incoming streams to the corresponding equilibrium value. Each additional equilibrium stage changes concentrations again and a sufficient number of equilibrium stages, applied to enrich one component, leads to a previously specified concentration of the component.

The concentration steps achievable in one equilibrium stage depend on equilibrium and on concentration of the incoming streams, which again depend on the amount of the countercurrently flowing streams. This will be illustrated below.

## 8.2.2 General Equations

For the concept of theoretical stages, Eqs. 8.1 to 8.4 may be transformed to the following equations [24]. The meaning of the variables may be taken from Fig. 8.3, the legend below Eq. 8.9 and the legend to Eq. 8.1 to 8.5.

Equilibrium relationships:

$$V_{iP} = \frac{K_{ip}\, V_p}{L_p}\, L_{ip}.$$

8.6

Mass balances:

$$L_{ip} + V_{ip} - L_{ip+1} - V_{ip-1} - F_{ip} = 0,$$

8.7

$$\sum_i L_{ip} = L_p \quad \text{and} \quad \sum_i V_{ip} = V_p.$$

8.8

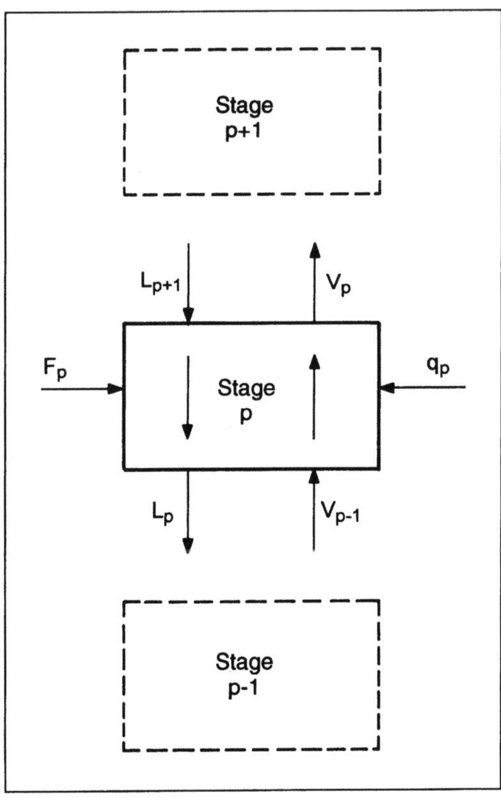


**Fig. 8.3.** Concept of theoretical stages in countercurrent gas extraction for the definition of the variables in Eqs. 8.6 to 8.9.

Enthalpy ($H$) balances:

$$L_p H_{L_p} + V_p H_{V_p} - L_{p+1} H_{L_{p+1}} - V_{p-1} H_{V_{p-1}} - F_p H_{F_p} - q_p = 0, \qquad\qquad 8.9$$

with the following indices:
$p$    = theoretical stage $p$;
$i$     = component $i$;
$L, V$ = liquid or gaseous phase;
$F$    = feed.

For $m$ components and n theoretical stages, there are $n(2m + 3)$ equations. These equations in general are a system of nonlinear equations. Strategies for a solution are available, but will not be further discussed. The main difficulty, with respect to gas extraction, is to calculate the equilibrium distribution coefficients with sufficient accuracy to represent the separation factor and the two phase region.

Phase equilibrium in context with countercurrent processes is usually written as distribution coefficient or $K$-factor:

$$K_i = y_i/x_i. \qquad\qquad 8.10$$

In gas extraction, $K_i$ is strongly dependent on pressure and temperature and may well be strongly dependent on concentration.

$$K_i = f(P,\ T,\ x_i,\ x_j,\ y_i,\ y_j). \qquad\qquad 8.11$$

Values for $K_i$ can be calculated with the available methods for phase equilibrium calculations, as discussed in Chapter 3, but can also be taken from experimental results and used in the analysis of the separation problem without employing phase equilibrium calculation.

The solution of these equations has been treated, for example, by Moricet [27] for the example of separation of monoglycerides from a mixture of glycerides, by Brignole [1] for the separation of ethyl alcohol from aqueous solutions, and by Colussi et al. [15] for multicomponent mixtures.

If the equilibrium distribution coefficients can be provided by calculation or by empirically determined relationships, then procedures in commercially available process calculation programs may be used for modeling countercurrent gas extraction. So far the main problem resides with providing the equilibrium relationships for a sufficient and appropriate number of components of the feed mixture. Since equilibrium data are usually scarce, the reduction of multicomponent mixtures to equivalent mixtures containing far less components will be discussed below. For these equivalent mixtures then simplified methods, like the McCabe-Thiele method or the Ponchon-Savarit method, can be applied for an approximate solution of the separation problem.

## 8.2.3 Reduction of Multicomponent and Complex Mixtures to Equivalent Binary, Ternary or Quaternary Systems

The complications with detailed modeling of separations for a complex mixture have led to simplifications of the problem. The main simplification is the reduction of the number of components of a mixture to an equivalent binary or ternary mixture. A binary mixture is sufficient for representing the separation problem, if the influence of the solvent on the mass flow of the phases can be neglected, i.e., solubility of the supercritical solvent in the liquid phase and solubility of the low volatile components in the supercritical solvent are approximately constant over the concentration range under consideration. This corresponds to a linear relation between the countercurrently flowing phases or a linear operating line (see below).

To obtain a representative binary system, two neighboring compounds are selected from the mixture of substances, between which the separation shall take place. These are called "key components". If they are the major components they can be treated as a binary mixture representing the total complex or multicomponent mixture. The binary system of key components is then an equivalent binary system. If it is not possible to define key components, then pseudo-components may be defined, as discussed in Section 3.7.

It is supposed that all other components distribute in the countercurrent separation with one of the key components to the top or with the other to the bottom according to their equilibrium distribution caused by supercritical carbon dioxide under operating conditions. In a distillation process this means that the substances with a vapor pressure higher than the lighter key component are all transported to the top of the distillation column. In countercurrent gas extraction, the compounds with a higher distribution coefficient than the key component, which is enriched at the top of the column, are also enriched in the gaseous phase and are transported to the top of the column. The same reasoning applies for the components with a distribution coefficient lower than the key component, which is enriched at the bottom of the column: they are transported to the bottom of the separation column.

The distribution factor may change with concentration to such an extent that the direction of enrichment in the column is reversed for a component, especially if an entrainer is employed [3]. This is analogous to azeotropy in distillation, or solutropy in a liquid-liquid extraction.

For determining the key components, the equilibrium distribution coefficients must be known. From these distribution coefficients the separation sequence can be directly concluded and key components selected. Usually equilibrium distribution coefficients are experimentally determined by measuring phase equilibrium. These measurements are not easy and are therefore tedious and expensive. Another possibility for determining key components and the distribution of the other components of a complex mixture is by chromatographic analysis. If the chromatographic method separates according to the same molecular properties as countercurrent gas extraction, then the separation sequence of the components of a complex mixture for gas extraction can directly be concluded from the chromatogram.

The easiest case to understand is if the same supercritical solvent is used in chromatography as mobile phase and in countercurrent extraction as solvent. With an unpolar stationary phase in chromatography, it is not farfetched to expect the same sequence of substances during elution by chromatography (SFC) as is obtained by an equilibrium distribution in the supercritical solvent. Then, the chromatogram can be used for defining key components, without knowledge of phase equilibrium. An example is given in Section 8.4 for the separation of tocopherols.

In general, additional equilibrium measurements are necessary, since solubility and equilibrium distribution coefficients are needed for the analysis of the separation problem. But it may be possible that a relationship can be found between the equilibrium distribution and retention time in chromatography and equilibrium solubility in the gaseous phase and peak area, so that equilibrium measurements may be reduced to a few calibrating experiments.

If the amounts of countercurrently flowing phases is not constant due to changes in concentration and solubility along a separation cascade, then the influence of the solvent must also be taken into consideration. This leads to an equivalent ternary system, consisting of the two key compounds and the supercritical solvent. This case is treated in Sections 8.2.7 and 8.2.8.

An additional component to the supercritical solvent and two components which are to be separated, an entrainer, modifier or just a third compound which cannot be represented by the key components, leads to an equivalent system of four components (a quaternary system). This case is discussed in Section 8.5.

### 8.2.4 Number of Theoretical Stages for a Binary or Equivalent Binary System

For an equivalent binary system consisting of component 1 and component 2, the set of equations for modeling gas extraction can be transformed to Eqs. 8.12 to 8.15. The process is schematically shown in Fig. 8.4. Several assumptions have been made: In a separation cascade (enriching section or stripping section) mass flows to and from the cascade are only allowed at the ends. The process is carried out at constant pressure and temperature.

Equilibrium:

$$y_1 = f(x_1). \tag{8.12}$$

A constant separation factor $\alpha$, which can be a medium value, can be written as:

$$y_1 = \frac{\alpha_{12}\, x_1}{1 + (\alpha_{12} - 1)\, x_1}. \tag{8.13}$$

Mass balance, enriching section are:

$$V_p y_{1p} - L_{p+1} x_{1p+1} = V_n y_{1n} - R_n x_{1R_n},\qquad 8.14$$

$$V_p - L_{p+1} = V_n - R_n.\qquad 8.15$$

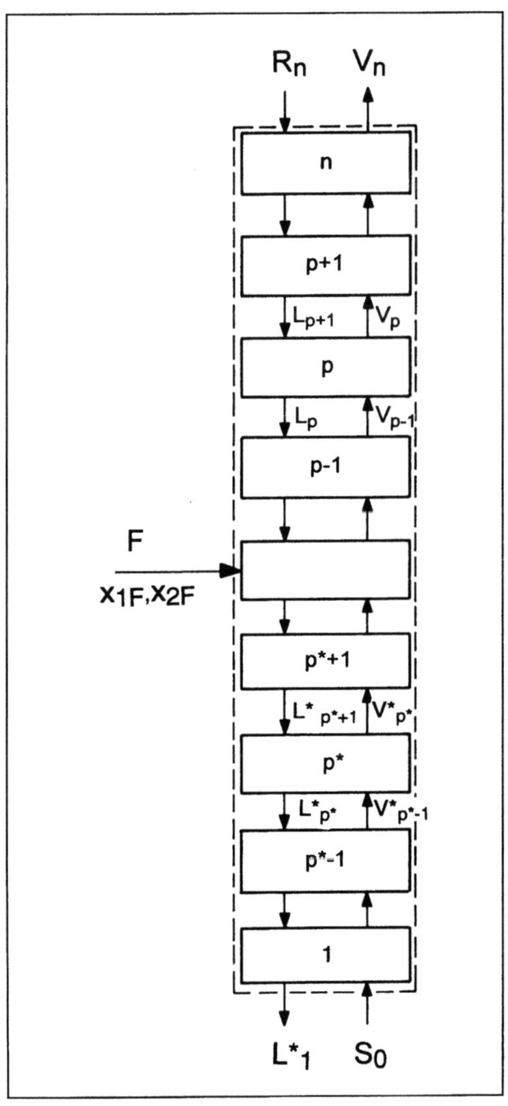


**Fig. 8.4.** Flow scheme of a separation for an equivalent binary system.

Mass balance, stripping section are:

$$V_{p*}^* y_{1p*} - L_{p*+1}^* x_{1p*+1} = S_0 \cdot y_{1S_0} - L_1^* \cdot x_{11},$$ 8.16

$$V_{p*}^* - L_{p*+1}^* = S_0 - L_1^*.$$ 8.17

Total mass balance is:

$$F + S_0 + R_n = V_n + L_1^*,$$ 8.18

with:

$\alpha_{12}$ = separation factor between components 1 and 2 at given conditions of temperature and pressure;

$y_1, x_1$ = concentrations of component 1 in the gaseous and liquid phase;

$y_{1p}, x_{1p}$ = concentrations of component 1 in the gaseous and liquid flow from equilibrium stage $p$;

$V_p$ = gaseous flow from stage $p$;

$L_p$ = liquid flow from stage $p$;

$R_n$ = reflux to stage $n$ of the enriching section;

$S_0$ = solvent to the first stage of the stripping section;

$F$ = feed mixture (equivalent binary mixture);

$L_1^*$ = liquid flow leaving stripping section (raffinate phase);

$V_n$ = gaseous flow leaving enriching section (extract phase).

Energy effects have been neglected, therefore no enthalpy balances have to be considered. If energy flows are important, due to concentration changes or heating or cooling, the set of equations has to be completed with enthalpy balances.

The above set of equations can be programmed and solved by computer. But it is possible to solve these equations graphically, which has the advantage of more clarity and that it can be carried out manually. The method is well known as the McCabe-Thiele method. If equations for equilibrium and mass balance are linear in the $y,x$-system of coordinates, which may well be at low or high concentrations, an analytical solution, known as the Kremser-Souders-Brown equation (KSB) is applicable.

## 8.2.5 Analysis of Countercurrent Gas Extraction According to the McCabe-Thiele Method

For the McCabe-Thiele method the following assumptions are made: The composition in both the countercurrent flows can be represented by one concentration and the amount of flows is constant along a separation cascade.

Phase equilibrium and material balances are plotted in a $y,x$-diagram, where the ordinate represents concentrations of component 1 in the gaseous phase and the abscissa represents concentrations of component 1 in the liquid phase (Fig. 8.5). Equilibrium can be represented by Eq. 8.13, if the separation factor $\alpha$ can be considered constant, otherwise the equilibrium line must be defined by individual experimental points or correlation functions based on experimental data.

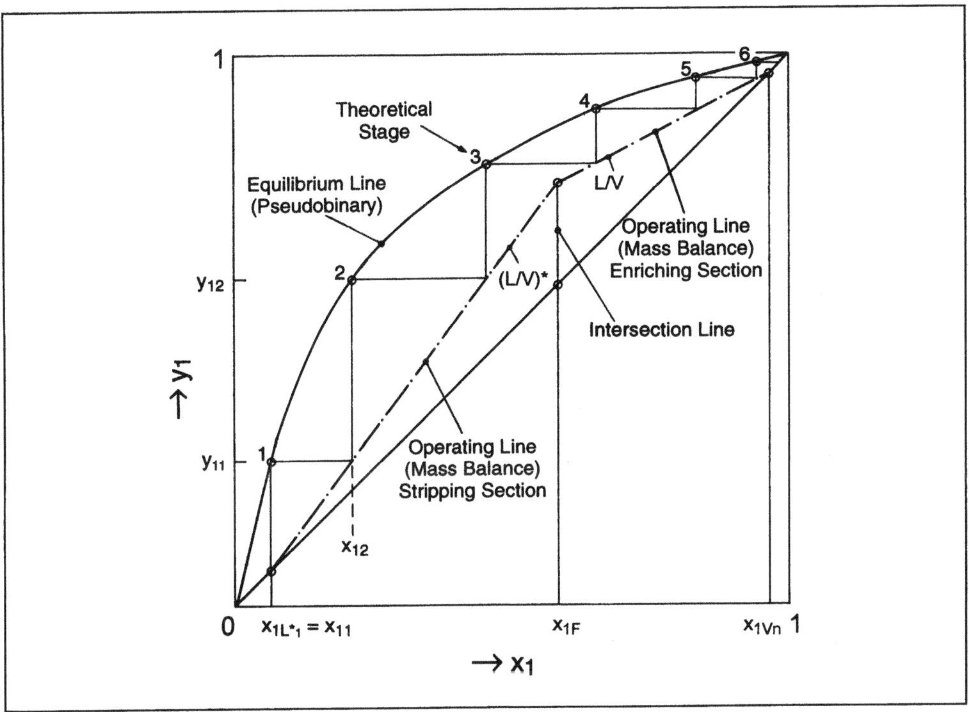


**Fig. 8.5.** $y,x$-diagram of an equivalent binary system in gas extraction for application of the McCabe-Thiele method.

### 8.2.5.1 Determination of Operating Lines

The mass balance is a linear equation, since $V$ and $L$ are constant. For the enriching section the mass balance is given by Eq. 8.19:

$$y_{1p} = (L_{p+1}/V_p)\, x_{1p+1} + (V_n\, y_{1n} - R_n\, x_{1R_n})/V_p. \qquad 8.19$$

The second term on the right side of Eq. 8.19 is a constant and the first term represents the gradient of the line. Since flows are assumed to be constant, $L_n = R_n$. The gradient of the mass balance line, or operating line, is determined by $R$ (at given $V$), the liquid reflux flow introduced at the top of the column. The operating line for the enriching section begins at the concentration of component 1 leaving the column at the top. The equivalent concentration of component 1 in the gaseous flow leaving the column is the same as the equivalent concentration of component 1 in the liquid reflux, therefore $x_{1R_n}$ on the diagonal line $x = y$ is one point of the operating line. This line can be drawn with the gradient $R/V$, the ratio of the countercurrent flows in the enriching section. $R$ is determined by the amount of liquid reflux introduced at the top of the column. $V$ is on one hand determined by $E$, the amount of extract removed, plus the amount of reflux $R: V = E + R$. On the other hand $V$ is determined by the equilibrium solubility of components 1 and 2 (the key components) in the supercritical solvent, as given by the total solubility of the complex mixture in the supercritical solvent times the fraction of component 1 and 2 (equilibrium loading assumed).

Note that variables are valid for an equivalent binary system. Therefore, concentrations and flows are calculated on a solvent-free basis.

The same reasoning holds for the stripping section. One point of the operating line is determined by the concentration of the liquid flow (raffinate) removed from the lower end of the column on the diagonal. The gradient of the stripping section operating line is determined by the ratio of the flows in the stripping section $L^*/V^*$. $L^*$ is the amount of the equivalent binary system of the liquid raffinate phase removed at the lower end, $R_0$, plus $V^*$, the amount dissolved by the incoming supercritical solvent at the bottom and transported upwards in the stripping section: $L^* = R_0 + V^*$.

For the stripping section the operating line is given by:

$$y_{1p^*} = (L^*_{p^*+1}/V^*_{p^*})x_{1p^*+1} + (S_0 y_{1S_0} - L^*_1 x_{11})/V^*_{p^*}. \qquad 8.20$$

The feed distributes between the liquid and the gaseous flow according to its conditions of state. If the feed is introduced as liquid saturated with the supercritical gas, then only the liquid flow is changed and $L^* = L + F$. But there may be cases where the feed is gaseous or only partly vaporized. Then both flows are changed at the feed introduction point according to the ratio of gas to liquid. The intersection of the operating line of the enriching and stripping section is given by either the sum or the difference of these equations, taking the same concentration variables for both sections and omitting the indices for the intermediate stages:

Enriching section:   $Vy_1 = Lx_1 + (V_n y_{1n} - R_n x_{1R_n})$ $\qquad$ 8.21

Stripping section:   $V^* y_1 = L^* x_1 + (S_0 y_{1S_0} - L^*_1 x_{11})$ $\qquad$ 8.22

---

$(V - V^*)y_1 = (L - L^*)x_1 + (V_n y_{1n} - R_n x_{1R_n}) - (S_0 y_{1S_0} - L^*_1 x_{11})$ $\qquad$ 8.23

With the assumption that $V - V^* = V_F$ and $L^* - L = L_F$ and considering that the second term on the right side is equal to $Fx_{1F}$, the total amount of component 1 in the feed, Eq. 8.23 is transformed to:

$$V_F y_1 = L_F x_1 + Fx_{1F}. \qquad 8.24$$

The gradient of the intersecting line depends on the ratio $- L_F/V_F$, with $L_F$ the amount of liquid key components and $V_F$ the amount of gaseous key components in the feed $F$. If the feed is introduced as saturated liquid, the gradient of the intersecting line is $\infty$. If the feed is introduced as saturated gas, the gradient of the intersecting line is 0. An unsaturated liquid (subcooled) as feed has an intersection line with a gradient between 1 and $\infty$, an unsaturated gas (superheated) has an intersection line with a gradient between 0 and $- 1$. The intersection line crosses the diagonal at the total concentration of key component 1 in the feed.

### 8.2.5.2 Determination of Theoretical Stages

The number of equilibrium stages can be determined by starting at one end of the column, e.g., at the lower end of the separation column at the concentration $x_{11}$. This concentration is specified by the amount of key component 1 allowed in the bottom product and has been used already to determine the lower end of the stripping section operating line. The equilibrium concentration corresponding to $x_{11}$ is $y_{11}$, the equilibrium concentration of the gaseous equivalent binary mixture leaving the first equilibrium stage. This concentration is given by the corresponding point on the equilibrium line or the solution of the equilibrium equation for $x = x_{11}$. For the gas leaving stage 1 and the liquid flowing to stage 1, the mass balance is represented by the operating line. The corresponding point to $y_{11}$ on the operating line or the solution of the equation for the operating line for $y_{11}$ yields the composition $x_{12}$ of the liquid flow to stage 1.

In the next step the equilibrium gas composition corresponding to $x_{12}$ is determined, yielding $y_{12}$ the equilibrium composition of the gas leaving stage 2, and so on. Alternate solutions of the equilibrium equation and the appropriate equation for the operating line lead to a sequence of equilibrium stages. The sequence can be terminated when the specified concentration for key component 1 in the top product is obtained. Graphically, the number of equilibrium stages is represented by the points on the equilibrium line.

The number of equilibrium stages resulting from this binary analysis of a multicomponent separation problem is the lower limit. Further stages may be necessary to separate additional compounds from the product.

The introduction of unsaturated mass flows into the separation device is of influence on the operating line in a separation cascade. If the supercritical solvent or the reflux is fed unsaturated with either the low volatile components or the gas to the countercurrently operated column, then the first equilibrium stage can be assumed to achieve saturation. The effect on the operating line is determined by the amount of mass transferred. Since solubility of the gas in the liquid phase is relatively high, the effect of adding unsaturated liquid may be substantial and the operating line for this area of the column nonlinear. At the lower end of a cascade, where the supercritical solvent is introduced, the effect can be small if solubility of the low volatile components in the supercritical solvent is low and the concentration change in the first theoretical stage is small compared to the total concentration change in the separation cascade. Then, the operating line can be considered to be linear. On the other hand, concentration changes in one theoretical stage may be great, causing substantial variation of the countercurrently flowing phases. Then, it may be necessary to construct a curved operating line or switch to the Ponchon-Savarit method, presented in Section 8.2.8.


### 8.2.5.3 Limiting Operating Conditions for the Amount of Countercurrent Flows

A separation column with countercurrent flows can be operated at or near two limiting cases: In the first case, all the product at the end of a separation column is reintroduced into the column. No material is removed or added after a first filling. This mode of operation results in the minimum number of theoretical stages necessary for a certain separation, specified by top and bottom product concentration of component 1. In the second case, flows in the column are minimized. Since the ratio of liquid phase flow to vapor phase flow determines the gradient of the operating line, the minimum gradient of the enriching section operating line determines the minimum liquid flow possible. When the operating line touches the equilibrium line, the liquid phase flow is at its lowest possible value for operating the column. Then, the number of theoretical stages is infinite.

**Minimum equilibrium stages.** The degree of separation in a countercurrent separation column increases with the ratio of liquid to gas flowing through the column. The best separation is achieved if all the top product, after being removed from the gaseous solvent, is again introduced to the column as reflux. Then, the product quantity is zero and the reflux ratio is infinite. The operating line is represented in this case by the diagonal in a $y,x$-diagram. Since the distance between operating line and equilibrium line has reached the maximum value, which is possible, the number of equilibrium stages is minimal for a specified separation. This operating mode can be used for starting up a column and for comparison of separation efficiencies.

**Minimum reflux.** At lower reflux ratios than infinite, the operating line approaches the equilibrium line. The reflux may be lowered until the operating line touches the equilibrium line. Then, the number of equilibrium stages becomes infinite and the reflux ratio is the minimum reflux ratio for a specified separation. These limiting conditions are illustrated in Fig. 8.6.

The real reflux ratio must be chosen in between the limiting values. The actual value depends also on economical considerations. At low reflux ratios (but above the minimum reflux ratio), the column will be high and thin. At high reflux ratios, the column will be low and thick. Since for pressure vessels it is more expensive to increase diameter than length, the optimum reflux ratio will be near the minimum reflux ratio.


### 8.2.5.4 Consideration of Equilibrium Data

Some consideration must be given to the equilibrium data used in the analysis of the separation. The equivalent binary separation factor is determined from the $K$-fac-

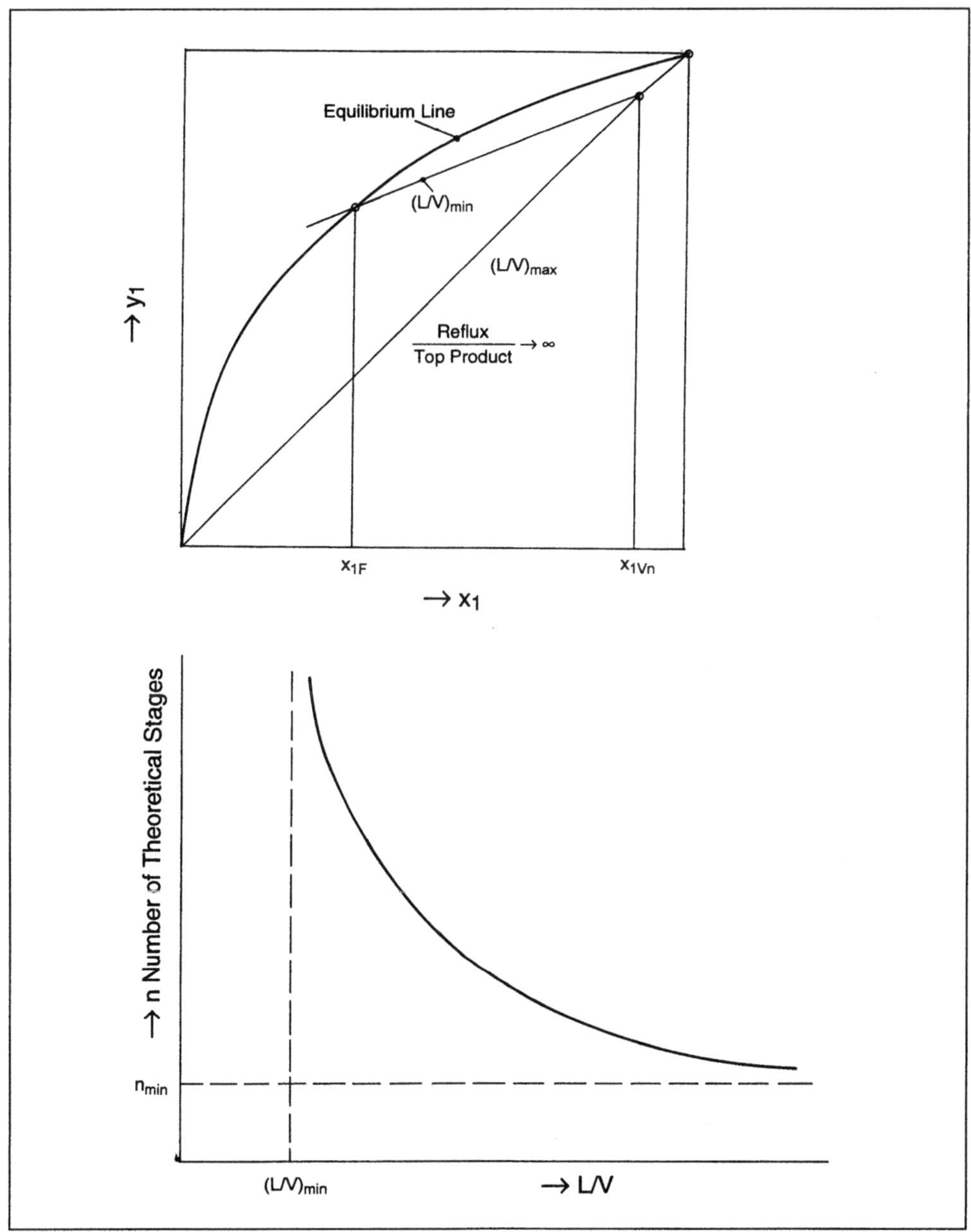



**Fig. 8.6.** Limiting conditions for number of equilibrium stages and reflux for an equivalent binary separation.

tors of an equivalent binary system. In a binary system, equilibrium is not dependent on the amount of phases. The separation factor is the same for the first gas bubble formed and the last drop of liquid evaporating. For mixtures containing more than two components, the separation factor depends on the amount of phases being in contact at equilibrium. Let us consider the example of a ternary system of component 1, component 2, and a solvent $S$, as illustrated in Fig. 8.7. If solvent $S$ is added to the initial feed mixture $F$, the mixing point lies on a straight line connecting $F$ and $S$. The first gas bubble is formed if the amount of solvent is high enough for a mixing point just inside the two-phase area, but practically on the equilibrium curve (phase boun-

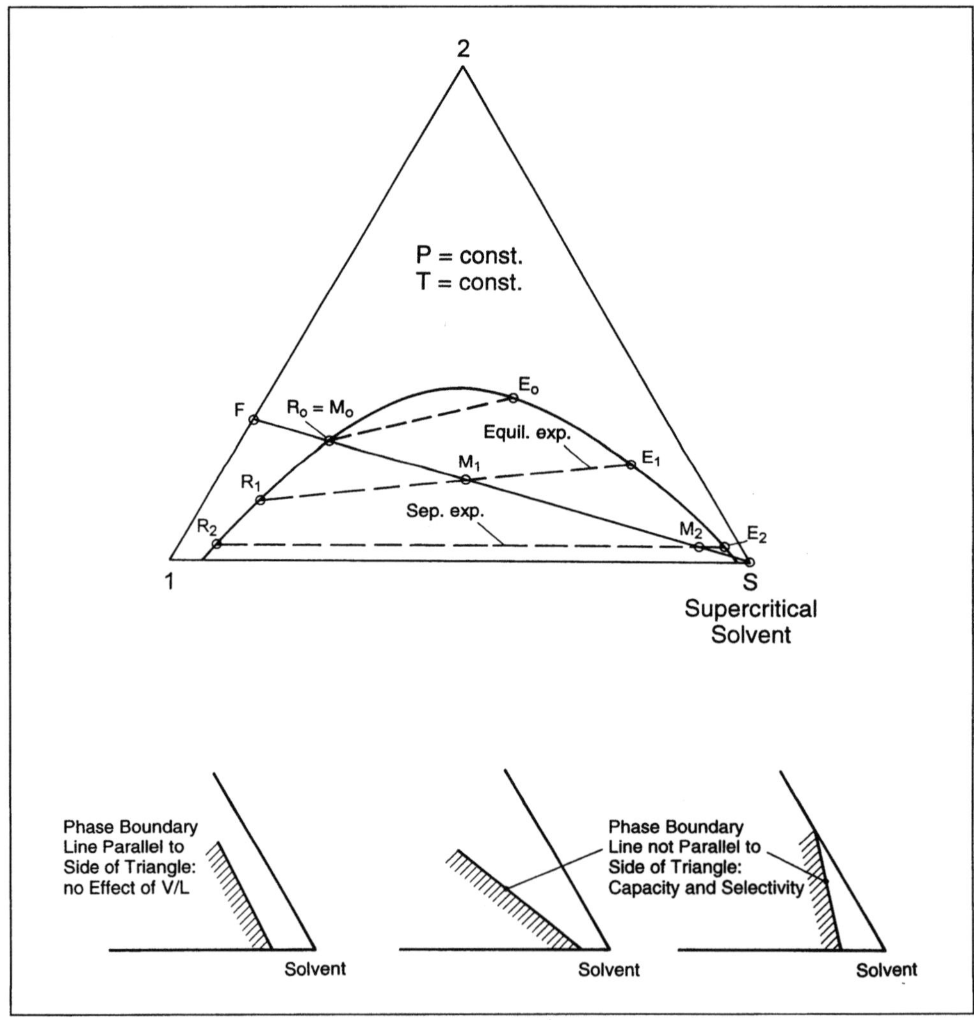


**Fig. 8.7.** Illustration of the effect of the ratio of gaseous to liquid phase in a phase equilibrium experiment and in a countercurrent separation.

270

dary line) on the feed side at $M_o$. $M_o$ is in equilibrium with $E_o$. Increasing the amount of solvent $S$ leads to mixing point $M_1$. $R_1$ is in equilibrium with $E_1$, representing the equilibrium composition of the gas phase on the solvent side of the phase boundary line. A higher amount of solvent results in a mixing point $M_2$. Phase equilibrium is represented by a tie line through $M_2$. This tie line crosses the phase boundary line in $R_2$ and $E_2$, representing liquid and gaseous equilibrium phases which are different from $M_1$ and $E_1$. If the tie line through $M_2$ is not parallel to the tie line through $M_1$, then the separation factor is also different in both cases. Tie lines normally are not parallel, but change their gradient.

Equilibrium data used in the analysis of countercurrent gas extraction are often measured by the static analytical method (see Section 3.8.2). The amounts of gaseous and liquid phases in this experiment are of the same order of magnitude. In most cases the gas phase may be two to three times larger in volume than the liquid phase. Masses of the phases may be about equal, and the number of moles will be most probably higher in the gaseous phase than in the liquid phase. The ratio of the two phases is quite different in countercurrent gas extraction. Typically, the amount of the gaseous phase is at least an order of magnitude higher than the amount of the liquid phase (compare the examples presented in Sections 8.3 and 8.4), regardless of whether on a volume, mass or molar basis. Therefore, loading and separation factor for countercurrent gas extraction are different from the equilibrium values determined at the same temperature and pressure in the equilibrium autoclave. The magnitude and the importance of this difference has to be studied before equilibrium loading of the gaseous phase and the separation factor for an equivalent binary analysis can be specified.

The effect is greatest for a ternary type-I equilibrium with phase boundary lines which are not parallel to the triangular side lines. For phase boundary lines parallel to the triangular side lines, or ternary type-II equilibrium, the effect can be negligible.


## 8.2.6 Analysis with the KSB-Equations

The KSB-equations allow to determine the number of theoretical stages as a function of the degree of separation and the equilibrium distribution coefficient. These equations are applicable if the equilibrium concentration in the gaseous phase is a linear function of the concentration in the liquid phase and the operating line is linear.

These conditions are not met for a separation covering an extended concentration range. For an approximation, the separation can be divided into sections, for which the assumptions are approximately valid.

The KSB equation for the removal of a compound is given by the following equation:

$$\frac{x_{1E} - x_{1E}{}^*}{x_{1R} - x_{1R}{}^*} = \frac{1 - (L/(mV))}{1 - (L/(mV))^{n+1}}, \qquad 8.25$$

with

$x_{1E}$ = concentration of component 1 in the liquid flow leaving the separation cascade;

$x_{1R}$ = concentration of component 1 in the liquid flow entering the separation cascade;

$x_{1E}^*$ = equilibrium concentration of component 1 in the liquid flow, corresponding to $x_{1E}$;

$x_{1R}^*$ = equilibrium concentration of component 1 in the liquid flow, corresponding to $x_{1R}$;

$L$ = liquid phase flow;

$V$ = gaseous phase flow;

$m$ = gradient of the equilibrium line $y = mx + b$;

$n$ = number of theoretical stages.

The number of theoretical stages $n$ can be determined, if the above equation is rearranged for $n$:

$$n = \frac{\ln\{[1 - L/(mV)][(x_{1R} - x_{1R}^*)/(x_{1E} - x_{1E}^*)] + L/(mV)\}}{\ln[(mV)/L]}. \qquad 8.26$$

## 8.2.7 Including the Supercritical Solvent into the Analysis of the Separation Process

In the analysis of the separation process, the supercritical solvent has to be taken into account if the mutual solubility of liquid and gas changes with composition in the range covered by the separation. If the feed mixture can be represented by an equivalent binary system, then the analysis can be carried out on an equivalent ternary system consisting of the supercritical solvent, key component 1 and key component 2. Both key components can be pseudocomponents. The equations for the countercurrent separation process for such a ternary system are given by Eqs. 8.6 to 8.9.

Energy effects (the enthalpy balances) are neglected, since temperature of the column is assumed to be constant. Enthalpy of mixing effects is considered to be of second order. A temperature gradient in the column can be taken into account by the equilibrium distribution coefficients (separation factor) and the solubilities included in the mass balance equations.

For the solution of these equations, an appropriate number of parameters must be specified in order to solve the equations for the other parameters. Usually, composition and condition of the feed mixture, conditions of state and equilibrium distribution coefficients, the concentration and the amount of one of the key components in one of the streams leaving the separation device is known.

The equations can be solved by assuming a fixed number $n$ of equilibrium stages and carrying out the calculation for different values of $n$. Then, the set of equations has to be solved simultaneously for several times, until a solution is obtained where the specified separation can be achieved.

The other possibility for solving the set of equations is by a stage-to-stage calculation, starting at one end of the column. In this case, composition of the flows at one end of the separation column must be known, which serves as the starting point for the calculation. An alternate solution of equilibrium equations and mass balances leads to the number of equilibrium stages in one calculation sequence.

For the simplified three-component analysis of a separation problem, the stage to stage calculation should always be sufficient. This calculation can be carried out numerically or graphically. The graphic solution is convenient for understanding the method and will be discussed subsequently.

The graphical solution can be carried out in any diagram where the three components of the mixture can be represented. The Gibbs-triangular diagram is well known and often used in liquid-liquid extraction. For a ternary system in gas extraction, the gaseous phase in a Gibbs-diagram is compressed to a small region near the solvent corner. Concentration changes in the gaseous phase cannot be determined accurately. Therefore, a Jänecke-diagram is employed in which these problems can be avoided. It is explained in the following section.


## 8.2.8 Ponchon-Savarit Method for an Equivalent Ternary System

The Ponchon-Savarit [36, 41] method is applicable for separation problems of countercurrent gas extraction, which can be reduced to an equivalent ternary system consisting of the supercritical solvent and the two key components. The mutual solubility of the phases may change and even the case of a limited two-phase area (type I system) can be covered. The separation factor may vary with concentration.

The equivalent ternary mixture is represented in a rectangular diagram, with the solvent ratio on the ordinate and the concentration of one of the key components, on a solvent free basis, on the abscissa (Fig. 8.8). Such a diagram is known as a "Jänecke-diagram" [22]. In this diagram the solvent plays the same role as enthalpy in an enthalpy-concentration diagram. The upper line represents the composition of the gaseous phase, the lower one that of the liquid phase. Since the solvent ratio in the liquid phase is of the order of $10^0$, the liquid phase in most cases will coincide with the abscissa. Tie lines may be included into the diagram (broken lines in Fig. 8.8). A point on the phase envelope lines represents a saturated phase, while a point ($M$) in between these lines represents a mixture of two phases, which at equilibrium splits into equilibrium phases determined by the tie line through $M$.

If two mass flows $A$ and $B$ are mixed, the mixing point $M$ lies on the straight line $AB$, the length of $AM$ representing the quantity of $B$, and the length of $MB$ representing the quantity of $A$.

The mass balance for component 1 in the enriching section is given by:

$$V_p y_{1p} - L_{p+1} x_{1p+1} = V_n y_{1n} - R_n x_{1Rn} = E y_{1E}, \qquad 8.27$$

with $E$ as the amount of extract withdrawn at the top of the column (including the solvent) and $y_{1E}$ the concentration of component 1 in the extract.

Since we assume a steady state process, $Ey_{1E}$ = constant. Since $V_p$ and $L_{p+1}$ represent gaseous and liquid flows at any cross-section of the enriching section, Eq. 8.27 represents a straight line in the Jänecke-diagram. All lines, representing mass balances for different cross-sections of a separation cascade, meet at one point, $P_E$, given by $Ey_{1E}$. The abscissa of this point is the concentration of component 1 in the top product on a solvent free basis. The ordinate is the amount of solvent removed at the top of the column and fed into the gas cycle. This amount of solvent is given by $V_n y_{Sn} - R_n x_{SRn}$, the total amount of solvent in the upward gas flow minus the amount of solvent reintroduced with the reflux. For use in the diagram, these quantities have to be adjusted to the units plotted on the coordinates.

The same reasoning holds for the stripping section. The mass balance equation for component 1 is:

$$V^*_{p*}y_{1p*} - L^*_{p*+1}x_{1p*+1} = S_0 y_{1S_0} - L^*_1 x_{11} = Bx_{1B}, \qquad 8.28$$

$B$ = bottom product, including the dissolved solvent.

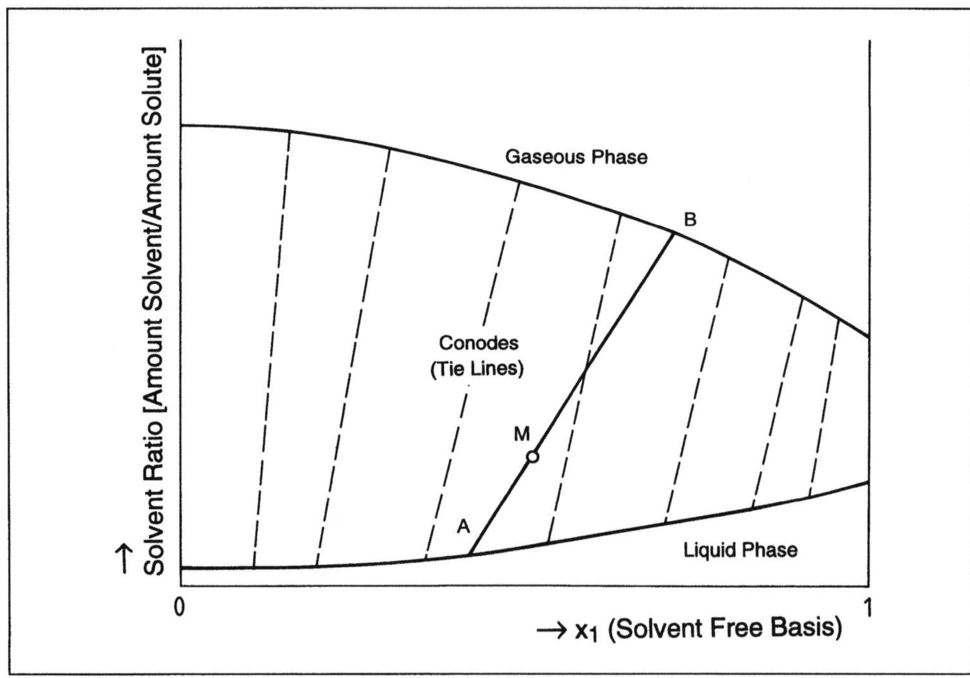


**Fig. 8.8.** Jänecke-diagram for representing an equivalent binary separation with varying solubility in both phases.

Since $V_{p*}^*$ and $L_{p*+1}^*$ represent gaseous and liquid flows at any cross-section of the stripping section, Eq. 8.28 represents a straight line in the Jänecke-diagram. All lines, representing mass balances for different cross-sections meet at one point, $P_B$, given by $Bx_{1B}$. The abscissa of this point is the concentration of component 1 in the bottom product on a solvent free basis, the ordinate is the amount of solvent added at the bottom of the column. This amount of solvent is given by $S_0 y_{SS0} - L_1^* x_{S1}$, the total amount of solvent in the solvent stream entering the column at the bottom minus the amount of solvent in the downward liquid flow. For use in the diagram, these quantities have to be adjusted to the units plotted on the coordinates.

The total mass balance is given by $F = E + B$. Therefore, in the Jänecke-diagram these points lie on a straight line.

The solution procedure for a separation problem is as follows:
– determine $x_{1B}$ and $B$ from a mass balance;
– determine $P_E$ from $y_{1E}$ and the amount of solvent removed;

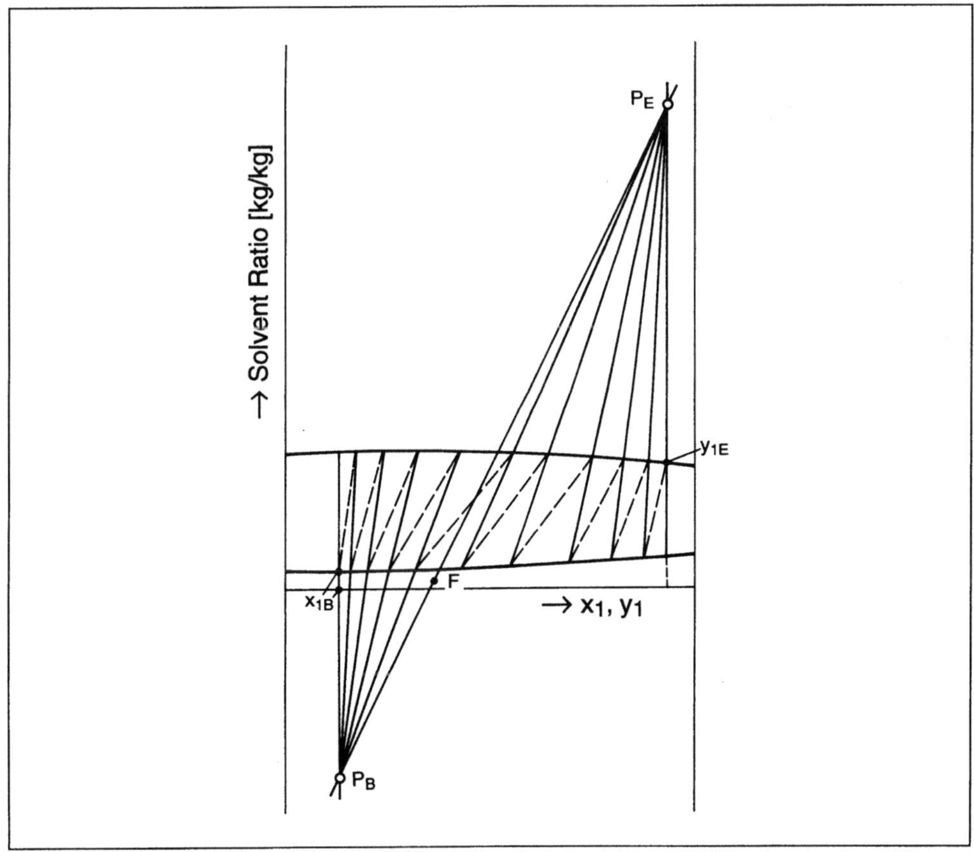



**Fig. 8.9.** Determination of number of theoretical stages in a Jänecke-diagram with the method of Ponchon-Savarit.

- $P_B$ can then be determined by connecting $P_E$ and $F$ with a straight line and continuing this line until it intersects with $x_{1B}$.
- Start at $x_{1B}$ on the line for the liquid phase. The phase in equilibrium with $x_{1B}$ is obtained via the tie line through $x_{1B}$. Connect this phase with $P_B$. From the intersection of this line with the liquid phase, the corresponding tie line leads to the next gas phase in equilibrium. Continue this procedure until $F$ is reached, then choose $P_E$ as the new point of intersection and continue the procedure until the concentration $y_{1E}$ in the top product is reached. The number of tie lines applied is the number of theoretical stages.

If no product is removed (total reflux), then $P_E$ and $P_B$ are at infinite distance, and the lines for the mass balances are vertical.

An example of a calculation of number of theoretical stages by the Ponchon-Savarit method is shown in Fig. 8.9. It may be the graphical representation of a computer calculation employing the above discussed equations.

## 8.2.9 The NTU-HTU Method

In separation columns with countercurrently flowing phases, concentrations of the phases change gradually rather than in equilibrium steps. Therefore, the NTU-HTU method, taking into account the monotonous variation along the column, is sometimes preferred to the theoretical stage model.

A mass balance for the amount of component 1 transferred from one phase to the other in an infinitesimal section of the column $dh$, related to the area $A$ of the cross-section of the column is then:

$$\frac{V}{A}\frac{dy_1}{dh} = -\frac{L}{A}\frac{dx_1}{dh}. \qquad 8.29$$

The amount of component 1 transferred from the liquid phase can be expressed by the equation for mass transport:

$$L dx_1 = k_G A_{ex}(y_1^* - y_1), \qquad 8.30$$

with
$A_{ex}$ = area for mass transfer between the phases;
$k_G$ = overall mass transfer coefficient for component 1, related to the gaseous phase;
$y_1^*$ = equilibrium concentration of component 1, related to the concentration of component 1 in the liquid phase.

With $a = A_{ex}/(A\,dh)$, the specific area for mass transfer between the phases ($m^2$ transfer area per $m^3$ volume of the column), Eq. 8.29 is transformed to

$$\frac{V}{A}\frac{dy_1}{dh} = k_G a(y_1^* - y_1). \qquad 8.31$$

From Eq. 8.31 it can be seen that the variation of $y_1$ with column height depends on the difference between the equilibrium concentration and actual concentration of component 1 in the gaseous phase. This concentration difference is the driving potential for the mass transfer.

The necessary height of a column for achieving a certain concentration change between incoming flow ($y_{1in}$) and outcoming flow ($y_{1out}$) can be obtained by integration.

$$h = \int_0^h \mathrm{d}h = \frac{V}{A} \int_{y_{1in}}^{y_{1out}} \frac{\mathrm{d}y_1}{k_G a(y_1^* - y_1)}. \qquad 8.32$$

Integration of Eq. 8.32 is simple if the product $k_G a$ is constant along the column. This is usually assumed. Then, Eq. 8.32 is transformed to

$$h = \underbrace{\frac{V}{k_G a A}}_{HTU} \underbrace{\int_{y_{1in}}^{y_{1out}} \frac{\mathrm{d}y_1}{(y_1^* - y_1)}}_{NTU}. \qquad 8.33$$

The first term on the right side of Eq. 8.33 is the **Height of a Transfer Unit**, the second term corresponds to the **Number of Transfer Units**. The height of a column can then be calculated from these quantities, which are characterized by indices $o$ for overall mass transfer including both phases and $G$ for the gaseous phase, to which the driving potential is related:

$$h = HTU_{oG}NTU_{oG}. \qquad 8.34$$

The height of a transfer unit is determined by mass transfer and therefore dependent on mass transfer equipment. The number of transfer units only depends on equilibrium properties. A separation problem can therefore conveniently be split into an equilibrium-dependent and an equipment-dependent part, as in the theoretical stage model.

## 8.2.10 Analysis of Countercurrent Gas Extraction

*Definition of the separation problem*

- Specify the separation problem: Composition of products, yield of certain components, feed quantity;
- Determine composition of the feed mixture;
- Determine phase equilibria, by experiment if necessary (feed, other compositions and products if possible, variation of gas/liquid-ratio, pressure, temperature, gas mixtures, entrainers);

– Correlate phase equilibria, if possible, to reduce necessary number of experiments;
– Determine separation factors (if necessary from experimental data only) for the compounds of the feed mixture;
– Decide on which components are to be separated (in most cases, determined by the specification of the separation problem);
– Decide on characterizing the feed mixture for analyzing the separation:
    – Multicomponent mixture with practically all components;
    – Representation of the multicomponent mixture by "continous thermodynamics";
    – Multicomponent mixture with a reduced number of pseudo-components;
    – Reduction to an equivalent binary system on the basis of "key components";
    – Reduction to an equivalent ternary system on the basis of "key components," including the supercritical solvent as one component.
– Determine equilibrium data according to the chosen method of representing the mixture.


*Simulation of the separation*

– Selection of method for determining number of theoretical stages or number of transfer units;
– Determination of limiting values (minimum reflux ratio, minimum number of theoretical stages);
– Variation of gas/liquid ratio or reflux-ratio;
– Determination of possible separation with a limited number of theoretical stages;
– Determination of number of theoretical stages necessary for a given separation;
– Determination of concentration profiles.

In the following, two examples will be presented, the separation of fatty acid methylesters according to chain length and to degree of saturation, and the separation of tocopherols from a condensate from water vapor distillation carried out to purify edible oils.


# 8.3 The Separation of Fatty Acids According to Chain Length and Saturation

## 8.3.1 Background

The separation of saturated and unsaturated fatty acids is a problem that has been treated many times and with different separation processes in order to selectively recover unsaturated fatty acids. Vacuum distillation is an established separation technique, but countercurrent gas extraction may be an alternative that avoids high temperatures and improves yields.

The interesting fatty acids have a carbon chain length of 14 to 24, with the major amounts concentrating from C16 to C22. The effects of unsaturation on physical properties are superimposed by the effects of chain length. Furthermore, free fatty acids tend to dimerize to a certain extent. Fatty acids therefore are treated mostly as methyl- or ethyl-esters.

The separation of $\omega$-3-fatty acids from fish oil is one of the incentives to study the possibility of separating fatty acids according to chain length and degree of saturation. The separation of a model mixture of methylesters of fatty acids with pure $CO_2$ and carbon dioxide – propane mixtures, was investigated by van Gaver [18]. In the following, results with pure carbon dioxide will be treated in some detail.

## 8.3.2 Feed

The model feed mixture was prepared from methyl esters of different carbon chain length. Since they were of natural origin, other components were present in minor quantities. They have been neglected for the purposes of the investigation. The composition of the feed mixture for phase equilibrium measurements and separation experiments are listed below in Table 8.2. Since two preparations of the feed mixture were necessary, the concentrations for the mixtures used for equilibrium and separation experiments differ slightly. The composition of the mixtures is listed in connection with the method of analysis employed, since in multicomponent mixtures the composition determined is somewhat dependent on the method of analysis.

**Table 8.2.** Main components of mixtures of methyl esters of fatty acids used in the investigation of separating saturated and unsaturated fatty acids (after van Gaver [18]). The figures have been rounded off compared to the original literature.

| Component | Phase equilibrium experiments | Separation experiments |
|---|---|---|
| | Composition (wt.-%) | |
| C14:0 | 1.0 | 0.7 |
| C16:0 | 18.5 | 14.4 |
| C16:1 | 1.1 | 0.4 |
| C18:0 | 10.8 | 15.5 |
| C18:1 | 34.3 | 32.7 |
| C18:2 | 32.1 | 32.3 |
| C18:3 | 2.2 | 4.0 |
| | 100.0 | 100.0 |

### 8.3.3 Analytical Conditions

The analysis of the fatty acid esters was carried out by capillary gas chromatography. During the investigation the column was changed from a DB 225 from J&W Scientific to a Omegawax TM 320 from Supelco. A comparison of the results on the different columns did not show a significant difference. The analytical conditions are summarized in Table 8.3.

**Table 8.3.** Analytical conditions for the gas chromatographic analysis of fatty acid methyl esters (after van Gaver [18]).

|  | DB 225 | Omegawax TM 320 |
| --- | --- | --- |
| Gas chromatograph | hp 5890 series II | hp 5890 series II |
| Detector | FID, 220 °C | FID, 250 °C |
| Stationary phase | DB 225, 0.25 μm | Omegawax TM 320 0.25 μm |
| Column | J&W Scientific | Supelco |
|  | 30 m, 0.32 mm | 30 m, 0.32 mm |
| Column temperature | 50 – 140 °C (40 °C/min) | 50 – 140 °C (7.5 °C/min) |
|  | 140 – 220 °C (3 °C/min) |  |
| Injector temperature | 50 – 250 °C (200 °C/min) | 50 – 250 °C (200 °C/min) |
| Mobile phase | He | He |
| Injected quantity | 1 μl | 1 μl |

### 8.3.4 Phase Equilibria

Phase equilibria have been measured by the synthetic method, using a view cell to acquire data on the extent of the two phase region, and by the static analytical method, in order to determine equilibrium concentrations and the influence of pressure and temperature on composition. Density of the phases was determined with an oscillator-type densitometer. In Fig. 8.10 the equilibrium properties are summarized, showing the limit of the two-phase area, the density of the coexisting phases at different temperatures, the equilibrium concentration of the model mixture in the coexisting phases, and the total solubility of the model mixture in supercritical $CO_2$.

The strong increase of the loading of the gaseous phase requires a very good control of the operating pressure.

In Fig. 8.11. the distribution coefficients and separation factors are shown for different components and varying conditions. The $K$-factors of individual components are a strong function of the total concentration in the gaseous phase up to a loading of about 4 wt.-%. Above this value $K$-factors vary only slightly with loading and approach equal distribution ($K = 1$).

$K$-factors reveal that a separation according to chain length and degree of saturation is possible. The shorter chain fatty acid esters and the unsaturated esters tend to enrich in the gaseous phase. It is therefore possible to define the C16- and the C18-

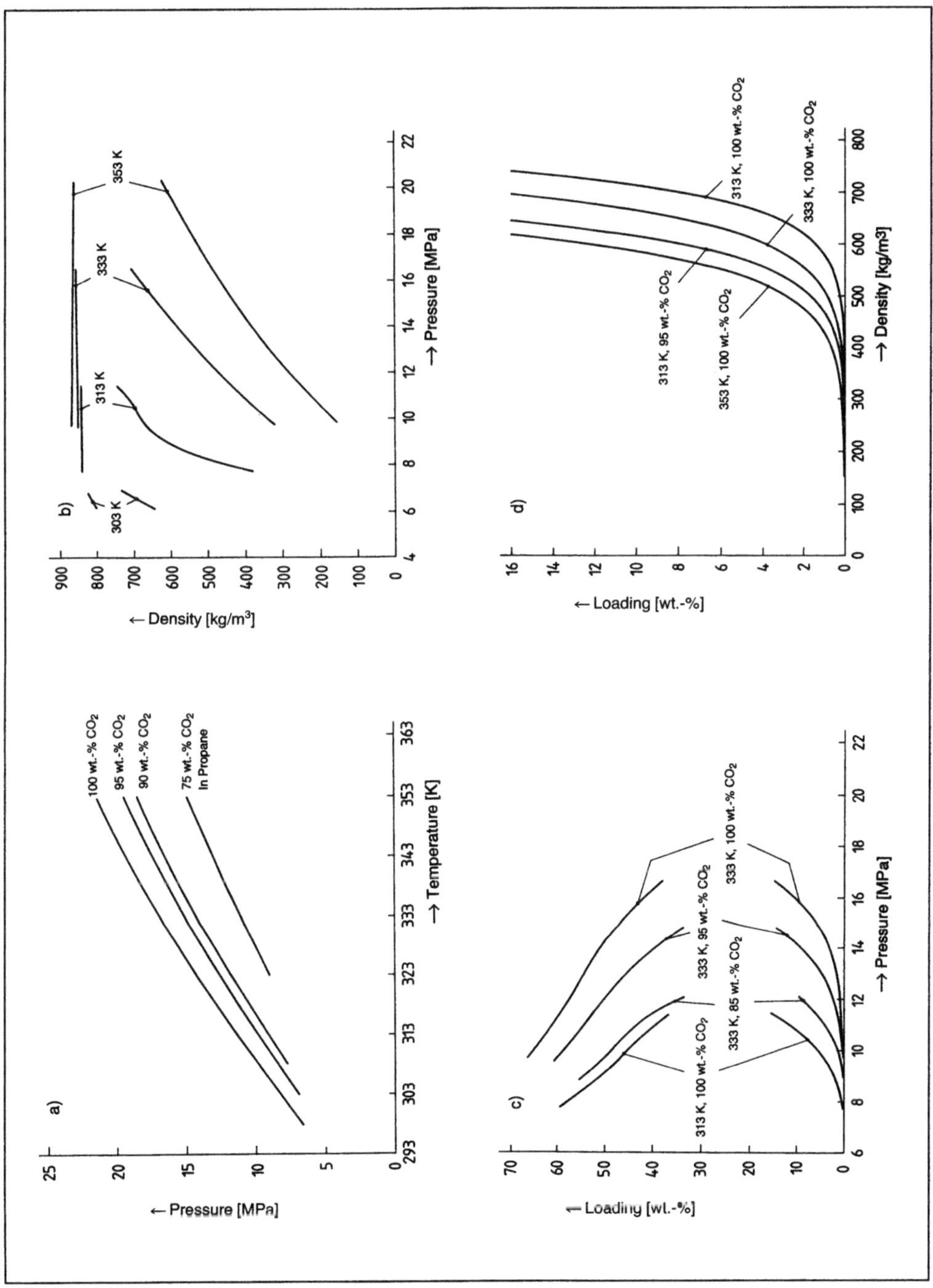

**Fig. 8.10.** Phase equilibria of a model mixture of C16/C18 fatty acid methyl esters (after van Gaver [18]).
a)  Phase boundary line for the two phase area;
b)  density of coexisting phases for different temperatures;
c)  equilibrium concentration of the model mixture in the coexisting phases on a solvent free basis;
d)  total solubility of the model mixture in supercritical $CO_2$ in dependence on density.

281

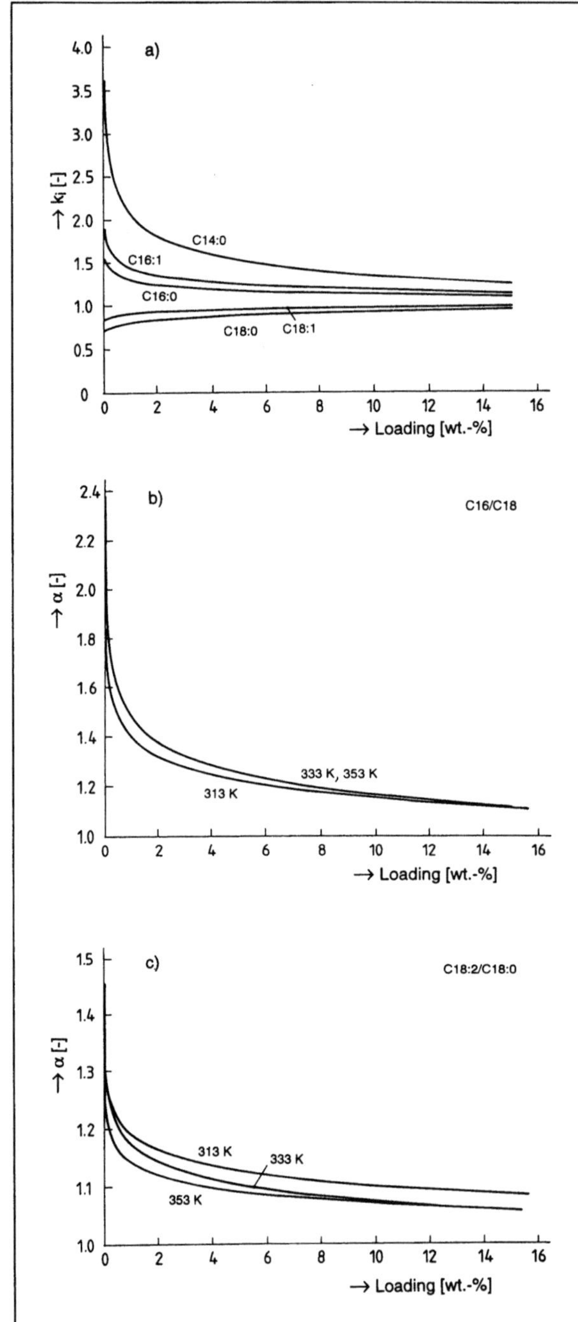



**Fig. 8.11.** Distribution coefficients and separation factors for the model mixture of C16/C18 fatty acid methyl ester in supercritical $CO_2$ (after van Gaver [18]).
a) *K*-factors of individual components as a function of the total concentration in the gaseous phase;
b) separation factor between C16 and C18 for different temperatures;
c) separation factor between C18:2 and C18:0 for different temperatures.

282

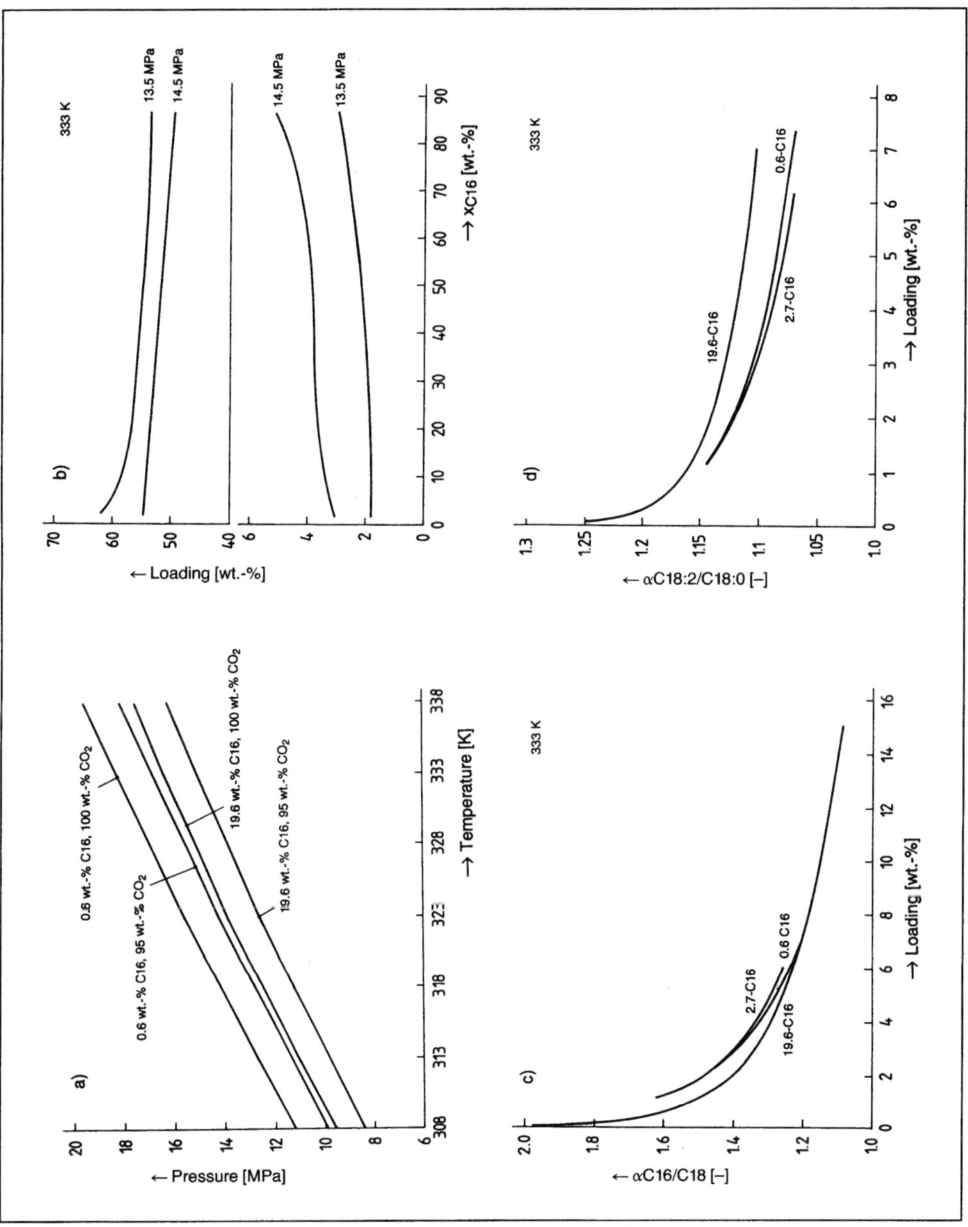



**Fig. 8.12.** Influence of concentration on equilibrium properties of a model mixture of fatty acid methyl-esters (after van Gaver [18]).

a) Phase boundary line for different concentrations of C16-ester;

b) concentration of esters in the coexisting phases (solvent free basis) as a function of C16:0-ester concentration;

c) separation factor of C16/C18 as a function of solubility in the gaseous phase for different concentrations of C16-ester;

d) separation factor of C18:2/C18:0 as a function of solubility in the gaseous phase for different concentrations of C16-ester.

esters as pseudocomponents for the separation according to chain length and to use these pseudocomponents as key components. For the separation according to degree of saturation, linoleic acid methylester (C18:2) and stearic acid methylester (C18:0) are selected as the key components.

Separation factors behave similar to $K$-factors with respect to total concentration of the fatty acid methylesters in the gaseous phase, or loading of the gaseous phase. For the C16/C18-separation, separation factors start at a value of about 2 and approach 1.2 at 4 wt.-% loading. For the separation according to saturation (C18:2/C18:0), separation factors start at about 1.3 and decrease to 1.1 at a loading of 4 wt.-%. The influence of temperature on the separation factors is small.

The feed mixture yields only one point on the equilibrium curve with respect to concentration. In general, separation factors are not constant with composition. It is therefore necessary to also determine phase equilibria for other compositions. For a model mixture this may be achieved by different preparations. For real mixtures, appropriate material has to be prepared during the first separation experiments.

For the model mixture of fatty acid methyl esters the influence of varying composition is shown in Fig. 8.12. With higher content of the shorter chain fatty acid esters, the two-phase boundary line shifts to lower pressures due to the increased mutual solubility. Separation factors were found to be independent on concentration in the range covered by the investigation (see Table 8.4). The scattering of the data may be attributed to the limited accuracy of the analytical method.

**Table 8.4.** Separation factors for the model mixture of C16/C18-esters in dependence on the concentration of C16 (after van Gaver [18]).

| C16-ester | $\alpha_{C16/C18}$ | $\alpha_{C18:2/C18:0}$ |
|---|---|---|
| wt.-% | 60 °C, 13.5 MPa | |
| 1.8 | 1.39 | 1.35 |
| 20.4 | 1.43 | 1.34 |
| 55.0 | 1.23 | 1.19 |
| 74.5 | 1.24 | 1.21 |
| 86.5 | 1.39 | 1.34 |

## 8.3.5 Selection of Process Conditions

Selection of process conditions is a process of optimization. Investment and operating costs have to be considered. Therefore, the optimized set of parameters can only be achieved by iteration and interactively with experts, since, so far, sufficient and detailed information on countercurrent gas extraction processes for a pure computational solution is not available. For the separation experiments in a laboratory or pilot plant, process parameters can be chosen without optimization with respect to economy.

Enriching of unsaturated free fatty acids comprises two separation steps: the separation according to carbon chain length and the separation according to degree of saturation. For the first separation step a loading of 3 to 5 wt.-% (pressure about 14 MPa) at a temperature of 60 °C was selected. The main aspect for preferring a process temperature of 60° to 40 °C is the steep increase of loading with pressure at the lower temperature, requiring a very precise control of pressure in order to achieve a constant separation factor. The separation factor for the separation of C16/C18-fatty acid methylesters at these conditions is 1.30 to 1.40 (see Fig. 8.11).

For the separation of saturated and unsaturated fatty acid methyl esters the separation factor is lower than for the separation according to chain length. Therefore, an operating point with a lower loading of 2 wt.-% at 60 °C was chosen. The separation factor at these conditions is $\alpha = 1.15$.

## 8.3.6  Results of Separation Experiments

Separation experiments were carried out in a pilot plant with a 13-m high column of 70-mm inner diameter, equipped with a regular packing of the type Sulzer CY. The

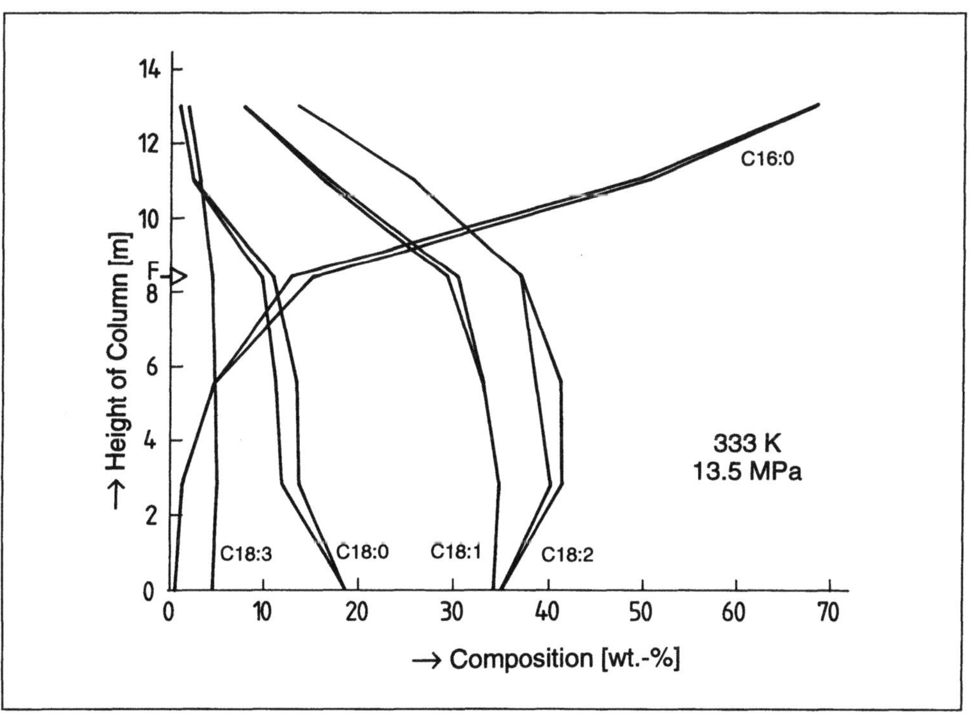


**Fig. 8.13.** Separation of fatty acid methylesters according to chain length. Concentration profiles (gas and liquid) of a separation experiment in a pilot plant column (after van Gaver [18]).

experiments were carried out at constant temperature and variations of reflux and point of feed introduction were investigated.

With the initial mixture (see Table 8.2) first experiments for a separation according to chain length were carried out. An example of the result of such an experiment is shown in Fig. 8.13. The C16-esters can be enriched at the top of the column and removed from the bottom product. The maximum obtainable gas flow was 62000 kg/(m²h), due to the limiting capacity of the gas pump. Flooding was not observed.

For the separation experiments according to degree of saturation, the bottom product of the first separation step was used, containing about 1.1 wt.-% C16-methylester and 77.5 wt.-% unsaturated C18-methylesters. The rest of the mixture is C18:0-methylester and some longer chain methylesters. The maximum obtainable gas flow was 64000 kg/(m²h). Flooding was not observed. An example of the concentration profiles for this separation step is shown in Fig. 8.14.

In the experiments, ω-3-fatty acid methylesters were enriched to about 70 wt.-%, unsaturated fatty acid methylesters were concentrated to > 95 wt.-%. The conditions for a further increase in concentration of ω-3-fatty acids will be discussed in the analysis of the separation steps in Section 8.3.7.

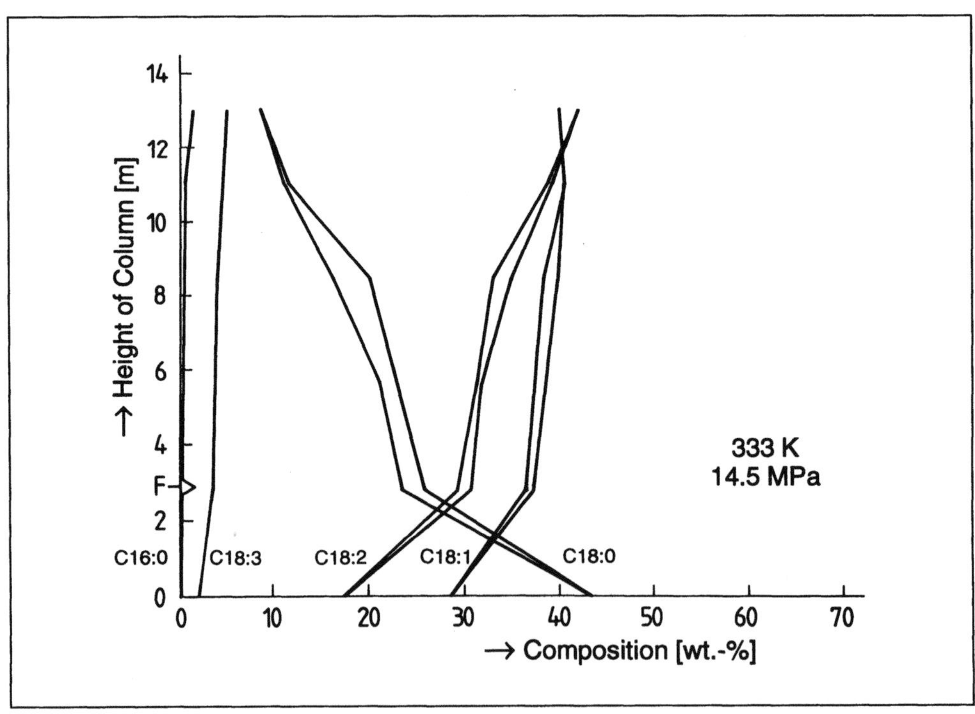


**Fig. 8.14.** Separation of fatty acid methylesters according to degree of saturation. Concentration profiles (gas and liquid) of a separation experiment in a pilot plant column (after van Gaver [18]).

The possibility of the separation of saturated and unsaturated fatty acid methylesters by countercurrent gas extraction with supercritical carbon dioxide was clearly demonstrated.

## 8.3.7 Analysis of the Separation

The analysis of the separation comprises the interpretation of the experimental results, i.e. the determination of theoretical stages or number of transfer units for the separation achieved. From that value, the height of a theoretical stage or transfer unit can be obtained from the theoretical stages achieved in the experiment and the length of the column. With these data, the separation can be further analyzed for the necessary theoretical stages, and for the resulting height of a column in dependence of reflux ratio.

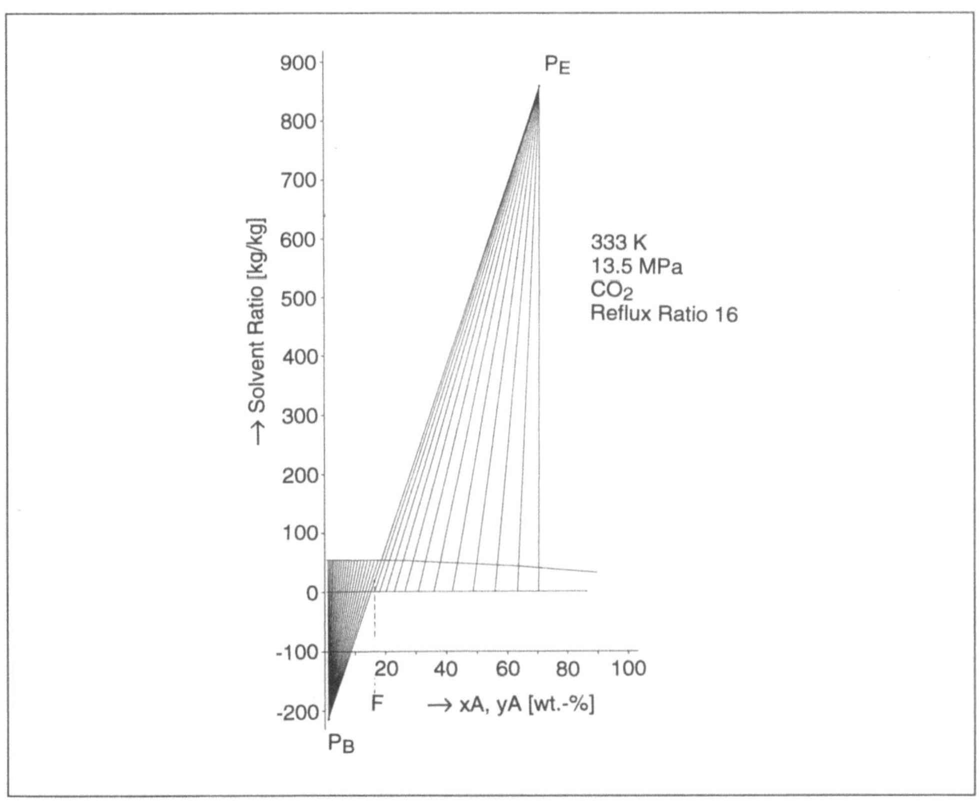



**Fig. 8.15.** Determination of theoretical stages for the separation of fatty acid methylesters (separation according to chain length). (After van Gaver [18]).

Before any separation experiment is carried out, from an analysis of the separation based on phase equilibrium, the limiting values for theoretical stages and reflux ratio can be derived in dependence of an anticipated separation. If the number of theoretical stages of a column, operated with the feed mixture, are known, e.g., from previous experiments or results on similar systems, the separation experiment can be simulated beforehand. A reasonable reflux ratio can be selected, and experiments can be planned to verify the simulation results. Experimental results may then be used to modify the data for the simulation. Such an analysis enables to carry out experiments near optimum operating points and will lead to a far lower number of experiments than have to be carried out otherwise.

Numbers of theoretical stages were determined by reducing the multicomponent model mixture to an equivalent binary system. For the separation according to carbon chain length the C16-components and the C18-fraction were defined as key pseudo-components. For separations achieved with the initial feed mixture, about 30 theoretical stages were obtained by three different methods: McCabe-Thiele, Ponchon-Savarit and NTU. The height of a theoretical stage was determined for this separation to be about 0.43 m. In Fig. 8.15 an example with the Jänecke method is presented. For the separation according to degree of saturation, phase equilibrium data were not available so that the separational analysis could not be carried out.

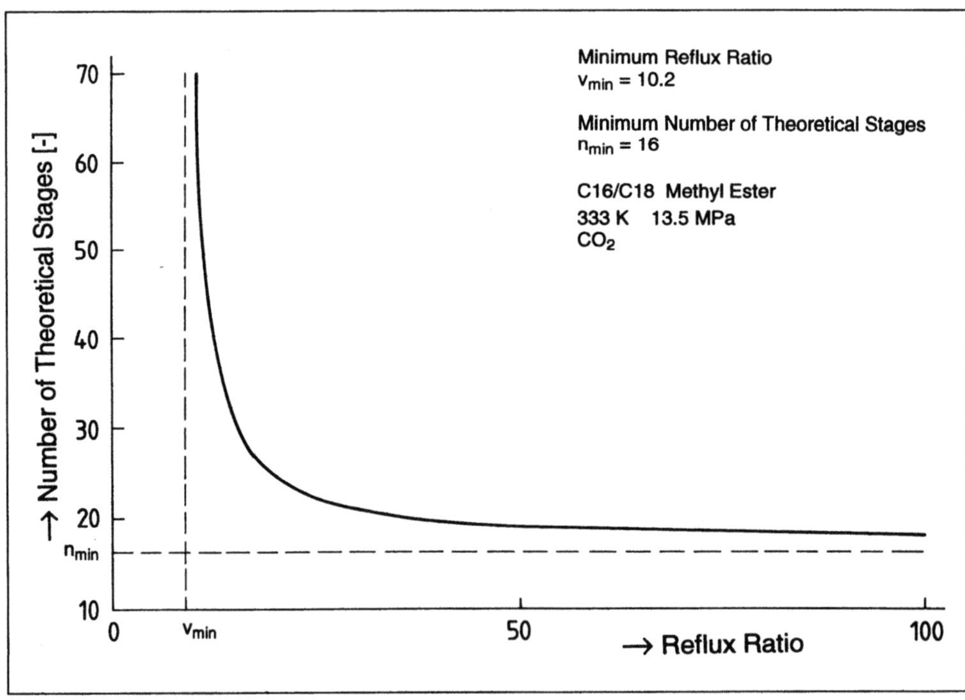



**Fig. 8.16.** Number of theoretical stages as a function of the reflux ratio for the separation of fatty acid methylesters according to chain length. (after Riha [38]). Data from van Gaver [18].

The limiting values have been determined and the influence on the number of theoretical stages evaluated for a certain experimental run. The result is presented in Fig. 8.16. For a separation with 16.3 wt.-% of C16-fatty acid methylester in the feed,

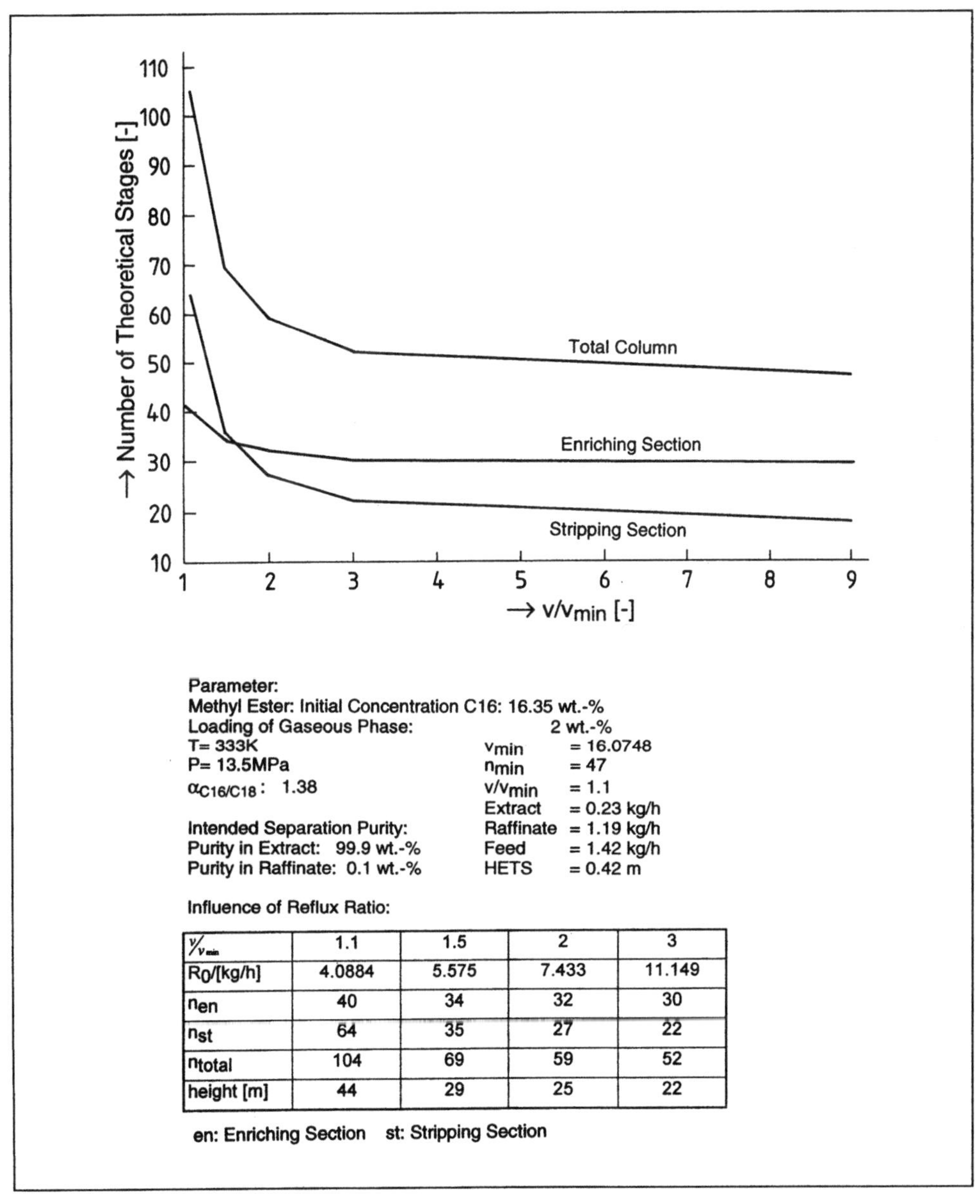


| $v/v_{min}$ | 1.1 | 1.5 | 2 | 3 |
|---|---|---|---|---|
| R$_0$/[kg/h] | 4.0884 | 5.575 | 7.433 | 11.149 |
| n$_{en}$ | 40 | 34 | 32 | 30 |
| n$_{st}$ | 64 | 35 | 27 | 22 |
| n$_{total}$ | 104 | 69 | 59 | 52 |
| height [m] | 44 | 29 | 25 | 22 |

**Fig. 8.17.** Determination of the necessary number of theoretical stages for a certain separation. Separation of fatty acid methylesters according to chain length (after Riha [39]).

77.2 wt.-% in the top product, and 0.6 wt.-% in the bottom product, the minimum reflux ratio is 10.2 and the minimum number of theoretical stages is 16.

Since the number of theoretical stages (determining the height of a separation column) is of less influence on investment costs than reflux ratio (determining the diameter of a column), the reflux ratio for a commercial separation will be chosen at a relatively low value above the limiting minimum reflux. For an experimental column, where demonstration of the possibility of a separation and providing of samples is the main incentive, reflux ratio may be chosen at a relatively high value.

The necessary number of theoretical stages for separations up to high concentrations can be determined, assuming that the values, obtained from the experimental separation, are valid also for higher purities of the top and bottom product. For the C16/C18-separation a result obtained with these assumptions is shown in Fig. 8.17. A separation was assumed, in which 99.9 wt.-% of C16 was the upper limit in the bottom product. The number of theoretical stages determined for this separation amounts to 69, for a reflux ratio of 24, which is 1.5 of the minimum reflux ratio. The separation could be realized in a column with $\approx$ 30 m effective packing-height, if effectivity of the mass transfer equipment also can be assumed to remain constant at high concentrations of one pseudo-component.

# 8.4 Tocopherols

## 8.4.1 Background

Tocopherols are organic substances of antioxidative activity. They occur in many plants. Tocopherols are easily oxidized and thereby prevent the formation of toxic peroxides in biological systems. The main natural source for tocopherols is edible oils, since they are produced in huge quantities. Raw edible oils contain about 1000 ppm (0.1 wt.-%) tocopherols. During the refining process the tocopherol content of the oil is reduced to about 700 ppm. A good part of the tocopherols is removed in the deodorization process, a water vapor distillation in which free fatty acids and other unwanted substances are removed, but also part of the valuable compounds, like the tocopherols. The amount depends on processing conditions. From the water vapor distillation a condensate is formed (deodorizer condensate) which consists of an aqueous and an organic phase. The organic phase contains most of the substances removed from the oil. Tocopherols and other substances are enriched in this condensate, which can be further processed to obtain fractions rich in these substances. The usual processing is carried out by vacuum distillation. Thereby a fraction containing about 50 wt.-% tocopherols is obtained. For further enrichment, countercurrent gas extraction could be the right process, since operating temperature is low and degradation by temperature zero. Therefore, the separation of tocopherols from enriched deodorizer condensate has been investigated with supercritical carbon dioxide as solvent in a countercurrent multistage separation [14, 19].

## 8.4.2 Feed Mixture (Initial System of Components)

The initial feed mixture used in this investigation is a commercially available product, which contains about 55 wt.-% tocopherols. The main components are squalene,

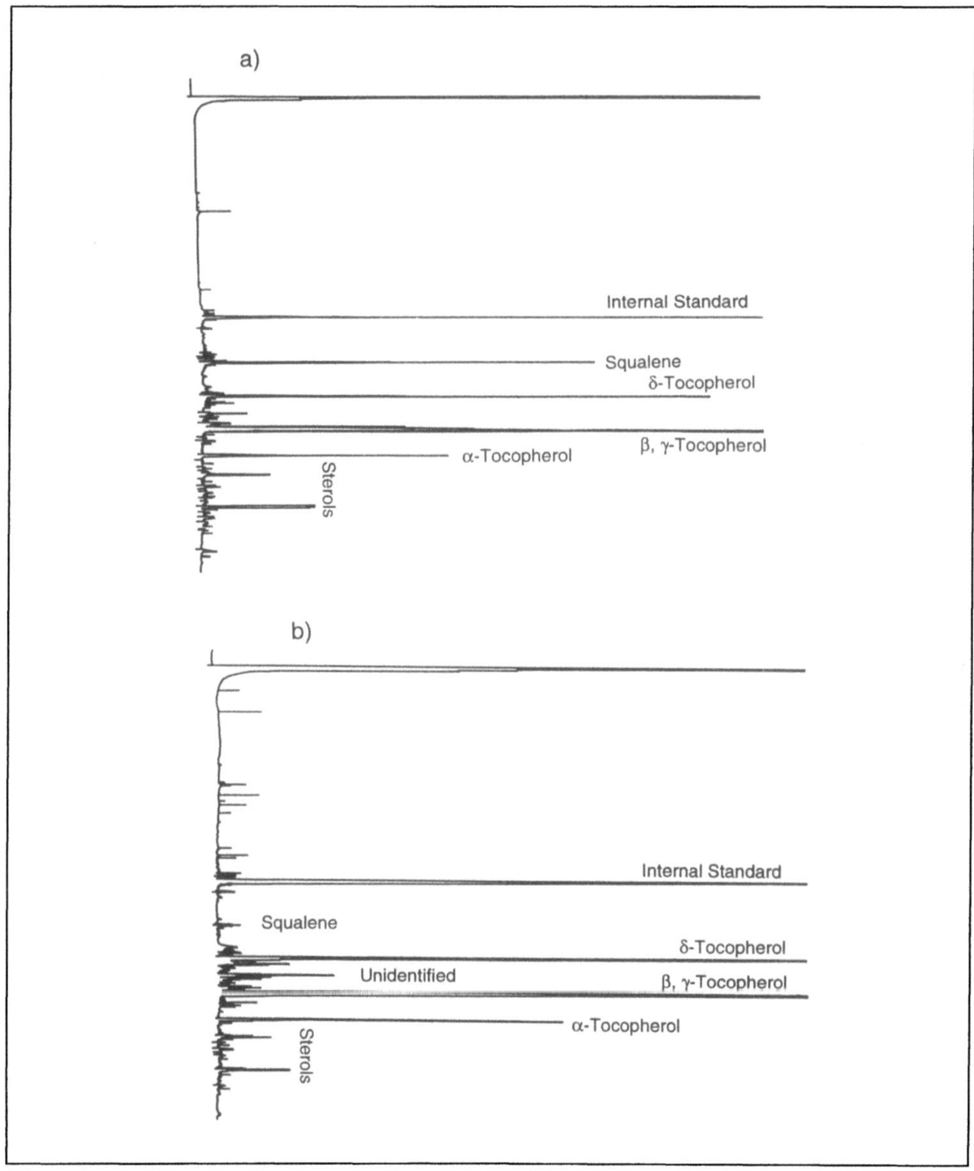



**Fig. 8.18.** Gas chromatogram of tocopherol feed mixture. a) Initial feed mixture; b) Tocopherol-enriched feed mixture [19].

the tocopherol isomers, sterols, sterolesters, triglycerides and about 10 to 15 wt.-% unknown substances; in total there may be more than 100 individual components. For the separation experiments, two additional mixtures have been employed to cover a wider concentration range. The mixtures were prepared by countercurrent separation (tocopherol enriched mixture) and by mixing the initial feed mixture with squalene (squalene enriched mixture). The approximate feed concentrations of the different mixtures used in the separation experiments are listed in Table 8.5. Gas chromatograms of two of the feed mixtures are presented in Fig. 8.18. Only the major components have been identified.

**Table 8.5.** Main components of tocopherol feed mixture.

| | Squalene wt.-% | Tocopherol isomers, wt.-% | Sterols wt.-% |
|---|---|---|---|
| Initial feed mixture | 9 | 55 | 5 |
| Tocopherol-enriched feed mixture | 0.5 | 65–75 | 6–10 |
| Squalene-enriched feed mixture | 86.5 | 8 | 1 |

Concentrations do not add up to 100 %, due to unidentified components.

Chemically, the tocopherols are isoprenoidic compounds and belong to the lipids. Tocopherols are viscous, oily liquids of orange color. The chemical structure is shown in Fig. 8.19. The tocopherols have asymmetric centers in the 2, 4', and 8' position.

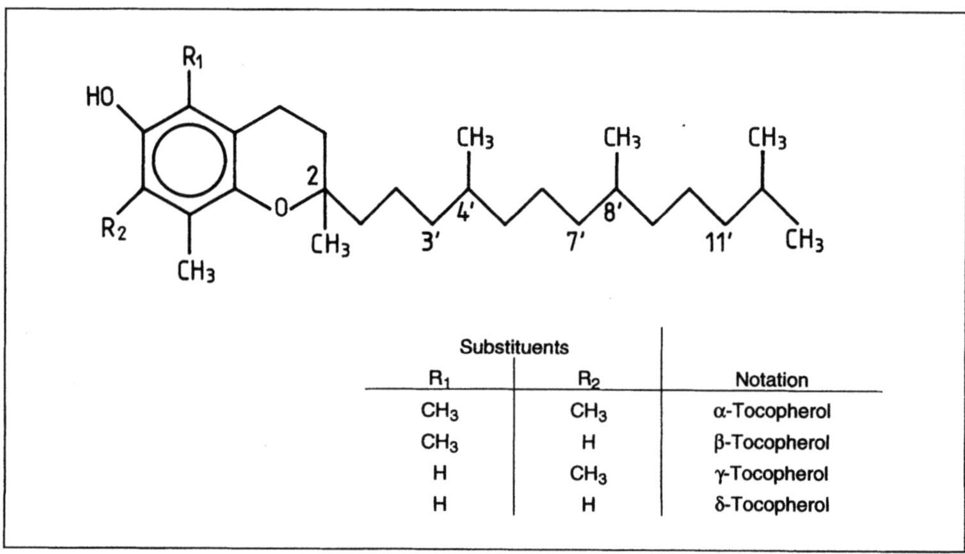


| Substituents | | |
|---|---|---|
| R1 | R2 | Notation |
| CH3 | CH3 | α-Tocopherol |
| CH3 | H | β-Tocopherol |
| H | CH3 | γ-Tocopherol |
| H | H | δ-Tocopherol |

**Fig. 8.19.** Chemical structure of tocopherols.

Therefore stereoisomers are possible. For more details on tocopherols and related substances in edible oils the reader is referred to the literature.

## 8.4.3 Analytical Conditions

Conditions for the chromatographic analysis were chosen to quantify the main components, which are important for the separation. A total separation of all the minor components was not intended. In separate HPLC experiments it was ensured that the GC-peaks of the major components did not contain other coumpounds [19]. Analytical conditions are listed in Table 8.6.

**Table 8.6.** Analytical conditions for the gas chromatographic analysis of the feed mixture for the tocopherol separation (after Gottschau [19]).

|  | DB 225 |
| --- | --- |
| Gas chromatograph | hp 5840 A |
| Detector | FID, 603 K |
| Stationary phase | DB 5, 0.1 µm |
| Column | J&W Scientific |
|  | 30 m, 0.25 mm |
| Column temperature | 340 – 480 K (20 K/min) |
|  | 480 – 530 K (2 K/min) |
| Injector temperature | 593 K |
| Mobile phase | He |
| Injected quantity | 2 µl |

## 8.4.4 Phase Equilibria

Phase equilibria have been determined with supercritical carbon dioxide and the initial feed mixture, containing 50 wt.-% tocopherols, and with two other mixtures, one containing about 70 wt.-% tocopherols and only 0.5 wt.-% squalene, the other containing about 85 wt.-% squalene [14, 19]. In Fig. 8.20, the loading of the gaseous phase with low volatile components is shown for a constant density of $CO_2$ of 655 kg/m$^3$ and for different concentrations of squalene. At a low squalene content the amount dissolved in the gas is up to 10 g/kg (1 wt.-%), at a high squalene content, the solubility increases up to 40 g/kg (4 wt.-%).

Separation factors are shown in Fig. 8.21 and Fig. 8.22. They can be calculated from the measured concentrations of equilibrium phases. For the figures, separation factors at a constant density of supercritical carbon dioxide of 655 kg/m$^3$ have been selected.

Separation factor $\alpha$ for the components $\alpha$-tocopherol – campesterol is in the range of 2.6 to 1.5 in the concentration range of 60 – 80 wt.-% $\alpha$-tocopherol. $\alpha$ declines rapidly with increasing $\alpha$-tocopherol content. In general, the higher values for $\alpha$ are obtained at lower pressure (20 MPa), the lower values for $\alpha$ at higher pressures. This corresponds with the effect that selectivity declines with increasing capacity of a solvent. Experimental uncertainties are the cause for a substantial scatter of the data.

The separation factor for squalene – $\alpha$-tocopherol runs from $\alpha \approx 4$ at low squalene concentrations, to values of $\alpha \approx 1$ at high squalene concentrations.

The result is that a separation of the tocopherol fraction from the other components is possible by extraction with carbon dioxide. The results of the phase equilibrium measurements can now be used to analyze the two separation steps necessary and to plan the separation experiments.

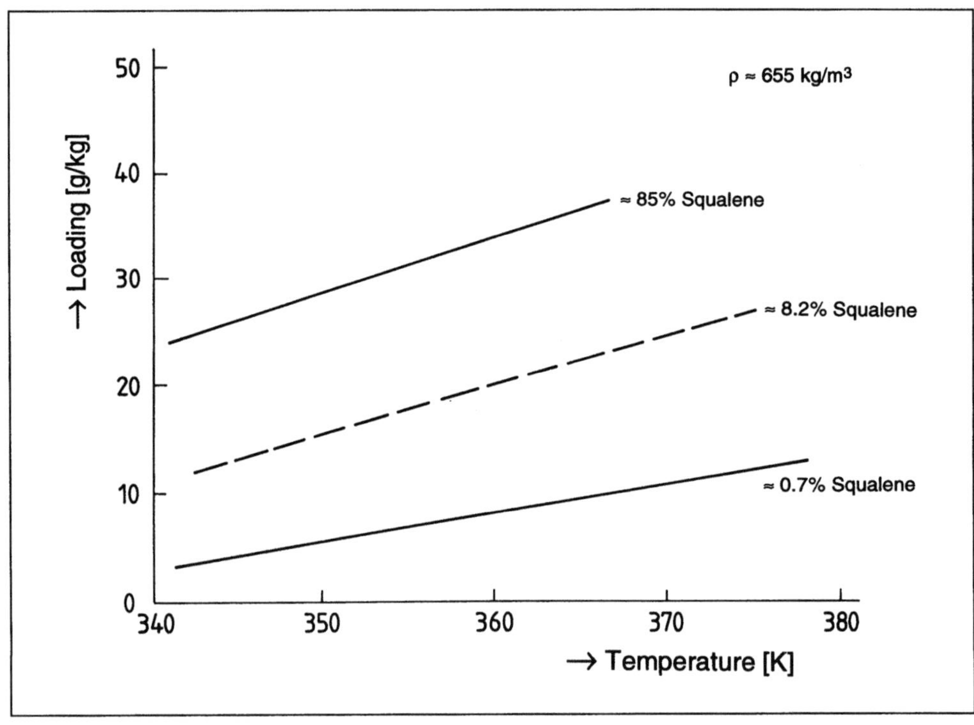


**Fig. 8.20.** Loading of gaseous phase with tocopherol fraction [19].

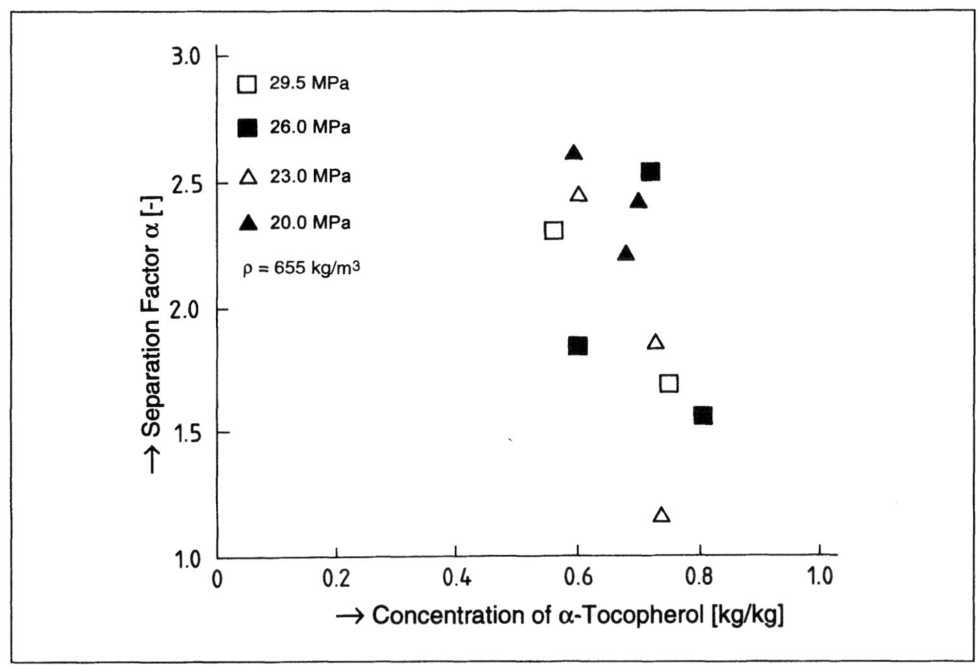

Fig. 8.21. Separation factor for the components $\alpha$-tocopherol – campesterol [19].

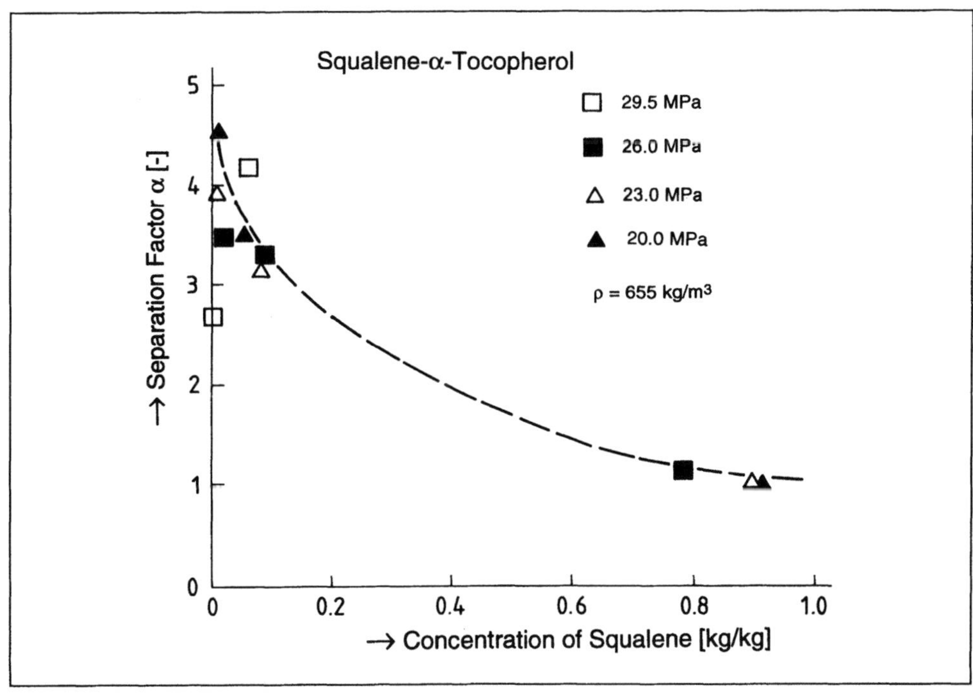


Fig. 8.22. Separation factor for the components squalene – $\alpha$-tocopherol [19].

295

## 8.4.5 Reduction of Number of Components, Selection of Key Components, Separation Sequence and Process Conditions

The chromatographic analysis of the feed mixture with an unpolar column separates the components analogue to the unpolar solvent $CO_2$. The components are eluted in the chromatographic separation in the same sequence as is obtained from the separation factor, due to equilibrium distribution in carbon dioxide. Therefore, the chromatograms of Fig. 8.18 can be taken to select the separation cuts and the key components. In the case of the tocopherols the cuts can be laid between the components squalene and δ-tocopherol for the separation of the more volatile components and between α-tocopherol and campesterol for the separation of the tocopherols from the heavier ends. For the calculation of the separation factors between tocopherols and squalene, as shown above, α-tocopherol was used as a reference substance, because of scattering analytical data for δ-tocopherol. These components can be taken as key components and the separation analysis may be carried out using these keys as equivalent binary system.

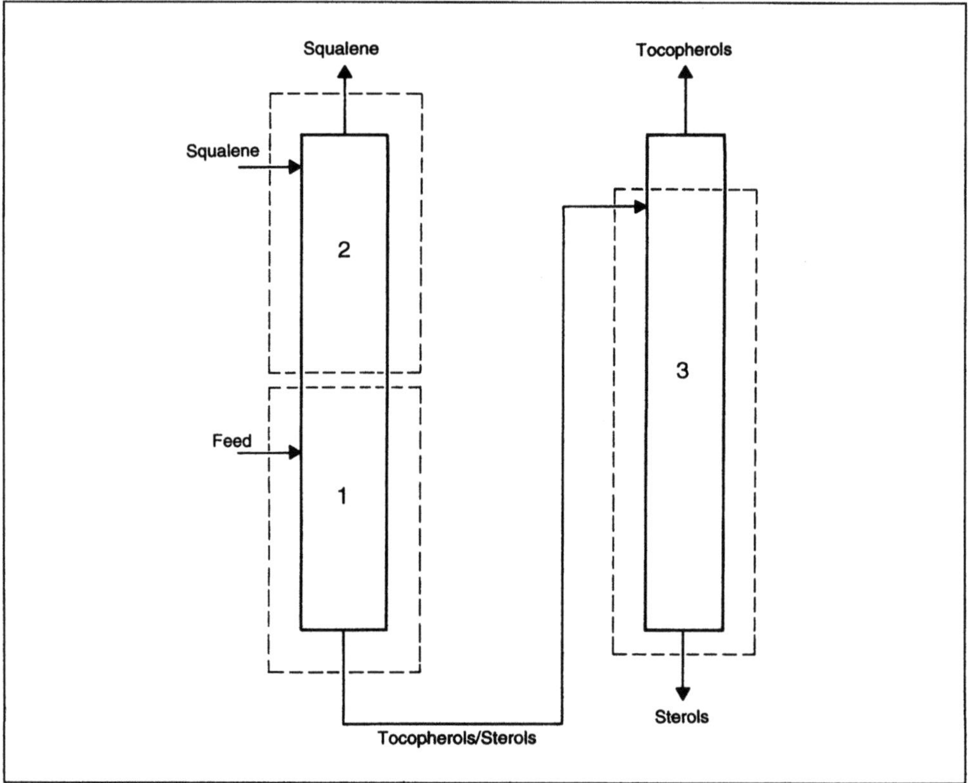



**Fig. 8.23.** Separation sequence for carrying out the separation experiments in a laboratory column.

296

Process conditions for carrying out the separation experiments are chosen as a compromise between high separation factor and high loading. Since experience with separations of this type is not abundant, a wider range of operating conditions will actually be necessary in order to learn about the behavior of the system in a column, than would be necessary if enough information would be available for countercurrent gas extraction. After a separation analysis carried out with equilibrium data beforehand, then only a few separation experiments should be required.

For the experimental procedure of enriching the tocopherol fraction, the separation sequence schematically shown in Fig. 8.23 was chosen. Squalene was separated in a first separation as top product from the tocopherol/sterol mixture. In a second separation, the tocopherol fraction was separated as top product from the sterol fraction. Since the length of the column was limited to 7 m, the separation steps were further divided into an enriching and a stripping step. For the squalene/tocopherol separation in a first step the squalene was removed, using the column as a stripping section to remove the squalene. In this section, an appreciable amount of tocopherols leaves the top of the stripping section together with squalene. In order to bring down the tocopherols to the bottom, in a second sequence of runs, the column was operated as an enriching section, with pure squalene as reflux. In such a way, the two conditions of a separation, purity of the products and yield, could be demonstrated in a limited column. In a third sequence of runs, the removal of the sterols from the tocopherols was demonstrated in an enriching section, with the tocopherol fraction as top product. The total removal of the tocopherols from the sterol fraction was not investigated, since the main goal was to achieve a pure tocopherol fraction. Examples of the results are presented below.

## 8.4.6 Results of Separation Experiments

In Figs. 8.24 and 8.25 the achieved separation is plotted against the ratio of liquid to gas flowing countercurrently through the column. About three theoretical stages have been verified in the column of 17 mm inner diameter with 5 mm spirals as packing at a maximum throughput of gas of about $20\,000\ kg/(m^2h)$. Above this value flooding was observed.

A tocopherol fraction of at least 85 wt.-% tocopherols could be produced. A chromatogram of this product is shown in Fig. 8.26. The components on both sides of the tocopherols have been removed. But between the tocopherol peaks, there are several components which cannot be removed by the same type of separation. A further purification of the total tocopherol mixture with countercurrent gas extraction would be possible by separating the tocopherol isomers, but this separation is not reasonable with gas extraction due to the low separation factors.

The separations have been analyzed with equivalent binary systems acccording to the method described above. An example is given in Fig. 8.27, where graphical solutions according to the Ponchon-Savarit method in a Jänecke-diagram and according to McCabe-Thiele in an $y,x$-diagram are presented.

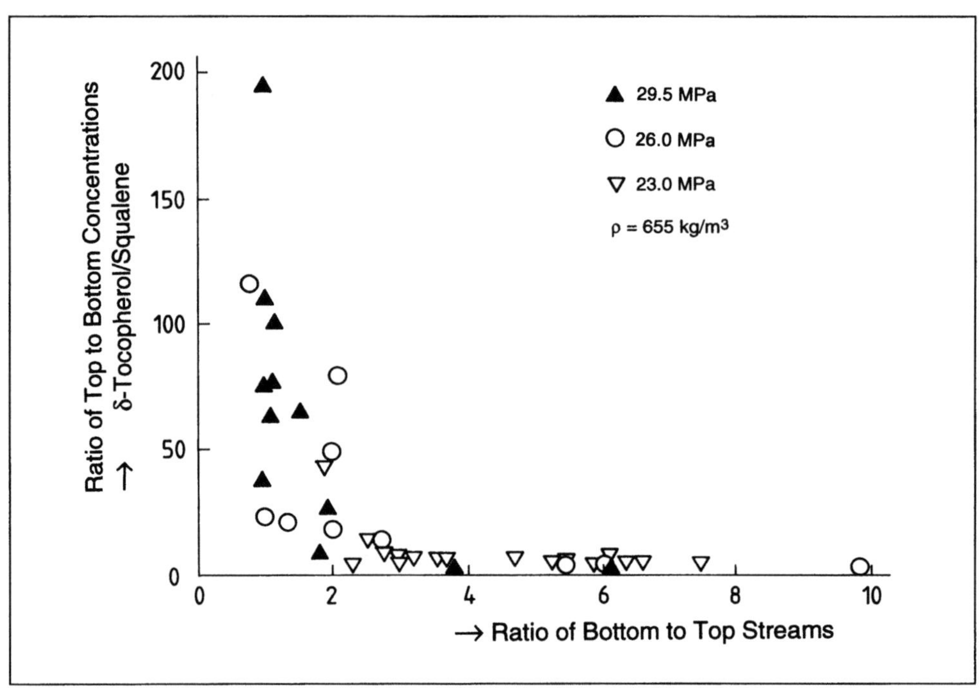

**Fig. 8.24.** Separation of squalene and δ-tocopherol [19].

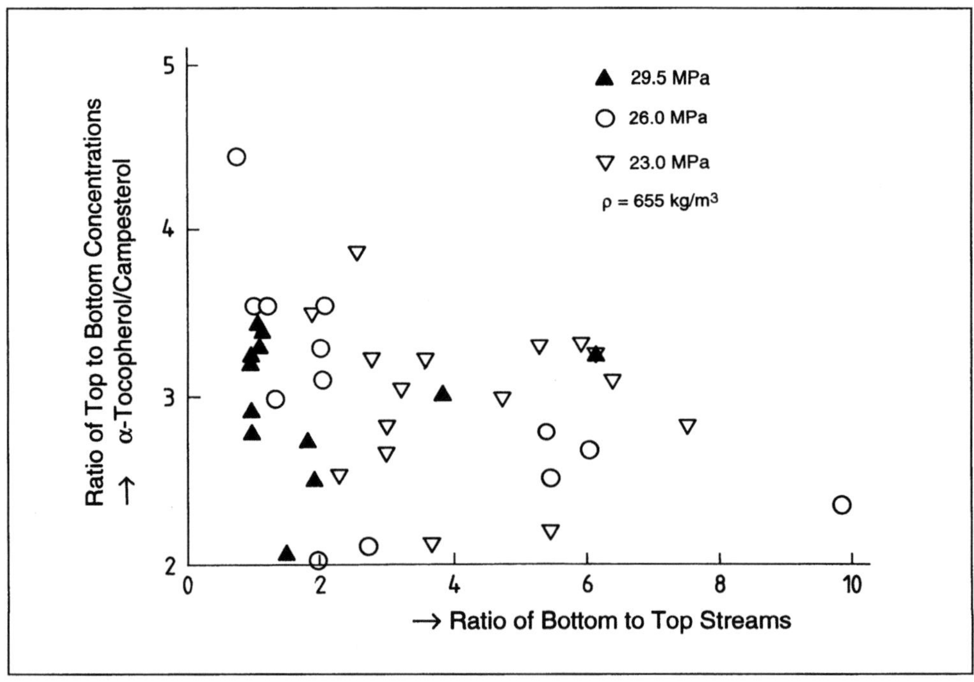

**Fig. 8.25.** Separation of α-tocopherol and campesterol [19].

298

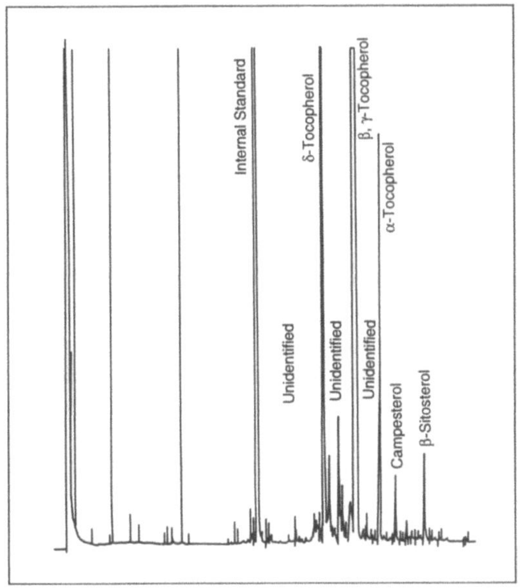



**Fig. 8.26.** Chromatogram of top product of tocopherols.

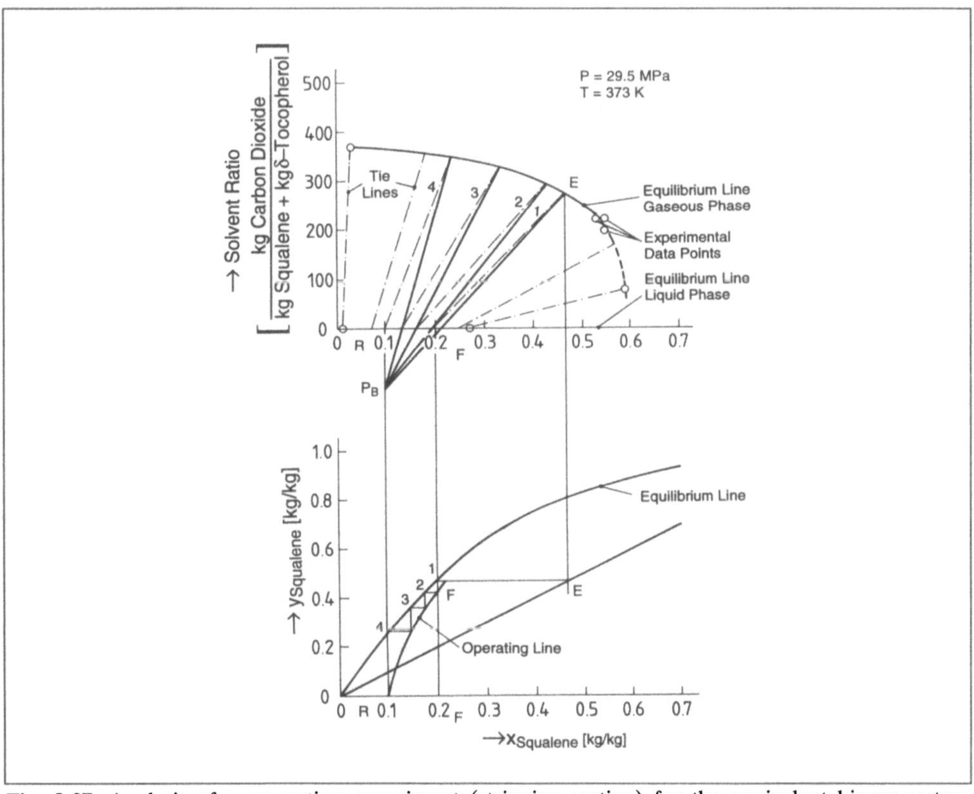



**Fig. 8.27.** Analysis of a separation experiment (stripping section) for the equivalent binary system squalene/δ-tocopherol with the Ponchon-Savarit method (upper diagram) and the McCabe-Thiele method (lower diagram).

# 8.5 Countercurrent Gas Extraction with a Co-Solvent

## 8.5.1 Basic Considerations

The solubility of low volatile substances in supercritical solvents may be too low for carrying out a countercurrent gas extraction. In a countercurrent separation the extract must be transported with the supercritical solvent through the column. The diameter of the column is determined by the necessary gas flow. The amount of low volatile material transported is proportional to the gas flow and to the equilibrium concentration of the low volatile material in the gaseous phase. The downward flow in the enriching section of a column is split from the extract and carried by the gaseous phase to the top of the column. For the operating of the mass transfer equipment, a minimum quantity of liquid flow is necessary. Therefore, beside operating as near as possible at the flooding point, it is useful to increase the solubility of the low volatile components in the supercritical solvent. As a rule of thumb, the minimum solubility in the gaseous phase is about 1 wt.-%, but a reasonable concentration is 2 to 5 wt.-%.

Concentration of the low volatile components in the gaseous phase can be enhanced by several means:

I) Density of the gas can be enhanced by either increasing pressure, lowering the temperature, or adding a co-solvent. With increasing density of the gaseous phase, densities of both phases approach one another. Since a certain density difference is necessary to enable two-phase flow in the field of gravity, increasing density of the gas is limited. As a rule of thumb, the minimum density difference can be considered to be $100 \ kg/m^3$.

II) Temperature can be enhanced, increasing vapour pressure of the low volatile substances. This is also advantageous for mass transfer rates. Solubility of low volatile substances in the gaseous phase increases with temperature only at relatively high pressures. Phase equilibria must be considered.

III) A co-solvent can be added, which increases density and can further enhance concentration in the supercritical solvent by molecular interactions.

To achieve an increase of concentration of low volatile substances in the gaseous phase, the binary system supercritical solvent – co-solvent must be supercritical at operating conditions. Then, the ternary system of supercritical solvent – co-solvent – low volatile components is of ternary type I. In Figs. 8.28 and 8.29 two examples are shown: a liquid (ethanol) and a gas (propane) as co-solvents. Solubility in the gaseous phase in these cases is significantly enhanced by the co-solvent at higher concentrations of co-solvent.

The separation factor in general decreases with increasing solubility of the low volatile components in the gaseous phase. With increasing concentration of co-solvent an increasing separation factor is possible, especially with components with specific interactions. For a detailed discussion of the effect of co-solvents on phase equilibrium, see also Section 3.5.

A separation process becomes more complicated by using a solvent mixture of supercritical component and co-solvent. The co-solvent has to be removed from raf-

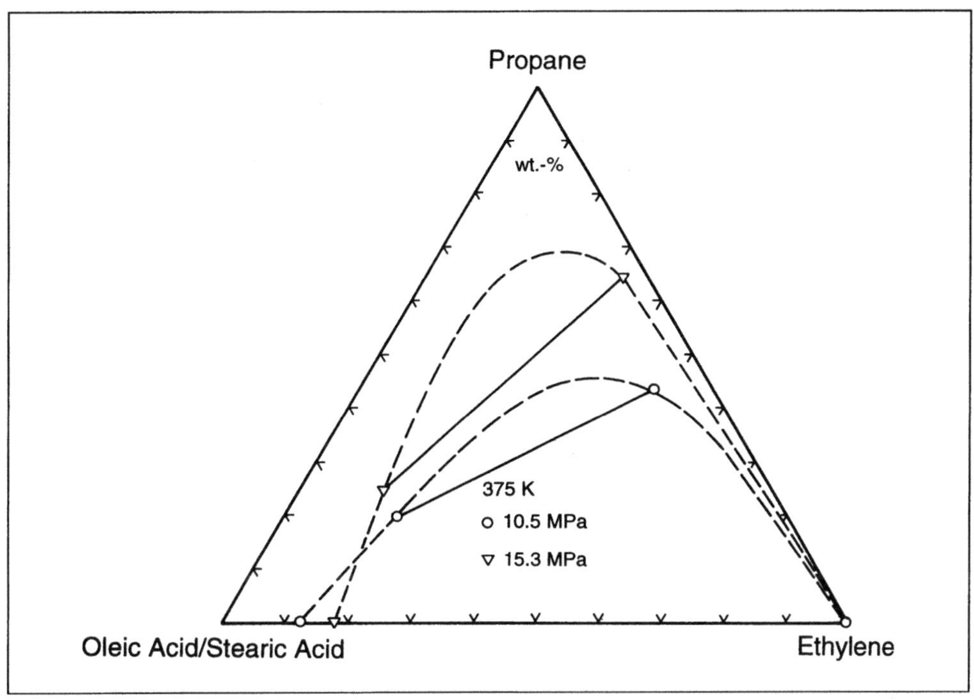



**Fig. 8.28.** The role of a co-solvent: A gas (at STP) as co-solvent. Oleic acid/stearic acid (low volatile pseudo-component) – propane (co-solvent) – ethylene (supercritical solvent).

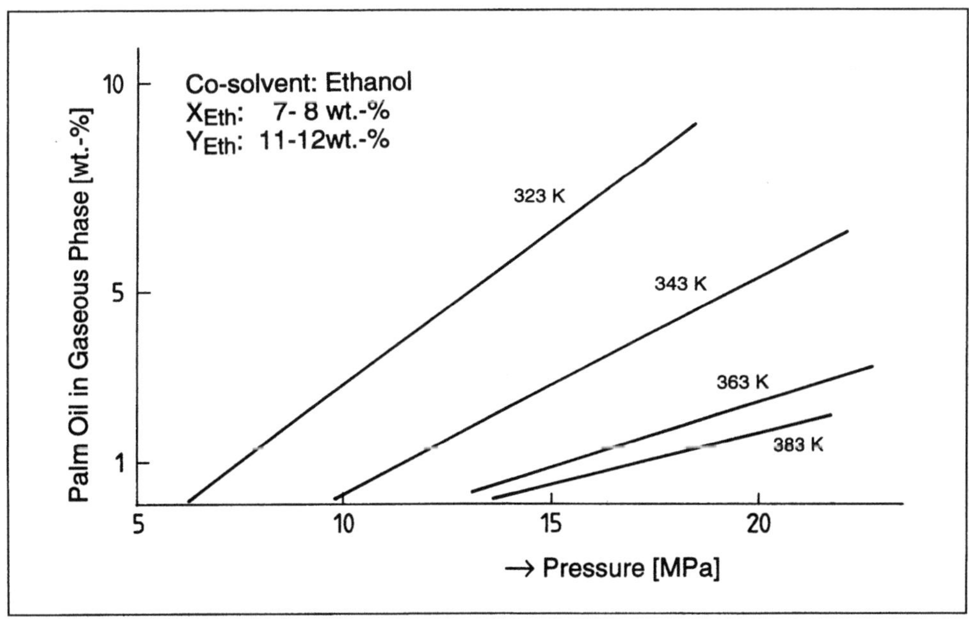



**Fig. 8.29.** The role of a co-solvent: A liquid solvent (at STP) as co-solvent. Palm oil (low volatile pseudo-component) – ethanol (co-solvent) – carbon dioxide (supercritical solvent).

finate and extract. This can easily be achieved by expansion or by stripping with the supercritical component, but it requires an additional process step. Furthermore, the co-solvent can distribute differently between gaseous and liquid phase, thus changing solvent composition. Since solubility and selectivity can strongly depend on co-solvent concentration, the composition of the solvent mixture has to be controlled within narrow limits.

Even if there is no intention to employ a co-solvent, the feed mixture may contain the co-solvent in itself. Components of medium volatility will act as a co-solvent. Therefore it is useful to understand phase behavior and the operation of a countercurrent column with a co-solvent.

In a gas extraction with a co-solvent at least four components are involved: two components which are separated, the co-solvent, and the supercritical component. Solubility and separation factor depend on pressure, temperature and composition of the system. Operating conditions are chosen so that solubility and separation factor enable the intended separation. Countercurrent gas extraction itself is then carried out at constant pressure and, in general, also constant temperature.

## 8.5.2 Countercurrent Gas Extraction with a Co-Solvent as Quasi-Ternary Mixture

In the context of this chapter the different behavior of a gas extraction system with a co-solvent is of interest. Therefore, low volatile components are considered as one pseudo-component. The process can then be treated for a quasi-ternary system in a triangular diagram, as shown in Fig. 8.30.

An extraction apparatus for a separation with a co-solvent consists of the main separation column (1), where the intended separation between the components of the feed mixture takes place, and two separation devices (2), (3), where the supercritical solvent is separated from extract and raffinate. A certain part of the extract is reintroduced as reflux (4) to the separation column, if it is operated as fractionated extraction with two separation cascades. The other components of the plant serve to adjust pressure and temperature and the concentration of the solvent, or to maintain mass flows, and they are not discussed within this context.

The feed mixture ($F$) is fed to the separation column and is split within the column into an extract (top product) $P_E$ and a raffinate (bottom product) $R_n$. Within the column, mass transfer takes place between the countercurrently flowing gaseous and liquid phase, in such a way that an extract $E_1$ leaves the column. Extract $E_1$ is split into the extract $E'$ and the cycle gas by pressure reduction, temperature increase or other measures. $E'$ is further split into the top product $P_E$ and the reflux $R_0$. Reflux $R_0$ is fed to the top of the column and serves as downflowing liquid in the countercurrent separation. $E'$, $P_E$, and $R_0$ are of the same composition. From $P_E$ the solvent has to be removed in a separator. The co-solvent may be removed by stripping with pure supercritical solvent.

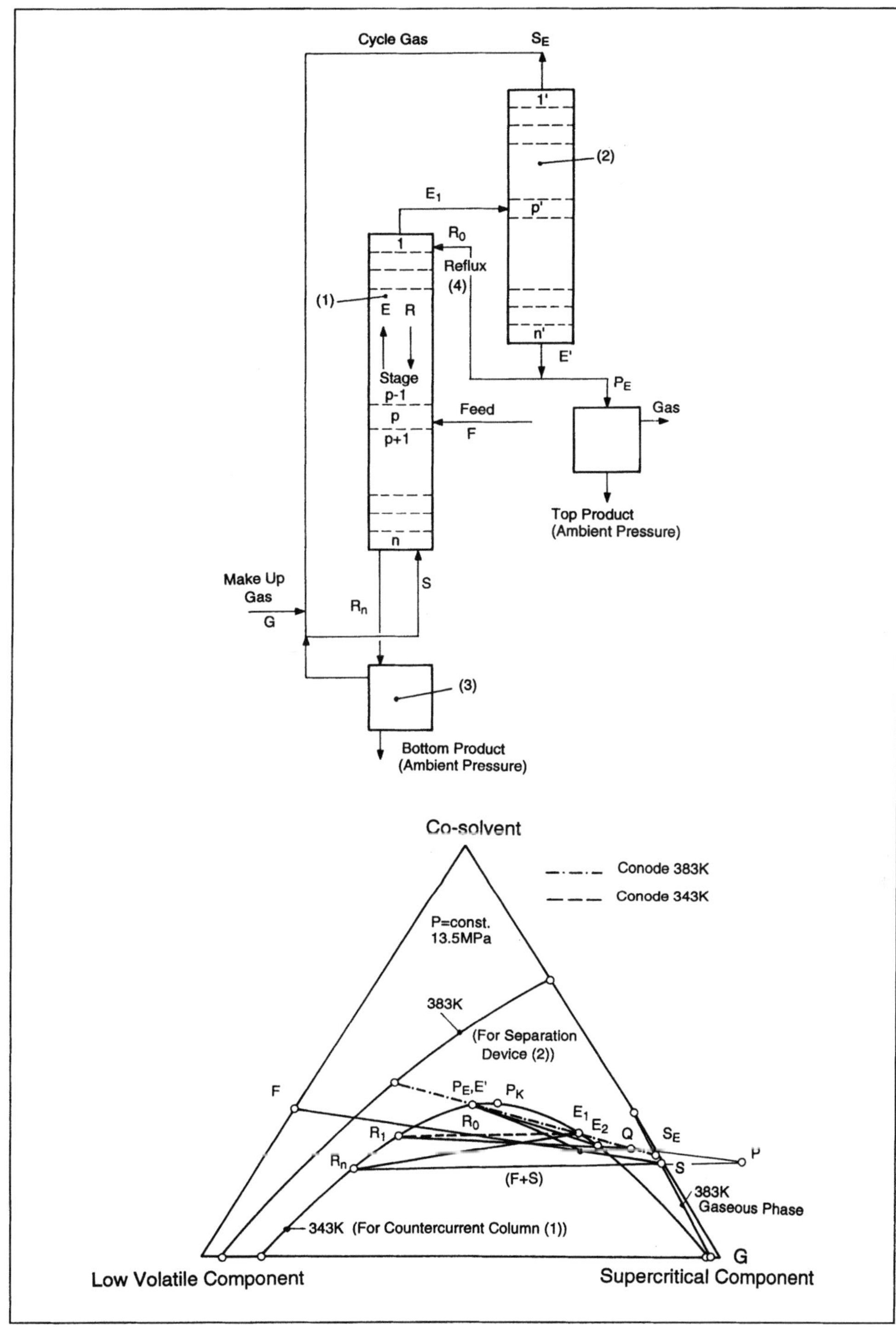



**Fig. 8.30.** Schematic graphic representation of a countercurrent gas extraction with a co-solvent.

303

The following discussion of a countercurrent separation in a ternary system is written with the symbols for a co-solvent. If for the co-solvent, another low volatile substance is taken, the analysis is the same for the separation of two low volatile substances (without co-solvent).

The net mass flow $Q$ from the top of the separation column is given by (the meaning of the symbols is illustrated in Fig. 8.30):

$$Q = E_1 - R_0. \qquad\qquad 8.35$$

A mass balance around the top until stage number $p$ is given by:

$$E_{p+1} + R_0 = R_p + E_1, \text{ or} \qquad\qquad 8.36$$

$$E_{p+1} - R_p = Q = E_1 - R_0. \qquad\qquad 8.37$$

Because of Eq. 8.37, $Q$ is the net mass flow towards the top of the separation column for all stages from 1 to $n - 1$. In a triangular diagram, representing the gas extraction system, all mass balances of the enriching section of the column will pass through $Q$. The number of theoretical stages for the variation of the concentration of the quasi-ternary system can thus be determined by alternately using conodes and drawing mass balance lines through $Q$. From a mass balance one obtains:

$$\frac{R_p}{Q} = \frac{\overline{E_{p+1}\,Q}}{\overline{R_p E_{p+1}}}, \qquad\qquad 8.38$$

with

$$\overline{R_p E_{p+1}} + \overline{E_{p+1}Q} = \overline{R_p Q}. \qquad\qquad 8.39$$

The ratio of the internal flows $R_p/E_{p+1}$ can be obtained from:

$$\frac{R_p}{E_{p+1}} = \frac{\overline{E_{p+1}Q}}{\overline{R_p Q}} = \frac{x_{AEp+1} - x_{AQ}}{x_{ARp} - x_{AQ}} \qquad\qquad 8.40$$

The ratio of the internal flows $R_p/E_{p+1}$ is different from stage to stage. At the top of the column the ratio is $R_0/E_1 = \overline{E_1 Q}/\overline{R_0 Q}$. For the determination of $Q$, first the composition and amount of the cycle gas $S_E$ has to be found. For this a mass balance of the separator at the top has to be considered. Extract flow $E_1$ enters, and cycle gas $S_E$ leaves this separator. Then the mass balance is given by: $E_1 = S_E + E'$. $E'$, $R_0$ and $P_E$ are on the same straight line as $S_E$, $E_1$ and $Q$. In addition, $S_E$ has to be inside the triangular diagram. If $S_E$ contains only a small amount of co-solvent, it coincides practically with the base line of the triangle.

The mass balance is given by: $E_1 = S_E + P_E + R_0$. Because of $E = Q + R_0$, $S_E + P_E = Q$ is obtained. $Q$ therefore is on the straight line $\overline{S_E\,E'}$ between $E_1$ and $S_E$. Then the reflux ratio can be calculated:

$$\frac{R_0}{P_E} = \frac{(x_{AE_1} - x_{AQ})}{(x_{AP_E} - x_{AE_1})} \frac{(x_{AP_E} - x_{AS_E})}{(x_{AQ} - x_{AS_E})}. \tag{8.41}$$

$Q$ must be chosen on the line $\overline{S_E P_E}$ in such a way that the reflux ratio is achieved. Since the distance between $S_E$ and $Q$ is very small, a graphical construction will not be helpful.

For the stripping section, a mass balance yields $E_{p*+1} - R_{p*} = P$, and a total mass balance $Q = P + F$. Therefore $P$ lies on the straight line connecting $F$ and $Q$ at the crossing point with the line $\overline{R_n S}$. After obtaining $P$, the pole point for the stripping section, the number of theoretical stages in the stripping section can be determined in the same way as for the enriching section.

From a total mass balance for the plant $G + F = R_n$ with $G = S - S_E$, the quantity of make up gas can be determined:

$$G x_G = R_n x_{GRn} + P_E x_{GPE}. \tag{8.42}$$

With $x_G = 1$ the quantity of make up solvent $G$ is obtained:

$$G = R_n x_{GRn} + P_E x_{GPE}. \tag{8.43}$$

$G$ is only a small part of the total cycle gas flow $S_E$. Therefore, $S$ is very close to $S_E$. Then, $P$ is also very close to $S_E$ and can be assumed to be identical with $S_E$ for an approximate graphical solution.

For carrying out the separation process, the mixing point of $E_2 + R_0$ must lie in the two-phase area. During operating of the separation column, $E_1$ and $S_E$ have to be controlled for the concentration of co-solvent.


# 8.6 Laboratory Plants and Equipment

The main part of a laboratory plant for countercurrent gas extraction is the separation column, where the countercurrent contact of liquid and gaseous phase takes place. Equally important is the equipment for establishing the flow of the phases through the column. Other components depend on the size of the column, whether the supercritical solvent is recycled or not, and on the degree of automatization. In the following, two examples of laboratory installations are described. They are non-commercial installations used at our laboratory for separation experiments on edible oils, fats and related compounds, and on mineral oil fractions.

### 8.6.1 Laboratory Plant 1

The plant is designed for separation experiments on edible oil mixtures, fat components, or similar compounds, using pure carbon dioxide or carbon dioxide with an entrainer as a solvent. The column diameter is small, so that experiments can be run with about 0.1 kg of feed material. Maximum operating pressure is 35 MPa, maximum operating temperature is 420 K ($\approx$ 150 °C), minimum operating temperature (at supercritical conditions for $CO_2$) is about 320 K ($\approx$ 50 °C). Automatization is kept low. The separation column can be operated as stripping section, as enriching section, or with both sections. A schematic drawing of the plant is shown in Fig. 8.31.

The separating column consists of several sections of a standard high pressure tube of 25.4 mm (1″) outer diameter and 17.5 mm (3/4″) inner diameter, connected by standard cross fittings (316 SS or 1.4401, Autoclave Engineers, Erie, PA, USA). Total column length is 7 m, of which 6 m were filled with mass transfer packing. For achieving mass transfer, different types of packing were used, first wire spirals (Braunschweiger Wendeln) of 3 and 5 mm diameter (VVF, Ransbach-Baumbach, 1.4404) and then a regular wire mesh packing (Sulzer EX). Individual sections of the packing were 300 to 500 mm high, see Fig. 8.32. Between the packing sections, the liquid phase was redistributed by special insets with a conical drilled hole.

At the lower end of the column a view cell (NOVA Swiss) permits observation of the liquid filling, and towards the upper end a second view cell helps to detect flooding. At the lower end of the column 500 mm were free of packing and served as liquid reservoir. In the free zone at the upper end liquid and gaseous phase could separate. The separation column is heated in four sections by heating wires. All heated parts are insulated. Control of temperature was achieved by two point controllers and Pt100 temperature sensors. Feed reservoir, feed line, and bottom of the column were heated separately. Temperature within the column was determined by Ni-CrNi thermocouples (TIR 1 to TIR 5), pressure was measured by pressure transducers (PIR 1 and PIR 2). The solvent was passed through an electrically heated preheater filled with spirals. For removal of the extract from the supercritical solvent a precipitator of 1 l volume, for controlling pressure, a buffer autoclave of 4 l volume and for removing liquid or solid particles from the gas stream a filter autoclave of 0.5 l was employed. The different parts of the plant were connected with standard type tubing of 1/4″ (6.35 mm) outer diameter and 3.09 mm inner diameter. Rupture discs secured the plant against overpressure.

Feed material was pumped with a metering piston pump (4) of 1000 cm$^3$/h capacity (Bran & Lübbe, Norderstedt, FRG) from a heated reservoir (3) of 5 l capacity. Feed flow was adjusted by stroke length of the piston of the pump. Feed could be injected at three locations into the column (5, 6, 7), depending on the mode of operation of the column. For example, location 7 was used for feed injection for running the column as stripping section. The bottom product was removed from the column by a metering valve (8). The extract phase, consisting of supercritical solvent and extract, was expanded through a metering valve (10) into the precipitatior. The expanded gas was led to the bottom of the autoclave in order to avoid short cut flow. The gas leaving the precipitator was further expanded (valve 12), passed through the filter autoclave, filled with active charcoal, and then led to the suction side of the cycle membrane

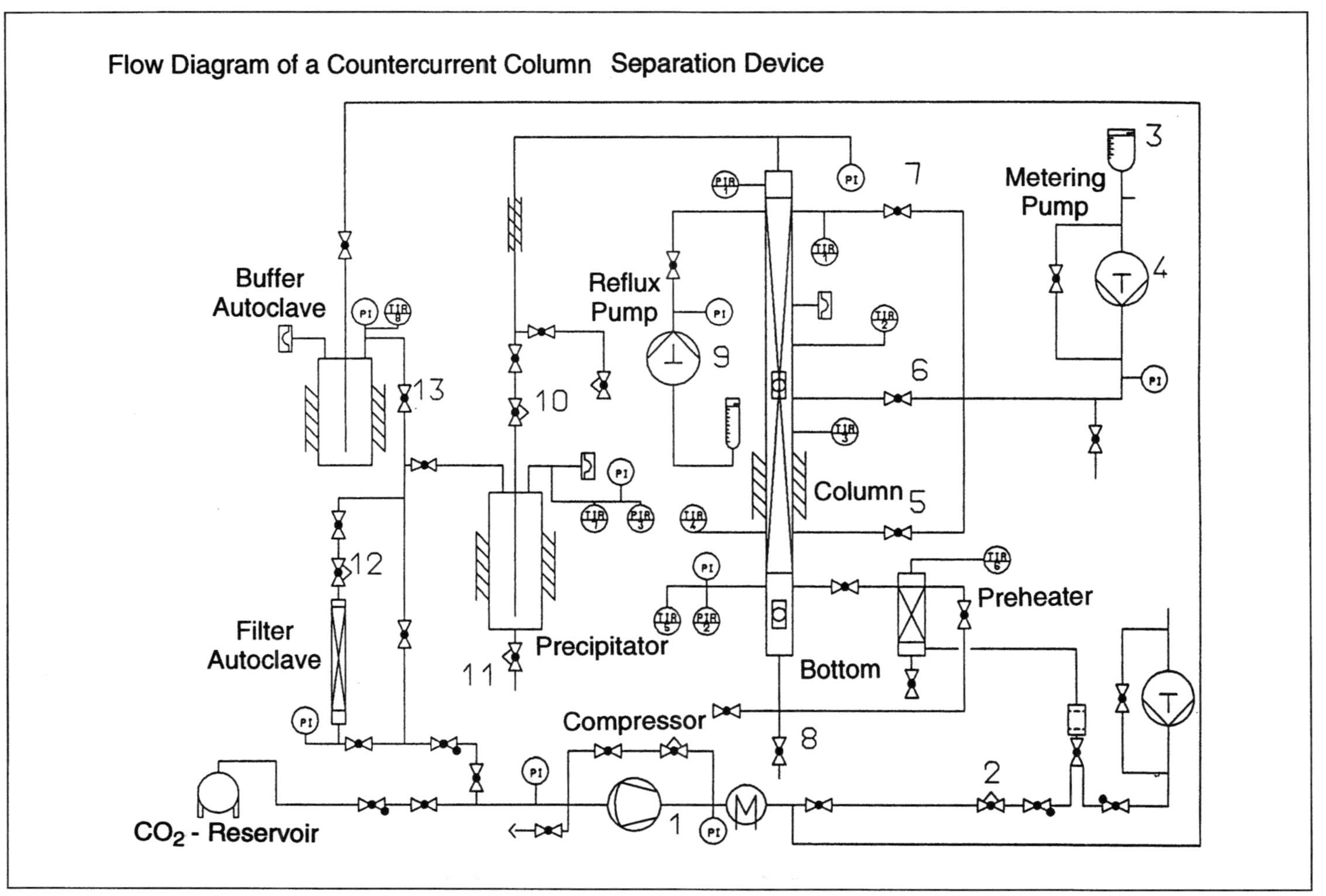


**Fig. 8.31.** Flow diagram of laboratory plant 1.

 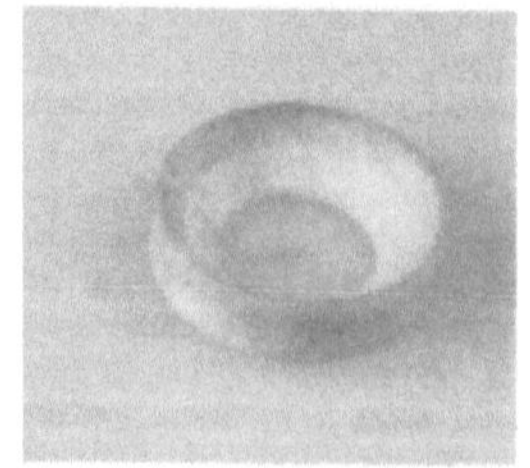 

**Fig. 8.32.** Packings of laboratory plant 1.

compressor (1) (max. flow 20 kgCO$_2$/h at 50 MPa; Andreas Hofer, Mühlheim/Ruhr, FRG). In the suction line, a coriolis mass flow meter (Rheonik, Maisach, FRG) measured solvent flow rate. Pulsation caused by the compressor was dampened by a metering valve (2) on the high pressure side of the compressor. Losses of solvent during the operation of the plant were compensated from the buffer autoclave. Liquid reflux was fed to the column by a HPLC-pump (9) (Milton-Roy) from a heated reservoir. Measured values were registered and processed with a computer (Atari, MEGA-ST).

### 8.6.2 Laboratory Plant 2

Laboratory plant 2 was designed for separation experiments on heavy mineral oil fractions. The setup is similar to laboratory plant 1. Therefore the components are only mentioned briefly. A flow scheme is presented in Fig. 8.33. The operation of the plant is described below.

The plant is designed for a maximum pressure of 35 MPa and a maximum temperature of 420 K ($\approx$ 150 °C). Length of the separation column is 4 m, inner diameter is 20 mm. At the lower end a view cell is installed for observing the liquid phase. The column is electrically heated by heating wires. The temperature is kept constant throughout the column and controlled by controllers and thermocouples. Feed is pumped from a heated reservoir (B1) with the feed pump (P1) and via a preheater (W1) to the column. The location of the feed injection depends on the mode of operation of the column as stripping or enriching section only, or with both sections. The gaseous solvent is pumped through a buffer autoclave (B2) from the common gas reservoir. After passing preheater (W4) it enters the column at the bottom and flows upward as the gaseous extract phase countercurrently to the liquid phase through the separation column. The extract phase is removed from the top, passes heat exchanger (W2) and an expansion valve, and is then led into the first precipitator (A1). From there the extract is further expanded into the second precipitator (A2), from which the extract is removed. The gaseous solvent leaves precipitator A1, is cooled in heat exchanger W3 and liquefied. The liquefied gas is pumped by cycle plump P2 to the heat exchanger W4, where it is heated and evaporated. After evaporation the gaseous solvent is recycled to the column. The raffinate phase is removed at the bottom of the column through precipitator A3.

For starting up the plant, the heating is switched on first. Then the gaseous solvent ($CO_2$ or $CO_2/C_3H_8$) is pumped to the plant until the operating pressure is reached. The cycle pump is started up and the solvent cycled with the intended flow rate. The gas cycle is operated without introducing feed until pressure and temperature are

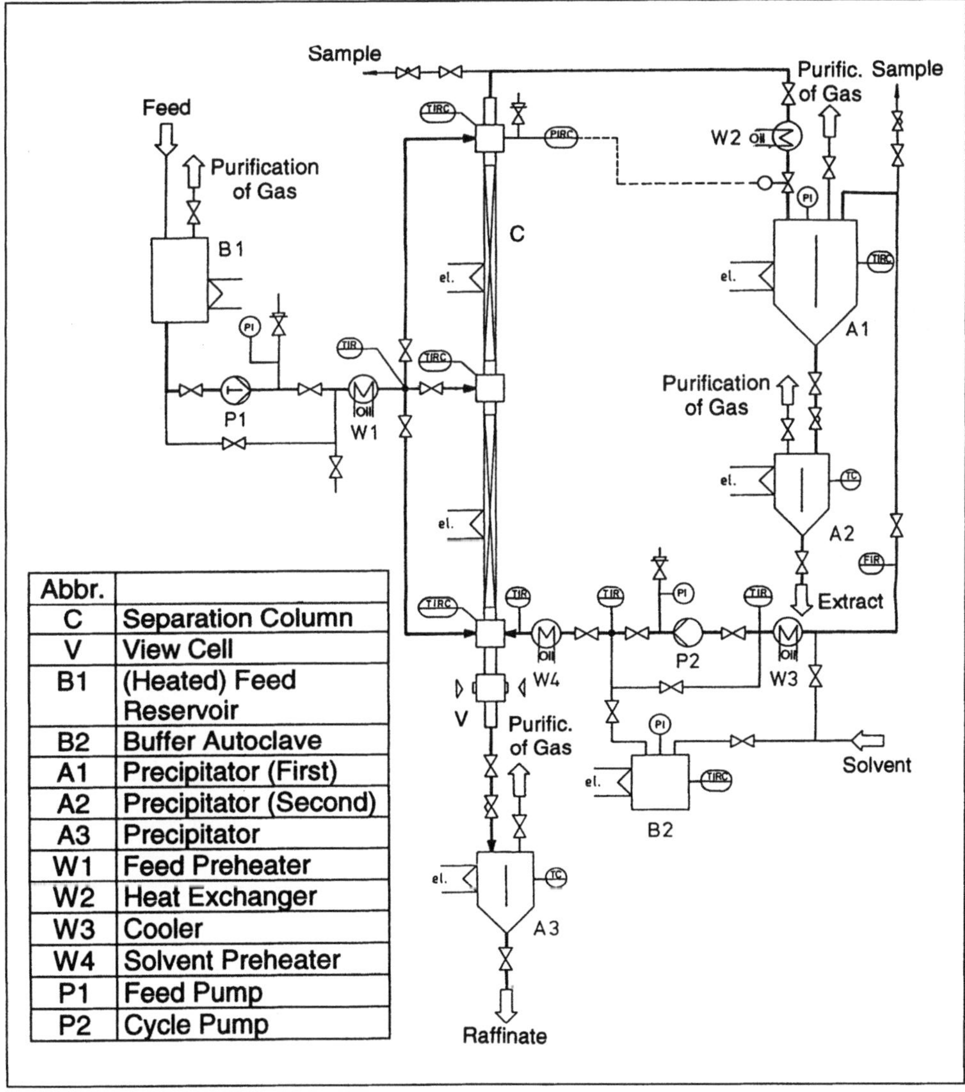


| Abbr. | |
|---|---|
| C | Separation Column |
| V | View Cell |
| B1 | (Heated) Feed Reservoir |
| B2 | Buffer Autoclave |
| A1 | Precipitator (First) |
| A2 | Precipitator (Second) |
| A3 | Precipitator |
| W1 | Feed Preheater |
| W2 | Heat Exchanger |
| W3 | Cooler |
| W4 | Solvent Preheater |
| P1 | Feed Pump |
| P2 | Cycle Pump |

**Fig. 8.33.** Flow scheme of laboratory plant 2.

constant and the plant is in a steady state, which is obtained after 30 to 45 min. Then, the cycle pump is shut off and the feed pump started up. Feed is fed into the column until at the bottom the first liquid droplets appear. Then, the cycle pump is started up again and entrainer is fed to the system if required. Time of experiment is counted from this point. The plant is now operated for about 2.5 h in a steady state. Product flows are continuously removed at the top and at the bottom. Reflux is fed to the column, if required by the mode of operation. After that time, steady state operation of the plant with respect to composition of gaseous and liquid phase is assumed (and has been experimentally determined). The plant is operated for a further 1.5 h, during which, at intervals of 30 min samples are drawn from top and bottom product. These samples are analyzed and the result is taken as the separation result under given conditions. After an operation time of 5 h, the plant is shut down, emptied, and cleaned, if necessary.

## References


1. Brignole EA, Andersen PM, Fredenslund A (1987) Supercritical fluid extraction of alcohols from water. Ind Eng Chem Res, 27: 254 – 261
2. Brunner G, Peter S, Riha R (1974) The properties of compressed gases and their applicability for extractive distillation in plate columns. Preprints VTG/AICHE-Joint Meeting, München 1974, Vol II, D 3-4
3. Brunner G (1978) Phasengleichgewichte in Anwesenheit komprimierter Gase und ihre Bedeutung bei der Trennung schwerflüchtiger Stoffe. Habilitationsschrift, Universität Erlangen-Nürnberg
4. Brunner G, Peter S, Retzlaff B, Riha R (1979) The separation of low volatile substances using compressed gases. In: Timmerhaus KD, Barber MS (eds) High-Pressure Science and Technology. Proc 6th AIRAPT conf, Boulder 1977. Plenum, New York, pp 565 – 572
5. Brunner G, Peter S (1979) Die Trennung schwerflüchtiger Stoffe mit Hilfe komprimierter Gase in Anwesenheit von Schleppmitteln. Ber Bunsenges Phys Chem 83: 1137
6. Brunner G, Peter S (1980) Verfahrenstechnische Grundlagen und Möglichkeiten der Anwendung komprimierter Gase zur Stofftrennung. Chem Ing Tech 52: 445; VDI-Berichte Nr. 391: 561 – 571
7. Brunner G, Peter S (1982) On the solubility of glycerides and fatty acids in compressed gases in presence of an entrainer. Separation Science and Technology, 17: 199 – 214
8. Brunner G, Peter S (1982) State of art of extraction with compressed gases (Gas Extraction). German Chem Eng 5: 181 – 195
9. Brunner G, Kreim K (1985) Zur Trennung von Ethanol-Wasser-Gemischen mittels Gasextraktion. Chem IngTech 57: 550 – 551; MS 1271/85, VDI Berichte pp 1569 – 1590
10. Brunner G, Kreim K (1986) Separation of ethanol from aqueous solutions by gas extraction. Ger Chem Eng 9: 246 – 250
11. Brunner G (1986) Anwendungsmöglichkeiten der Gasextraktion im Bereich der Fette und Öle. Fette Seifen Anstrichmittel 88: 464 – 474
12. Brunner G (1987) Stofftrennung mit überkritischen Gasen (Gasextraktion). Chem Ing Tech 59: 12 – 22
13. Brunner G (1988) Extraction with supercritical fluids. In: Ullmann's Encyclopedia of Industrial Chemistry, Vol B3, chapter 7.3, Verlag Chemie Weinheim
14. Brunner G, Malchow T, Stürken K, Gottschau T (1991) Separation of tocopherols from deodorizer condensates by countercurrent extraction. J Supercritical Fluids, 4: 72 – 80
15. Colussi IE, Fermeglia M, Gallo V, Kikic I (1992) Supercritical multistaged multicomponent separation: Process simulation. Computers chem Engng 16/1: 211 – 224
16. Czech B (1991) Das Betriebsverhalten einer Gegenstromkolonne bei der Erzeugung von Diglyceriden mit Hilfe der nahekritischen Extraktion. Dissertation, Universität Erlangen-Nürnberg

17. Ender U (1989) Der Einfluß der Druckpulsation auf die überkritische Fluidextraktion von Monoglyceriden. Dissertation, Universität Erlangen-Nürnberg

18. Gaver D van (1992) Fractionatie van vetzuuresters met supercritische extractie. Dissertation, Universiteit Gent

19. Gottschau T (1994) Untersuchungen zur Anreicherung von Tocopherolen durch Hochdruckgegenstromextraktion mit überkritischem Kohlendioxid. Dissertation, Technische Universität Hamburg-Harburg

20. Haan AB de, Supercritical fluid extraction of liquid hydrocarbon mixtures. Dissertation, University of Delft, Netherlands

21. Hodel M (1990) Untersuchungen zum Einsatz der on-line "Supercritical Fluid Chromatography" für die Messung von Hochdruck-Phasengleichgewichten und zur Bestimmung des Betriebsverhaltens einer SF-Gegenstromextraktionskolonne, Dissertation ETH Nr. 9301, Eidgenössische Technische Hochschule Zürich

22. Jänecke E (1906) Z anorg Chemie, 51: 132

23. Kelly RM (1987) General processing considerations. In: Rousseau RW (ed) Handbook of Separation Process Technology. John Wiley & Sons, New York

24. King CJ (1980) Separation Processes, 2nd edn, McGraw-Hill, New York

25. Kohmann H (1982) Betriebsverhalten einer überkritischen Fluidextraktion im Gegenstrom, untersucht am System Stearinsäureglyceride – Aceton – $CO_2$. Dissertation, Universität Erlangen-Nürnberg

26. Kreim K (1983) Zur Trennung des Gemisches Ethanol – Wasser mit Hilfe der Gasextraktion. Dissertation, Technische Universität Hamburg-Harburg

27. Moricet M (1982) Simulierung von Gasextraktion in Bodenkolonnen am Beispiel der Abtrennung von Monoglycerid aus einem Ölsäureglyceridgemisch sowie der freien Fettsäure aus Palmöl. Dissertation, Universität Erlangen-Nürnberg

28. Peter S, Brunner G, Riha R (1973) Zur Trennung schwerflüchtiger Stoffe mit Hilfe fluider Phasen. Dechema-Monographien, 73: 197

29. Peter S, Brunner G, Riha R (1974) Phasengleichgewichte bei hohen Drücken und Möglichkeiten ihrer technischen Anwendung. Chem Ing Tech 46: 623

30. Peter S, Brunner G, Riha R (1976) Zur Abtrennung des Monoglycerids der Ölsäure aus einem Glyceridgemisch mit Hilfe von komprimiertem Kohlendioxid in Gegenstromkolonnen. Fette Seifen Anstrichmittel 78: 45

31. Peter S, Brunner G, Riha R (1977) Wirtschaftliche Aspekte der Stofftrennung mit Hilfe komprimierter Gase. Chem Ing Tech 49: 61

32. Peter S, Brunner G, Riha R (1978) Economic aspects of the separation of substances by means of compressed gases in countercurrent processes. Ger Chem Eng, 1: 26

33. Peter S, Brunner G (1978) Die Trennung schwerflüchtiger Stoffe mit Hilfe komprimierter Gase in Gegenstromprozessen. Angewandte Chemie 90: 794

34. Peter S, Brunner G, Riha R (1973) Verfahren zur Trennung von Stoffgemischen durch extraktive Destillation. DOS 23 40 566; 10. 08. 73

35. Peter S, Brunner G, Riha R (1982) Process for separating mixtures of substances of low volatility, US Patent 4,345,976 (Aug 24)

36. Ponchon M (1921) Tech Mod, 13:20. Cited from King [24]

37. Retzlaff B (1979) Zur Trennung von Glyceriden und Fettsäuren mit Hilfe komprimierter Gase in einer Füllkörperkolonne. Dissertation, Universität Erlangen-Nürnberg

38. Riha R (1976) Die Trennung schwerflüchtiger Stoffe mit Hilfe komprimierter Gase in Bodenkolonnen dargestellt am Beispiel der Trennung von Ölsäure-Monoglycerid und Diglycerid. Dissertation Universität Erlangen-Nürnberg

39. Riha V (1993) Trenntechnische Analyse zur Anreicherung von ungesättigten Fettsäureestern. Private communication, Technische Universität Hamburg-Harburg

40. Sattler K (1988) Thermische Trennverfahren: Grundlagen, Auslegung, Apparate. VCH, Weinheim, Basel, Cambridge, New York

41. Savarit R (1922) Arts Metiers, pp 65, 142, 178, 241, 266 and 307. Cited from King [24]

42. Tiegs C (1985) Die Trennwirksamkeit verschiedener Einbauten und der Stoffübergang bei der Extraktion mit dichten Gasen in Gegenstromkolonnen. Dissertation, Universität Erlangen-Nürnberg

43. Weidner E (1985) Die Trennung von Lecithin und Sojaöl mit Hilfe eines überkritischen Extraktionsmittels. Dissertation, Universität Erlangen-Nürnberg

# 9 Chromatography with Supercritical Fluids (Supercritical Fluid Chromatography, SFC)

## 9.1 Chromatographic Fundamentals

### 9.1.1 Chromatography as Separation Process

In a chromatographic separation a mixture of substances is transported by a carrier, called mobile phase, along a not moving surface, called stationary phase. Chromatography is therefore a separation based on two mass separating agents: the mobile phase and the stationary phase. Between the two phases mass transfer processes take place, which lead to a different transport velocity along the surface of the stationary phase for the different components of a mixture. The components reach the end of the stationary phase at different times and can be detected and collected separately.

The stationary phase can be inside a tube as a packing or a thin layer (column chromatography) or on a flat surface. The interactions of the components of the mixture depend on the nature of the stationary phase. Adsorption on surfaces, solvation in a liquid layer or solid layer (membrane), and reversible chemical reaction between ionic parts of molecules are examples of interaction mechanisms.

A feed mixture $F$ of substances, dissolved in the mobile (carrier) phase is fed to the set up for chromatographic separation at one end of the stationary phase. According to the interaction mechanism and the equilibrium distribution values at the actual conditions of state, the compounds in the mixture distribute differently between mobile and stationary phase. Due to the continuous flow of the mobile phase along the stationary phase, the equilibrium distribution is disturbed. If, for example, the mobile phase after the first introduction of the feed mixture contains less of $F$, the compounds distributed to the stationary phase tend to redistribute to the mobile phase. The less the interaction forces between $F$ and the stationary phase and its molecules, the greater the tendency of $F$-molecules to again enter the mobile phase. If, on the other hand, the feed mixture in the mobile phase meets a section of the stationary phase free of components of $F$, the components of $F$ tend to distribute to the stationary phase. These processes result in a different transport velocity along the stationary phase for different molecules. Molecules having weak interaction forces with the stationary phase are transported quickly, others with strong interactions are transported slowly. Beside the interactions with the stationary phase, the solvent power of the mobile phase determines the distribution of the components. Since supercritical gases have a high solvent power and exert this solvent power at low temperatures,

SFC appears to be the alternative to gas chromatography, which needs high temperatures for the evaporation of the feed mixture and for liquid chromatography, where liquid solvents may be replaced. In this chapter SFC is compared to GC and LC. The modes of operation and the different sizes of chromatographic separations and the use of SFC for on-line analysis and for measuring component properties are discussed.

## 9.1.2 Mode of Operation of Chromatographic Separations

A chromatographic separation can be carried out with intermittent injections (discontinuous operation) or with continuous feeding of the mixture (continuous operation). Most chromatographic separations are carried out batchwise, with different elution procedures, called elution chromatography, frontal chromatography or displacement chromatography.

**Elution chromatography.** A certain quantity of the mixture is dissolved in the mobile phase and transported with it through the chromatographic column. The transport velocity depends on the distribution of the components between mobile and stationary phase, determined by solvent power of the mobile phase and the interactions of the components with the stationary phase. The components reach the end of the chromatographic column at different times after injection and can be detected (identified) and collected. A schematic representation of the operational mode in elution chromatography is shown in Figs. 9.1 and 9.2.

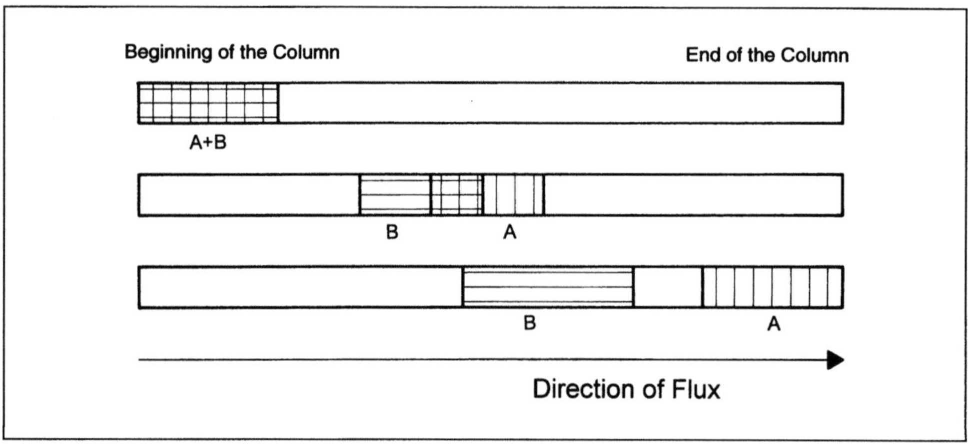


**Fig. 9.1.** Elution chromatography. The mixture of components *A* and *B* is separated in the column into two pure fractions *A*, *B*, and reaches the end of the column as a function of time as shown in the chromatogram of Fig. 9.2.

314

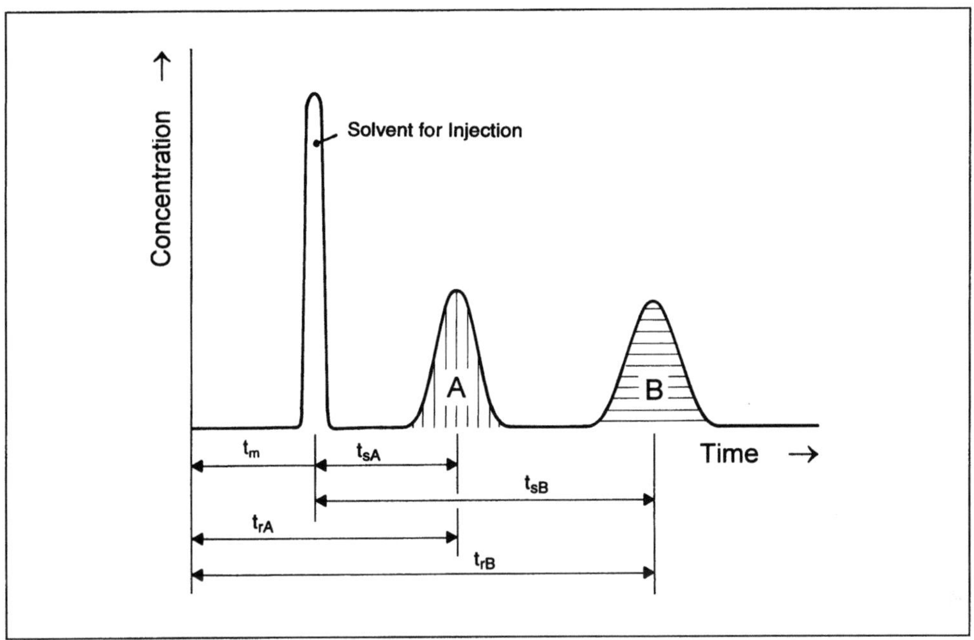



**Fig. 9.2.** Chromatogram of a two-component mixture separated by elution chromatography.

Advantages of elution chromatography are the total separation of the injected quantity of substances, even for more than two substances. Therefore, analytical separations are preferably carried out in this mode. Disadvantageous is the intermittent injection.

Due to several effects, the molecules of a substance are not all transported with the same velocity through the chromatographic column. Therefore, the response signal for the quantity of a substance after elution, called peak, is a distribution, very similar to a Gauss-distribution. The Gauss-distribution represents the peak the closer, the lower the concentration of substances to be separated in the mobile phase. The peaks of a chromatogram can be characterized by the times needed for elution. The time needed for eluting the maximum concentration (equivalent to the maximum of the peak) is called retention time (see also Section 9.3).

**Frontal chromatography.** The feed mixture $F$ is fed continuously to the chromatographic column. In the column the component with the weakest interactions with the stationary phase is transported fastest. It forms a frontal zone in which, at the end of the column, it can be removed as pure compound, until the compound with the next stronger interactions is eluted. From this time on, only a mixture of substances can be eluted. Figure 9.3 illustrates the mode of operation in frontal chromatography for a binary mixture of components $A$ and $B$, and in Fig. 9.4 a corresponding chromatogram is shown.

An advantage of frontal chromatography is that the component at the front can be collected for a certain time, which may be a greater amount than can be achieved with one or several injections in elution chromatography. Disadvantages are the incomplete yield of the component at the front and the loss of the ability to separate more than two components in one column.

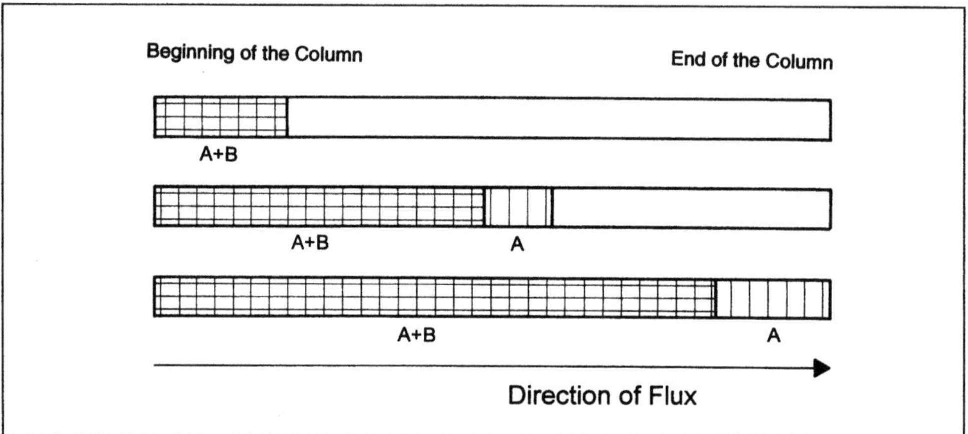


**Fig. 9.3.** Mode of operation for frontal chromatography.

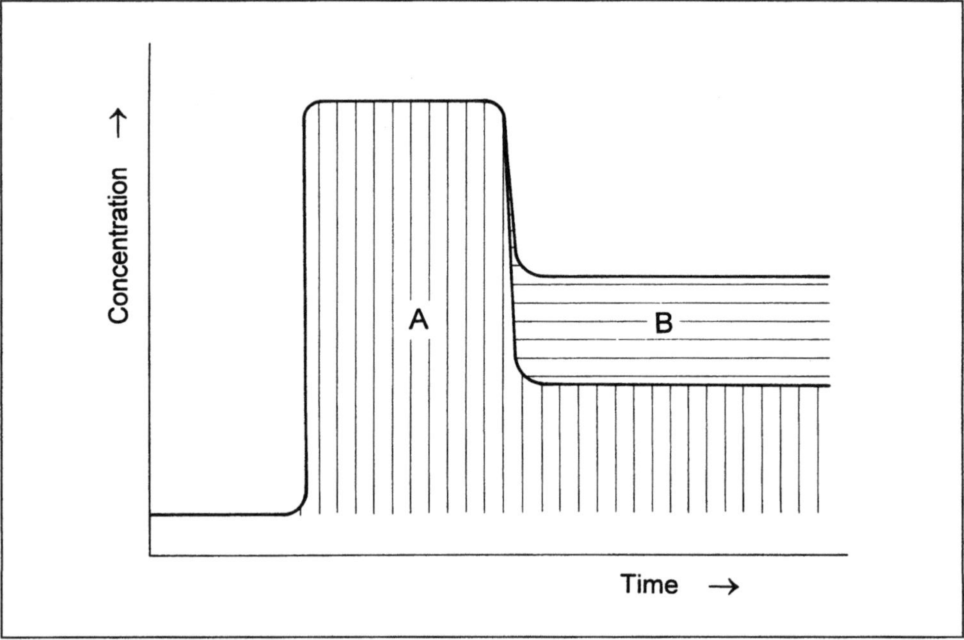


**Fig. 9.4.** Chromatogram of a separation by frontal chromatography.

**Displacement chromatography.** A certain quantity of the feed mixture *F* is injected to the chromatographic column. *F* completely adsorbs on the stationary phase in the column. Then, a substance is added to the mobile phase which has stronger interactions with the stationary phase than any of the substances of the feed mixture. This substance displaces the components of the feed mixture and the chromatogram

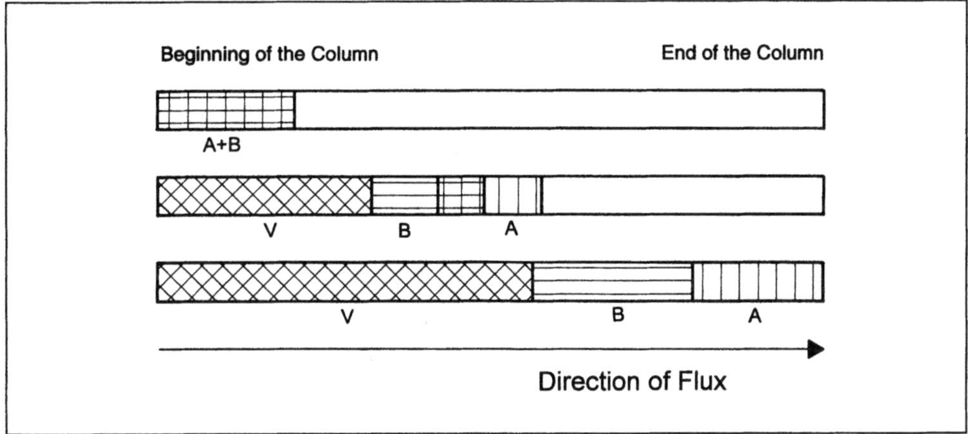



**Fig. 9.5.** Separation of a binary system *AB* in displacement chromatography. Component *V* is the displacement agent.

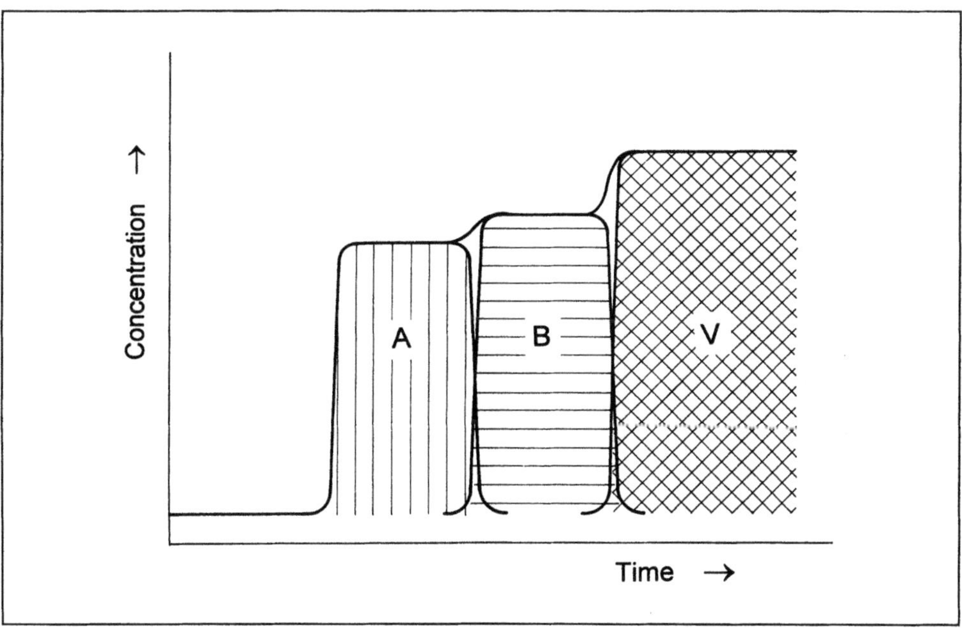



**Fig. 9.6.** Chromatogram of a separation with displacement chromatography.

develops at the end of the column according to the schematic representation in Figs. 9.5 and 9.6 for a binary mixture. It is also possible to use different substances for displacing different compounds. After the separation the column has to be cleaned from the displacing agent before a new injection can be started.

An advantage of displacement chromatography is that higher concentrated fractions than in elution chromatography can be achieved. Dilute substances can be adsorbed on the stationary phase and then eluted in high concentration with the displacement agent. Disadvantageous is the down time needed for regeneration of the chromatographic column.

**Continuous chromatography.** Continuous chromatography is a countercurrent separation with a chromatographic system. The mixture is fed continuously to the

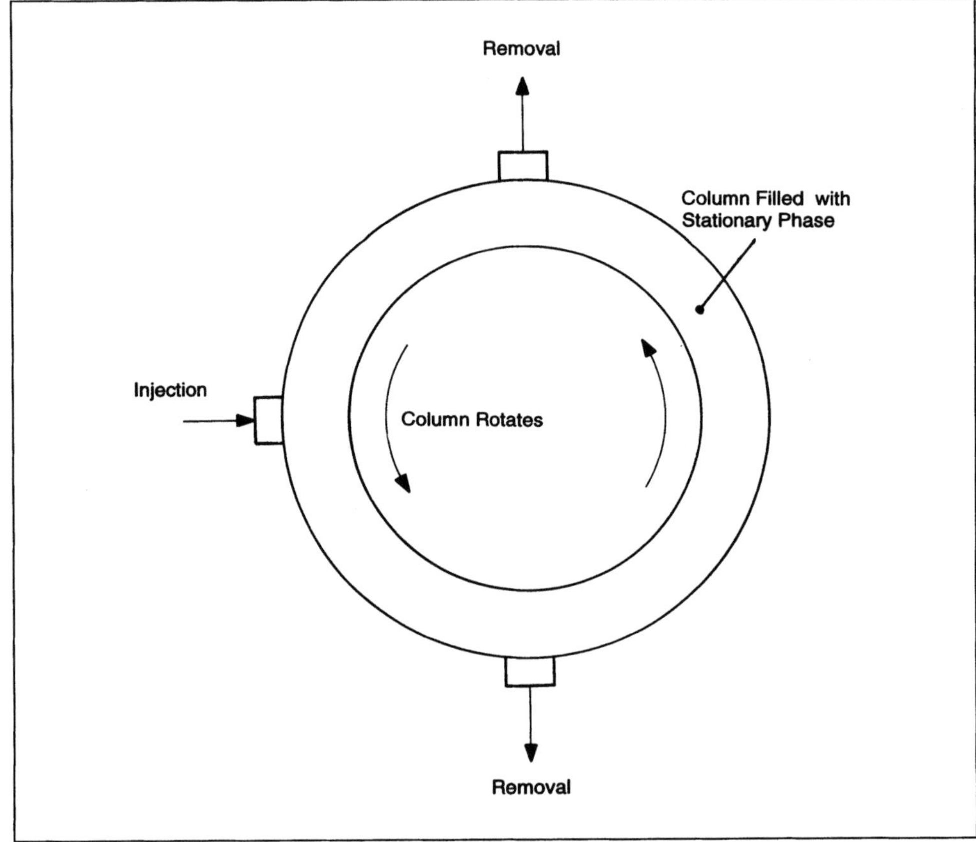


**Fig. 9.7.** Continuous chromatography with a rotating column.

separation column. In order to make possible a continuous chromatographic separation employing a stationary phase, either the stationary phase or the point of injection of $F$ and the point of removing the separated components must be moved. Moving the stationary phase can be achieved, for example, by a rotating column, leaving points of injection and removal fixed. A schematic drawing of such a chromatographic system is shown in Fig. 9.7. Construction of such an apparatus is mechanically complicated. Injection and removal points can be moved by switching valves, which open or close different sections of a chromatographic column. A schematic representation of such a chromatographic system is given in Fig. 9.8. Advantages and disadvantages of continuous chromatography are discussed in connection with scaling up of chromatography.

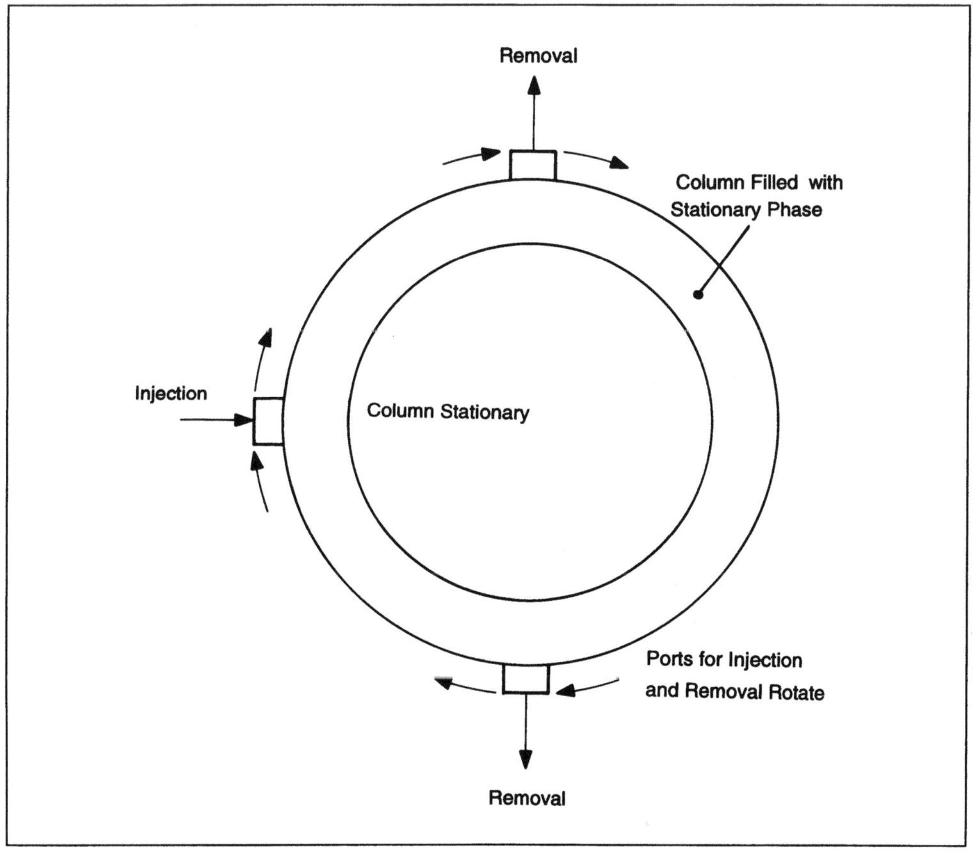


**Fig. 9.8.** Continuous chromatography with a valve switching system for employing different sections of the chromatographic column.

### 9.1.3 The Mobile Phase

In supercritical fluid chromatography (SFC) the mobile phase is a supercritical gas or a near critical liquid. Compared to gas chromatography (GC), where a gas is under ambient pressure, and liquid chromatography (LC), where a liquid is used as mobile phase, the solvent power of the fluid mobile phase in SFC can be varied by density, e.g., by pressure changes at constant temperature. Solubility increases in general with pressure under supercritical conditions of the mobile phase. Since temperature is near the critical temperature of the mobile phase, temperature sensitive compounds can be processed. The chromatographic separation can be carried out at constant pressure (isobaric operation) or with increasing pressure (pressure programmed). In addition, temperature can be varied. SFC has one more adjustable variable for optimization of elution than GC or LC.

A supercritical fluid has properties similar to a gas and also similar to a liquid. While density and solvent power may be compared to those of liquids, transport coefficients are more those of a gas. SFC, because of its mobile phase, can cover an intermediate region between GC and LC, as illustrated in Fig. 9.9 with respect to density and diffusion coefficient. For preparative and production scale operations, SFC has the advantage of easy separation of mobile phase from separated compounds. A disadvantage is that strongly polar and ionic molecules are not dissolved by supercritical gases, which can be advantageously used in SFC.

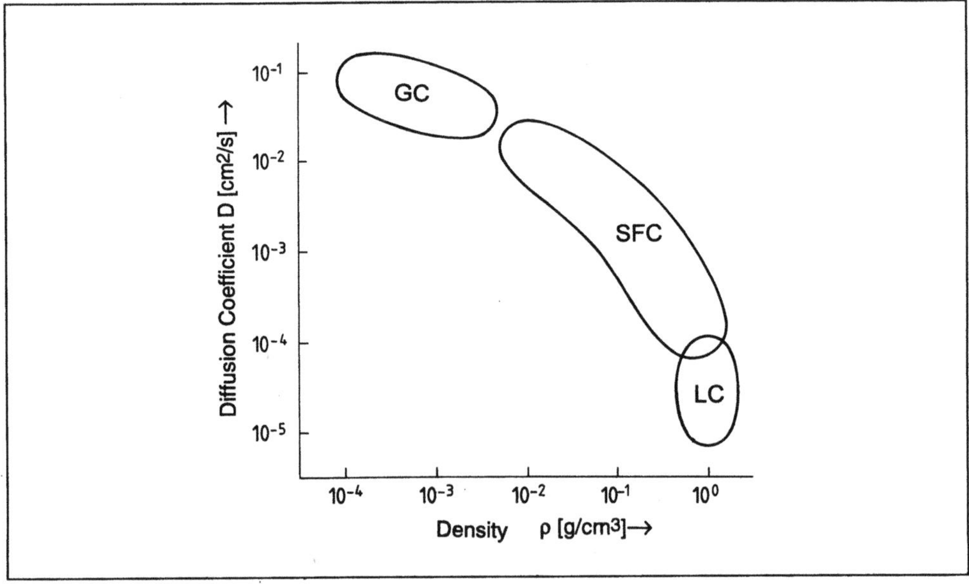


**Fig. 9.9.** Areas for the different mobile phases in chromatographic separations with respect to component properties [30].

320

Most gases which can be used in SFC are unpolar. Therefore, polar substances of a feed mixture can only be eluted by adding a polar modifier. Polar gases like ammonia of sulfur dioxide are reactive compounds under pressure and the equipment must be able to withstand corrosive conditions. On the other hand, carbon dioxide is easy to handle and safe. Polar modifiers, which are easier to handle than ammonia or sulfur dioxide may instead be applied. To make effective use of the possibilities of SFC, allowable pressures should be high. A maximum pressure of around 40 MPa is necessary. Therefore, equipment for HPLC, normally limited to 15 MPa, is not sufficient.

Composition of the mobile phase can substantially influence separation in SFC. Retention times of substances may be very much different due to polarity or other physico-chemical properties of the components of the mobile phase. In an investigation of this effect, Pickel [23] found large differences in the separation of aromatic hydrocarbons with $CO_2$, $N_2O$, $C_3H_8$, and $C_3H_6$, as shown in Fig. 9.10.

Gases applied in SFC are mostly unpolar. The polarity of carbon dioxide at low densities is comparable to that of n-hexane and at higher densities to that of methylene chloride. Nitrous oxide and the alkanes butane or pentane behave similar. Polar substances are eluted only after long retention times and in broad peaks or even not at all. In these cases, a polar modifier, added to the gaseous mobile phase, introduces the necessary polarity to the mobile phase. The modifier then determines the elution sequence, which can be changed by the amount and the type of modifier. Figure 9.11 shows an example for chromatograms of xanthines.

With increasing content of a modifier in the mobile phase, retention times become shorter. For polycyclic aromatic compounds, Leyendecker et al. [14] have investigated the influence of 1.4-dioxane as modifier on n-pentane as mobile phase. The result is illustrated in Fig. 9.12. At high dioxane-contents, no separation is achieved in the chromatographic column. All substances are eluted at the same time. An influence of the modifier on the capacity ratio can be observed at relatively low concentrations. This is due to a second effect of an entrainer. Polar sites of the stationary phase strongly adsorb polar components of the feed mixtures and thereby cause long retention times and broad peaks. Molecules of a polar modifier, which are available permanently during flow of the mobile phase, can occupy the polar sites of the stationary phase and thus render possible a separation without interference of the strongly polar sites.

Temperature and pressure can be employed in supercritical fluid chromatography as parameters for influencing separation characteristics. Temperature directly determines vapor-pressure of the feed components and density of the mobile phase and, indirectly, adsorption equilibrium. With higher temperature, vapor-pressures of the feed components increase exponentially. Density decreases proportionally to temperature if conditions are far from critical, but in the region of the critical point of the mobile phase, which is the main area of application of SFC, density varies dramatically with temperature. The solvent power of the mobile phase, which increases with density, is therefore changed substantially in this region. The influence on chromatographic separation depends on the relative importance of these two effects, if other conditions remain unchanged.

An example of the influence of temperature on chromatographic separation of polycyclic aromatic compounds with n-butane as mobile phase is given in Fig. 9.13.

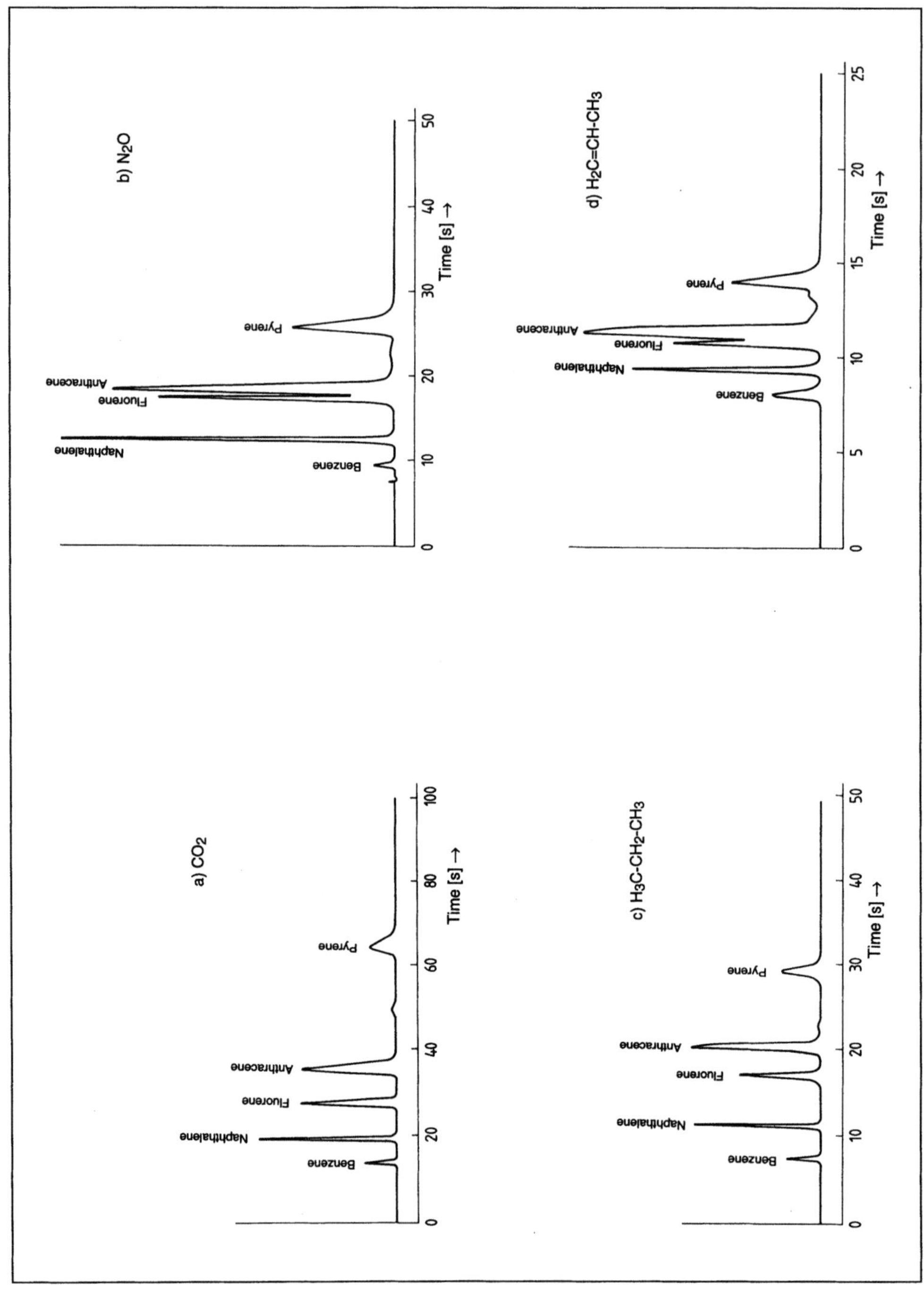

**Fig. 9.10.** Separation of aromatic hydrocarbons with different gases as mobile phase. Gases: a = carbon dioxide ($CO_2$); b = nitrous oxide ($N_2O$); c = propane ($C_3H_8$); d = propylene ($C_3H_6$); Column: 30 x 4.6 mm, unmodified silica gel. Initial pressure 12 MPa; Temperature 296.15 K; Flow rate 670 cm$^3$/min at STP (after Pickel [23]).

At temperatures far below the critical temperature of n-butane, retention times are long and elution peaks broad, due to the low vapor-pressures of the feed components. Just around the critical temperature, retention times are short, peaks are sharp and narrow, and separation of the components is sufficient. Above the critical temperature retention times increase. This is due to the large changes in density in the critical region. Just above the critical temperature, separation is best. Therefore, this re-

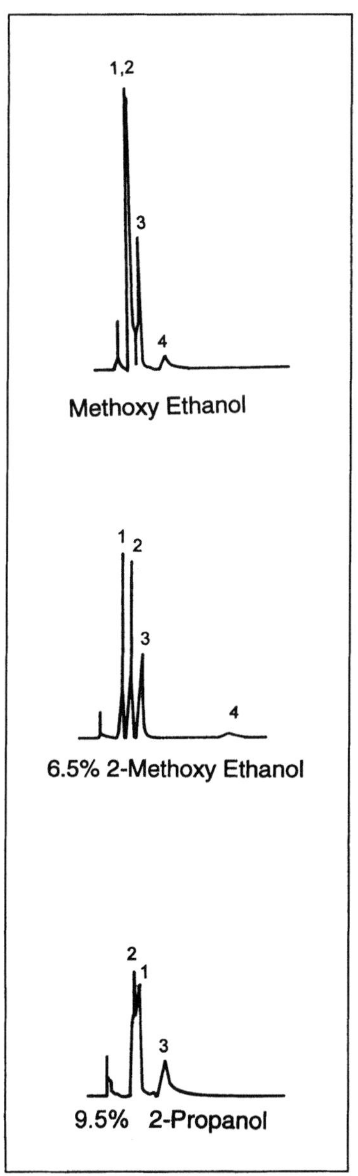



**Fig. 9.11.** Elution sequence of xanthines for different content and type of modifier. 1 = caffeine; 2 = theophylline; 3 = theobromine; 4 = xanthine (after Randall [24]).

gion may be used for mixtures which are difficult to separate. At further increased temperatures, retention times again become shorter due to the increasing vapor pressure of the feed components.

Pressure mainly influences density of the mobile phase. An example for the influence of pressure on separation in SFC is presented in Fig. 9.14. With increasing pressure, the influence of temperature is diminishing, since density varies less with temperature at higher pressures. For estimating the influence of pressure and temperature referring to a $P, V, T$-diagram is helpful.

SFC allows the variation of temperature and pressure for optimizing separation conditions as well as during the separation process itself. Such an operational mode is called pressure and temperature programing. Temperature programing is well known from gas chromatography, but is less common in SFC, since pressure programing can be very effective. Pressure- and temperature programing may be combined to

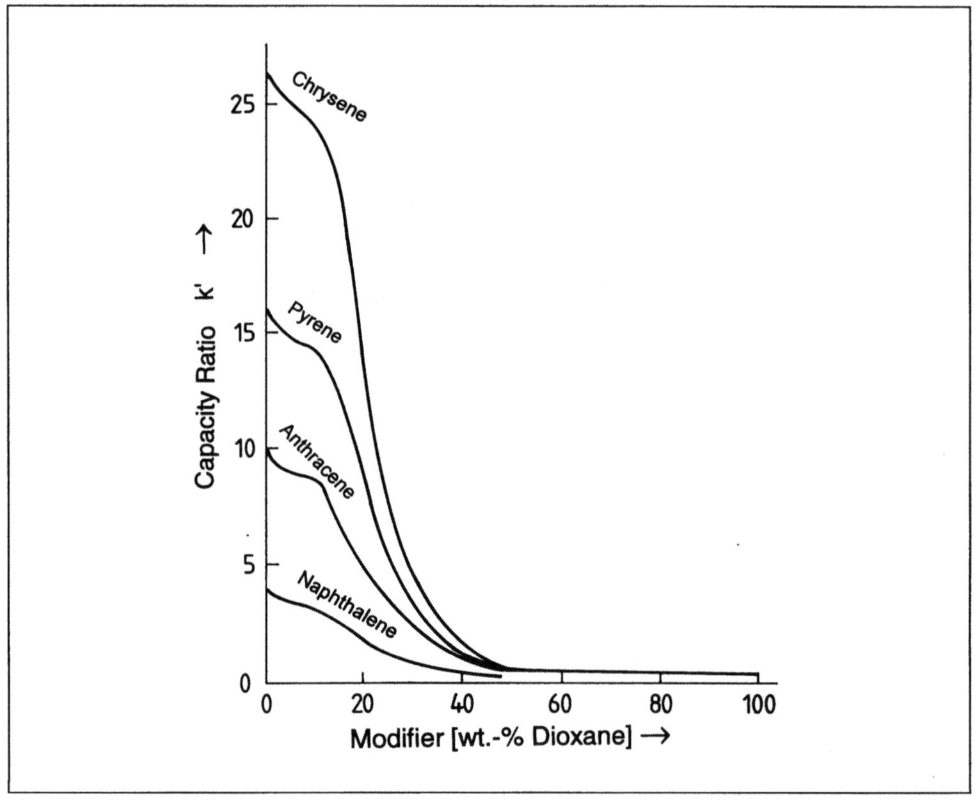


**Fig. 9.12.** Variation of capacity ratios of polycyclic aromatic compounds due to modifier concentration (1.4-dioxane) in the mobile phase (n-pentane). Pressure at the column outlet 3.6 MPa; Temperature 513.15 K (after Leyendecker et al. [14]).

324

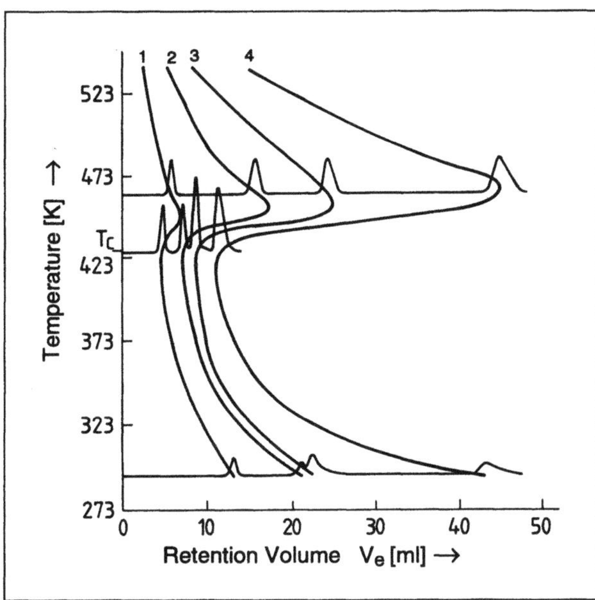



**Fig. 9.13.** Variation of retention times (retention volumes $V_e$) of polycyclic aromatic components in n-butane at 4.5 MPa with temperature. 1 = naphthalene; 2 = anthracene; 3 = pyrene; 4 = chrysene (after Klesper and Leyendecker [11]).

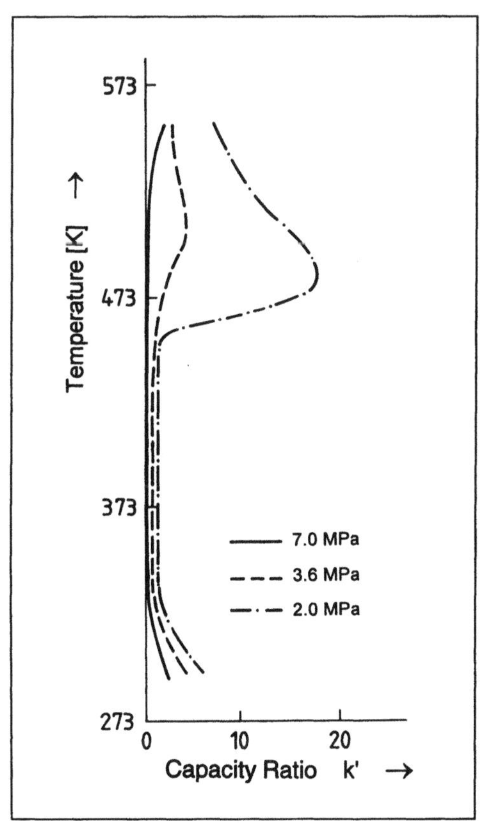



**Fig. 9.14.** Variation of retention times of chrysene with pressure. Mobile phase n-butane (after Klesper and Leyendecker [11]).

325

density-programing. An example for a pressure- and density-programed analytical separation is shown in Fig. 9.15.

In preparative chromatography, conditions are kept constant during separation, since feed mixtures of several injections may be on their way at the same time in the column. The elution of substances of different molecular weight in isobaric SFC separations is better than in isothermal GC, since vapor pressure is not so important in SFC. Compared to LC, the tendency of peak broadening is lower in SFC, since diffusion coefficients are far higher.

Flow rate of the mobile phase is a further important parameter which affects the number of theoretical stages in chromatographic separations. Due to the low viscosity of near critical mobile phases, flow rate in SFC can be high, and number of theoretical stages remains nearly constant over a wide range of flow rate (see below).

A more detailed discussion of chromatographic fundamentals and especially analytical applications of SFC can be found in the abundant literature on analytical SFC, e.g., in [8, 13, 31, 39, 40].

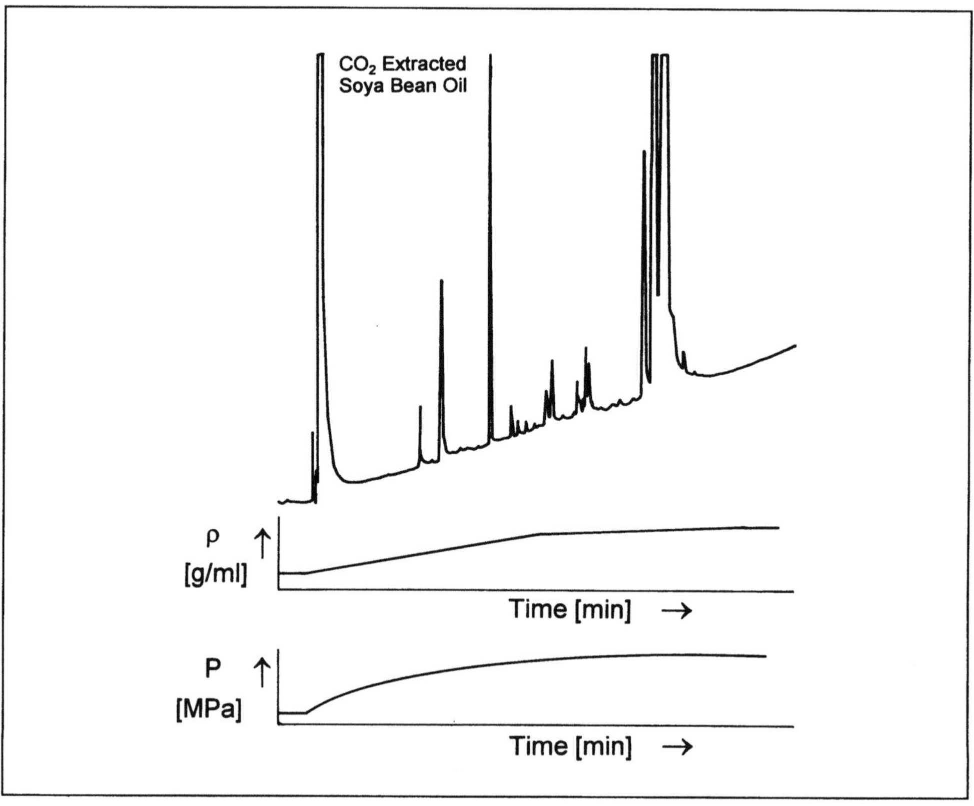


**Fig. 9.15.** Pressure and density programmed separation in SFC.

326

## 9.1.4 Scale of Chromatographic Separations

Chromatography is a separation process with a high separation potential. Separations can be carried out which are not possible with other separation processes, based only on one or no mass separation agent. On the other hand, a chromatographic system is characterized by thin layers and low capacities of mobile and stationary phase (the mass separation agents) which lead to low concentrations and small throughputs compared to separation processes with only fluids involved. In scaling up chromatography the main problem is to maintain the conditions of a chromatographic separation, i.e., plug flow of the mobile phase, no axial dispersion of the transported substances, and rapid equilibration between mobile and stationary phases in order to establish the same conditions for all molecules of one species of a mixture at one time. Otherwise, the substance will be distributed along the column during transport and eluted at low concentration over an extended period of time.

Chromatographic separations are carried out with very different amounts of feed mixtures in analytical chromatography, preparative chromatography, and production scale chromatography.

In *analytical chromatography* a mixture of substances is separated to enable the detection of the individual substances according to identity and quantity. A high power of separation is necessary in order to make possible direct determination of the quantity of a substance with detectors, after elution from the chromatographic column. The separated and eluted substances are not needed and are not collected but are instead disposed of. The recording of the detector signal, the chromatogram, is the product of analytical chromatography. The quantity of feed mixture is small, in the range of μg or mg, and is determined by detection limits. Chromatographic columns have diameters of fractions of a millimeter to several millimeters.

*Preparative chromatography* is applied for isolating one or more substances from a mixture for further use. Areas of application are the separation of intermediate and end products of a chemical or biochemical synthesis or the purification of natural substances. Products are used for determining physico-chemical properties or for activity tests, like in testing enzymes for their catalytic activity. The quantity of the substance which must be isolated depends on its further use. For the identification of a new chemical substance, a few mg may be enough, while for a new pharmaceutical active compound an amount of grams up to kilograms must be isolated and purified. Purity of the compound may be somewhat lower than in analytical chromatography. Concentrations of 95 to 99 % may be sufficient for most purposes. Optimization is worked out in direction of maximizing production rate. Chromatographic columns have diameters in the range of several mm to 100 mm.

*Production-scale chromatography* is the application of a chromatographic separation process for producing quantities of a substance which can be used in practical applications, mostly for commercial purposes. Chromatographic columns have large diameters, from up to 600 mm in liquid chromatography (HPLC) to several meters in ion exchange chromatography.

Chromatographic separations are costly separation processes if carried out for production of a substance. Therefore, chromatography is only applied for substances of high value.

## 9.1.5 Methods for Scaling up Chromatography

Scaling up chromatography can be achieved by several means: 1) increasing size of the chromatographic separation column, 2) employing more than one (many) chromatographic separation columns in parallel, and 3) increasing the amount of substance for one injection.

Increasing the size of a tube is a straightforward process, but maintaining chromatographic conditions is not. The larger the diameter of the column, the more difficult to maintain plug flow. Pressure drop becomes a problem, mainly in liquid chromatography, since the forces on large diameter columns are high. Therefore, par-

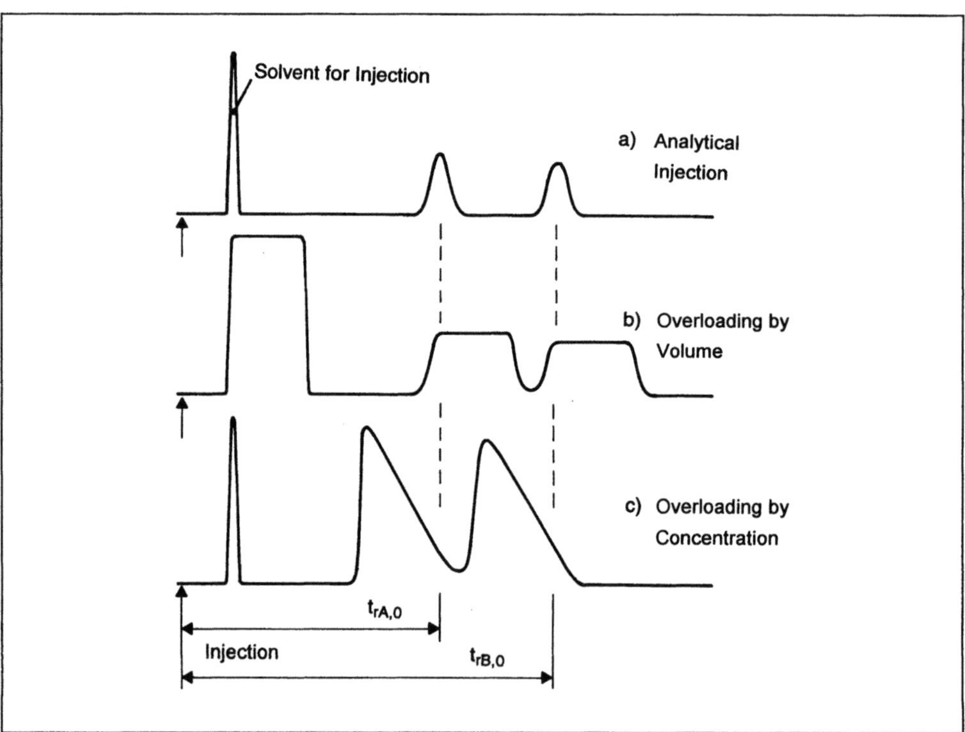


**Fig. 9.16.** Comparison of chromatograms under analytical conditions (part a), overloading by volume (part b), and overloading by concentration (part c).

328

ticle diameters for larger columns are about a factor of 10 greater than for analytical packings, about 50 μm, as compared to 3 – 5 μm in analytical applications. Large size packings of small diameter particles are difficult to produce without inhomogeneities.

Increasing the injected amount of substance can be achieved by either injection of a higher concentration or a higher volume of the solution. In the case of injecting a higher volume, broad peaks of a substance with an extended plateau are eluted. This mode is called overloading by volume. At higher concentrations of the feed mixture in the mobile phase, asymmetric peaks of a substance are eluted. The peak maximum is displaced against retention time under analytical conditions. Changes of the peak depend on the adsorption isotherms of the substance. The variations of the peaks with overloading are schematically shown in Fig. 9.16.

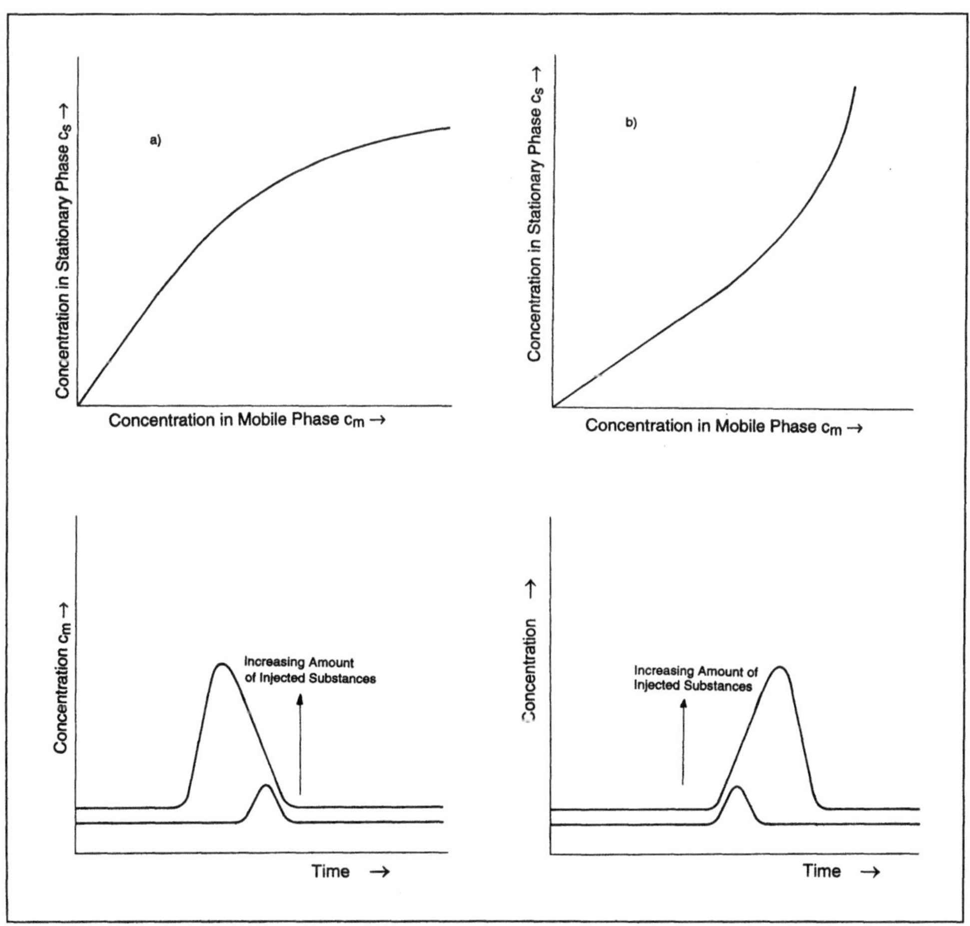


**Fig. 9.17.** Adsorption isotherms with nonlinear parts and the corresponding peaks of a chromatogram.

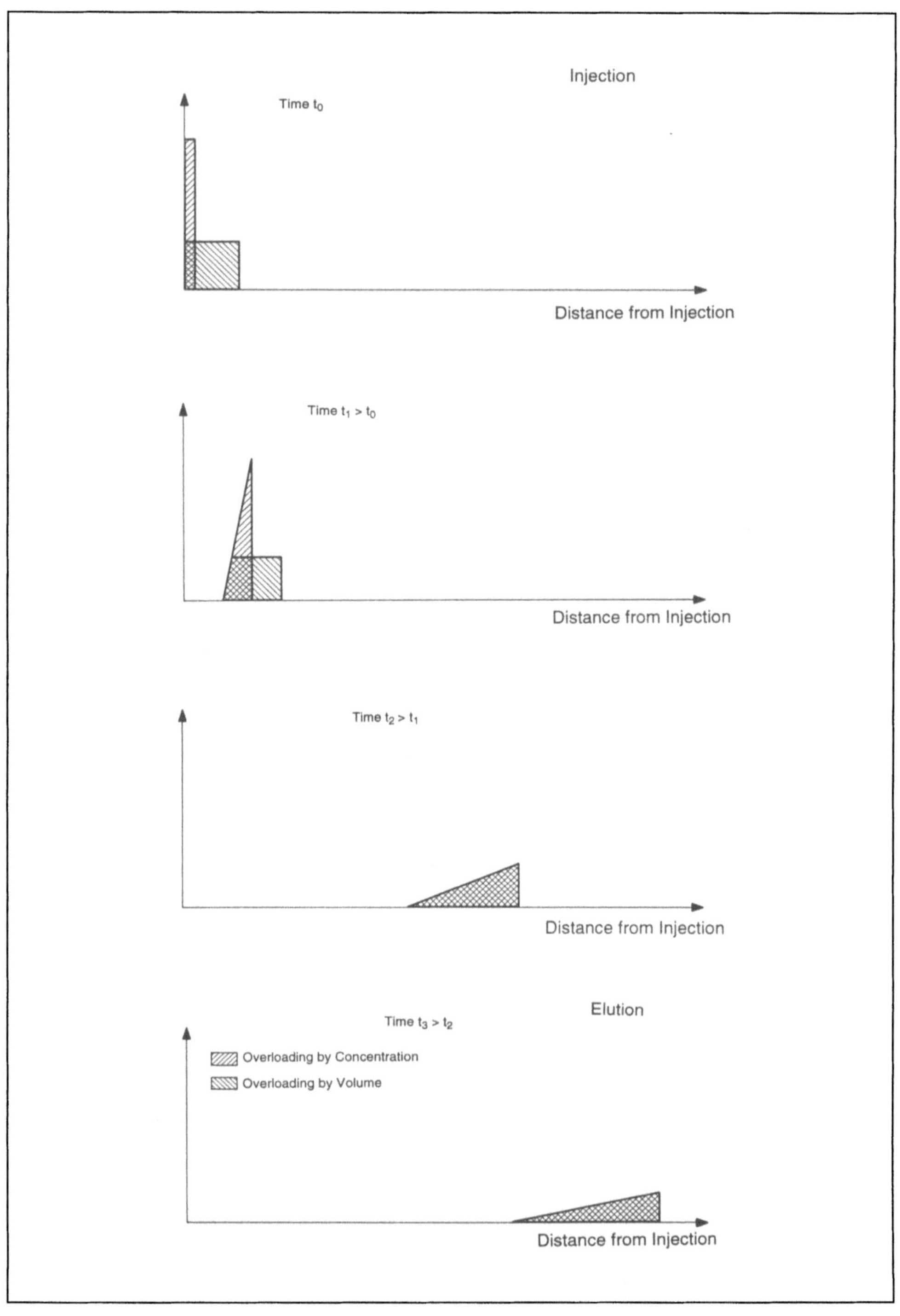



**Fig. 9.18.** Development of a peak under conditions of combined overloading by concentration and volume [38].

330

At low concentration each isotherm of equilibrium distribution between stationary and mobile phase is linear. Concentrations of a substance in the mobile and stationary phase are directly proportional. At higher concentrations the stationary phase gets saturated for many substances, since the adsorbens can adsorb only a limited quantity of molecules and the capacity of a liquid stationary phase is limited by solubility. These cases for nonlinear adsorption isotherms are illustrated in Fig. 9.17. Increased amounts of substances are transported faster through the column, since the quantities of the substance, which cannot be taken up by the stationary phase at one place are transported with the mobile phase without retardation to the next free place in the column. Therefore, peaks of the chromatogram are shifted to shorter retention times. The slope of the peaks is steep at the front and flat at the back. Concentration declines exponentially (Fig. 9.17.a).

For the reverse case, in which the concentration of a substance in the stationary phase increases overproportionally with concentration in the mobile phase, the substance is eluted in the form of leading peaks with an exponentially increasing front side of the peak and a steep slope at the back (Fig. 9.17.b).

Overloading by concentration and by volume can be combined. No additional peak broadening occurs if the injection volume is limited to about half the volumetric flow of the mobile phase in which the peak is eluted. The developing of the peak under combined overloading is illustrated in Fig. 9.18.

# 9.2 Practical Verification of SFC

## 9.2.1 Description of an SFC: General Design

An apparatus for chromatographic separations with a supercritical gas consists of the separation column as central part in a temperature controlled environment (1), (Fig. 9.19), the reservoir for the mobile phase (2), a unit for establishing, maintaining and controlling pressure (3), an optional unit for adding a modifier (4), the injection part for introducing the feed mixture (5), a measuring device (detector) for determining concentration of the eluted substances (6), a sample collection unit (7), a unit for processing the mobile phase (8) and another one for processing data and controlling the total apparatus (9).

The flow of the supercritical gas under pressure is maintained by long-stroke piston-pumps, reciprocating piston-pumps or membrane-pumps which deliver the mobile phase in liquefied form. The fluid is then heated to supercritical conditions before entering the column. Pressure and flow rate must be kept as constant as possible in order to maintain constant conditions for separation and to achieve a stable base line in the chromatogram. Oscillating pumps therefore can have three heads which deliver at different times or a pulsation dampener in order to minimize pulsation.

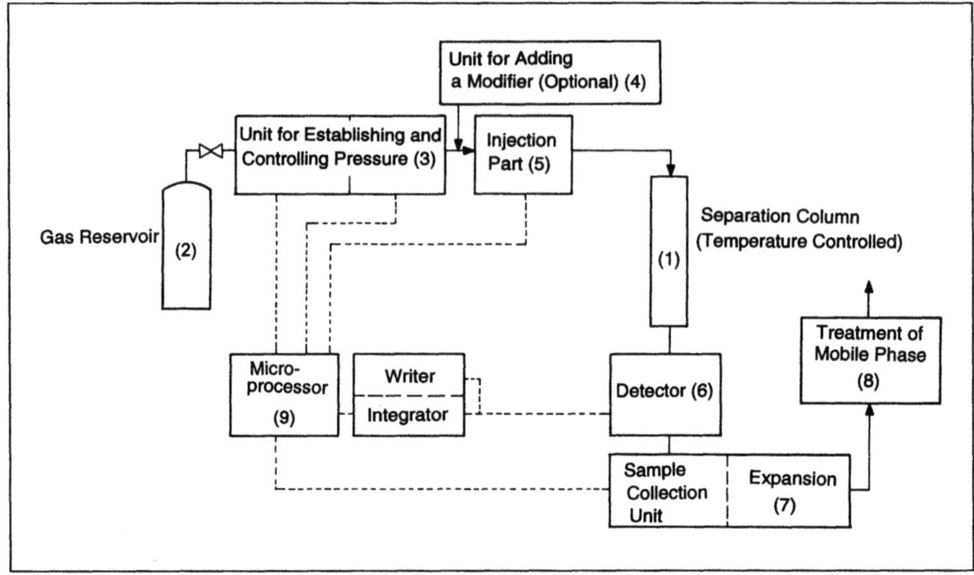



**Fig. 9.19.** Flow scheme of apparatus for supercritical fluid chromatography (SFC).

## 9.2.2 Injection Techniques

Injection of the mixture to be separated is accomplished for analytical purposes by sample loops which may be filled at ambient pressure and are injected into the flow of the mobile phase by switching a multiposition valve in the appropriate position. Such valves can be manufactured as linear moving or rotating valve, as shown in Fig. 9.20.

For preparative separations the feed is pumped by metering pumps into the flow of the mobile phase. Intensive mixing can be achieved in line by static mixers. Other possibilities comprise a column, where the mixture is placed under ambient pressure and is then eluted by the mobile phase and transported to the separation column, or a combination with a gas extraction unit. The extract of the gas extraction process can be directly passed through the chromatographic column. The separated substances can be collected. The extract from an extraction unit is diluted with respect to the interesting compounds. It can be collected on a column and after some time transported to the chromatographic separation. This operational mode is illustrated in Fig. 9.21.

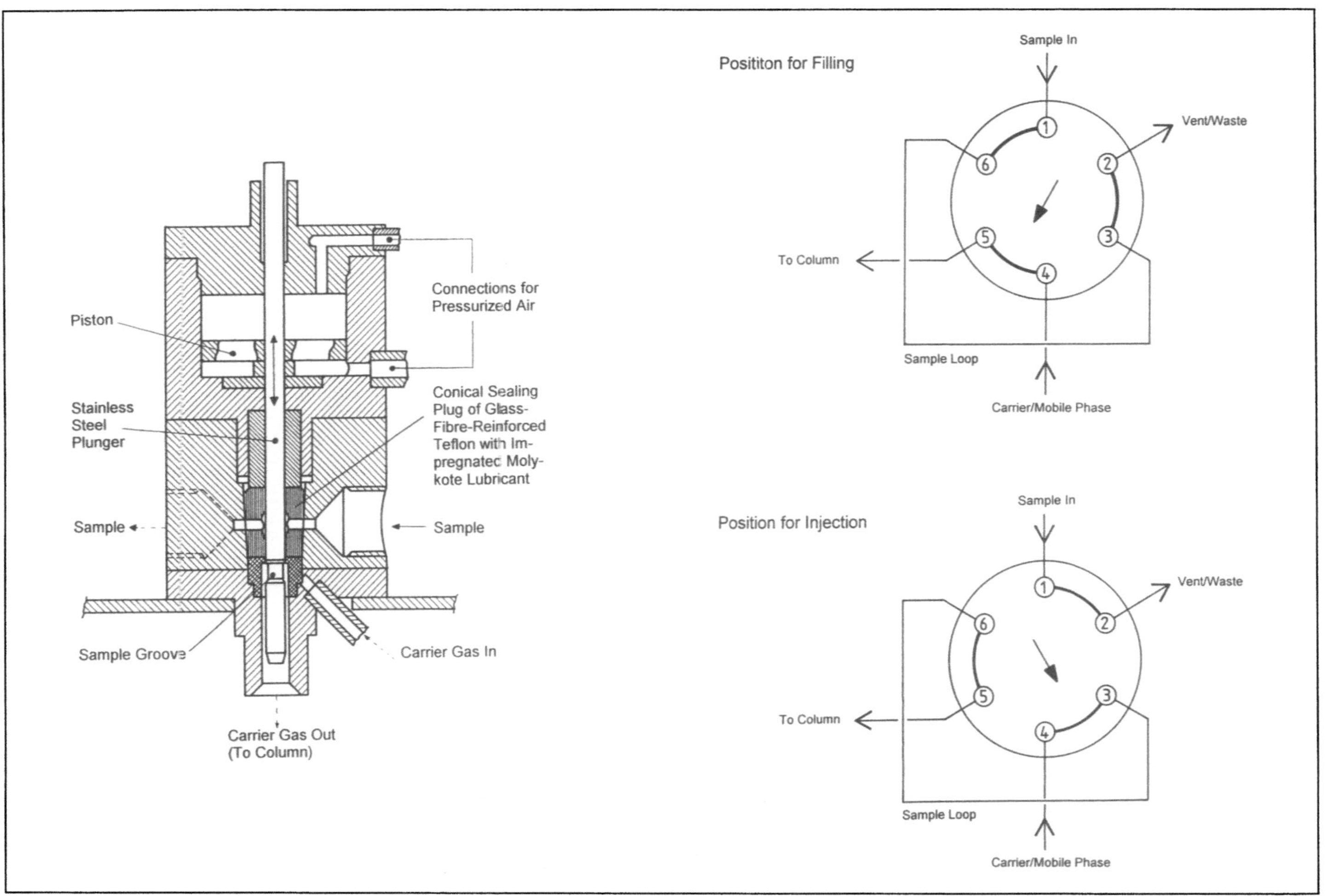


**Fig. 9.20.** Multiposition-valves for injection of samples into a SFC.

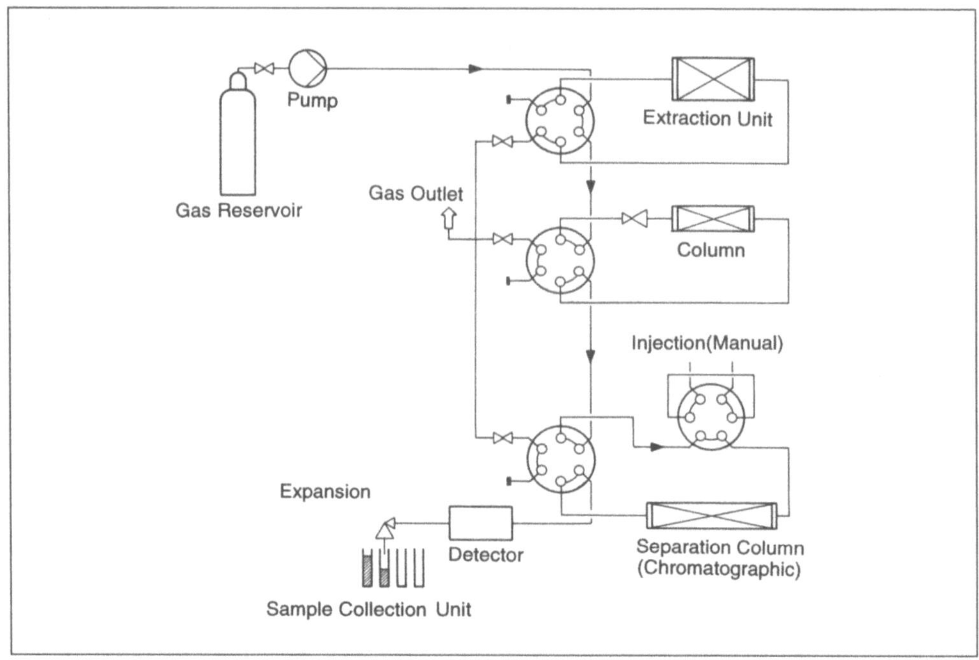



**Fig. 9.21.** Coupling of SFE with SFC. Concentration of compounds in a collecting column (after Yamauchi and Saito [42]).

## 9.2.3 Columns

Columns for chromatographic separation with supercritical gases are chosen, like other chromatographic columns, according to the needs of the separation. For analytical purposes the choice is between packed and capillary column. Capillary columns are used with a length between 10 m and 25 m. Pressure drop is low compared to liquid mobile phases. Therefore, capillary columns with inner diameters of 50 to 100 µm can be used and a high number of theoretical stages verified. Separations with capillary columns can be nearly as effective as in gas chromatography. While in early applications steel capillaries had been used in SFC, since 1980, fused silica capillary columns have replaced the steel capillaries. Stationary phases mostly stem from polysiloxanes and polyglycols. Frequently used stationary phases are listed in Table 9.1.

Under conditions of SFC, the compounds of the stationary phases may be slightly soluble in the mobile phase and are therefore fixed by linking them by chemical reactions. Numerous packed columns are available, many from HPLC applications. Normal phase chromatography (polar stationary phase, unpolar mobile phase) and reverse phase chromatography (unpolar stationary phase, polar mobile phase) are applied, but are not as important in SFC as in HPLC, since a polar mobile phase in

334

**Table 9.1.** Frequently used stationary phases in SFC.

| Composition | Trade name | Application |
|---|---|---|
| polysiloxane $(-O-\underset{R'}{\overset{R}{Si}}-)_n$ | | separation according to molecular weight |
| R, R': 100 % methyl | OV-1, SE-30 | |
| 95 % methyl, 5 % phenyl | OV-3, SE-52 | |
| 94 % methyl, 1 % vinyl, 5 % phenyl | SE-54 | |
| 25 % cyanopropyl, 50 % methyl, 25 % phenyl | OV-225 | |
| polyethylene glycol $(-CH_2-CH_2-O-)_n$ | Carbowax 20 M | separation according to polarity |

SFC involves a polar modifier. Most separations in SFC are carried out with unmodified silica gel or chemically modified silica gel as stationary phase, Fig. 9.22. For packed columns particles are available with diameters in the range of 3 to 100 μm. For analytical purposes particles in the range of 3 to 5 μm have a high separation power in a packing and enable a high linear velocity of the mobile phase leading to short retention times. For preparative purposes particles in the range of 20 to 100 μm are used. Special filling techniqes are necessary to ensure a homogeneous packing. Saito and Yamauchi [25, 26, 27] and Yamauchi and Saito [41, 42] applied columns of 7 to 20 mm diameter, Perrut [20, 21, 22] a column of 60 mm inner diameter and 600 mm length with particles of 10 – 25 μm, Alkio et al. [1] a 900-mm long column with 40 – 63 μm diameter particles.

The length of the column is dependent on the allowable pressure drop. Pressure drop usually is in the range of 1 to 4 MPa for 250 mm. In this range for the pressure

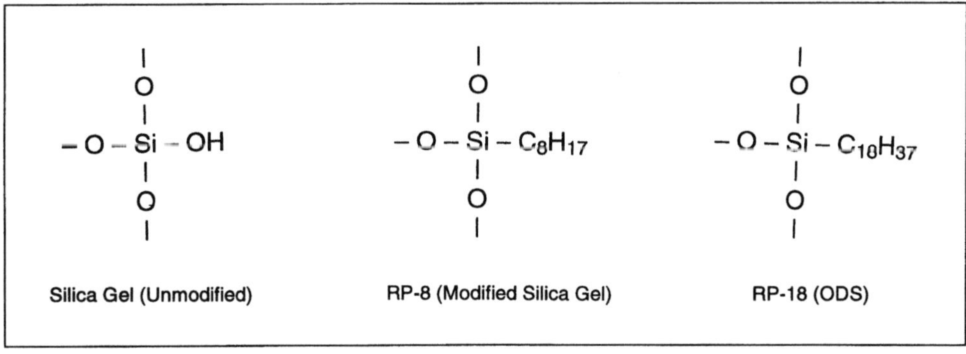

**Fig. 9.22.** Chemical structure of frequently used stationary phases in packed column SFC.

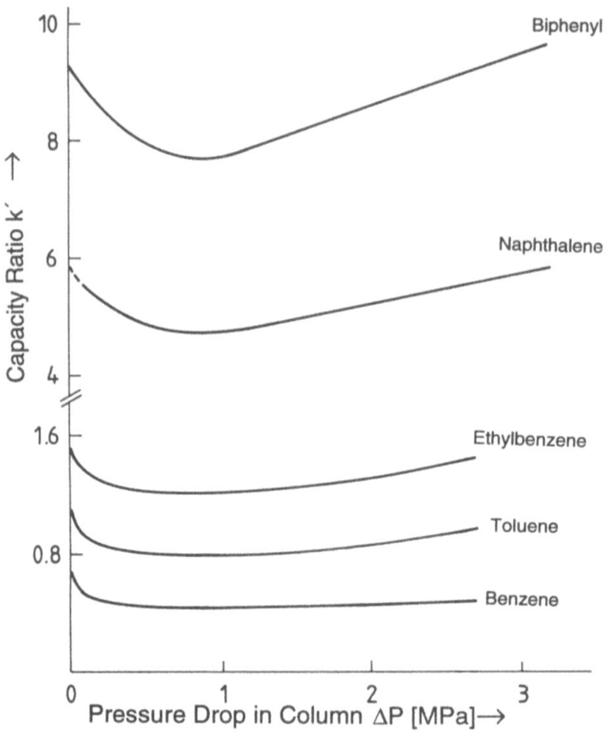

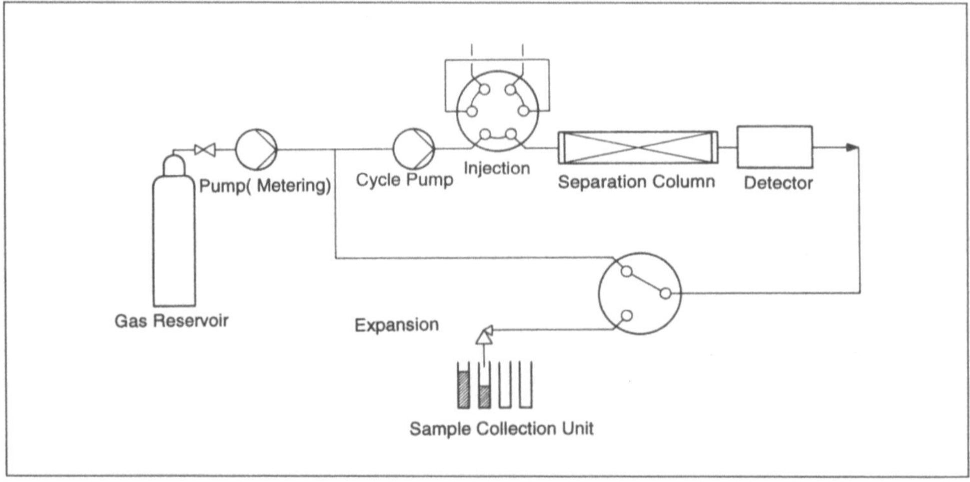


**Fig. 9.23.** Capacity ratios in SFC for various substances in dependence on pressure drop over the column (after Schoenmakers et al. [30]).



**Fig. 9.24.** Recycle chromatography after Saito et al. [25, 26, 28]. The substances of a mixture are passed several times through the chromatographic column until a sufficient separation is achieved.

336

drop capacity factors are nearly independent of pressure drop, as has been demonstrated by Schoenmakers et al. [29] (see Fig. 9.23).

To avoid unacceptable pressure drop, Saito et al. [25, 26, 28] applied a recycling technique, schematically shown in Fig. 9.24. A cycle pump transports the eluted substances several times to the beginning of the column. Thus, the separation power of the column can be enhanced, without increasing pressure drop. Peak broadening occurs due to the cyclic operations.

## 9.2.4 Detectors

Detection of a substance is necessary in analytical and preparative chromatography. In general, the same detectors are used as in gas and liquid chromatography. Selection of a detector depends on the quantity of substance available and the chemical

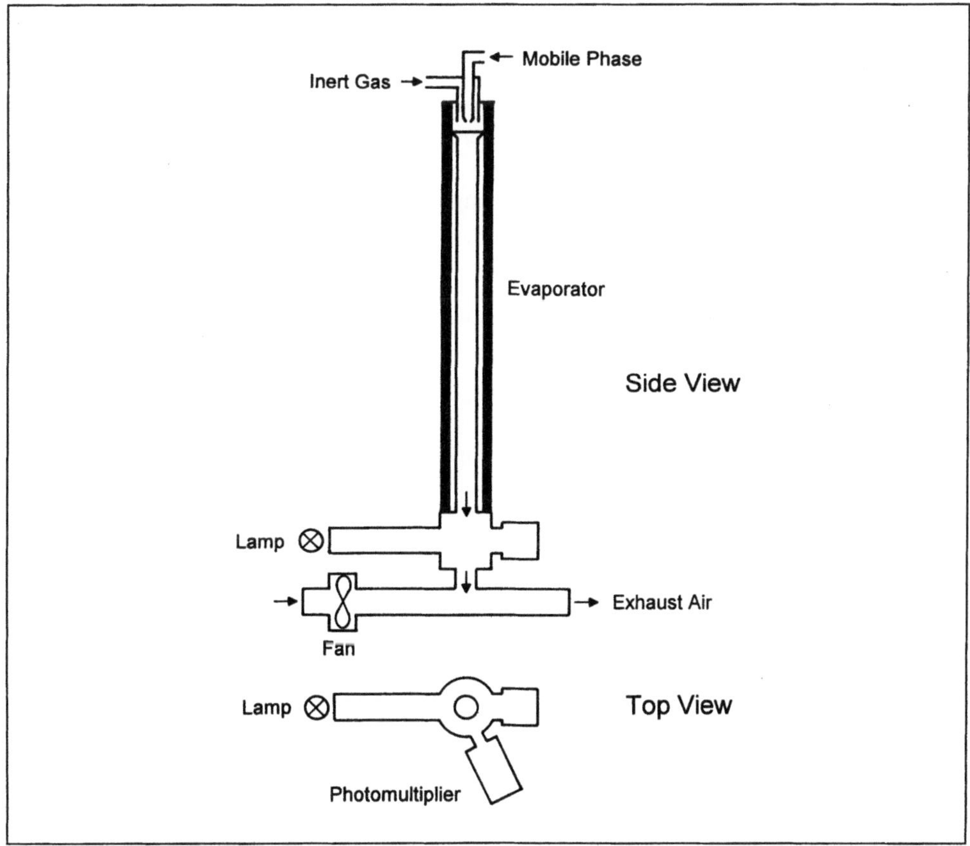


Fig. 9.25. Schematic drawing of a light scattering detector (Applied Chromatography Systems [2]).

nature of the compound. A flame ionization detector (FID) detects substances down to nanogram quantities. Between two electrodes a voltage of 300 V and a hydrogen flame are maintained. If a substance with at least one carbon-hydrogen bonding is eluted from the column to the detector, it is burned and ions are formed, which leads to a current between the electrodes. The current is amplified and processed as a signal for the concentration of the substance. Nearly all substances can be detected. Response factors mainly differ according to number of carbon atoms, therefore calibration is easy.

The ultraviolet spectroscopy detector (UV) is a nondestructive detector, which can be applied at column pressure. It is widely used, but is limited to substances with chromophoric groups. Saturated hydrocarbons, fatty acids and glycerides may be difficult to detect quantitatively. These substances may be detected with a refractive increment detector (RID), where the variation of refractive index of the mobile phase caused by dissolved substances is applied for detection. Other detectors are the fluorescence detector and the light-scattering detector.

A light-scattering detector is schematically shown in Fig. 9.25. In such a detector the mobile phase is intensively mixed with an inert gas and heated while flowing downward a tube [35, 36]. The inert gas and the temperature increase reduce solvent capacity of the mobile phase. The eluted substances precipitate and are carried as droplets or particles into the detection chamber. Into this chamber a tungsten lamp delivers visible light, which is dispersed by the droplets or particles. The dispersed light is detected by a photomultiplier under an angle of 60 degree. The signal is proportional to the mass of light-scattering particles. Therefore, the light-scattering detector acts as mass detector and its signal is independent on chromophoric groups. It can be applied for detection of chromophoric and non-chromophoric substances in a mixture, as, for example, in fatty acids and glycerides.

## 9.2.5 Expansion of the Mobile Phase and Sample Collecting System

In analytical SFC, the mobile phase is either expanded after or before detection. Downstream to a detector, which is operated under column pressure, expansion can be achieved by normal expansion valves. They can act as back pressure regulators and may be controlled by a central unit. An interesting alternative to an expansion valve was designed by Saito and Yamauchi [26], who use time-controlled opening and closing of an unrestricted tube for expansion. This has the advantage that blocking of the tube by precipitating substances is avoided. Another expansion technique was adapted from GC: A glass capillary is formed into a long, thin capillary, a so-called restrictor. Problems with blocking and difficulties with reproducible manufacturing of the restrictors are disadvantages of this solution.

In analytical SFC expansion techniques are determined by detection needs. The amount of substances is small and can easily be handled. In preparative, and more so in productive SFC, this is totally different. The quantity of the mobile phase is not small and it must be recycled. To avoid backmixing, the recycled mobile phase must be totally free from any dissolved substances. Concentrations of the dissolved substances are low. In most cases they will be in the range of 0.1 or even 0.01 %. Then, separation methods for the dissolved substances from the mobile phase become important. A discussion on methods for separating the extract from the supercritical solvent is given in Chapter 6. Here, one example proposed by Perrut [20, 21, 22] for chromatographic systems is presented in Fig. 9.26. After elution and detection, the mobile phase together with the dissolved substance is heated and expanded. By these means the solvent power of the mobile phase is reduced and the substance precipitates; it is collected in one of several collecting vessels, one for each substance. The substances are removed after sufficient quantities of each of the substances have accumulated after several injections. Before the expanded mobile phase can be recycled by a cycle pump, it is passed through an adsorbing bed, where remaining quantities of the dissolved substances and other unwanted substances (as, for example, water) are removed. As in any solvent cycle, make up gas must be added, and a small part of the solvent must be removed for disposal or for special cleaning.

In preparative SFC so far mostly extracts from plants like lemon peel oil, tocopherols from wheat germ or ubichinones have been treated. Unsaturated fatty acids from fish oil, mostly processed as esters, is a subject investigated heavily in recent years. More specialized applications deal with polymers or the fractionation of coal tar. Berger and Perrut [3] have reviewed the applications of preparative SFC.

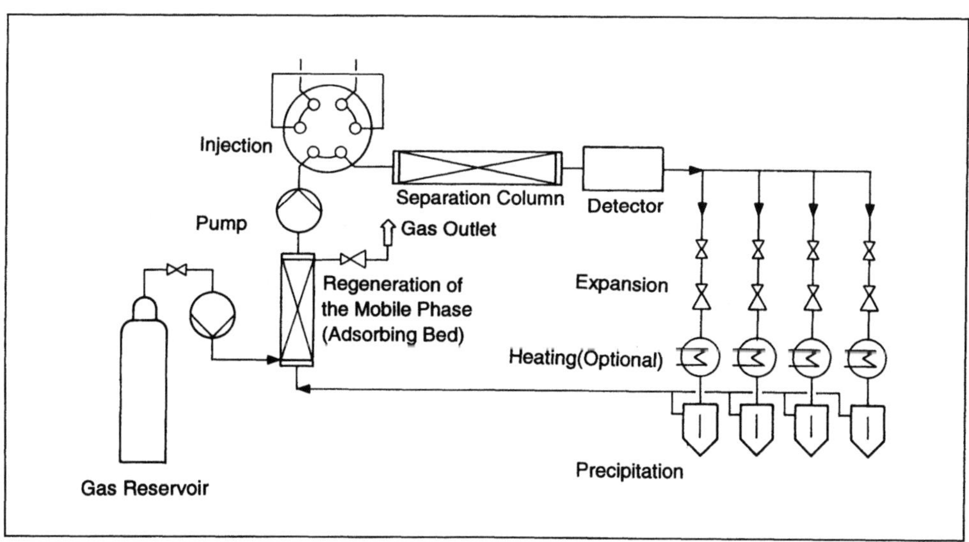


**Fig. 9.26.** Flow scheme of a preparative SFC with recyle of the mobile phase (after Perrut [20, 21, 22]).

# 9.3 Theoretical Description of SFC with Special Consideration of Scale-up

Chromatography is a separation process with two mass separation agents, the mobile and the stationary phase. The substances to be separated distribute between these two phases. As in other separation processes, the chromatographic process can be analyzed assuming repeated equilibrium distribution between the phases. Thermodynamics is then the appropriate framework to analyze chromatographic separation under equilibrium conditions. Mass transfer considerations allow conclusions to the modification of elution characteristics. From molecular properties conclusions can be drawn on molecular interactions between the two phases and the substances of the injected mixture and, therefore, on the possibility of its separation.

## 9.3.1 Fundamental Thermodynamic Equations

The following discussion follows the one given by Mollerup et al. [18].
The partition coefficient $K'$ in chromatography is defined as

$$K' = (c_{is}/c_{im}) \qquad \text{and} \qquad K' = \lim_{c_i \to 0} (c_{is}/c_{im}) \qquad\qquad 9.1$$

for finite concentrations $\qquad\qquad$ for infinite dilution,

with
$c_{is}$ = concentration, e.g., mole per volume of the solute in the stationary phase;
$c_{im}$ = concentration, e.g., mole per volume of the solute in the mobile phase.

The equilibrium partition coefficient normally used in thermodynamic analysis of mixtures is defined by the ratio of the mole fractions:

$$K'_e = (x_{is}/x_{im}), \qquad\qquad 9.2$$

with $x_{is}$ and $x_{im}$ the mole fractions of the solute in the stationary and the mobile phase. Note that the ratio of the mole fractions is inverse to conventional use in separation processes. $K'_e$ can be calculated from the ratio of the fugacity coefficients $\varphi_i$:

$$K'_e = \varphi_{is}/\varphi_{im}, \qquad\qquad 9.3$$

and the fugacity coefficient $\varphi_i$ can be calculated from:

$$RT \ln \varphi_i = (\partial A'(T,V,n)/\partial n_i)_{T,V} - RT \ln z, \qquad\qquad 9.4$$

with $n$ the number of moles of the mixture, $z$ the compressibility factor, $A'$ the residual Helmholtz-energy, which can be calculated from an equation of state:

$$A'\,(T,V,n) = -\int_{\infty}^{V}\left(P - \frac{nRT}{V}\right)\,\mathrm{d}V. \qquad 9.5$$

The difference between the two partition coefficients is also a question of the standard state. For the definition of $K,'$ the standard state is a solution, where the molar concentration is 1 or infinite dilution. This standard state depends on the solvent. The usual standard state in thermodynamics is that of a pure component and the chemical potential is written:

$$\mu_i\,(T,P,x) = \mu_i^{o}(T,P) + RT\ln(\gamma_i x_i), \qquad 9.6$$

with $\gamma_i$ = activity coefficient, which is normalized so that $\lim \gamma_i = 1$ for $x_i = 1$. Sometimes an infinite dilution standard state is defined with $\lim \gamma_i = 1$ for $x_i = 0$.

The capacity factor (capacity ratio) $k'$ is defined as the net retention time $t_s = t_r - t_m$ of the solute in the stationary phase, divided by the residence time in the mobile phase $t_m$. $t_r$ is the retention time of the solute (see Fig. 9.2). If equilibrium can be assumed, the number of moles of a solute in a phase is proportional to the residence time of the species in the phase:

$$k' = (t_r - t_m)/t_m = n_{is}/n_{im}. \qquad 9.7$$

$$k' = K\,\frac{V_s}{V_m} = \frac{K}{\beta}, \qquad 9.8$$

where $\beta$ is the volumetric phase ratio of $V_s$, the volume of the stationary phase, to $V_m$, the volume of the mobile phase.

$$k' = K'_e\,\frac{n_s}{n_m} = \frac{K'_e}{\beta_e}, \qquad 9.9$$

where $\beta_e$ is the molar phase ratio. With $v$ the molar volume of a phase and $V$ the total volume of a phase:

$$V_m = n_m v_m \text{ and } V_s = n_s v_s \text{ and } \beta = (\beta_e v_m)/v_s. \qquad 9.10$$

$$k' = K'_e\,\frac{v_m V_s}{v_s V_m} = \frac{K'_e}{\beta}\,\frac{v_m}{v_s}. \qquad 9.11$$

If the stationary phase does not swell, $V_s$ and $V_m$ are constant, and if the composition of the stationary phase does not change, $v_s$ is also constant. Then, the variation of $K'$ and $k'$ due to temperature, pressure, and density of the mobile phase can be calculated from known properties. The derivatives of $K'$ and $k'$ are equal, since the volumetric phase ratio is assumed to be constant. For the dependence on pressure and temperature the following equations can be derived:

*Temperature dependence at constant pressure:*

$$\left( \frac{\partial \ln k'}{\partial (1/T)} \right)_P = \frac{H_{im} - H_{is}}{R} - \frac{T^2}{v_m} \left( \frac{\partial v_m}{\partial T} \right)_P , \qquad 9.12$$

*Temperature dependence at constant density:*

$$\left( \frac{\partial \ln k'}{\partial (1/T)} \right)_\varrho = \left( \frac{\partial \ln k'}{\partial (1/T)} \right)_P - T^2 \left( \frac{\partial \ln k'}{\partial P} \right)_T \left( \frac{\partial P}{\partial T} \right)_\varrho \qquad 9.13$$

*Pressure dependence at constant temperature:*

$$\left( \frac{\partial \ln k'}{\partial P} \right)_T = \frac{1}{RT} (v_{im} - v_{is}) + \frac{1}{v_m} \left( \frac{\partial v_m}{\partial P} \right)_T \qquad 9.14$$

*Density dependence at constant temperature:*

$$\left( \frac{\partial \ln k'}{\partial \varrho_m} \right)_T = \left( \frac{\partial \ln k'}{\partial P} \right)_T \left( \frac{\partial P}{\partial \varrho_m} \right)_T . \qquad 9.15$$

For modeling capacity factors and partition coefficients an equation of state can be used. Phase ratio and molar volume of the stationary phase, if not known, can be treated as adjustable parameters. Mollerup et al. [18] were able to correlate capacity factors with good accuracy, as is shown in one example for a mixture of hydrocarbons in Fig. 9.27. Calculation of capacity factors for tocopherols and other interesting substances for a separation process has so far not been published.

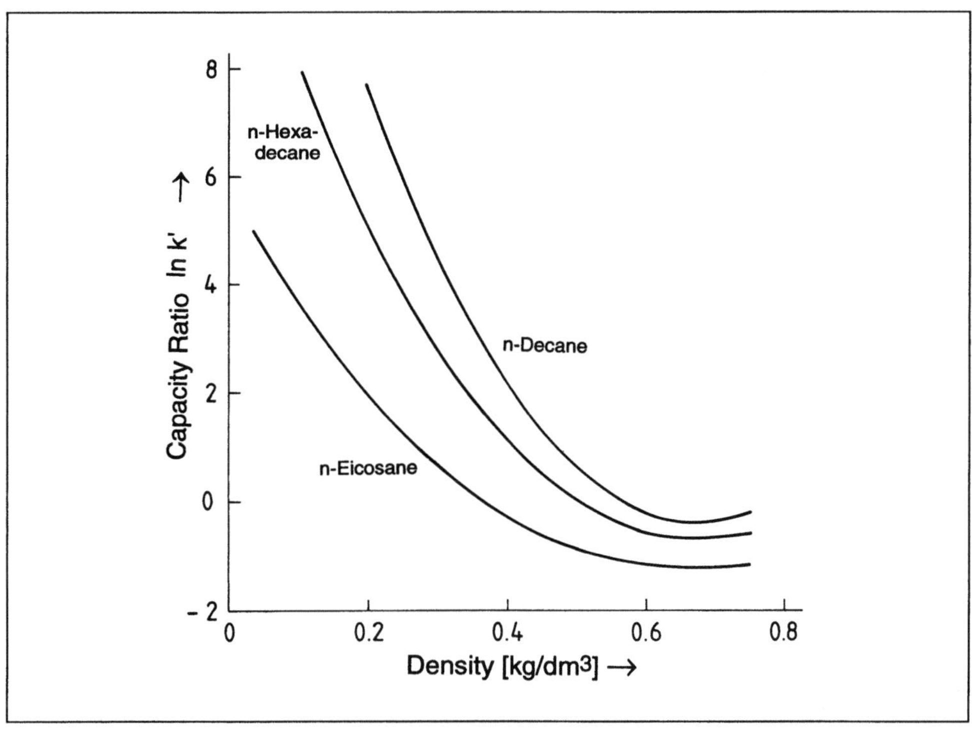


**Fig. 9.27.** Capacity factors of paraffines as a function of density (after Mollerup et al. [18]).

## 9.3.2 Number of Theoretical Stages in Chromatographic Separations

In this section an expression is derived for the number of theoretical stages for a chromatographic separation. The method is based on the assumption of multiple equilibrium distributions. The feed mixture is fed to the first equilibrium stage. Equilibrium is established between mobile and stationary phase. Then, the mobile phase is transported to the next stage, where again a distribution between mobile and stationary phase takes place. These steps are repeated while the components of a mixture migrate through the column. Due to the different distribution coefficients, different substances migrate with different speed and are therefore separated. This model is illustrated in Fig. 9.28. It consists of separate distribution vessels through which the mobile phase flows in sequence and transports the components to be separated from stage to stage. It is assumed that the stationary phase remains in each of the distribution vessels and the volume of the stationary phase in each of the vessels is equal. Thermodynamic equilibrium is established in each vessel. The indices used are $m$ for the mobile, and $s$ for the stationary phase.

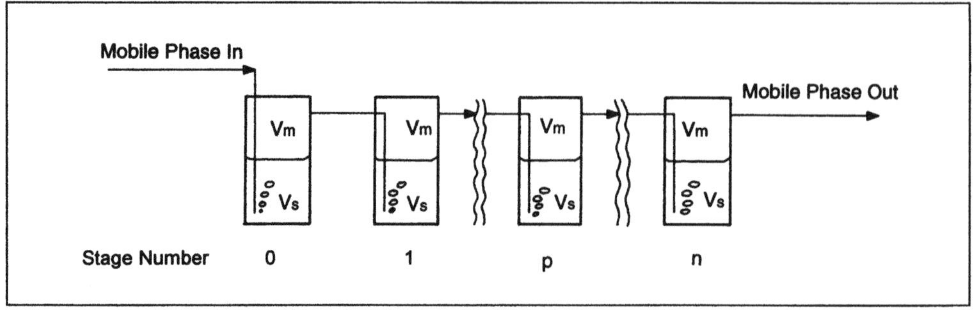


**Fig. 9.28.** Model for a chromatographic separation consisting of a sequence of vessels in which equilibrium distribution of the components between mobile and stationary phase takes place.

For stage $p$ equilibrium is determined by the following equation:

$$c_{imp} = K_i c_{isp}, \qquad\qquad 9.16$$

with $K_i$ the equilibrium distribution coefficient, $c_{imp}$ the concentration of component $i$, in moles per volume, in the mobile phase in stage $p$, and $c_{isp}$ the concentration of component $i$ in the stationary phase in stage $p$. Note that $K_i$ is defined in the usual way for separation processes, i.e., inverse to $K'$.

The change of the quantitiy of a component along the equilibrium stages can be derived from a mass balance. Since the components in the mobile phase are gases, the quantity of a substance is expressed by volume. The quantity transported from stage $p$-1 to stage $p$ minus the quantity transported with the mobile phase from stage $p$ must be equal to the change of quantity of component $i$ in the mobile and stationary phase:

$$c_{imp\text{-}1}\mathrm{d}V - c_{imp}\mathrm{d}V = V_m \mathrm{d}c_{imp} + V_s \mathrm{d}c_{isp}, \qquad\qquad 9.17$$

$$V_s \mathrm{d}c_{isp} = V_s \frac{\mathrm{d}c_{imp}}{K_i}, \qquad\qquad 9.18$$

$$\frac{\mathrm{d}c_{imp}}{\mathrm{d}V} = \frac{c_{imp-1} - c_{imp}}{V_m + V_s/K_i}. \qquad\qquad 9.19$$

This differential equation can be integrated with the initial conditions, that at zero time, at zero quantity of the mobile phase, a quantity $F$ of feed is injected to stage zero, containing $F_i$ moles of component $i$.

344

$$c_{im0}V_m + c_{is0} = c_{im0}(V_m + V_s/K_i)$$
$$c_{imp} = 0 \quad \text{for } p > 0 \Bigg\} \quad \text{for } V = 0. \tag{9.20}$$

Integration of Eq. 9.19, considering initial conditions yields:

$$c_{imp} = \frac{F_i}{V_m + V_s/K_i} \; \frac{e^{-v'_i} \, v'_i{}^p}{p!} \quad \text{with } v'_i = \frac{V}{V_m + V_s/K_i}. \tag{9.21}$$

The integrated equation is a Poisson-distribution. It is similar to a Gauss-distribution and can be approximated by it for large values of $p$. The approximation, using the Gauss-distribution is:

$$c_{imp} = \frac{F_i}{V_m + V_s/K_i} \; \frac{1}{\sqrt{2\pi p}} \, e^{-(v'_i - p)^2/(2p)}. \tag{9.22}$$

With this equation, the elution function, detected as "peak" in a chromatogram, can be determined as the concentration of component $i$ for a given stage $p$ as a function of the dimensionless cumulated volume of the mobile phase $v'_i$. Alternatively, the concentration of component $i$ along the chromatographic column for a constant cumulative volume of the mobile phase can be calculated.

If we substitute $n$, the number of theoretical stages, for $p$, the above equation is transformed to:

$$c_{im} = \frac{F_i}{V_m + V_s/K_i \sqrt{2\pi n}} \, e^{-(v'_i - n)^2/2n}. \tag{9.23}$$

This equation is plotted in Fig. 9.29. Characteristic values for the chromatographic separation can be obtained from the mathematical analysis of the function, Eq. 9.23.

Maximum of the peak:   $v'_i = n;$ $\hfill$ (9.24)

$$\text{Number of theoretical states: } n = v'_i \, \frac{V}{V_m + V_s/K_i}; \tag{9.25}$$

Points of inflection: $v'_{i,left} = n - \sqrt{n}$   and   $v'_{i,right} = n + \sqrt{n};$ $\hfill$ (9.26)

Points of intersection with the base line:

$$v'_{i,left} = n + 1 - 2\sqrt{n} \quad \text{and} \quad v'_{i,right} = n + 1 + 2\sqrt{n}; \tag{9.27}$$

Width of peak: $b = 4\sqrt{n}.$ $\hfill$ (9.28)

The cumulative dimensionless volume of the mobile phase $v'_i$ can be transformed to a time scale: $v'_i \sim t_i$, with $t_i$ for time, which is different for any component. With a coefficient $k_i$: $v'_i = k_i t_i$. With $v'_i = n_i$, the time $t_{ri}$ at which the peak maximum appears, is given by:

$$t_{ri} = (1/k_i)n_i \quad \text{or} \quad t_{ri} = k'_i n_i. \tag{9.29}$$

$t_{ri}$ is also called the retention time. From retention time and width of a peak the number of equilibrium stages can be calculated:

$$n_i = \left(\frac{4\,t_{ri}}{b_i}\right)^2. \tag{9.30}$$

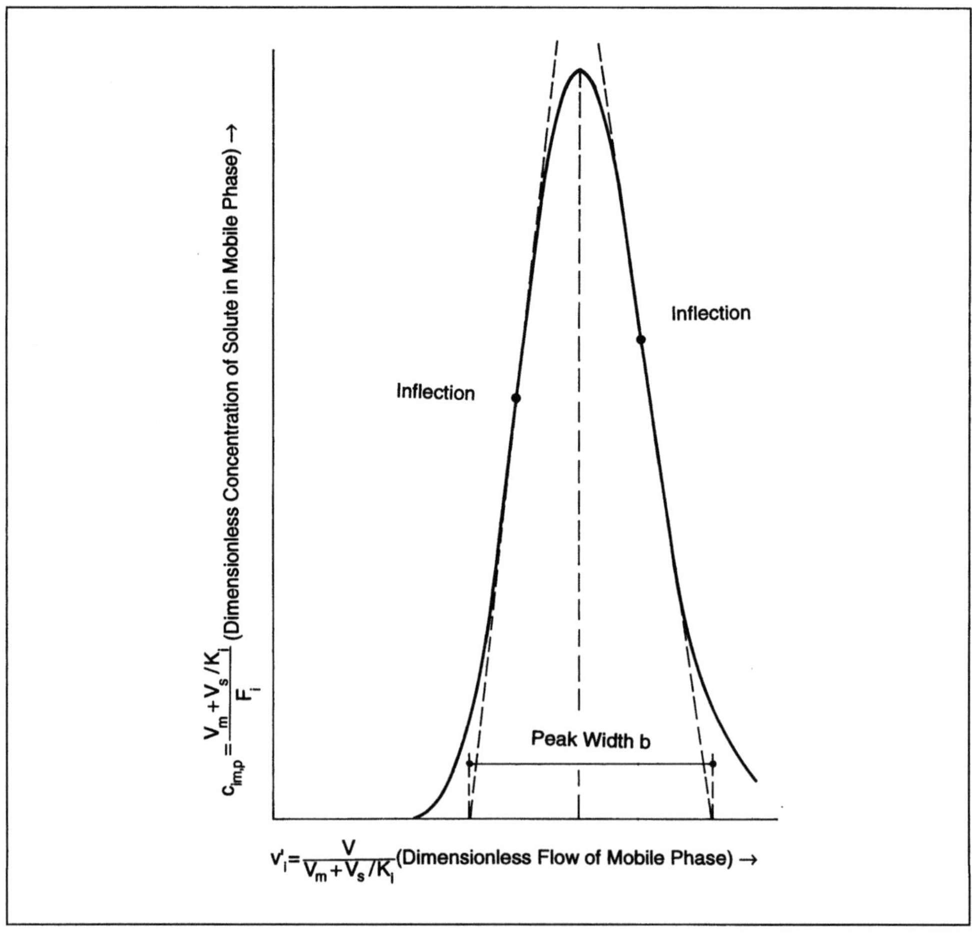


**Fig. 9.29.** Schematic drawing of a chromatographic peak.

346

The number of theoretical stages necessary for the separation of two components depends on the ratio of their capacity factors. This ratio is called selectivity $s$.

$$s_{ij} = k'_i / k'_j.$$ 9.31

In order to separate two components, a certain distance of the peak maxima compared to peak width must be obtained. This can be characterized by the resolution $R_{ij}$ between two peaks of components $i$ and $j$, which is defined by Eq. 9.32 in combination with Fig. 9.30.

$$R_{ij} = \frac{2\,(t_{rj} - t_{ri})}{b_i + b_j}.$$ 9.32

In terms of selectivity $s_{ij}$ and capacity factor $k'$, the resolution of two peaks of similar compounds with $n$ as an average number of theoretical stages can be written as:

$$R_{ij} = \frac{n^{1/2}}{4}\,\frac{s_{ij} - 1}{s_{ij}}\,\frac{k'}{1 + k'}.$$ 9.33

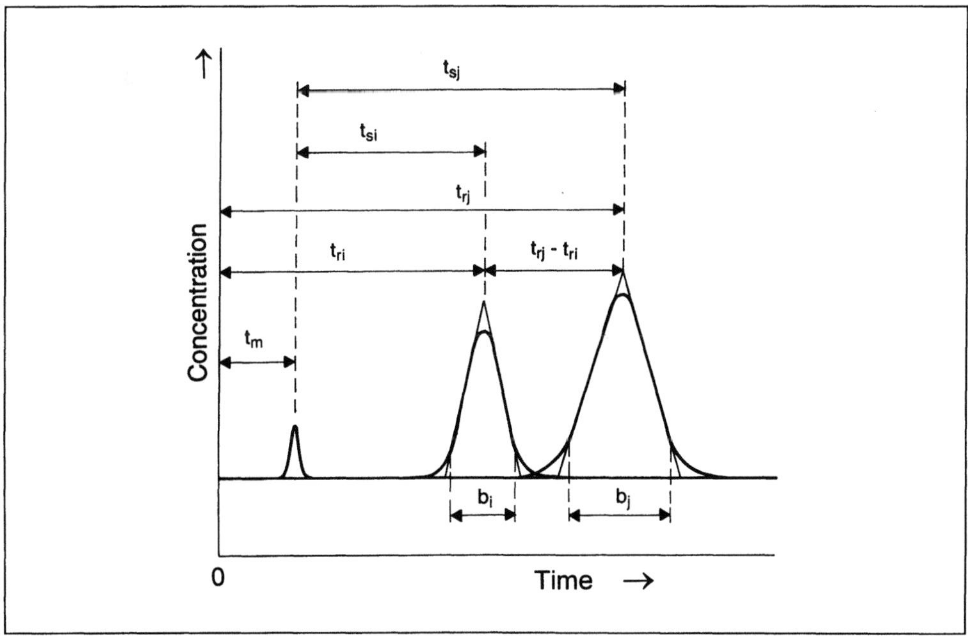


**Fig. 9.30.** Schematic chromatogram of three eluted components for defining resolution.

A satisfactory resolution is obtained for $R_{ij} \geq 1$, a baseline resolution is obtained for $R_{ij} \geq 1.5$.

If Eq. 9.33 is rearranged, the number of theoretical stages $n$ necessary for a separation of two components in dependence of resolution and selectivity can be calculated by Eq. 9.34. This equation is plotted in Fig. 9.31 for different values of $R_{ij}$ and $k'$.

$$n = 16\, R_{ij}^2 \left(\frac{s_{ij}}{s_{ij}-1}\right)^2 \left(\frac{1+k'}{k'}\right). \qquad\qquad 9.34$$

A capacity factor $k' = 2$ means that peak retention is twice that of holdup volume. From Fig. 9.31 it can be concluded, that minimum selectivity for SFC is about 1.03, for GC about 1.02, and for LC about 1.10.

With the number of theoretical stages, the equivalent height of a theoretical stage $H_s$ can be calculated with column length $l$ to:

$$H_s = l/n. \qquad\qquad 9.35$$

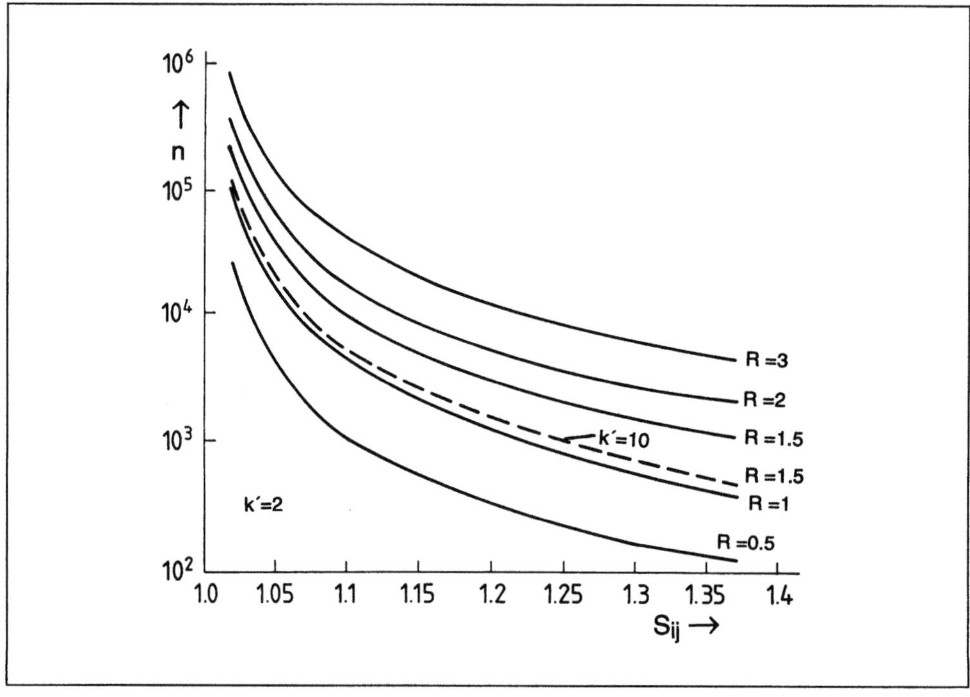


**Fig. 9.31.** Number of theoretical stages for a chromatographic separation in dependence of capacity ratio $k'$ and resolution $R_{ij}$.

The height of a theoretical stage $H_s$ for packed and capillary columns can be modeled by taking into account hydrodynamic characteristics of the flow in the column, as was proposed by van Deemter [6]:

$$H_s = 2\lambda d_p + \frac{2\gamma D_{im}}{u} + \frac{8\,k_i' d_F^2}{\pi^2\,(1 + k_i')\,D_{is}}\,u, \qquad\qquad 9.36$$

with
$\lambda$   = factor for the unequal distribution of particles in the packing;
$d_p$  = particle diameter;
$\gamma$   = tortuosity factor for channels in the packing;
$D_{im}$ = diffusion coefficient of component $i$ in the mobile phase;
$u$   = linear velocity of mobile phase;
$k_i'$  = capacity ratio of component $i$;
$d_F$  = film thickness of the stationary phase;
$D_{is}$ = diffusion coefficient of component $i$ in the stationary phase.

Equation 9.36 is schematically plotted in Fig. 9.32, where the influence of the different terms is illustrated. The first term takes into account the different length of paths, as the mobile phase flows in a packed column. This term becomes smaller with

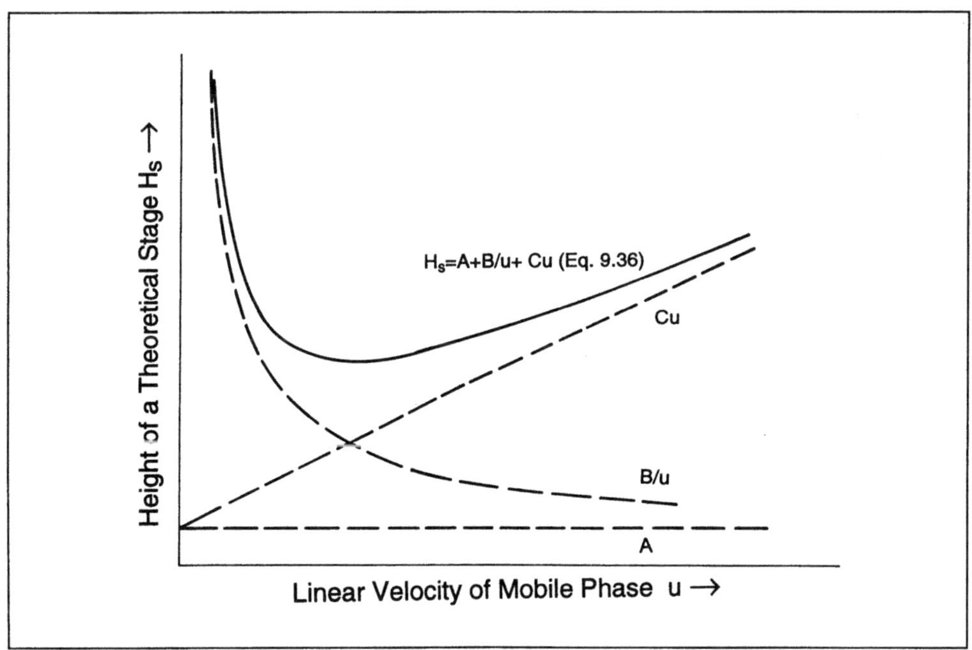


**Fig. 9.32.** Representation of the height of a theoretical stage $H_s$ by the van Deemter-equation [6].

smaller particles and more regular packings. The first term is zero for capillary columns. The second term considers axial diffusion in the column. Its influence decreases with increasing linear velocity. The third term deals with relaxation effects during distribution of feed components between mobile and stationary phase. It increases linearly with velocity of the mobile phase.

The three terms are added to the height $H_s$ of a theoretical stage in dependence of linear velocity of the mobile phase, shown in Fig. 9.32. At low velocities $H_s$ is high. The separational power of the column is not effectively used. $H_s$ decreases with increasing linear velocity of the mobile phase. But increasing relaxation effects lead to a minimum for $H_s$ and the increase of $H_s$ for higher linear velocities.

The minimum in the van Deemter-curve is about the optimum point for operating a chromatographic separation column. Yet actual operating conditions depend on the real shape of the curve. In Fig. 9.33 such curves are shown for SFC and HPLC for different particle diameters. For SFC, the minimum of $H_s$ is lower at higher linear velocities than for HPLC. Furthermore, the minimum is flat for SFC, thus allowing to choose an even higher linear velocity than that at the minimum, without losing much in separational effectivity.

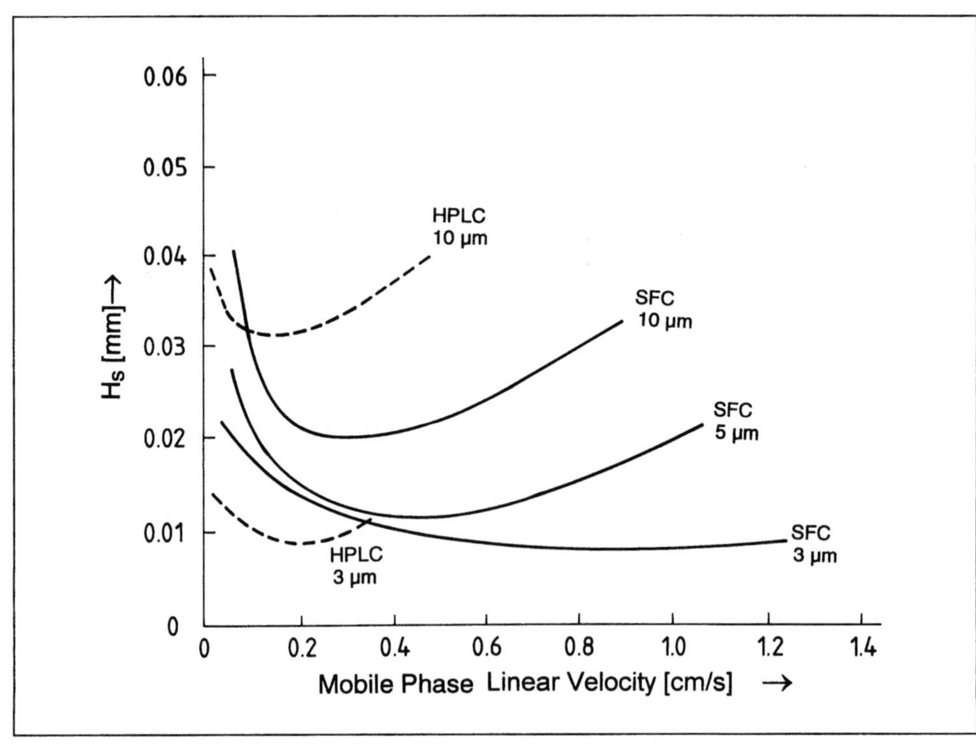


**Fig. 9.33.** Height of theoretical stage $H_s$ for SFC and HPLC for packed columns with different particle diameters (after Gere et al. [8]).

This derivation is valid for analytical concentrations and quantities. Scaling up chromatographic separations is not straightforward, since it depends on repeated distribution of components between (the surface of) a stationary phase and a mobile phase, for which it is difficult to maintain optimum conditions in a larger scale. In the mobile phase no backmixing and no short cut flows should disturb plug flow, in the stationary phase all interactions between the molecules of the stationary phase and one solute component in the mobile phase should be equal. Quantities of a substance distributing at one site should be small, since the distribution should be achieved for all molecules of one species at the same time. These conditions are difficult to achieve with increasing amounts of feed.

### 9.3.3 Theories for Scale-up

Theories for scaling up chromatographic separations with supercritical gases stem from HPLC, but should be applicable in SFC, especially if carried out in the near critical region. Two theories will be reviewed and applied to the separation of tocopherols and prostaglandins, which are discussed in more detail in the following section.

For scaling up, theories for the elution of substances must be modeled for overloading conditions, the number of theoretical stages has to be determined, and rules for maximizing throughput have to be defined.

The first theory to be discussed here, was presented by Knox and Pyper [12]. It is based on modeling the geometric shape of the eluted peaks. Overloading by volume and by concentration are treated separately.

*Peaks under conditions of overloading*

**Overloading by volume.** Under conditions of overloading by volume it is assumed that the feed at the starting point of the chromatographic separation can be described by a rectangular concentration profile. The width $b_{in}$ of this injected profile is then defined by:

$$b_{in} = \frac{V_{in}}{A_m\,(1 + k')}, \qquad\qquad 9.37$$

with $A_m$ for the free area of a cross-section in the chromatographic column, through which the mobile phase flows and $V_{in}$ the injected volume. For characterizing the concentration in the column, the injected quantity $Q$, divided by the area of a cross-section of the column $A$ is defined: $q = Q/A$. The initial concentration $c_0$ can then be calculated from the injected quantity and the length of the injected profile $z_{in}$:

$$c_0 = q/z_{in}. \qquad\qquad 9.38$$

For a peak in an analytical injection, the concentration of the eluted substance in the maximum of the peak, $c_{max}$, is:

$$c_{max} = q/\sqrt{2\pi H_s l}, \qquad\qquad 9.39$$

with $H_s$ as the height of a theoretical stage and $l$ column length measured from the point of injection. For the injection of higher quantities,

$$z_{in} > 8/(H_s l)^{0.5}, \qquad\qquad 9.40$$

corresponding to about 20 times the analytical injected profile length, peak maxima are plateaus with a maximum concentration of $c_0$.

**Overloading by concentration.** At higher concentrations, thermodynamic equilibrium must be considered, in addition to axial dispersion, since the capacity ratio of a component may not be constant within a peak. For the leading front and the tailing front of a peak different conditions may be valid. The capacity ratio, written with the volumes of mobile ($V_m$) and stationary ($V_s$) phases and the concentrations of component $i$ in the mobile ($c_{im}$) and stationary phase ($c_{is}$):

$$k'_i = (V_{is}c_{is})/(V_{im}c_{im}). \qquad\qquad 9.41$$

For low concentrations, the ratio of the concentrations $c_{is}/c_{im}$ is constant, if the ratio of the phases is assumed to be constant. This corresponds to the initial part of an isotherm, which is linear or approximately linear. At higher concentrations, the linear region of the adsorption isotherm is left and the ratio depends on concentration (see Fig. 9.34).

The capacity ratio of the peak front is characterized by Eq. 9.42, which corresponds to the secant in Fig. 9.34.

$$k'_{i,front} = c_{is}/c_{im}. \qquad\qquad 9.42$$

For the tail of the peak, a differential quotient is used, corresponding to the tangent in Fig. 9.34, since the capacity ratio changes faster with concentration than at the leading front:

$$k'_{i,tail} = dc_{is}/dc_{im}. \qquad\qquad 9.43$$

The maximum volume which can be injected without significant overlapping of peaks is calculated from the volume of mobile phase necessary for elution $V_{ir}$ and the standard deviation $\sigma_i = \sqrt{n}$, defined by a Gauss-distribution for the peak, eluted under analytical conditions. The peak width at the baseline is $4\sigma_i$:

$$V_{in,max} = (V_{ir} - V_{jr}) - \sqrt{2\pi}\,(\sigma_i + \sigma_j). \qquad\qquad 9.44$$

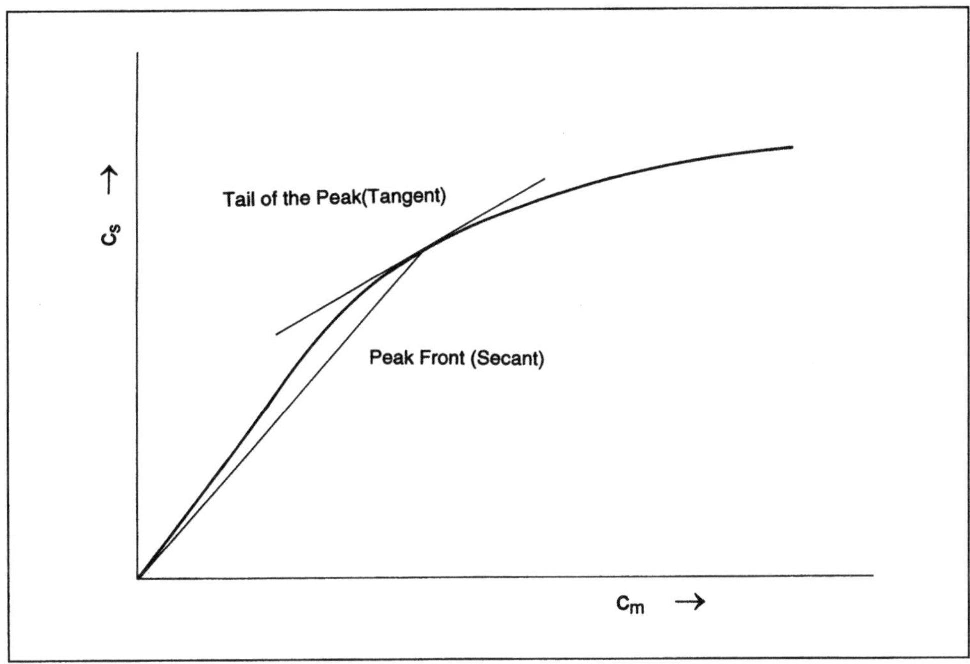


**Fig. 9.34.** Example of an adsorption isotherm. Capacity ratios $k'$ are different for the leading and the tailing front due to different concentrations.

In the next step, the equations for an isotherm must be connected with the peak front by means of the capacity ratio. The ratio of elution time for the front of a peak and the solvent is named the $R_F$-value. $R_{F0}$ is then the $R_F$-value for small (analytical) injection volumes. The elution profile as a function of the $R_F$-value is called $R\Delta$, the elution profile as a function of retention time or retention volume is called $k\Delta$, as illustrated in Fig. 9.35.

A Langmuir-type isotherm can be represented by the following equation:

$$c_{is} = \frac{k_0' \, c_{im}}{1 + a c_{im}}, \qquad\qquad 9.45$$

with $a$ the characterizing parameter of the *Langmuir*-isotherm and $k_0'$ the capacity ratio at analytical conditions.

The $R\Delta$-profile can be written as:

$$R_F = R_{F0} \left( 1 + 2 \, \frac{k_0'}{1 + k_0'} \, a \, c_{im} \right). \qquad\qquad 9.46$$

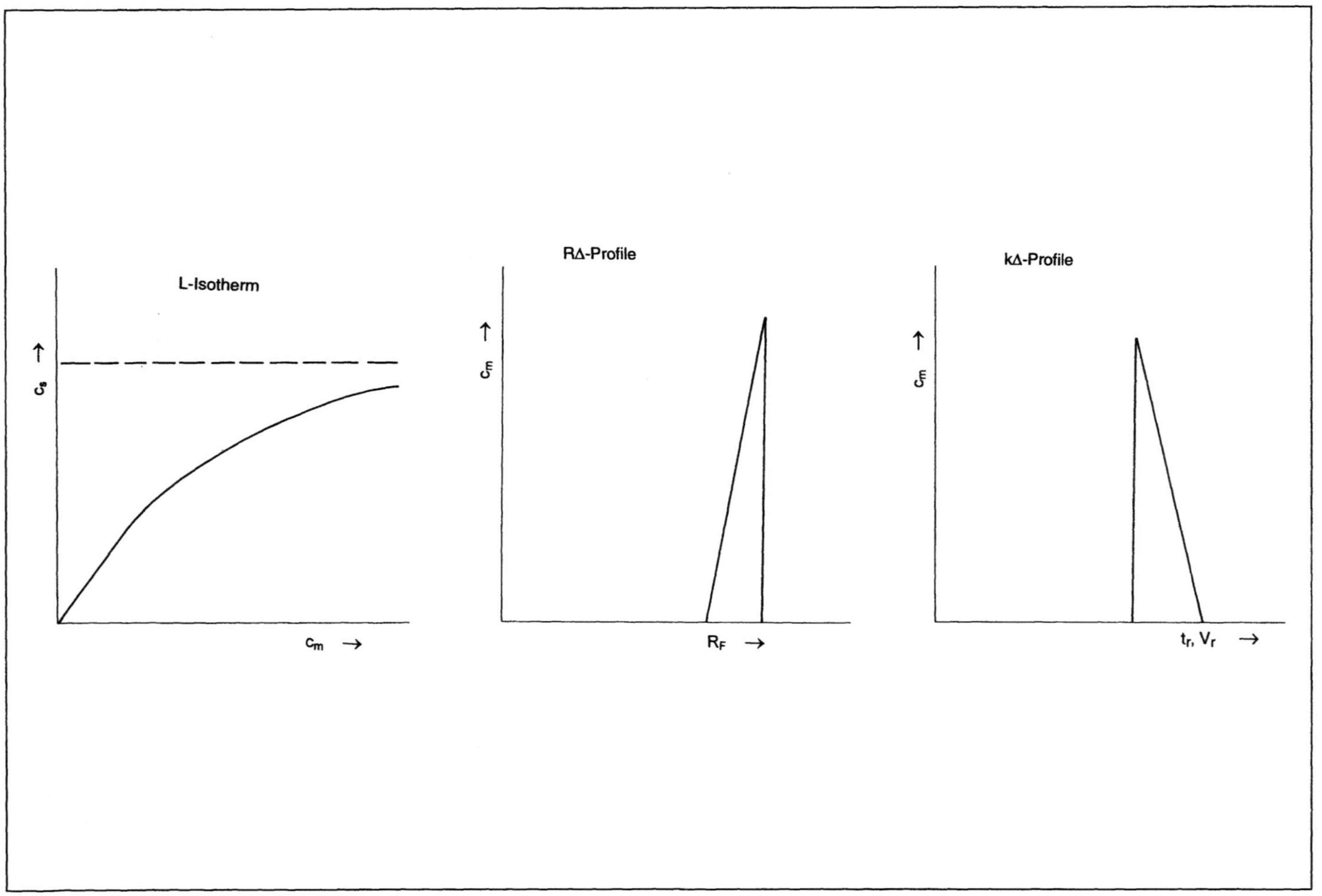


**Fig. 9.35.** Adsorption isotherm and elution profiles in scale-up theory of Knox and Pyper [12].

The $R_F$-value can be calculated from the capacity ratio:

$$k' = (1/R_F) - 1 \quad \text{and} \quad R_F = 1/(1 + k'). \tag{9.47}$$

The amount of substance in the stationary phase can be found by integrating Eq. 9.43:

$$c_{is} = \int_0^{c_{im}} k'\, dc_{im}. \tag{9.48}$$

The elution profiles can be approximated by developing the equation into a series, using the following terms:

$$\psi = k_0'/(1 + k_0') \quad \text{and} \quad R_{F0} = 1/(1 + k_0'). \tag{9.49}$$

Now, the isotherm, the $R\Delta$-, and the $k\Delta$-profile can be written as:

$$c_{is} = k_0' c_{im} (1 - ac_{im} + (4/3)\psi(ac_{im})^2 - 2\psi^2(ac_{im})^3 + \ldots), \tag{9.50}$$

$$R_F = R_{F0} (1 + 2\psi ac_{im}), \tag{9.51}$$

$$k' = k_0' (1 - 2ac_{im} + 4\psi(ac_{im})^2 - 8\psi^2(ac_{im})^3 + \ldots). \tag{9.52}$$

The quantity of a substance $Q$, represented by a peak, can be obtained by integrating the equation for the concentration, using $c_{*im}$ for the maximum concentration in the mobile phase and $l_F$ for the length, the peak front has been transported:

$$Q = 2 \int_0^{c_{*im}} [(c_{im} + c_{is})\, l_F R_{F0}\, ac_{im}]dc_{im} \tag{9.53}$$

Inserting the approximate series for $c_{is}$ and integrating results in:

$$Q = (l_F/ac_{im})(ac_{im}c_{*im})^2 [1 - (2/3)(ac_{im}c_{*im}) + (2/3)(ac_{im}c_{*im})^2$$
$$- (4/5)(ac_{im}c_{*im})^3 + \ldots]. \tag{9.54}$$

Since $Q$ is constant, $c_{*im}$ decreases with $l_F$. The peak gets broader and flatter. The width of the peak is $2\, ac_{im}c_{*im}l_0$, with $l_0$ the length, which a peak of analytical quantity is transported. The number of theoretical stages $n$ is then calculated from:

$$n = (f\, l^2)/(2ac_{im}c_{*im}l_0)^2, \tag{9.55}$$

with $f$ as a factor defining the shape of the peak and $l$ for the length of the column. For a Gauss-peak, $f = 16$, which is the best value also for triangular peaks. The height of a theoretical stage $H_s$ increases with quantity $Q$:

$$H_s = (1/4)\, ac_{im}\, R_{F0}\, Q. \qquad\qquad 9.56$$

The equation for the throughput $N$ can be written as:

$$N = \frac{4}{ac_{im}R_F} \left[ \frac{l}{d_p N^*} - h(v) \right] \frac{d_p \Delta p}{(L/d_p)^2\, \varphi\, \eta}, \qquad\qquad 9.57$$

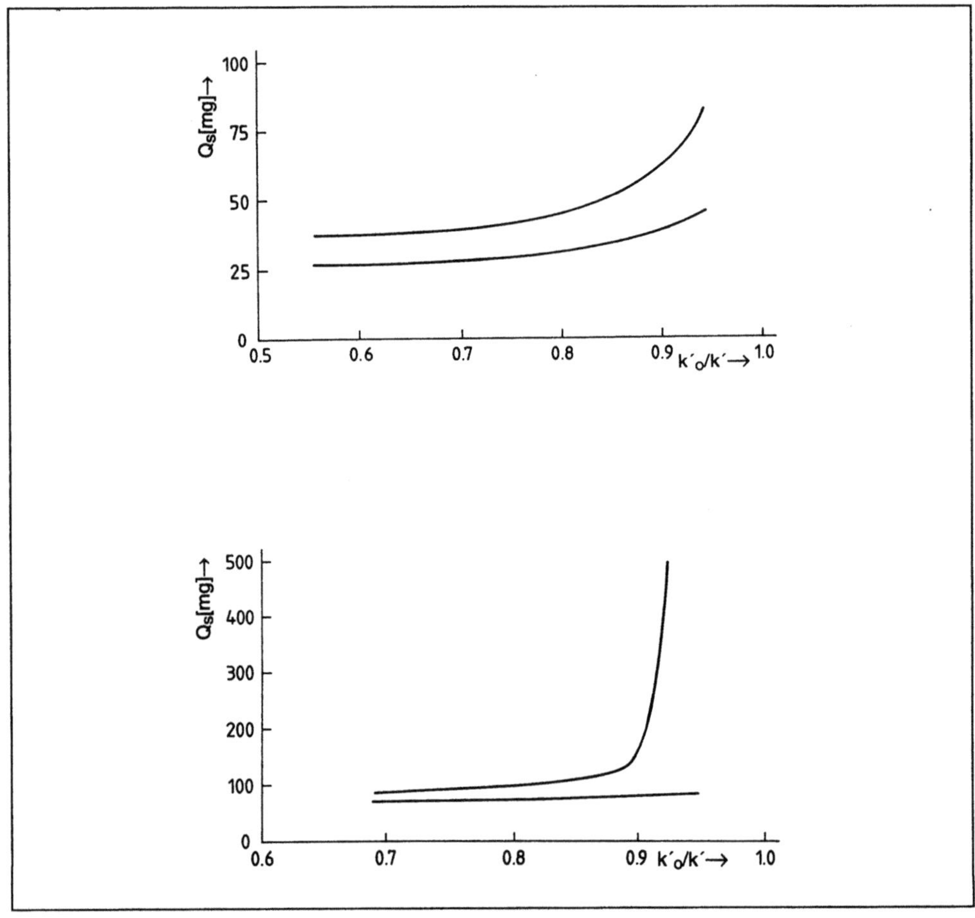

Fig. 9.36. Saturation capacity of a chromatographic column as a function of the capacity ratio under overloading conditions to the capacity ratio under analytical conditions [38].

356

with:
$l$ = column length, $d_p$ = particle diameter, $N^*$ = throughput for which the peaks are touching one another, $\Delta p$ = pressure drop across the column, $\eta$ = viscosity of the mobile phase, $\varphi$ = flow resistance coefficient, and $h(v)$ a modified van Deemter-equation:

$$h(v) = Av^{1/3} + B/v + Cv,\qquad\qquad 9.58$$

with

$$v = \frac{d_p^2\,\Delta p}{l\,D_m\varphi\,\eta},\qquad\qquad 9.59$$

where $D_m$ is the binary diffusion coefficient in the mobile phase. Since in preparative chromatography high linear velocities are applied, the above Eq. 9.58 can be approximated by: $h = Cv$.

Eble et al. [7] and Snyder et al. [32] published a theory on preparative chromatography, which allows to calculate the saturation capacity $Q_s$ of a column. This is the amount of a substance which can be injected to a column before the substance is eluted at the other end of the column.

$$Q_s = (2n)^{1/2}\,Q_i[k_0'/(1 + k')][k_0'/(k' - k_0')],\qquad\qquad 9.60$$

or:

$$Q_s = n_0(k_0'/(1 + k_0'))^2\,(Q_i/Q_n),\qquad\qquad 9.61$$

with $Q_i$ for the injected quantity of a substance and $Q_n$ as a loading parameter. For the dependence of the number of theoretical stages on the injected quantity $Q$, empirical Eq. 9.62 was derived from experimental data.

$$n/n_0 = 1/[1 + (3/8)Q_n]\qquad\qquad 9.62$$

$$\text{with } Q_n = n_0[k_0'/(1 + k_0')]^2\,\frac{Q_i}{Q_s}.\qquad\qquad 9.63$$

Results for $\alpha$-tocopherol are shown in Figs. 9.36 and 9.37. The saturation capacity is plotted against the capacity ratio for preparative and analytical conditions. For higher overloading (decreasing ratio of $k_0'/k'$), the resulting saturation capacity becomes constant and results of the two equations coincide within reasonable limits. For values of $k_0'/k' > 0.85$ the equations are not applicable. The saturation capacity has been determined for columns of different sizes. The saturation capacity (for $\alpha$-tocopherol) seems to increase linearly with the total volume of the column.

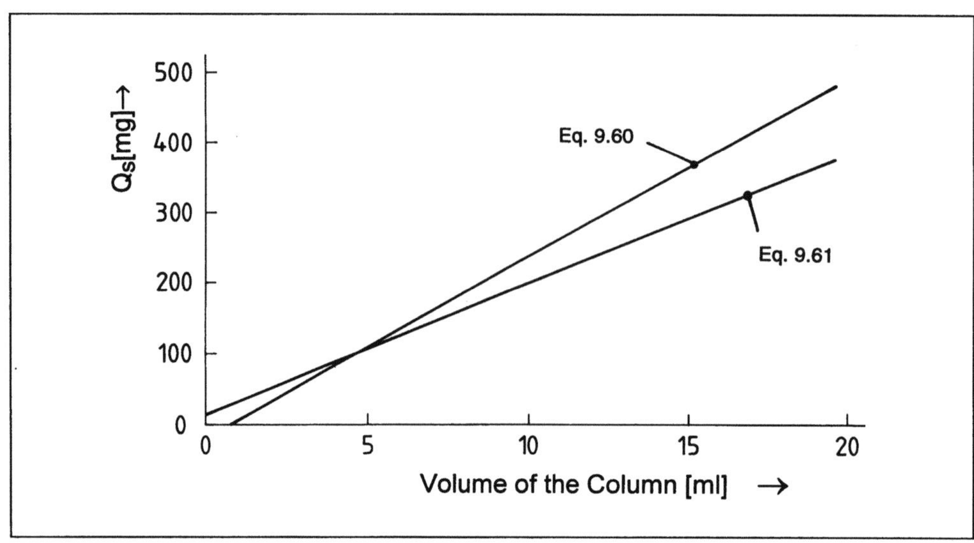



**Fig. 9.37.** Dependence of saturation capacity on total volume of a chromatographic column. Example: $\alpha$-tocopherol [38]. Column: LiChrosorb Si60; mobile phase: $CO_2$/methanol 0,5 wt.-%; average pressure: 15 MPa; temperature 293 K.

# 9.4 Supercritical Fluid Chromatography (SFC) on a Preparative Scale: Two Examples

## 9.4.1 General Approach

**Method.** Chromatographic separation on a preparative scale will be considered for the elution mode. For developing the separation, first conditions for an analytical separation are optimized and then under these conditions, separation with overloading is studied.

**Selection of feed mixtures.** The choice of a feed mixture depends on the intention of the investigation. For investigating general properties a model mixture of substances that is easy to handle and detect, and nonexpensive is sufficient. For proving the applicability of a process, a model mixture is not sufficient, since transfer of the results in general is not possible. A real mixture with all the difficulties of reality is more appropriate. For chromatographic separations the substances should be of some value, because chromatographic separations are expensive. In the following, scaling up of SFC is reviewed for tocopherols as a mixture of unpolar compounds and prostaglandins as example for polar substances, after Upnmoor [38].

**Apparatus.** For the investigation a SFC apparatus was employed which was assembled from commercially available components. Its flow scheme is presented in Fig. 9.38. Components for delivering and controlling the mobile phase comprised mod-

358

ules for pressurization, pressure control, and expansion. They were supplied by NWA (New Ways of Analytic, Lörrach, FRG). In the pressure enhancement module the mobile phase is liquefied in a cooled reservoir and pressurized by a pneumatically driven pump and delivered to the pressure control module, where the liquefied-gas flow is expanded in a controlled pressure reducer to operating pressure. This expansion simultaneously acts as pulsation dampener. Modifiers were pumped by an HPLC-pump (type LC-6 A, Shimadzu) and were mixed in-line with the mobile phase. The mobile phase was then heated to operating temperature in a coiled tube (1.65 mm inner diameter) of 2 m length. Samples were injected to the chromatographic column with a multiposition valve (Rheodyne 7010). The sample loop of the valve was filled with a µl-syringe. Substances were detected with a UV/VIS detector (SPD-6 AV, Shimadzu). The mobile phase was expanded in two stages within the expansion module. Modifiers and dissolved compounds precipitate during expansion and are separated from the gas in a gravitational separator. The liquid formed by expansion is removed after certain time intervals. The gas then flows to a mechanically operated gas flow controller, which acts as a second expansion step and allows to maintain a constant flow of gas through a metering valve, independent of the pres-

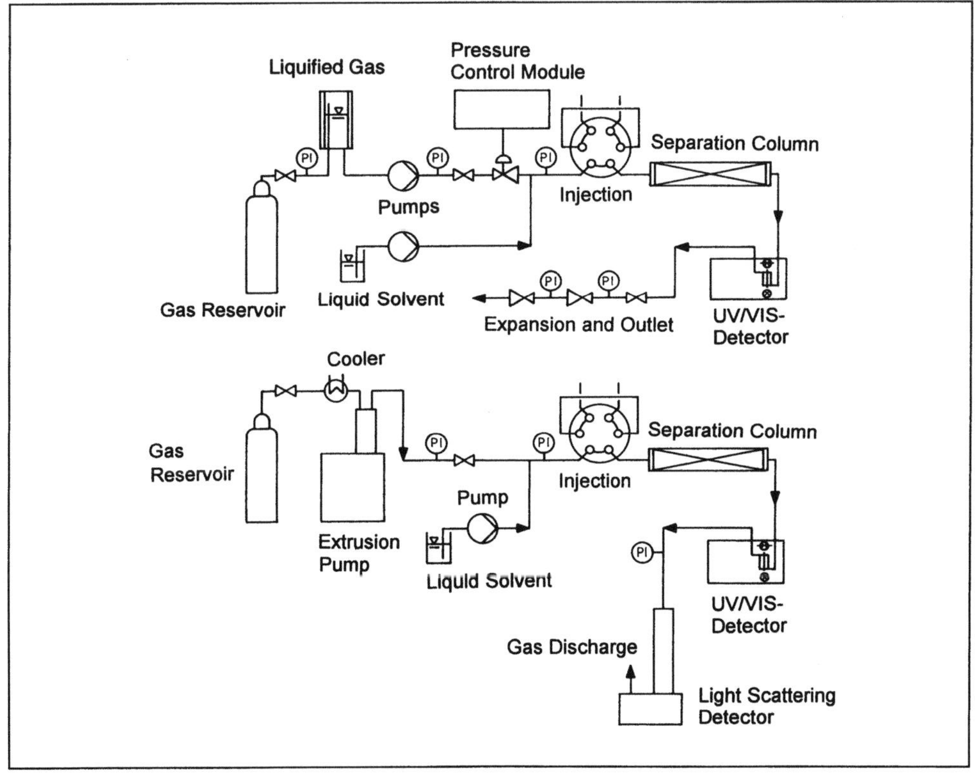



**Fig. 9.38.** Flow scheme of SFC apparatus [38].

sure on the high pressure side (type 8802, Brooks Instruments). All parts in contact with the mobile phase under pressure were manufactured from stainless steel or teflon.

Chromatographic columns are listed in Table 9.2. All columns were commercially available HPLC packed columns. LiChrosorb is an unmodified silica gel with irregular particles. Columns with RP-8 and RP-18 are modified on the surface by octyl- or octadecyl-groups. Ultrasphere ODS is an octadecyl modified silica gel with spherical particles. RP-select B is an octyl modified silica gel in the form of spherical particles with a low activity against basic substances.

**Table 9.2.** Chromatographic columns used for analytical and preparative chromatographic separations of tocopherols and prostaglandins [38].

| Length [mm] | Inner diameter [mm] | Particle diameter [mm] | Packing |
|---|---|---|---|
| 125 | 4 | 5 | LiChrosorb Si 60 |
| 250 | 4 | 5 | LiChrosorb Si 60 |
| 250 | 4.6 | 10 | LiChrosorb Si 60 |
| 250 | 10 | 7 | LiChrosorb Si 60 |
| 250 | 4 | 5 | LiChrosorb RP-8 |
| 250 | 4 | 5 | LiChrosorb RP-18 |
| 250 | 4 | 5 | LiChrospher RP-Select B |
| 250 | 4.6 | 5 | Ultrasphere ODS |

## 9.4.2 Separation of Prostaglandins

Prostaglandins are unsaturated hydroxy(keto)carboxylic acids with 20 carbon atoms, characterized by a ring with five carbon atoms (Fig. 9.39). The active substance is the $\alpha$-cis-isomer. Prostaglandins were available "pure" (reprodin, 88 % $\alpha$-cis-, 10 % B-isomer) and as mixture of isomers with 61 % B-, 18 % $\alpha$-cis-, < 1 % $\alpha$-trans- and 17 % $\beta$-cis-isomer. Prostaglandins were dissolved in ethylacetate for injection.

Prostaglandins are separated with HPLC on unmodified silica gel. Separation experiments with reversed phase packing materials were not successful. In analytical and preparative separations [10, 15, 16, 17] mobile phases based on hexane have been used. Such normal phase chromatographic separations can easily be transferred to SFC. In SFC, carbon dioxide is used instead of hexane as main component of the mobile phase. Figure 9.40 illustrates the SFC separation between the prostaglandin isomers. With increasing amount of modifier, capacity factors and therefore retention times decrease as well as selectivity (see Fig. 9.41). An increasing amount of modifier inactivates the surface of the silica gel by occupying sites of strong interaction and enhances interactions of the mobile phase with the solute by increasing

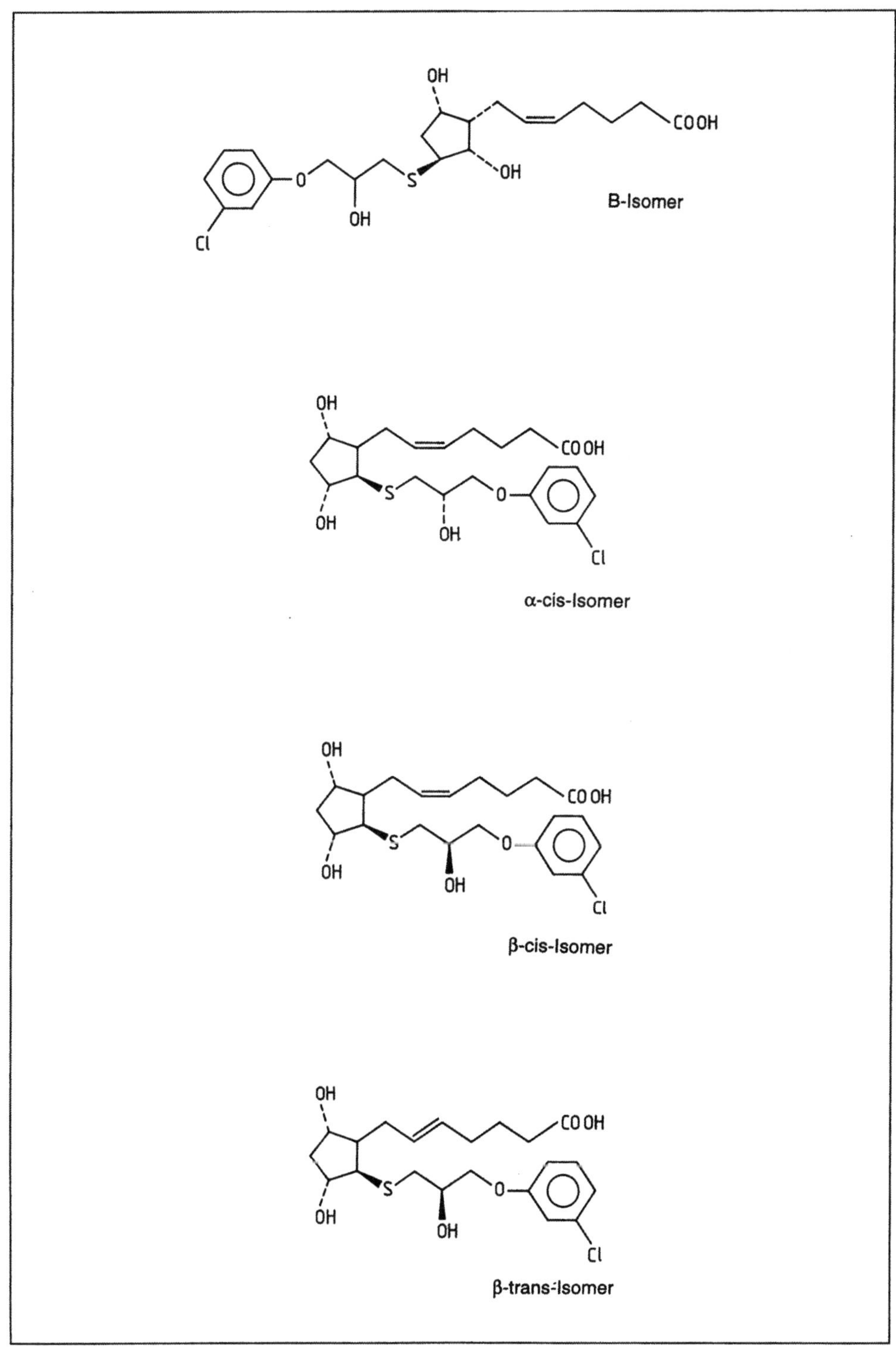



**Fig. 9.39.** Chemical structure of prostaglandins.

polarity. The second effect is dominant in the concentration range of 4 – 12 wt.-% methanol, used for the separation of the prostaglandins.

For the analytical separation on a 125 mm column, a linear velocity of 5 mm/s was chosen, resulting in a capacity factor of about 10. Selectivity under these conditions is

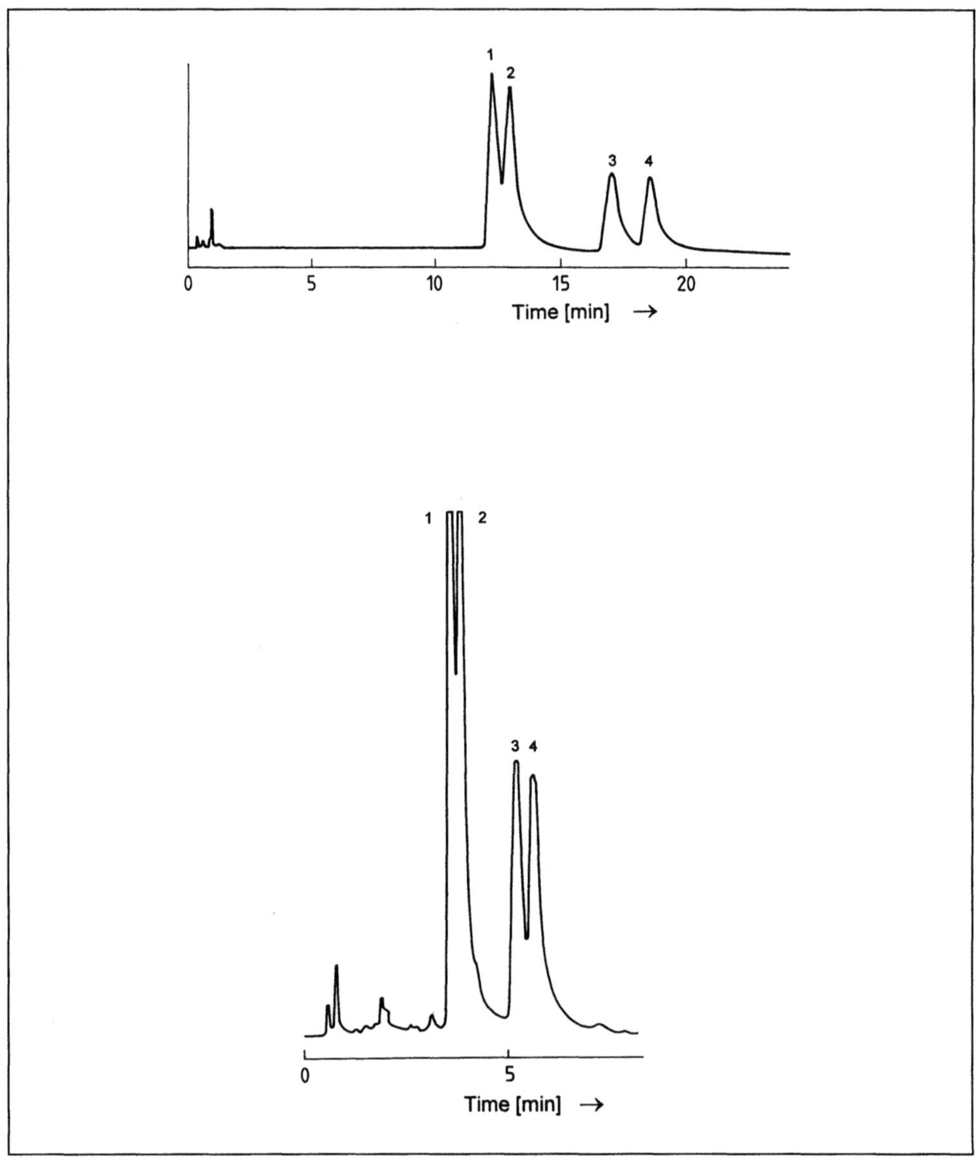

**Fig. 9.40.** Chromatograms of the separation of prostaglandins with SFC [38]. Temperature: 293 K; average pressure: 20 MPa; column: 125 x 4 mm; 5 μm LiChrosorb Si60; mobile phase: $CO_2$/methanol (top: 5.6 wt.%, bottom: 9.2 wt.%); 1, 2 = B-isomers; 3 = $\alpha$-cis; 4 = $\beta$-cis.

362

1.07, a low value for preparative separations. Therefore, the influence of pressure, temperature, and composition of the mobile phase on capacity factors and selectivity between the $\alpha$-cis- and the $\beta$-cis-isomers was determined, in order to improve analytical separation. With increasing pressure, selectivity is enhanced and capacity factors

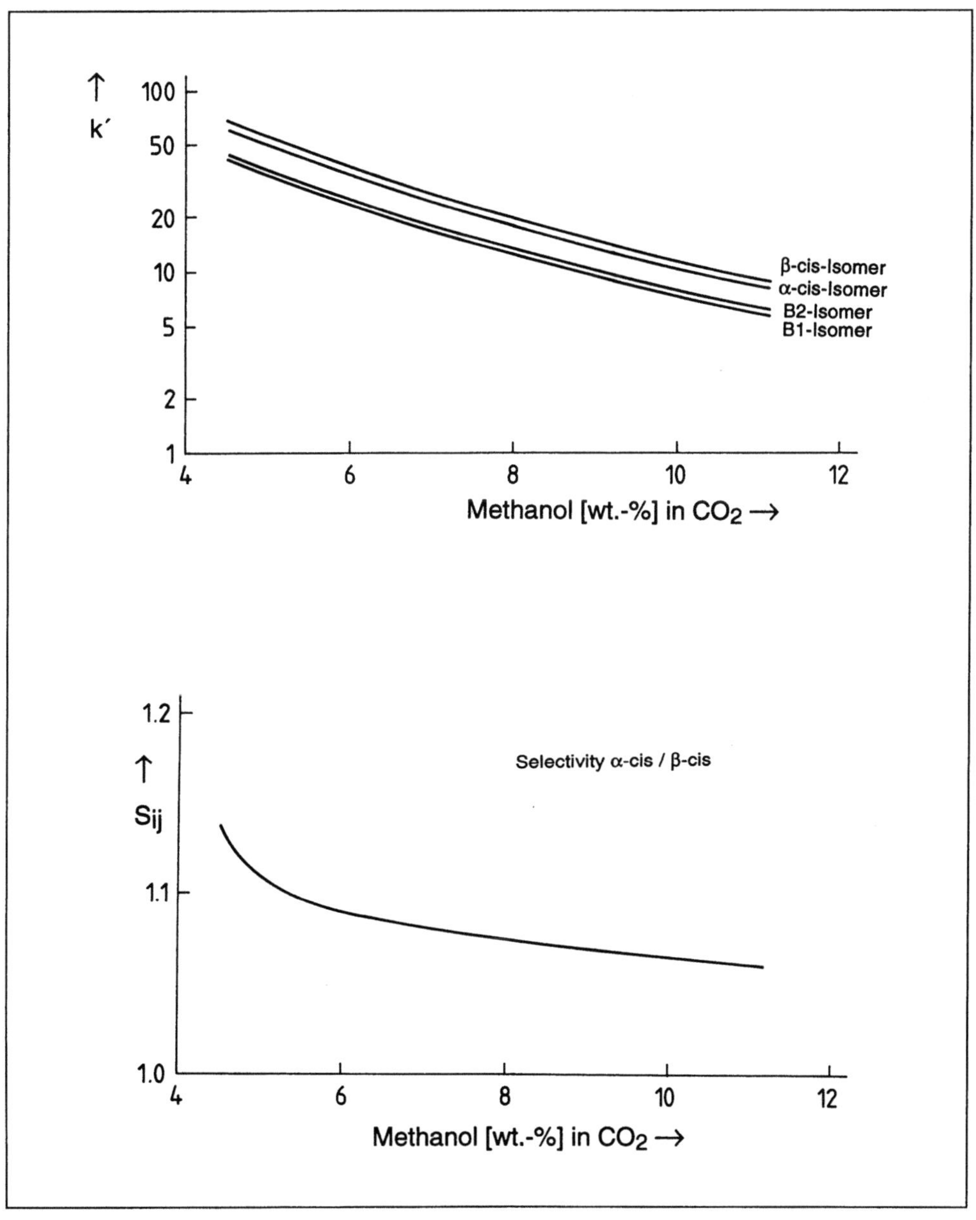



**Fig. 9.41.** Influence of amount of modifier on capacity factors and selectivity [37]. Average pressure: 20 MPa; temperature: 293 K; column: 125 x 4 mm; 5 µm LiChrosorb Si 60.

are lowered. Temperature within the applicable range has no significant influence on selectivity, only capacity factors increase slightly with temperature (Figs. 9.42 and 9.43).

Other modifiers can enhance selectivity. With 2-propanol, selectivity is significantly higher than with primary alcohols (Fig. 9.44). But the asymmetry of the peaks caused by 2-propanol leads to overlapping peaks, which allows to collect only a small part of the $\alpha$-cis-isomer. Therefore, methanol, in spite of the low selectivity, is the best modifier and was used in the scale-up experiments.

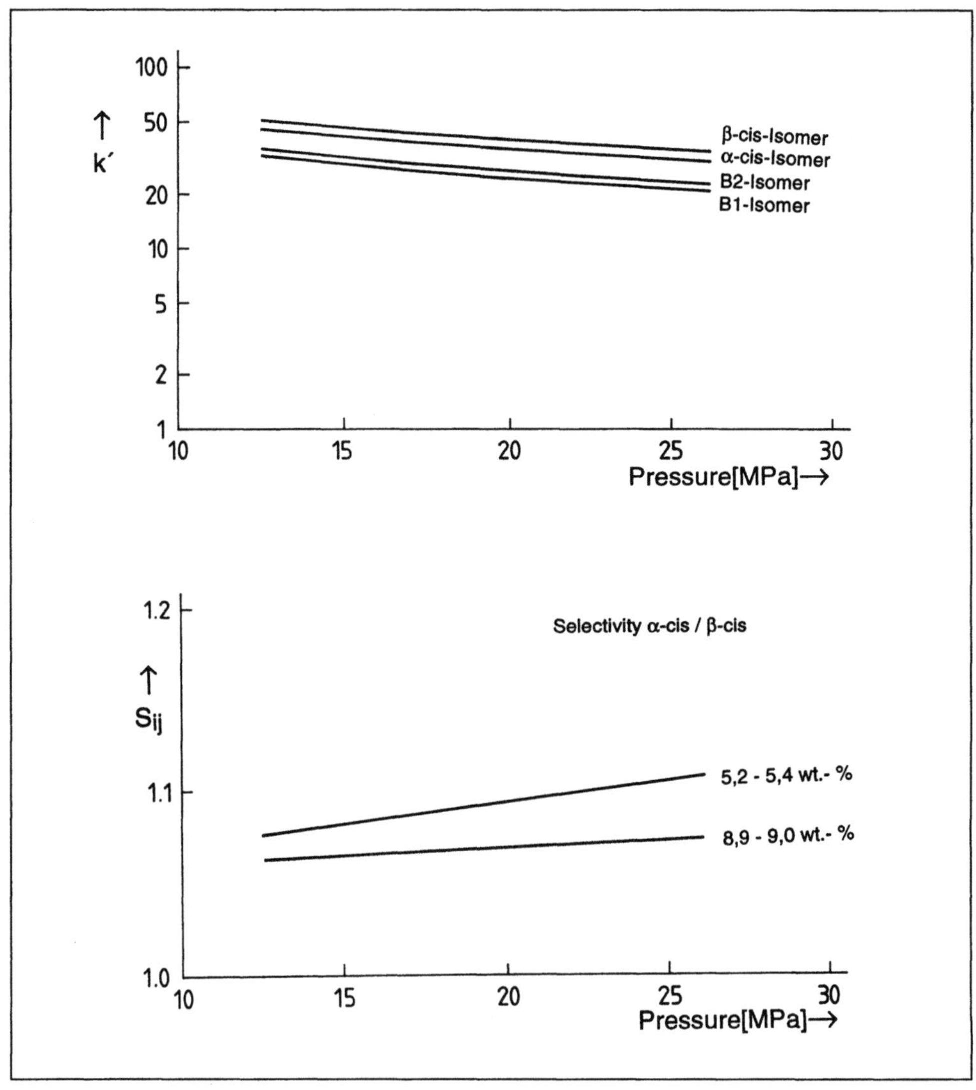


**Fig. 9.42.** Influence of pressure on capacity factors (upper diagram) and selectivity (lower diagram) [38].

Preparative separation experiments were carried out on a separation column of 250 mm length and 10 mm diameter, filled with silica gel particles (LiChrosorb Si 60) of 7 µm diameter. Injection volume was 100 µl with 15 mg prostaglandins dissolved in

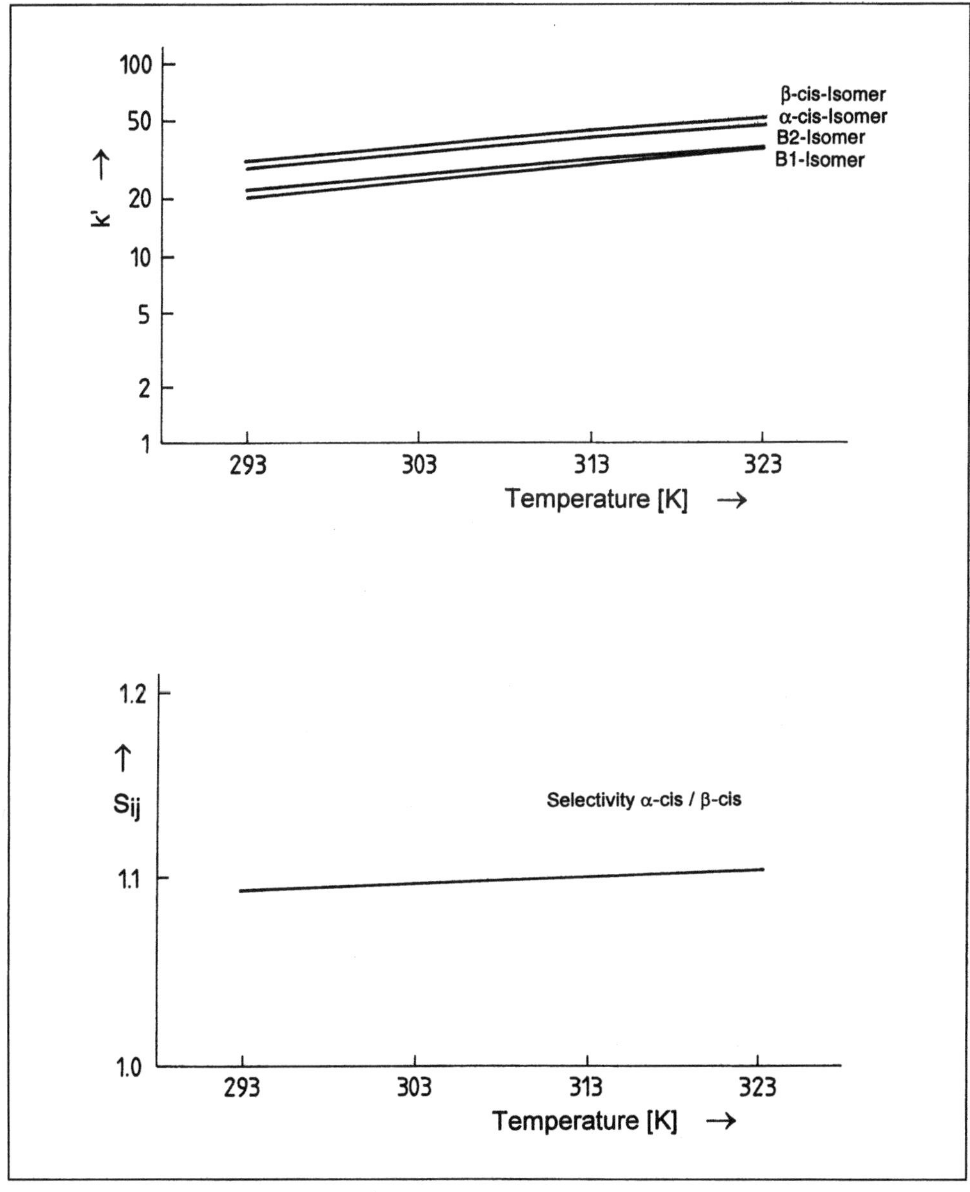



**Fig. 9.43.** Influence of temperature on capacity factors (upper diagram) and selectivity (lower diagram) [38]. Average pressure: 20 MPa; mobile phase: $CO_2$/methanol (5.3 wt.%); column: 125 x 4 mm; 5 µm LiChrosorb Si 60.

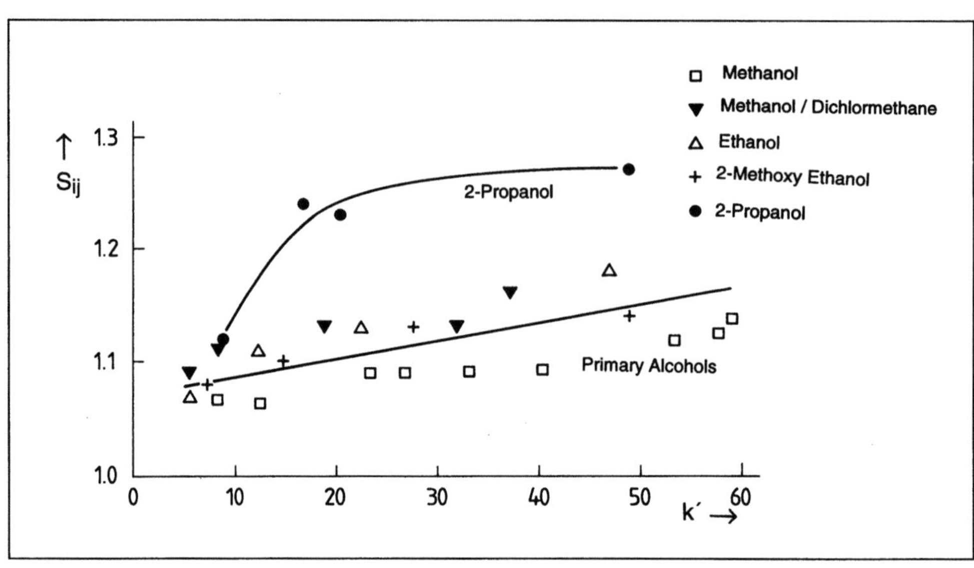



**Fig. 9.44.** Selectivity in dependence on capacity factor for different modifiers [38]. Selectivity between isomers $\alpha$-cis and $\beta$-cis in dependence on capacity ratio of the $\alpha$-cis isomer. Mobile phase: $CO_2$ with indicated modifiers; average pressure: 20 MPa; temperature: 292 K; column: 125 x 4 mm; 5 $\mu$m LiChrosorb Si 60.

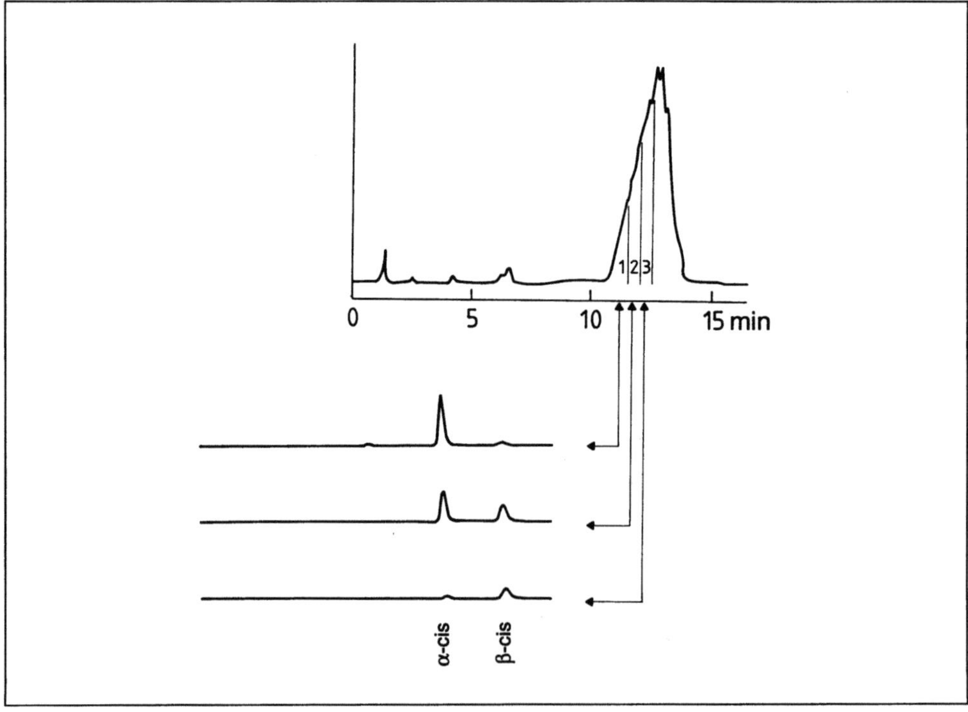



**Fig. 9.45.** Chromatogram of preparative separation of prostaglandins. Selected fractions are shown in corresponding HPLC-chromatograms [38]. Feed without B-isomers. Mobile phase: $CO_2$/methanol (10 wt.%); average pressure: 20 MPa; temperature: 292 K; column: 250 x 10 mm; 7 $\mu$m LiChrosorb Si 60.

366

ethylacetate. In Fig. 9.45 a chromatogram of a preparative separation is shown, together with the HPLC-chromatograms of the collected fractions. The first fraction contains about 80 % of $\alpha$-cis-isomer, the active compound.

The influence of increasing injection volume on the shape of the peaks is shown for different amounts of modifier in Fig. 9.46.

If these results of preparative SFC are compared to the achieved throughput in production liquid chromatography, then SFC has only about 1/10 of the capacity of LC. This disadvantage is not outweighed by the advantage in SFC of easy separation of the prostaglandins from the solvents. But this separation is not optimal. For taking SFC into consideration for the separation of such polar molecules, selectivity must be enhanced, probably by a different supercritical compound in the mobile phase.

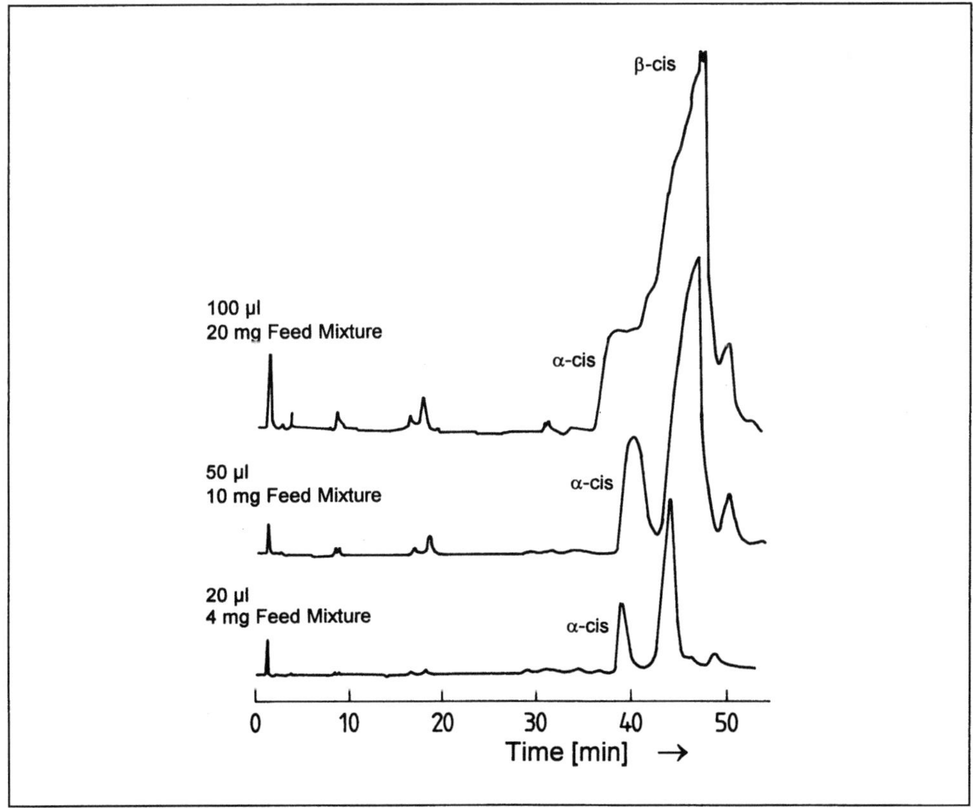


**Fig. 9.46.** Influence of injection volume on separation [38]. Feed without B-isomers. Mobile phase: $CO_2$/methanol (5.3 wt.%); average pressure: 24.2 MPa; temperature: 290 K; column: 250 x 10 mm; 7 µm LiChrosorb Si 60. Indicated are injected quantity and volume (Feed mixture: solution of 394 mg isomers in 2 µl ethylacetate).

## 9.4.3 Separation of Tocopherols

The chemical structure and some general information on tocopherols has been presented in Chapter 8, where the enrichment of tocopherols in a countercurrently oper-

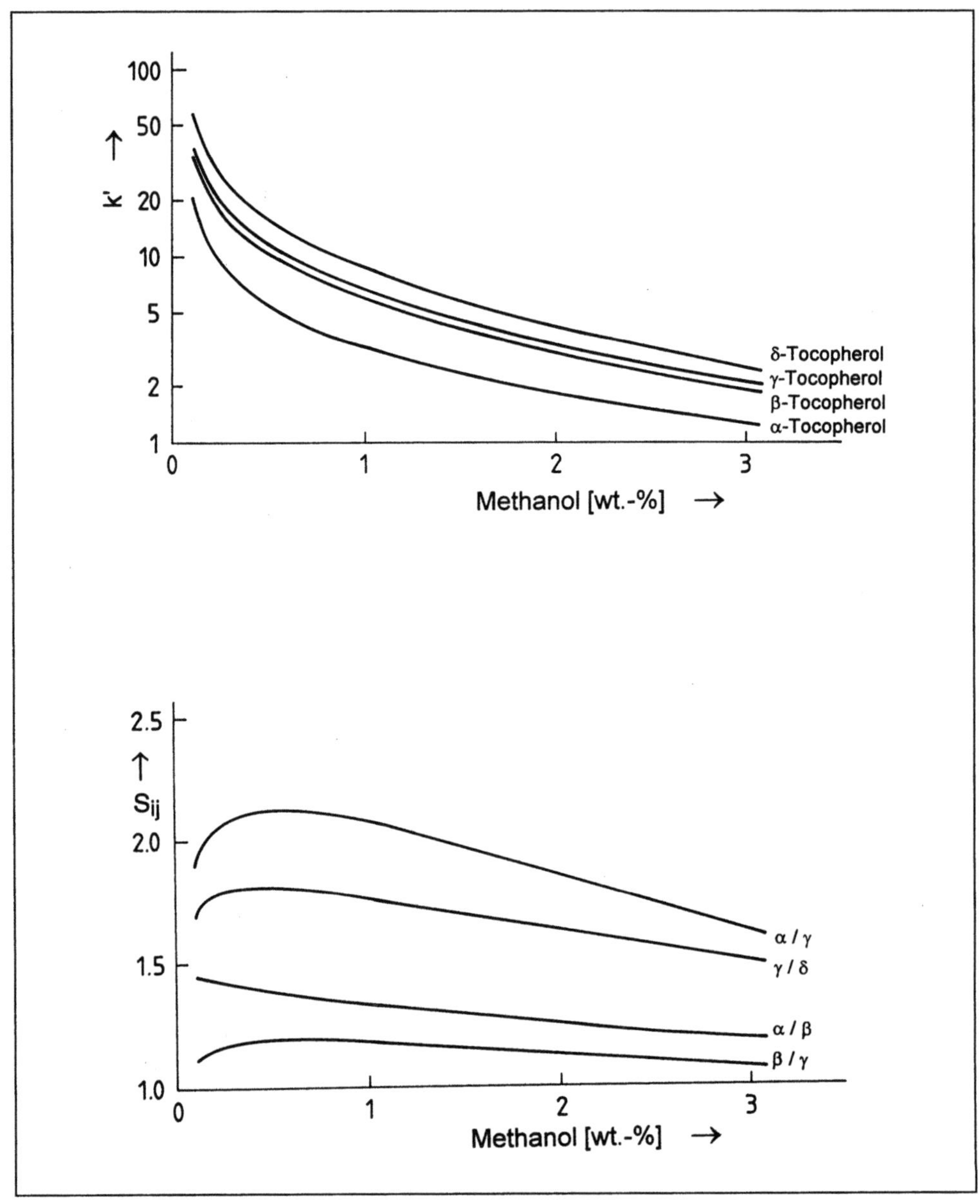


**Fig. 9.47.** Separation of tocopherols with SFC at analytical conditions [38]. Mobile phase: $CO_2$/methanol; average pressure: 15 MPa; column: 125 x 4 mm; 5 µm LiChrosorb Si60.

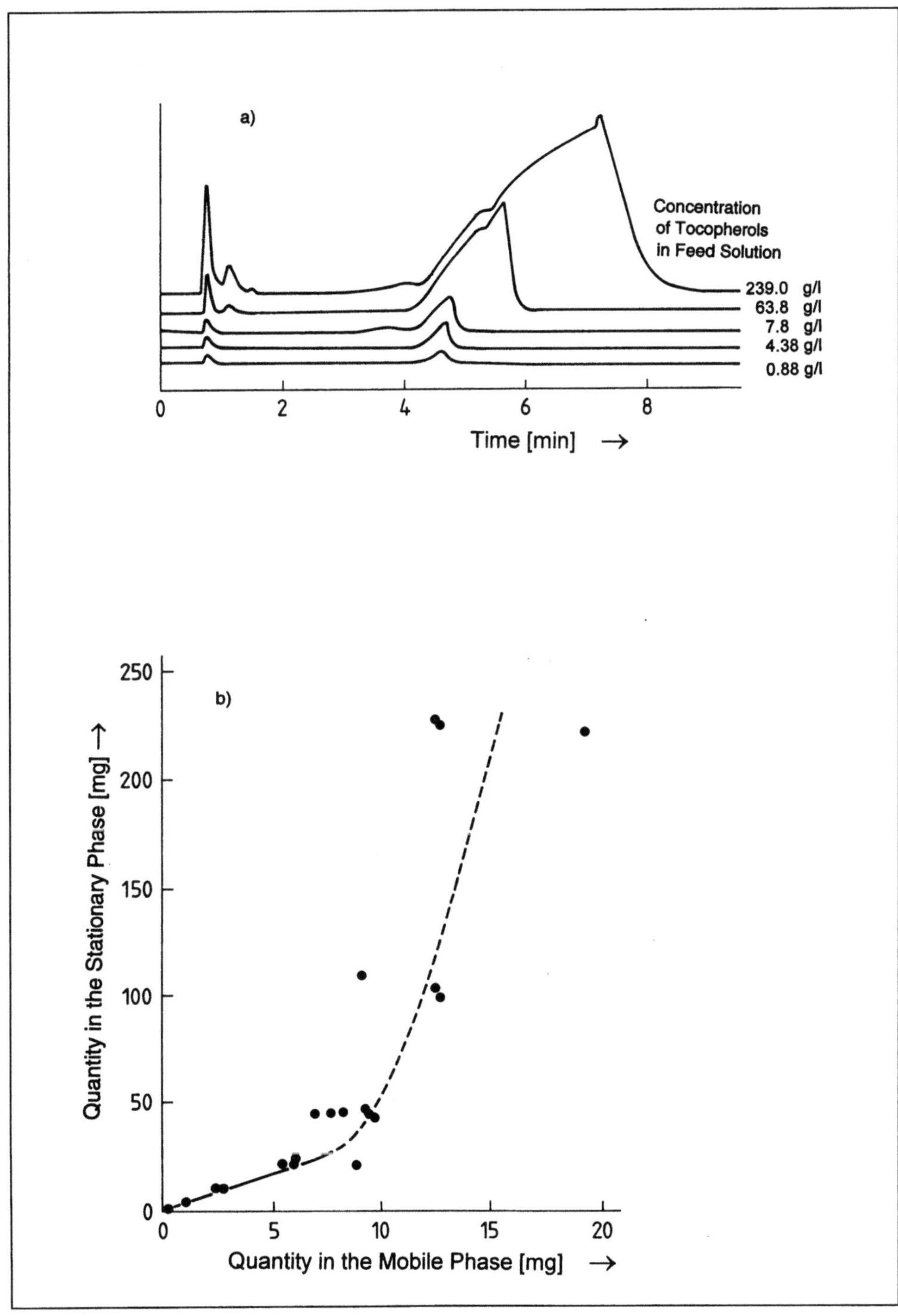


**Fig. 9.48.** a) Peaks of $\alpha$-tocopherol under overloading conditions; b) adsorption isotherm of $\alpha$-tocopherol on unmodified silica gel, as used for chromatographic separations of tocopherols [38].

ated column with a supercritical gas as solvent was reviewed. The maximum achievable tocopherol concentration was limited by components with nearly equal separation factors as the tocopherols. Those compounds are eluted in a chromatographic separation between the different tocopherol isomers. In contrast to a countercurrent extraction, a chromatographic separation can produce pure tocopherol isomers from such a fraction. Therefore, scale up of chromatographic separation is reviewed.

Analytical separation of tocopherols was carried out with HPLC by Coors and Montag [5], using silica gel as stationary and hexane-dioxane (94:6 parts by volume) as mobile phase. Bruns et al. [4] studied preparative separations of tocopherols from an enriched oil fraction on a 100 mm column of 400 mm length, using hexane-t-butyl-methylether (97:3 parts by volume) as eluent. Quantity of one injection was 15 g.

In the investigation which is reviewed here, a mixture rich in tocopherols (about 55 wt.-%), similar to that used by Bruns et al. was applied, containing 5 – 6 wt.-% $\alpha$-,

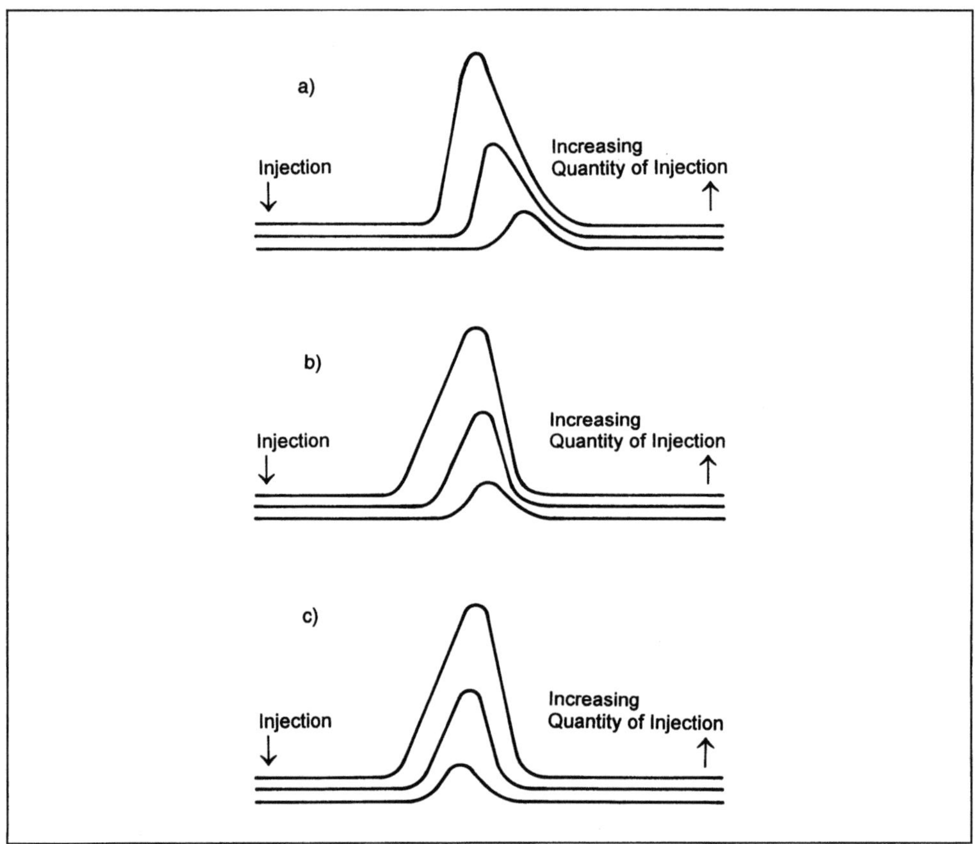



**Fig. 9.49.** Shapes of peaks under overloading conditions and with different modifiers. a) Langmuir-type adsorption isotherm; b) equal adsorption of modifier and eluted substance; c) stronger adsorption of modifier than eluted substance; for anti Langmuir-type isotherm ($\alpha$-tocopherol) see Fig. 9.48.

370

traces of $\beta$-, 35 – 44 wt.-% $\gamma$- and 12 – 14 wt.-% $\delta$-tocopherol besides 10 – 12 wt.-% squalene, 1 – 3 wt.-% sterols and unidentified components, probably sterolesters and others. Fundamental experiments on overloading were carried out with $\alpha$-tocopherol, a synthetic DL-mixture received as "Vitamine E purum". The tocopherols were dissolved in chloroform for injection.

The method for investigating scale up of chromatographic separation of tocopherols was the same as for prostaglandins: First, capacity factors in analytical separation, then, using $\alpha$-tocopherol as model substance, the influence of parameters on capacity factors, and then separation factors under overloading conditions were investigated. Some important experiments were repeated with the tocopherol-mixture.

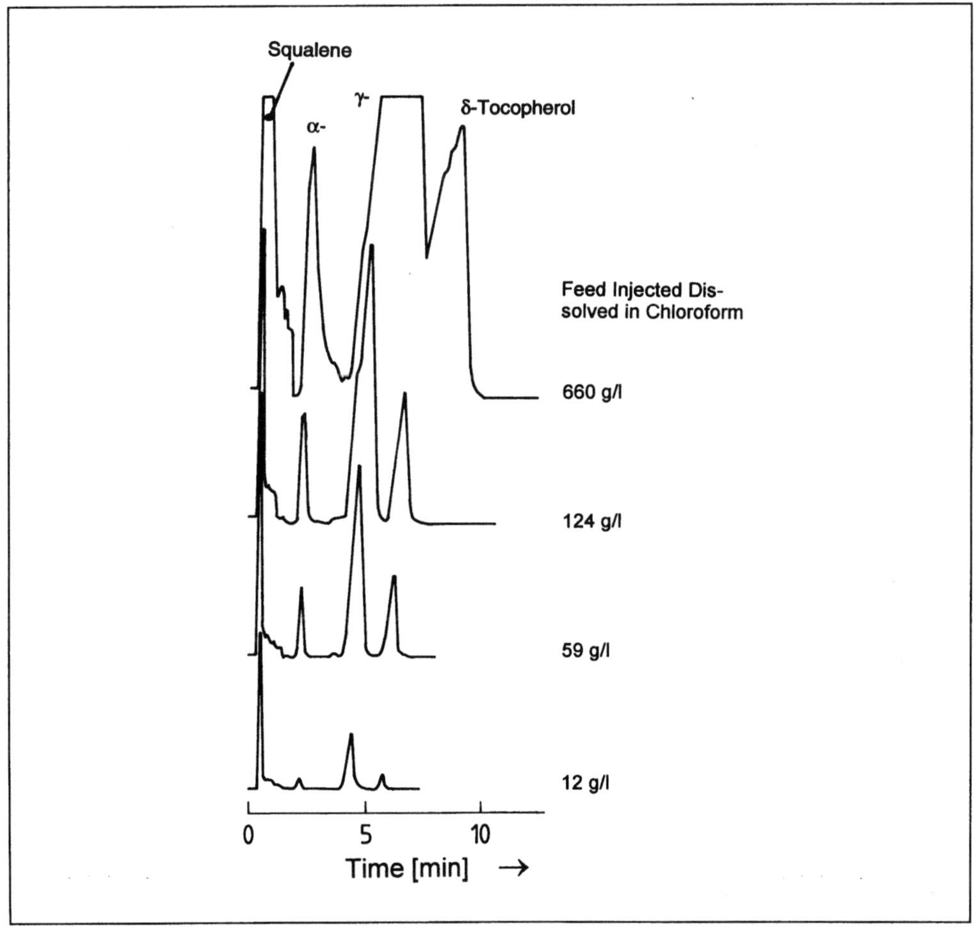


**Fig. 9.50.** Chromatograms of a tocopherol mixture under overloading conditions [38].

Capacity factors and selectivity with methanol as modifier are presented in Fig. 9.47. With increasing amount of modifier, capacity factors decrease, as do selectivities in general, with a maximum at a modifier content of about 0.5 wt.-% methanol. With other modifiers, like diethylether and acetonitrile, no separation was possible.

Overloading of $\alpha$-tocopherol as model substance results in peaks with a flat slope on the front side of the peak, so called "leading" peaks (Fig. 9.48, part a). With increasing concentration of injected tocopherols, the peak maximum is shifted to longer retention times, which is due to the anti Langmuir-type of adsorption isotherm (Fig. 9.48, part b). This is somewhat unusual. If a Langmuir-type adsorption isotherm is determining adsorption equilibrium, then peak maxima are shifted to shorter retention times, as could be shown by Golshan-Shirazi and Guiochon [9]. Modifiers also change the shape of a peak or elution characteristics. A modifier with equal strong adsorption on the stationary phase, as the dissolved substance, does not change peak maxima, while a modifier, with stronger interactions, induces peak maxima at higher retention times as is shown schematically in Fig. 9.49. Chromatograms of a mixture of tocopherols are presented in Fig. 9.50 under overloading conditions. At about 660 g/l (g tocopherol mixture per l solution) a limit is reached with respect to the possibility of collecting pure $\alpha$-tocopherol.

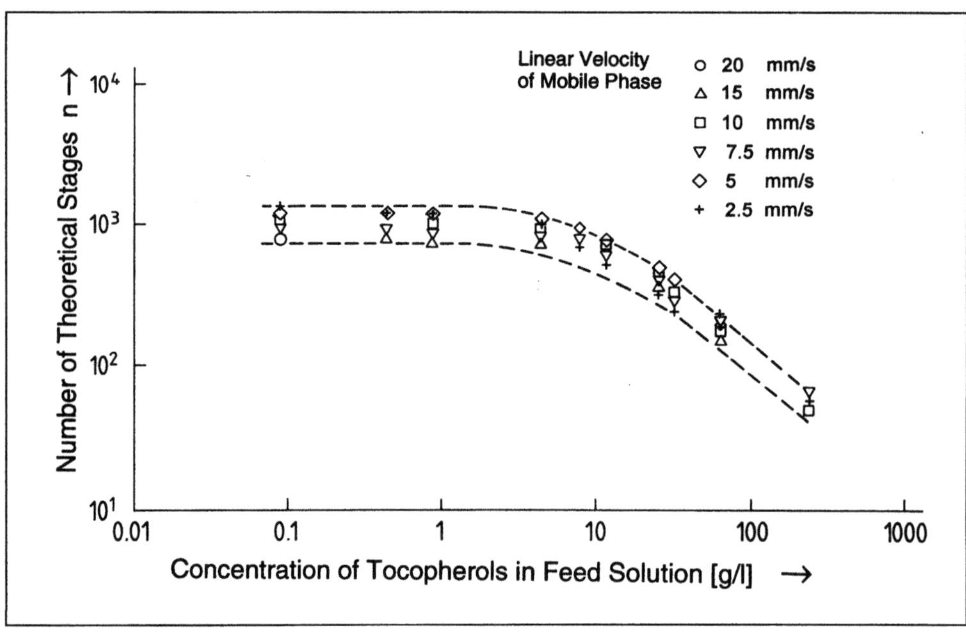


**Fig. 9.51.** Influence of linear velocity and injected quantity on the number of theoretical stages [38]. Solutions of $\alpha$-tocopherol in chloroform. Injected volume: 10 $\mu$l; mobile phase: $CO_2$/methanol (0.5 wt.%); average pressure: 15 MPa; temperature: 292 K; column: 125 x 4 mm; 5 $\mu$m LiChrosorb Si 60.

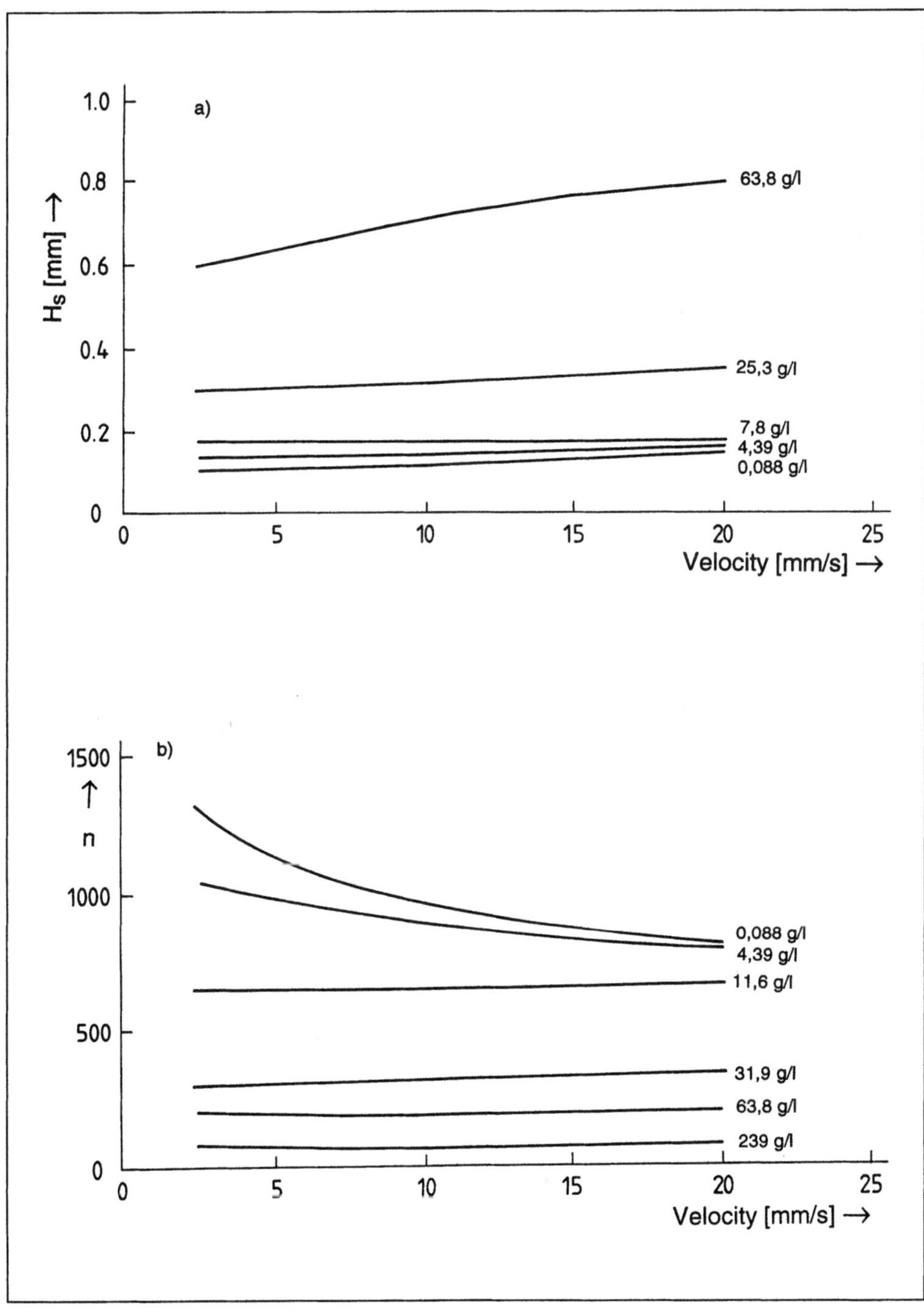



**Fig. 9.52.** Influence of linear velocity of the mobile phase on a) height of a theoretical stage (plate); b) number of theoretical stages [38]. Solutions of $\alpha$-tocopherol in chloroform. Injected volume: 10 µl; mobile phase: $CO_2$/methanol (0.5 wt.%); average pressure: 15 MPa; temperature: 293 K; column: 125 x 4 mm; 5 µm LiChrosorb Si 60.

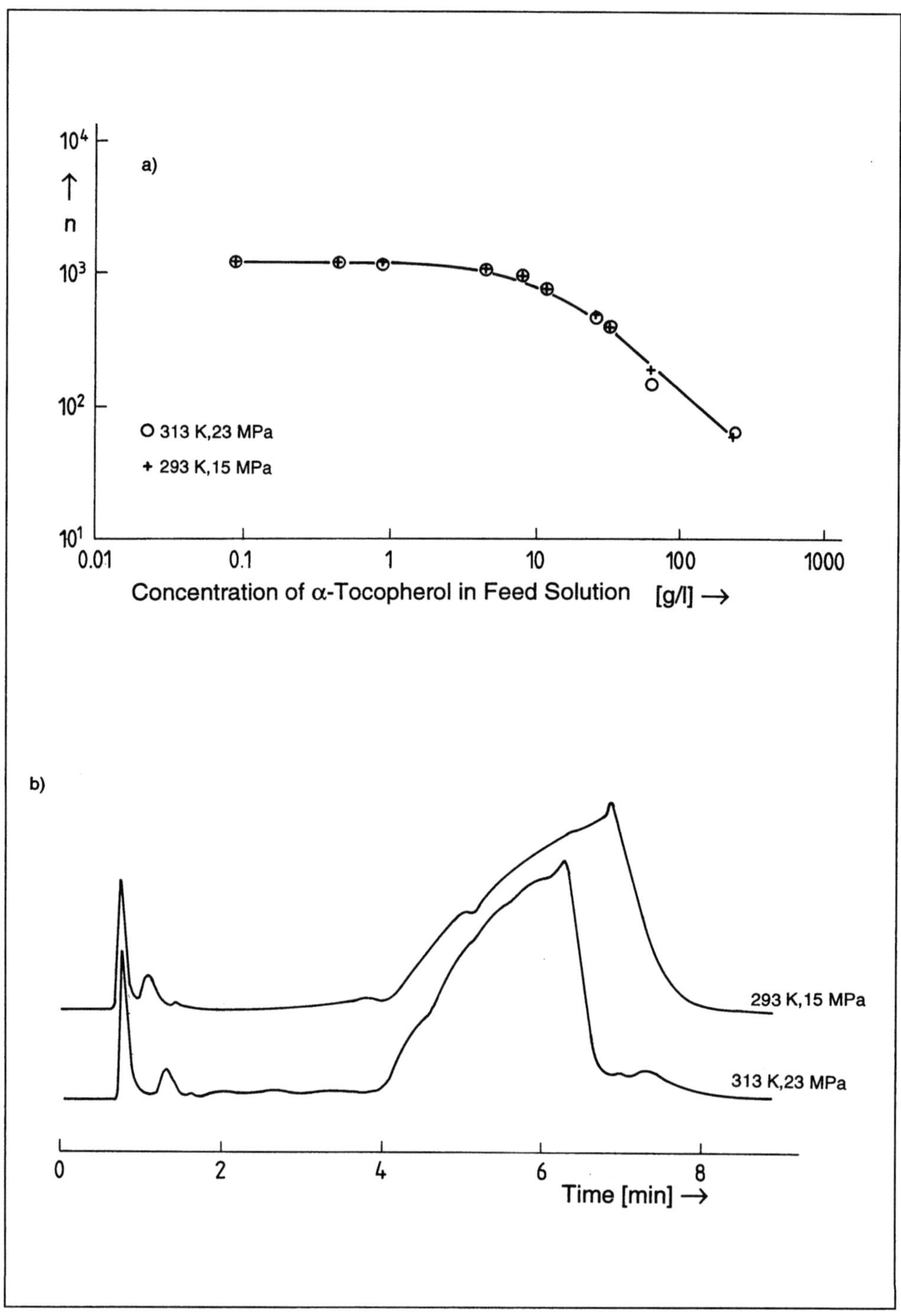

**Fig. 9.53.** Influence of temperature on a) number of theroretical stages; b) shape of the eluted α-tocopherol peak [38].

374

In the following, important parameters for scaling up chromatography are reviewed for the tocopherol mixture:

**Linear velocity** of the mobile phase is important because retention time is inverse proportional and throughput is direct proportional. The number of theoretical stages of a chromatographic separation column is slightly dependent on concentration of the injected tocopherol mixture at analytical concentrations (Fig. 9.51), with more stages at a lower velocity. But at higher concentrations, from about 5 g/l upwards, influence of linear velocity on number of stages is negligible. Peak broadening due to high linear velocity seems to be dominated by peak broadening due to higher concentrations. The number of theoretical stages decreases with concentration, from about the same concentrations at which the influence of velocity vanishes. These effects can be seen in Fig. 9.52.

With an increase in linear velocity, pressure drop over the column increases. At the outlet of the column conditions of the mobile phase may have changed markedly. This may be an additional limitation for a chromatographic separation. For the tocopherol separation, pressure drop is not important, since it has been carried out in a region where density changes moderately with pressure.

**Temperature.** At higher (supercritical) temperatures density is lower at the same pressure than in the subcritical region, and therefore retention times are longer. To achieve the same retention times, density must be chosen equal, i.e. pressure must be

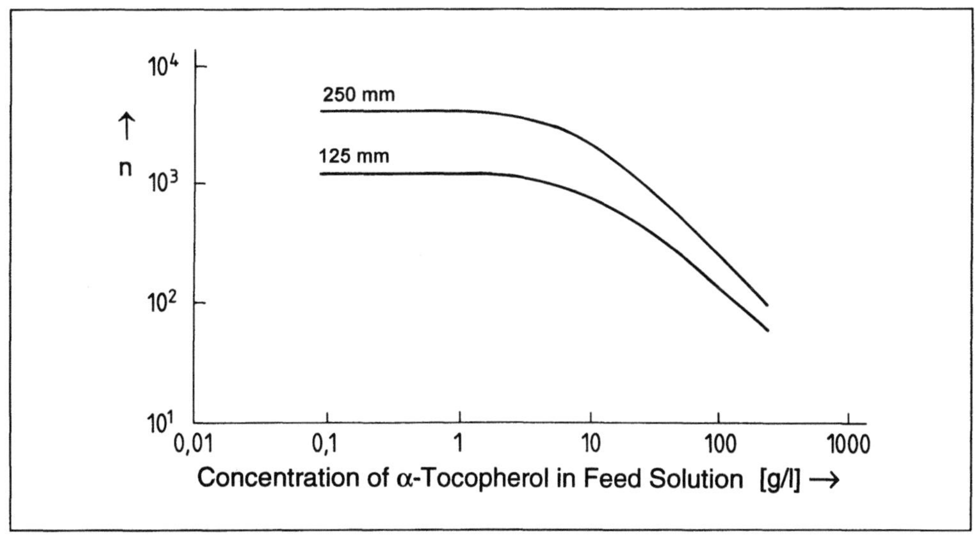



**Fig. 9.54.** Influence of column length on number of theoretical stages [38]. Solutions of α-tocopherol in chloroform. Injected volume: 10 µl; mobile phase: $CO_2$/methanol (0.5 wt.%); average pressure: 15 MPa; temperature: 293 K; column: 125 and 250 x 4 mm; 5 µm LiChrosorb Si 60.

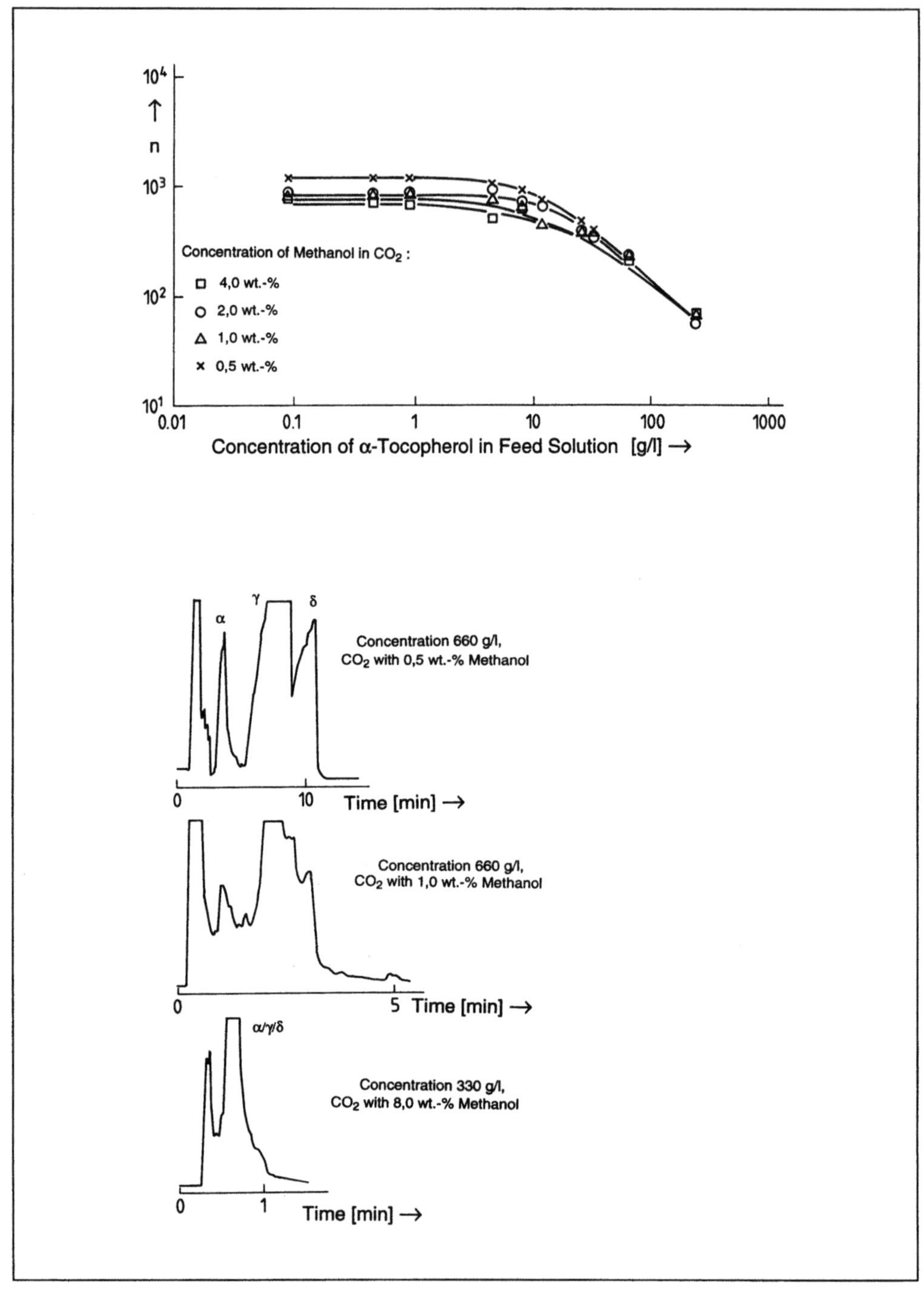


**Fig. 9.55.** Influence of modifier concentration on a) number of theoretical stages of a model substance (α-tocopherol); b) separation of a tocopherol mixture [38]. Solutions of α-tocopherol in chloroform. Injected volume: 10 µl; mobile phase: $CO_2$/methanol (as indicated); average pressure: 15 MPa; temperature: 293 K; column: 125 x 4 mm; 5 µm LiChrosorb Si 60.

376

increased. In the sub- and supercritical region, no influence on number of theoretical stages could be found (Fig. 9.53). But the peak at the higher, supercritical temperature is narrower, which is advantageous for preparative separations.

**Column length.** Influence of column length is great at low concentrations, but declines at higher concentrations, Fig. 9.54. It seems not to be useful for a better separation to use a longer column, since retention times and pressure drop are enhanced.

**Concentration of modifier.** At concentrations relevant for preparative separations, the influence of modifier becomes negligible on the number of theoretical stages for a model substance. But influence of modifier concentration on the separation of the tocopherols remains, as is shown in Fig. 9.55. Separation of the tocopherols becomes more difficult with increasing modifier concentration.

Results of separation experiments on a 4.6 mm diameter column were transferred to a 10-mm column, using 7-$\mu$m particles instead of 5-$\mu$m particles and a length of 250 mm instead of 125 mm. The results are compared to HPLC-separations published by Bruns et al. [4]. They achieved a throughput of 15 g in 60 min on a 100-mm column. If SFC results on the tocopherol mixture are extrapolated from the 10 mm column to a 100-mm column, then 24 g can be processed in 60 min. In this case, SFC has a higher capacity than liquid chromatography. Again it should be noted that separation conditions are not optimized. Therefore, an additional increase in capacity can be expected from SFC.

# 9.5 On-line Analysis with SFC

Supercritical fluid chromatography (SFC) is a suitable method for on-line analysis. Monitoring of chemical production and separation processes can be achieved in cases where substances can be eluted and separated by a mobile phase containing mainly a supercritical component. In the following, SFC as a method for on-line analysis is presented by an example, the determination of caffeine during the extraction from raw coffee beans or black tea. In this case, on-line analysis allows monitoring of the concentration in the gaseous solvent and the calculation of extracted caffeine quantity for determining the necessary time of extraction. This can otherwise only be taken from experience and confirmed later by off-line analysis.

A chromatographic system for on-line analysis consists 1) of the chromatographic separation and detection unit, 2) a device for taking samples from the process, handling samples, and injecting selected samples into the chromatographic system, and 3) a controlling device for automatic operation. In contrast to off-line analysis, the on-line anlysis system must be very reliable and stable over an extended period of time and cycle time must be short in comparison to changes in the process. A schematic drawing of the sampling part is given in Fig. 9.56 and of the chromatographic part in Fig. 9.57.

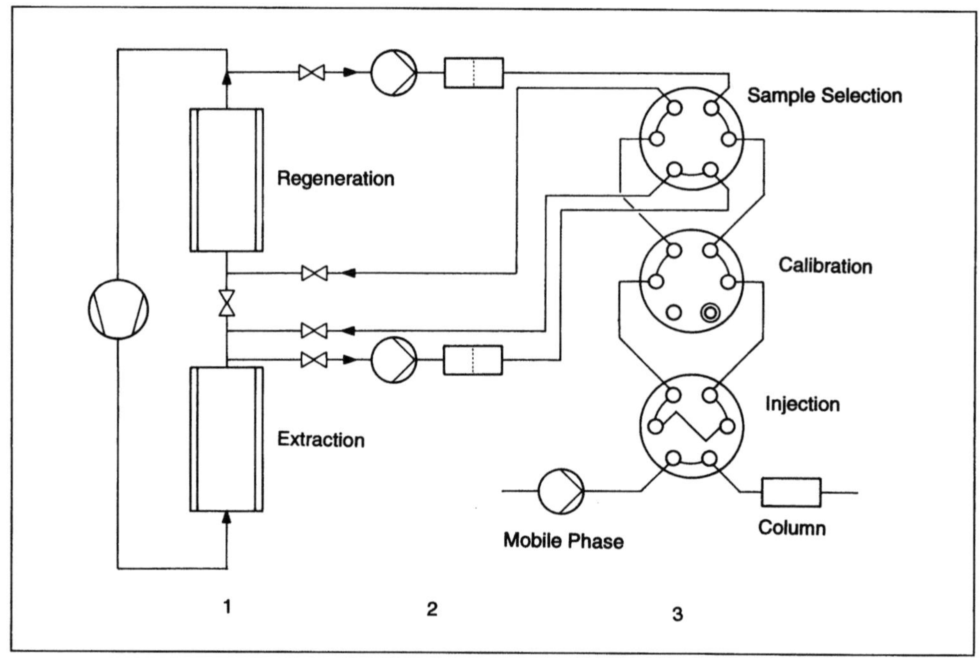



**Fig. 9.56.** Schematic drawing of the coupling device of a gas extraction plant with on-line-SFC. 1) Process unit; 2) Sample handling part; 3) Sample delivery for injection and calibration [34].

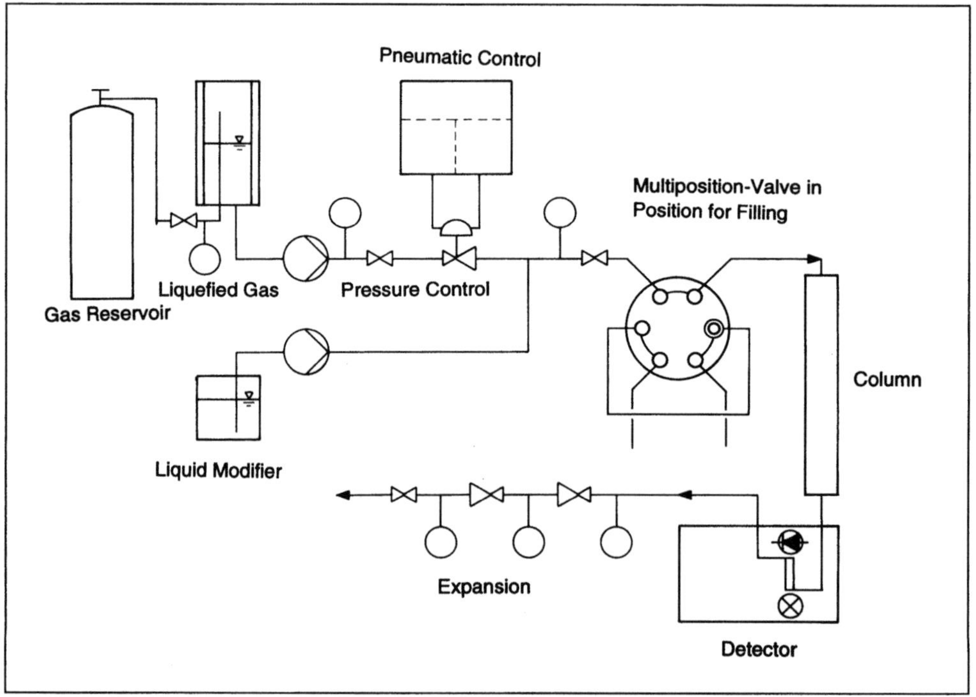



**Fig. 9.57.** Flow scheme of on-line SFC [34].

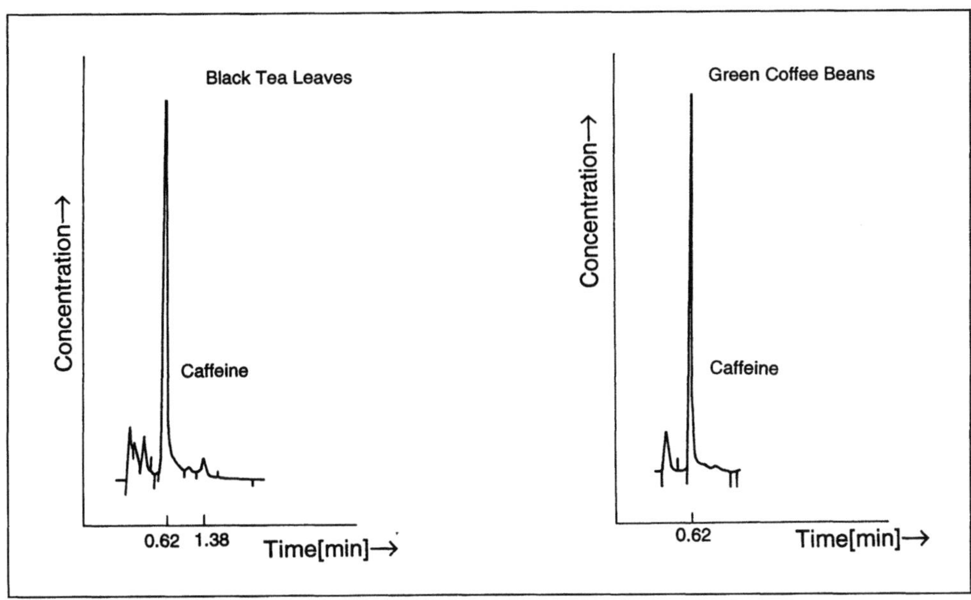



**Fig. 9.58.** Chromatograms of the extract from black tea (left) and raw coffee beans (right) with carbon dioxide as supercritical solvent. For chromatographic conditions, see text. Detection: UV/VIS: 222 nm (left), 272 nm (right) [34].

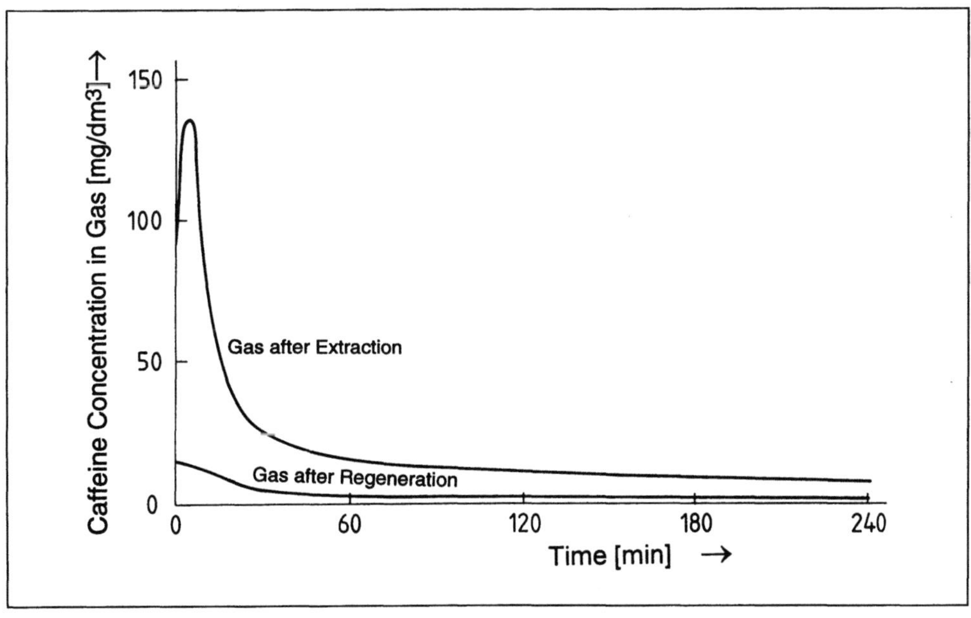



**Fig. 9.59.** Caffeine concentration after the extraction vessel (upper curve) and after the regeneration of the supercritical solvent (carbon dioxide) by absorption in water (lower curve) [34].

For monitoring the caffeine concentration after the extraction and the regeneration of the supercritical solvent, a small part of the solvent flow is removed at the outlet of both vessels, and pumped without significant decompression over a filter (2 µm), through the first of a series of three multiport valves and back again to the outlet line, nearby to the point of removal [34]. This is necessary to achieve a representative sample for the injection into the chromatographic column. There may be cases, where it is necessary to bypass the valve as long as no sample is actually injected, to avoid accumulation of substances by adsorption. The pump in the sample cycle ensures a reproducible and constant flow of sampled solvent. The filter avoids blocking of the sensitive multiport-valves. Size of the filter must be designed for operation of an extended time. For sample injection a series of three six port valves is useful. The first one takes a sample and delivers it to a second one, which is normally passed by, but can be used for introducing calibrating solutions after a certain number of analyses. The third valve is the actual injection valve, where the sample is injected from the sample loop of the valve to the mobile phase and then transported to the chromatographic column.

Chromatographic separation conditions can be taken from off-line analysis [33, 34, 35]. For achieving a short cycle time, a packed column may be advantageous. In our example, a packed column of 3.1-mm inner diameter and 50-mm length was selected, which allowed cycle times for the caffeine determination of about 2 min. Packing material was LiChrosorb Si60 (E. Merck, Darmstadt, FRG) of 5-µm diameter. Columns were prepared by a suspension method. The column was used for more than 5 months without changes in separation efficiency. The mobile phase consisted of carbon dioxide with 15 wt.-% methanol as modifier (tea) and 5.7 wt.-% for coffee. Linear velocity was 4.5 mm/s (tea) or 6.5 mm/s (coffee), initial pressure was 25 MPa with a pressure drop over the column of 3 MPa. Different conditions were necessary for coffee and tea, because of different caffeine concentrations and different composition of the extract. Examples of the chromatograms are shown in Fig. 9.58 and the monitoring of an experiment is presented in Fig. 9.59. On-line analysis in this case is of great advantage compared to off-line analysis. It is tedious to take samples from a compressed gas, and accuracy and reproducibility are low if the procedure is not carried out very carefully.

## References


1. Alkio M, Harvala T, Komppa V (1988) Preparative scale supercritical fluid chromatography. In: Perrut M (ed) Proc 2nd Int Symposium on Supercritical Fluids, pp 389 – 396
2. Applied Chromatography Systems (1987) Instruction manual for evaporative analyzer model 750/14
3. Berger C, Perrut M (1988) Purification de Molecules d'interet biologique par Chromatografie Preparative Avec Eluant Supercritique. Technoscope Biofutur 25: 3 – 8
4. Bruns A, Berg D, Werner-Busse A (1988) Isolation of tocopherol homologues by preparative high-performance liquid chromatography. J Chromatogr 450: 111 – 113
5. Coors U, Montag A (1988) Untersuchungen zur Stabilität des Tocopherolgehaltes pflanzlicher Öle. Fat Sci Technol 90: 129 – 136

6. Deemter JJ van, Zuiderweg FJ, Klinkenberg A (1965) Longitudinal diffusion and resistance to mass transfer as causes of nonideality in chromatography. Chem Eng Sci 5: 271 – 289

7. Eble JE, Grob RL, Antle PE, Snyder LR (1987) Simplified description of high-performance liquid chromatographic separation under overload conditions, based on the Craig distribution model. III. Computer simulations for two co-eluting bands assuming a Langmuir isotherm. J Chromatogr 405: 1 – 29

8. Gere D, Board R, Mc Manigill D (1982) Parameters of supercritical fluid chromatography using HPLC columns. Hewlett-Packard Publ 43-5953-1647, Avondale, Pennsylvania, pp 7 – 15

9. Golshan-Shirazi S, Guiochon G (1989) Theoretical study of system peaks and elution profiles for large concentration bands in the case of a binary eluent containing a strongly sorbed additive. J Chromatogr 461: 1 – 18

10. Kern JR, Lokensgard DM, Manes LV, Matsuo M, Nakamura K (1988) Separation of the stereo-isomers of an allenic E-Type prostaglandin. J Chromatogr 450: 233 – 240

11. Klesper E, Leyendecker D (1986) Supercritical fluid chromatography: Retention and resolution. Intl Lab 16(9): 18 – 30

12. Knox JH, Pyper HM (1986) Framework for maximizing throughput in preparative liquid chromatography. J Chromatogr 363: 1 – 30

13. Lee ML, Markides KE (eds) (1990) Analytical Supercritical Fluid Chromatography and Extraction. Chromatography Conferences, Provo, Utah

14. Leyendecker D, Leyendecker D, Schmitz FP, Klesper E (1986) Chromatographic behaviour of various eluents and eluent mixtures in the liquid and in the supercritical state. J Chromatogr 371: 93 – 107

15. Lohmus M, Kirjanes I, Lopp M, Lille Ü (1988) Solvent selectivity in the resolution of some regioisomeric and diastereomeric prostaglandin intermediates on silica. J Chromatogr 449: 77 – 94

16. Lohmus M, Lopp M, Lille Ü, Veisserik J (1988) Preparative liquid chromatographic separation of prostacyclin carba-analogues and their intermediates. J Chromatogr 450: 105 – 109

17. Markides KE, Fields SM, Lee ML (1986) Capillary Supercritical Fluid Chromatography of Labile Carboxylic Acids. J Chromatogr Sci 24: 254 – 257

18. Mollerup J, Staby A, Sloth HC (1992) Thermodynamics of supercritical fluid chromatography. 9th Int Symp on Prep and Ind Chromatography "PREP-92", Nancy

19. Martire DE, Boehm RE (1987) Unified molecular theory of chromatography and its application to supercritical fluid mobile phases. 1. Fluid-liquid (absorption) chromatography. J Phys Chem 91: 2433 – 2446

20. Perrut M (1982) Demande de brevet francais, No. 82 09 649, 3 juin 1982

21. Perrut M (1983) Brevet europeen No. 00 99 765, 25. 05. 1983

22. Perrut M (1984) US patent, No. 44 78 720, Oct. 23, 1984

23. Pickel KH (1986) Chromatografische Untersuchungen mit hochkompressiblen mobilen Phasen. Dissertation, Universität Erlangen-Nürnberg

24. Randall LG (1984) Carbon dioxide based supercritical fluid chromatography. Column efficiencies and mobile phase solvent powers. In: ACS Symposium Series 250: 135 – 169

25. Saito M, Yamauchi Y, Hondo T, Senda M (1988) Laboratory Scale Preparative Supercritical Fluid Chromatography in Recycle Operation: Instrumentation and Applications. In: Perrut M (ed) Proc 2nd Int Symposium on Supercritical Fluids, pp 381 – 388

26. Saito M, Yamauchi Y (1988) Recycle chromatography with supercritical carbon dioxide as mobile phase. HRC & CC 11: 741 743

27. Saito M, Yamauchi Y, Inomata K, Kottkamp W (1989) Enrichment of tocopherols in wheat germ by directly coupled supercritical fluid extraction with semi-preparative supercritical fluid chromatography. J Chromatogr Sci 27: 79 – 85

28. Saito M, Yamauchi Y (1990) Isolation of tocopherols from wheat germ oil by recycle semi-preparative supercritical fluid chromatography. J Chromatogr 505: 257 – 271

29. Schoenmakers PJ, Verhoeven FCCJG (1986) Effect of pressure on retention in supercritical fluid chromatography with packed columns. J Chromatogr 352: 315 – 328

30. Schoenmakers PJ, Uunk LG (1987) Supercritical fluid chromatography – recent and future developments. Eur Chromatogr News 1.3: 14 – 22

31. Smith RM (ed) (1988) Supercritical Fluid Chromatography. Roy Soc Chem, London

32. Snyder LR, Cox GB, Antle PE (1987) A simplified description of HPLC separation under overload conditions. A synthesis and extension of two recent approaches. Chromatographia 24: 82 – 96

33. Upnmoor D, Brunner G (1989) Investigation of retention behaviour on packed column SFC. Ber Bunsenges Phys Chem 93: 1009 – 1015
34. Upnmoor D, Brunner G (1989) Anwendungen der Fluidchromatographie (SFC) in der on-line Analytik. GIT Fachz Lab 33: 311 317
35. Upnmoor D, Brunner G (1989) Retention of acidic and basic compounds in packed column supercritical fluid chromatography. Chromatographia 28: 449 – 454
36. Upnmoor D, Brunner G (1992) Packed column supercritical fluid chromatography with light scattering detection. I. Optimization of parameters with a carbon dioxide/methanol mobile phase. Chromatographia 33: 255 – 260
37. Upnmoor D, Brunner G (1992) Packed column supercritical fluid chromatography with light scattering detection. II. Retention behaviour of squalene and glucose with mixed mobile phases. Chromatographia 33: 261 – 266
38. Upnmoor D (1992) Untersuchungen zur präparativen Fluidchromatographie am Beispiel von Tocopherolen und Prostaglandinen. Dissertation, Technische Universität Hamburg-Harburg
39. Wenclawiak B (ed) (1992) Analysis With Supercritical Fluids: Extraction and Chromatography. Springer, Berlin Heidelberg New York London Paris Tokyo Hong Kong Barcelona Budapest
40. White CM (ed) (1988) Modern Supercritical Fluid Chromatography. Hüthig, Heidelberg Basel New York
41. Yamauchi Y, Saito M, Hondo T, Senda M, Milet JL, Castiglioni E (1988) Coupled supercritical fluid extraction – supercritical fluid chromatography using pre-concentration/separation column: Application to fractionation of lemon peel oil. In: Perrut M (ed) Proc 2nd Int Symposium on Supercritical Fluids, pp 381 – 388
42. Yamauchi Y, Saito M (1990) Fractionation of lemon-peel oil by semi-preparative supercritical fluid chromatography. J Chromatogr 505: 237 – 246

# 10 Conclusion

Many interesting topics, which are related to gas extraction and supercritical gases as solvents have not been treated in this book, since the length of this book is necessarily limited. Among those topics that are interesting fields of research and application are:

- formation of small solid particles;
- formulation of pharmaceutical products;
- aerogels;
- impregnation of wood and dying of textile fibres;
- volume enhancement and disruption of plant materials;
- preparation of thin films.

An especially interesting area is the application of supercritical water for extraction and reaction.

In chemical reactions, the solvent power of supercritical components is a vast and rewarding area of investigation, in addition to the influence of a supercritical atmosphere on reaction kinetics and reaction rate. Four topics are mentioned explicitly:

- supercritical fluids as solvents during chemical reactions in general;
- supercritical fluids as diluents;
- supercritical fluids and solid catalysts;
- enzymatic reactions in a supercritical atmosphere.

# Subject Index